Fractional Calculus: Theory and Applications

Fractional Calculus: Theory and Applications

Editor: Gerald Paterson

NY RESEARCH PRESS
New York

Published by NY Research Press
118-35 Queens Blvd., Suite 400,
Forest Hills, NY 11375, USA
www.nyresearchpress.com

Fractional Calculus: Theory and Applications
Edited by Gerald Paterson

International Standard Book Number: 978-1-63238-632-8 (Hardback)

Cataloging-in-Publication Data

Fractional calculus : theory and applications / edited by Gerald Paterson.
p. cm.
Includes bibliographical references and index.
ISBN 978-1-63238-632-8
1. Fractional calculus. 2. Calculus. 3. Mathematics. I. Paterson, Gerald.

QA314 .F73 2019
515.83--dc23

Contents

Preface

It is often said that books are a boon to mankind. They document every progress and pass on the knowledge from one generation to the other. They play a crucial role in our lives. Thus I was both excited and nervous while editing this book. I was pleased by the thought of being able to make a mark but I was also nervous to do it right because the future of students depends upon it. Hence, I took a few months to research further into the discipline, revise my knowledge and also explore some more aspects. Post this process, I begun with the editing of this book.

Fractional calculus falls under the domain of mathematical analysis. It focuses on the study of the differentiation operator in terms of real and complex number powers. Fractional calculus has various applications across other scientific fields such as polymer chemistry through structural damping models, in quantum theory through the analysis of Schrödinger's equation, etc. This book elucidates new techniques and their applications in a multidisciplinary manner. The chapters provide significant information of this field to help develop a good understanding of fractional calculus and related fields. It also attempts to understand the multiple branches that fall under this discipline and how such concepts have practical applications. It aims to serve as a resource guide for students and experts alike as well as contribute to the growth of the discipline.

I thank my publisher with all my heart for considering me worthy of this unparalleled opportunity and for showing unwavering faith in my skills. I would also like to thank the editorial team who worked closely with me at every step and contributed immensely towards the successful completion of this book. Last but not the least, I wish to thank my friends and colleagues for their support.

Editor

1

Sequential optimality conditions for multiobjective fractional programming problems

Bhawna Kohli

Abstract In this paper, our aim is to develop necessary and sufficient optimality conditions in the absence of any constraint qualification for multiobjective fractional programming problem using the powerful combination of conjugate analysis and ε-subdifferential calculus. Furthermore, as an application of these conditions we derive sequential duality results for this class of problems.

Keywords Multiobjective fractional programming problem · ε-Subdifferentials · Sequential optimality conditions · Sequential duality

Introduction

In this paper, we consider the following multiobjective fractional programming problem:

$$\text{(P)} \quad \text{minimize } \phi(x) = \left(\frac{f_1(x)}{g_1(x)}\right), \ldots, \frac{f_p(x)}{g_p(x)}$$
$$\text{subject to } h_j(x) \leq 0, \quad j = 1, 2, \ldots, m$$

where f_i, g_i, h_j: $R^n \rightarrow R$, $i = 1, 2, \ldots, p$, $j = 1, 2, \ldots, m$. $f_i(.)$, $i = 1, 2, \ldots, p$, and $h_j(.)$, $j = 1, 2, \ldots, m$ are continuous, convex functions and $g_i(.)$, $i = 1, 2, \ldots, p$ are continuous, concave functions such that $f_i(x) \geq 0$ and $g_i(x) > 0$, $i = 1, 2, \ldots, p$ for all $x \in R^n$.

Let $E = \{x \in R^n: h_j(x) \leq 0, j = 1, 2, \ldots, m\}$ denote the feasible set for problem (P).

The study of multiobjective optimization problems has been a subject of great interest since multiobjective decision models can be widely applied to many practical problems which appear in the field of economics, management, medicine, etc. An important class of such problems is multiobjective fractional programming problems. Many authors studied optimality conditions and solution concepts for multiobjective optimization problems such as Chaoo and Atkins [3], Coladas et al. [4], Geoffrion [6], Gerth [7], Kaliszewski [13] and Li and Wang [14]. Though, generally we deal with exact optimal solutions but in many situations the concept of exact optimal solution cannot be applied but an approximate solution is required because from the computational point of view only approximate solutions can be obtained. So, in this article, we consider ε-approximate solutions defined as follows:

$\bar{x} \in E$ is an ε-weak efficient solution of (P) if there does not exist any feasible solution $x \in E$ such that

$$\frac{f_i(x)}{g_i(x)} < \frac{f_i(\bar{x})}{g_i(\bar{x})} - \varepsilon_i, \quad i = 1, 2, \ldots, p$$

where $\varepsilon = (\varepsilon_1, \varepsilon_2, \ldots, \varepsilon_p)$ with $\varepsilon_i \geq 0, i = 1, 2, \ldots, p$, when $\varepsilon = 0$, an ε-weak efficient solution is weak efficient solution of (P). For the notion of ε-optimal solution for scalar optimization problem one can refer to Bai et al. [1].

In the field of optimization, ε-optimality conditions have been discussed by many researchers like Loridan [15], Loridan and Morgan [16], Strodiot et al. [19], Yokoyama [22], Gajek and Zagrodny [5], Li and Wang [14], Lui [17], Tanaka [20], Yokoyama [24], Li and Wang [25], etc. Yokoyama [22] in 1992 obtained ε-optimality conditions for convex programming problem via exact penalty functions. In 1994, Yokoyama [23] extended the above results to vector minimization problem. Li and Wang [25] in 1998 introduced the concept of ε-proper efficiency and studied necessary and/or sufficient conditions for an ε-efficient solution (an ε-properly efficient solution, an ε-weak

B. Kohli (✉)
Department of Mathematics, University of Delhi,
Delhi 110007, India
e-mail: bhawna_kohli@rediffmail.com

efficient solution) for multiobjective optimization problem via scalarization and an alternative theorem.

Since study of multiobjective optimization problems is a subject of great importance, we have focused ourselves in this paper in developing optimality conditions for multiobjective fractional programming problem.

To derive necessary optimality conditions one needs to impose some kind of constraint qualification. But these qualifications may sometimes become cumbersome to verify and give rise to optimality conditions that are very difficult to trace from the view point of computation. In the absence of constraint qualifications (CQs.), Lagrange multiplier rules and Karush–Kuhn–Tucker (KKT) conditions may fail to hold. So, we need to develop optimality conditions without CQs. which would give more practical formulation of optimality conditions for multiobjective fractional programming problem (P). This motivates us to derive sequential optimality conditions for multiobjective fractional programming problem. Recently, work has been done in this direction for convex programming problems with cone convex constraints by Jeyakumar et al. [10, 12] and Bai et al. [1]. Jeyakumar et al. [10, 12] introduced the concept of sequential Lagrange multiplier rules for convex programs with cone convex constraints using the concept of epigraph of conjugate function in terms of ε-subdifferential computed at optimal solution. These conditions coincide with standard optimality conditions under the assumption of appropriate CQs. One of the main advantages of the ε-subdifferential which makes it a useful tool both in theory and practice is that for every $x \in \text{dom}f$, $\partial_\varepsilon f(x) \neq \phi$. Thibault [21] derived sequential optimality conditions using the subdifferential calculus for convex functions with cone convex constraints.

In this paper, our aim is to develop sequential optimality conditions for multiobjective fractional programming problem (P) via scalarization and using the concept of epigraph of conjugate function in terms of ε-subdifferentials computed at ε-weak efficient solution.

The paper is planned as follows: "Preliminaries" deals with some preliminary results that will be used in the sequel. In "Sequential optimality conditions", we derive sequential optimality conditions. Finally, in "Sequential duality results", sequential duality results have been obtained.

Preliminaries

In this section, we give some basic definitions and results which will be used in the sequel.

Let f: $R^n \to R \cup \{\pm\infty\}$.

The ε-subdifferential of f at $\bar{x} \in \text{dom}f$, $\partial_\varepsilon f(\bar{x})$ is defined as

$$\partial_\varepsilon f(\bar{x}) = \{\xi \in R^n : f(x) - f(\bar{x}) \geq \langle \xi, x - \bar{x} \rangle - \varepsilon, \forall x \in \text{dom}f\},$$

where $\varepsilon > 0$.

For detailed study on ε-subdifferentials one may refer to Hiriart-Urruty [8].

Remark 2.1 (Rockafellar and Wets [18]). If $\widehat{f}(x) = f(x) - \alpha$, $x \in R^n$, $\alpha \in R$, then

$$\widehat{f}^* = f^* + \alpha, \text{epi}\widehat{f}^* = \text{epi}f^* + (0, \alpha),$$

where f^* denotes the conjugate of function f and epif^*- denotes epigraph of f^*. For the definitions of conjugate and epigraph of a function one can see Bector et al. [2].

Remark 2.2 (Rockafellar and Wets [18]). For any scalar $\lambda > 0$,

$$(\lambda f)^* = \lambda * f^*, \quad (\lambda * f)^* = \lambda f^*$$

where $\lambda * f$ stands for epi-multiple. It satisfies

$$\text{epi}(\lambda * f) = \lambda \text{epi}f.$$

For details on conjugacy theory one may refer to Rockafellar and Wets [18].

Theorem 2.1 (Theorem 1.2.1, Bector et al. [2]). *A function f is a lower semicontinuous (lsc) function if and only if its epigraph is a closed convex set.*

For a set E, the indicator function δ_E is defined as

$$\delta_E(x) = \begin{cases} 0, & x \in E \\ +\infty & x \notin E \end{cases}$$

For a nonempty closed convex set E, δ_E is a proper, lsc, convex function.

Proposition 2.1 (Proposition 2.1, Jeyakumar et al. [10]). *Let f: $R^n \to R \cup \{\pm\infty\}$ be a proper, lsc, convex function and $\bar{x} \in \text{dom}f$. Then*

$$\text{epi}f^* = \bigcup_{\varepsilon \geq 0} \{(\xi, \langle \xi, \bar{x} \rangle + \varepsilon - f(\bar{x})) : \xi \in \partial_\varepsilon f(\bar{x})\}.$$

For $\bar{x} \in \text{dom}f$, $\partial_\varepsilon f(\bar{x})$ is nonempty and hence epif^* is nonempty.

Proposition 2.2 (Rockafellar and Wets [18], Jeyakumar et al. [12]) *For proper, lsc, convex functions f_1, f_2: $R^n \to R \cup \{\pm\infty\}$*

$$\text{epi}(f_1 + f_2)^* = \text{cl}(\text{epi}f_1^* + epif_2^*)$$

If $\text{dom}f_1 = R^n$

$$\text{epi}(f_1 + f_2)* = \text{epi}f_1 * + \text{epi}f_2 *$$

For any set $A \subset R^n$, we denote by coA and clcoA as convex hull and the closed convex hull of the set A, respectively.

Proposition 2.3 (Jeyakumar et al. [11]) *Let I be an arbitrary index set and f_i: $R^n \to R \cup\{\pm\infty\}$, $i \in I$, be proper, lsc, convex functions. Define $f(x) = \sup_{i\in I} f_i(x)$.*
Then, $\text{epi}f^* = \text{clco}\bigcup_{i\in I} \text{epi}f_i^*$.

Sequential optimality conditions

In this section, we prove sequential optimality conditions for multiobjective fractional programming problem (P).

We shall be using following Lemma on the lines of Lemma 5.1 [1] to prove our optimality conditions.

Lemma 3.1 *For (P), if* $E \neq \phi$. *Then*

$$\text{(i) epi } \delta_E^* = \text{cl}\left(\bigcup_{\mu_j \geq 0}\left(\sum_{j=1}^{m} \text{epi}(\mu_j h_j)^*\right)\right)$$

$$\text{(ii) } epi\left(\sum_{i=1}^{p} \lambda_i(f_i - v_i g_i) + \delta_E\right)^*$$

$$= \text{cl}\left(\text{epi}\left(\sum_{i=1}^{p} \lambda_i(f_i - v_i g_i)^*\right) + \bigcup_{\mu_j \geq 0}\left(\sum_{j=1}^{m} \text{epi}(\mu_j h_j)^*\right)\right)$$

where $v \in Rp$.

Since $f_i(.)$, $i = 1, 2, \ldots, p$, *are convex functions and* $g_i(.)$, $i = 1, 2, \ldots, p$, *are concave functions,* $f_i(.) - v_i g_i(.), i = 1, 2, \ldots, p$, *are convex functions.*

Theorem 3.1 $\bar{x} \in E$ *is an* ε-weak efficient solution of (P) if and only if there exist $\varepsilon' \geq 0$, $\{\varepsilon''_{nj}\}, \{\mu_{nj}\} \subset R_+, j = 1, 2, \ldots, m,$ $\lambda_i \geq 0, i = 1, 2, \ldots, p, \sum_{i=1}^{p} \lambda_i = 1$ *and* $\{\xi_i\} \subset R^n$, $i = 1, 2, \ldots, p$, $\{\xi''_{nj}\} \subset R^n$ $j = 1, 2, \ldots, m$ with $\sum_{i=1}^{p} \xi_i \in \partial_{\varepsilon'}\left(\sum_{i=1}^{p} \lambda_i(f_i(.) - \overline{v_i} g_i(.))\right)(\bar{x}), \xi''_{nj} \in \partial_{\varepsilon''_{nj}} h_j(\bar{x})$, $j = 1, 2, \ldots, m$ *such that*

$$\sum_{i=1}^{p} \xi_i + \sum_{j=1}^{m} \mu_{nj}\xi''_{nj} \to 0 \quad \text{as} \quad n \to \infty \tag{1}$$

and

$$\varepsilon' + \lim_{n\to\infty}\left[\sum_{j=1}^{m} \mu_{nj}\varepsilon''_{nj} - \left\{\sum_{j=1}^{m} \mu_{nj} h_j(\bar{x})\right\}\right] = \sum_{i=1}^{p} \lambda_i g_i(\bar{x})\varepsilon_i. \tag{2}$$

Assume that $f(x) < +\infty, g(x) < +\infty$, *for all* $x \in R^n$.
Here $\overline{v_i} = \frac{f_i(\bar{x})}{g_i(\bar{x})} - \varepsilon_i, i = 1, 2, \ldots, p.$

Proof Since $\bar{x} \in E$ is an ε-weak efficient solution of (P), there does not exist any feasible solution $x \in E$ such that

$$\frac{f_i(x)}{g_i(x)} < \frac{f_i(\bar{x})}{g_i(\bar{x})} - \varepsilon_i, \quad i = 1, 2, \ldots, p \tag{3}$$

Using parametric approach, problem (P) can be written as

$$\text{(P1)} \quad \text{minimize } (f_1(x) - v_1 g_1(x), \ldots, f_p(x) - v_p g_p(x))$$
$$\text{subject to } \quad h_j(x) \leq 0, \quad j = 1, 2, \ldots, m,$$

where $v \in R^p$ is the parameter.

By (3) we have that there does not exist any feasible solution $x \in E$ such that

$$f_i(x) < \left(\frac{f_i(\bar{x})}{g_i(\bar{x})} - \varepsilon_i\right) g_i(x), \quad i = 1, 2, \ldots, p$$

$$\Rightarrow f_i(x) - \left(\frac{f_i(\bar{x})}{g_i(\bar{x})} - \varepsilon_i\right) g_i(x) < 0 = f_i(\bar{x}) - \left(\frac{f_i(\bar{x})}{g_i(\bar{x})} - \varepsilon_i\right)$$
$$g_i(\bar{x}) - \varepsilon_i g_i(\bar{x}), \quad i = 1, 2, \ldots, p$$
$$\Rightarrow f_i(x) - \overline{v_i} g_i(x) < 0 = f_i(\bar{x}) - \overline{v_i} g_i(\bar{x}) - \varepsilon_i g_i(\bar{x}),$$
$$i = 1, 2, \ldots, p \tag{4}$$

$\Rightarrow \bar{x}$ is an $\bar{\varepsilon}$ - weak efficient solution of (P1) where $\bar{\varepsilon} = (\varepsilon_1 g_1(\bar{x}), \ldots, \varepsilon_p g_p(\bar{x}))$.

By weighted sum approach the problem (P1) can be converted to the following scalar optimization problem

$$\text{(P2)} \quad \text{minimize} \quad \sum_{i=1}^{p} \lambda_i(f_i(x) - v_i g_i(x))$$
$$\text{subject to} \quad h_j(x) \leq 0, \quad j = 1, 2, \ldots, m,$$
$$\text{where} \quad \lambda_i \geq 0, \quad i = 1, 2, \ldots, p, \sum_{i=1}^{p} \lambda_i = 1.$$

By (4), we have

$$f_i(x) - \overline{v_i} g_i(x) < 0 = f_i(\bar{x}) - \overline{v_i} g_i(\bar{x}) - \varepsilon_i g_i(\bar{x}),$$
$$i = 1, 2, \ldots, p$$

Since $\lambda_i \geq 0, i = 1, 2, \ldots, p$, multiplying above by $\lambda_i, i = 1, 2, \ldots, p$ and adding we get

$$\sum_{i=1}^{p} \lambda_i(f_i(.) - \overline{v_i} g_i(.))(x) < \sum_{i=1}^{p} \lambda_i(f_i(.) - \overline{v_i} g_i(.))(\bar{x})$$
$$- \sum_{i=1}^{p} \lambda_i g_i(\bar{x})\varepsilon_i$$

Hence $\bar{x}$ is an $\sum_{i=1}^{p} \lambda_i g_i(\bar{x})\varepsilon_i$—optimal solution of (P2).

Since $\bar{x}$ is an $\sum_{i=1}^{p} \lambda_i g_i(\bar{x})\varepsilon_i$—optimal solution of (P2), it is an $\sum_{i=1}^{p} \lambda_i g_i(\bar{x})\varepsilon_i$-optimal solution of the function $\sum_{i=1}^{p} \lambda_i(f_i(.) - \overline{v_i} g_i(.) + \delta_E)(x)$ on E, that is,

$$\sum_{i=1}^{p} \lambda_i(f_i(.) - \overline{v_i} g_i(.) + \delta_E)(x) \geq \sum_{i=1}^{p} \lambda_i(f_i(.) - \overline{v_i} g_i(.)$$
$$+ \delta_E)(\bar{x}) - \sum_{i=1}^{p} \lambda_i g_i(\bar{x})\varepsilon_i, \quad \text{for all } x \in E$$

$$\Rightarrow 0 \in \partial_{\sum_{i=1}^{p} \lambda_i g_i(\bar{x})\varepsilon_i}\left(\sum_{i=1}^{p} \lambda_i(f_i(.) - \overline{v_i} g_i(.)) + \delta_E\right)(\bar{x})$$

Hence

$$\left(0, -\sum_{i=1}^{p} \lambda_i(f_i(.) - \overline{v_i}g_i(.))(\bar{x}) + \sum_{i=1}^{p} \lambda_i g_i(\bar{x})\varepsilon_i\right)$$
$$\in \text{epi}\left(\sum_{i=1}^{p} \lambda_i(f_i(.) - \overline{v_i}g_i(.)) + \delta_E\right)^*$$

Using Proposition 2.2, we get

$$\left(0, -\sum_{i=1}^{p} \lambda_i(f_i(.) - \overline{v_i}g_i(.))(\bar{x}) + \sum_{i=1}^{p} \lambda_i g_i(\bar{x})\varepsilon_i\right)$$
$$\in \text{epi}\left(\sum_{i=1}^{p} \lambda_i(f_i(.) - \overline{v_i}g_i(.))\right)^* + \text{epi}\delta_E^*$$

Using Lemma 3.1(i) and Remark 2.2, we obtain

$$\left(0, -\sum_{i=1}^{p} \lambda_i(f_i(.) - \overline{v_i}g_i(.))(\bar{x}) + \sum_{i=1}^{p} \lambda_i g_i(\bar{x})\varepsilon_i\right)$$
$$\in epi\left(\sum_{i=1}^{p} \lambda_i(f_i(.) - \overline{v_i}g_i(.))\right)^*$$
$$+ \text{cl}\left[\bigcup_{\mu_j \geq 0}\left(\sum_{j=1}^{m} \mu_j \text{epi} h_j^*\right)\right]$$

Proposition 2.1 implies that there exist $\varepsilon' \geq 0$, $\{\varepsilon''_{nj}\}$, $j = 1,2,\ldots,m \subset R_+$ and $\sum_{i=1}^{p} \xi_i \in \partial_{\varepsilon'}\left(\sum_{i=1}^{p} \lambda_i (f_i(.) - \overline{v_i}g_i(.))\right)(\bar{x})$, $\xi''_{nj} \in \partial_{\varepsilon''_{nj}} h_j(\bar{x})$ such that

$$\left(0, -\sum_{i=1}^{p} \lambda_i(f_i(.) - \overline{v_i}g_i(.))(\bar{x}) + \sum_{i=1}^{p} \lambda_i g_i(\bar{x})\varepsilon_i\right)$$
$$= \left(\sum_{i=1}^{p} \xi_i, \left\langle \sum_{i=1}^{p} \xi_i, \bar{x}\right\rangle + \varepsilon' - \sum_{i=1}^{p} \lambda_i(f_i(.) - \overline{v_i}g_i(.))(\bar{x})\right)$$
$$+ \lim_{n\to\infty}\left[\left(\sum_{j=1}^{m} \mu_{nj}\xi''_{nj}, \left\langle \sum_{j=1}^{m} \mu_{nj}\xi''_{nj}, \bar{x}\right\rangle + \sum_{j=1}^{m} \mu_{nj}\varepsilon''_{nj} - \sum_{j=1}^{m} \mu_{nj}h_j(\bar{x})\right)\right]$$

Now, comparing both the sides we get

$$\sum_{i=1}^{p} \xi_i + \sum_{j=1}^{m} \mu_{nj}\xi''_{nj} \to 0 \quad \text{as} \quad n \to \infty$$

and

$$-\sum_{i=1}^{p} \lambda_i(f_i(.) - \overline{v_i}g_i(.))(\bar{x}) + \sum_{i=1}^{p} \lambda_i g_i(\bar{x})\varepsilon_i = \left\langle \sum_{i=1}^{p} \xi_i, \bar{x}\right\rangle$$
$$+ \varepsilon' - \sum_{i=1}^{p} \lambda_i(f_i(.) - \overline{v_i}g_i(.))(\bar{x})$$
$$+ \lim_{n\to\infty}\left[\left\langle \sum_{j=1}^{m} \mu_{nj}\xi''_{nj}, \bar{x}\right\rangle + \sum_{j=1}^{m} \mu_{nj}\varepsilon''_{nj} - \sum_{j=1}^{m} \mu_{nj}h_j(\bar{x})\right]$$

Using (1), we get

$$\varepsilon' + \lim_{n\to\infty}\left[\sum_{j=1}^{m} \mu_{nj}\varepsilon''_{nj} - \left\{\sum_{j=1}^{m} \mu_{nj}h_j(\bar{x})\right\}\right] = \sum_{i=1}^{p} \lambda_i g_i(\bar{x})\varepsilon_i.$$

Conversely suppose that (1) and (2) hold.

We have to show that $\bar{x}$ is an ε-weak efficient solution of (P).

Suppose on contrary that $\bar{x}$ is not an ε-weak efficient solution of (P). Then, as argued in the necessary part, we have that $\bar{x}$ is not a $\sum_{i=1}^{p} \lambda_i g_i(\bar{x})\varepsilon_i$-optimal solution of (P2). Then, there exists a feasible solution $x \in E$ such that

$$\sum_{i=1}^{p} \lambda_i(f_i(.) - \overline{v_i}g_i(.))(x) < \sum_{i=1}^{p} \lambda_i(f_i(.) - \overline{v_i}g_i(.))(\bar{x}) - \sum_{i=1}^{p} \lambda_i g_i(\bar{x})\varepsilon_i \tag{5}$$

Since $\sum_{i=1}^{p} \xi_i \in \partial_{\varepsilon'}\left(\sum_{i=1}^{p} \lambda_i(f_i(.) - \overline{v_i}g_i(.))\right)(\bar{x})$, $\xi''_{nj} \in \partial_{\varepsilon''_{nj}} h_j(\bar{x})$, $j = 1, 2, \ldots, m$ therefore, we have

$$\sum_{i=1}^{p} \lambda_i(f_i(.) - \overline{v_i}g_i(.))(x) - \sum_{i=1}^{p} \lambda_i(f_i(.) - \overline{v_i}g_i(.))(\bar{x})$$
$$\geq \left\langle \sum_{i=1}^{p} \xi_i, x - \bar{x}\right\rangle - \varepsilon'$$

$$h_j(x) - h_j(\bar{x}) \geq \left\langle \xi''_{nj}, x - \bar{x}\right\rangle - \varepsilon''_{nj}, \quad j = 1,2,\ldots,m \tag{6}$$

Since $\{\mu_{nj}\}$, $j = 1,2,\ldots,m \subset R_+$, therefore, multiplying (6) with μ_{nj}, $j = 1,2,\ldots,m$ and adding we get

$$\sum_{i=1}^{p} \lambda_i(f_i(.) - \overline{v_i}g_i(.))(x) - \sum_{i=1}^{p} \lambda_i(f_i(.) - \overline{v_i}g_i(.))(\bar{x})$$
$$+ \sum_{j=1}^{m} \mu_{nj}(h_j(x) - h_j(\bar{x})) \geq \left\langle \sum_{i=1}^{p} \xi_i, x - \bar{x}\right\rangle$$
$$+ \left\langle \sum_{j=1}^{m} \mu_{nj}\xi''_{nj}, x - \bar{x}\right\rangle - \left(\varepsilon' + \sum_{j=1}^{m} \mu_{nj}\varepsilon''_{nj}\right)$$

Taking limit as $n \to \infty$ and using (1) we get

$$\sum_{i=1}^{p} \lambda_i(f_i(.) - \overline{v_i}g_i(.))(x) - \sum_{i=1}^{p} \lambda_i(f_i(.) - \overline{v_i}g_i(.))(\bar{x})$$
$$+ \lim_{n\to\infty}\left[\sum_{j=1}^{m} \mu_{nj}h_j(x)\right]$$
$$\geq -\lim_{n\to\infty}\left[\varepsilon' + \sum_{j=1}^{m} \mu_{nj}\varepsilon''_{nj} - \sum_{j=1}^{m} \mu_{nj}h_j(\bar{x})\right]$$

Using (2) we get

$$\sum_{i=1}^{p} \lambda_i(f_i(.) - \overline{v_i}g_i(.))(x) - \sum_{i=1}^{p} \lambda_i(f_i(.) - \overline{v_i}g_i(.))(\bar{x})$$
$$+ \lim_{n\to\infty}\left[\sum_{j=1}^{m} \mu_{nj}h_j(x)\right] \geq -\sum_{i=1}^{p} \lambda_i g_i(\bar{x})\varepsilon_i$$

As x is feasible for (P), we have

$$\sum_{i=1}^{p}\lambda_i(f_i(.)-\overline{v_i}g_i(.))(x)-\sum_{i=1}^{p}\lambda_i(f_i(.)-\overline{v_i}g_i(.))(\bar{x})$$
$$\geq -\sum_{i=1}^{p}\lambda_i g_i(\bar{x})\varepsilon_i$$

which is contradictory to (5).

Hence our assumption was wrong, $\bar{x}$ is an ε-weak efficient solution of (P).

Corollary 3.1 *If in the above theorem we impose a constraint qualification, that is, the set* $\left[\bigcup_{\mu_j\geq 0}\sum_{j=1}^{m}\mu_j \mathrm{epi} h_j^*\right]$ *is closed, then the sequential optimality conditions reduce to the standard KKT conditions.*

We now illustrate above theorem with the help of the following example.

Example 3.1 Consider the problem

$$\min\left(\frac{f_1(x)}{g_1(x)},\frac{f_2(x)}{g_2(x)}\right)=(x^2+x,x)$$

subject to $h(x)=\begin{cases} x & x\geq 0\\ -x-1, & x<0\end{cases}\leq 0,$

Here f_i, g_i, $h{:}R\to R$, $i=1,2.$

Set of feasible solutions is given by

$E=\{x\in R:-1\leq x\leq 0\}.$

Here $\bar{x}=0$ is not an optimal solution as for $x=\frac{-1}{2}$ $\frac{f_1(x)}{g_1(x)}\ngeq\frac{f_1(\bar{x})}{g_1(\bar{x})}$ but it is an ε_1-optimal solution as for $\varepsilon_1=2$ $\frac{f_1(x)}{g_1(x)}\geq\frac{f_1(\bar{x})}{g_1(\bar{x})}-\varepsilon_1$, for all $x\in E$, and $\frac{f_2(x)}{g_2(x)}\ngeq\frac{f_2(\bar{x})}{g_2(\bar{x})}$ but it is an ε_2-optimal solution as for $\varepsilon_2=2$, $\frac{f_2(x)}{g_2(x)}\geq\frac{f_2(\bar{x})}{g_2(\bar{x})}-\varepsilon_2$, for all $x\in E$.

Now

$$\overline{v_1}=\frac{f_1(\bar{x})}{g_1(\bar{x})}-\varepsilon_1=-2,\quad \overline{v_2}=\frac{f_2(\bar{x})}{g_2(\bar{x})}-\varepsilon_2=-2,$$

Then there exist $\varepsilon'=1,\varepsilon''_{n1}=\frac{1}{n},\varepsilon''_{n2}=1+\frac{1}{n}$, $\lambda_1=\frac{1}{2},\lambda_2=\frac{1}{2}$, $\{\xi_{ni}\},\left\{\xi''_{nj}\right\}\in R$, $i=1,2,\ j=1,2$, with $\xi_1+\xi_2=1\in\partial_{\varepsilon'}(\lambda_1(f_1-\overline{v_1}g_1)+\lambda_2(f_2-\overline{v_2}g_2))(\bar{x})$ as $x\leq\frac{x^2}{2}+x+1$, for all $x\in R$ $\xi''_{n1}=1\in\partial_{\varepsilon''_{n1}}h(\bar{x})$, $\xi''_{n2}=-1\in\partial_{\varepsilon''_{n2}}h(\bar{x})$ as

$$\begin{cases} x\leq x+\varepsilon''_{n1}, & x\geq 0\\ -x\leq -x-1+\varepsilon''_{n2}, & x<0\end{cases},$$

and $\mu_{n1}=\frac{1}{n},\mu_{n2}=1$ such that

$$\xi_1+\xi_2+\sum_{j=1}^{2}\mu_{nj}\xi''_{nj}\to 0\quad\text{as}\quad n\to\infty$$

and

$$\varepsilon'+\lim_{n\to\infty}\left[\sum_{j=1}^{2}\mu_{nj}\varepsilon''_{nj}-\left\{\sum_{j=1}^{2}\mu_{nj}h_j(\bar{x})\right\}\right]$$
$$=\lambda_1 g_1(\bar{x})\varepsilon_1+\lambda_2 g_2(\bar{x})\varepsilon_2.$$

Theorem 3.2 *$\bar{x}\in E$ is an ε-weak efficient solution of (P) if and only if there exist $\{\varepsilon_n\},\{\varepsilon''_{nj}\},\{\lambda'_{ni}\},\{\mu_{nj}\}\subset R_+$, $i=1,2,\ldots,p$, $j=1,2,\ldots,m$, $\lambda_i\geq 0, i=1,2,\ldots,p$, $\sum_{i=1}^{p}\lambda_i=1$ and $\{\xi_{ni}\},\{\xi''_{nj}\}\subset R^n$, $i=1,2,\ldots,p$, $j=1,2,\ldots,m$ with $\sum_{i=1}^{p}\xi_{ni}\in\partial_{\varepsilon_n}\left(\sum_{i=1}^{p}\lambda_i(f_i(.)-\overline{v_i}g_i(.))\right)(\bar{x})$, $\xi''_{nj}\in\partial_{\varepsilon''_{nj}}h_j(\bar{x})$, $j=1,2,\ldots,m$ such that*

$$\sum_{i=1}^{p}\lambda'_{ni}\xi_{ni}+\sum_{j=1}^{m}\mu_{nj}\xi''_{nj}\to 0, \tag{7}$$

$$\sum_{j=1}^{m}\mu_{nj}h_j(\bar{x})\to 0$$

$\varepsilon_n\downarrow 0,\varepsilon''_{nj}\downarrow 0, i=1,2,\ldots,p,\ j=1,2,\ldots,m$ as $n\to\infty$. Assume that $f(x)<+\infty, g(x)<+\infty$, for all $x\in R^n$.

Proof Since $\bar{x}$ is an ε-weak efficient solution of (P), it is an $\sum_{i=1}^{p}\lambda_i g_i(\bar{x})\varepsilon_i$-optimal solution of (P2); hence it solves the following unconstrained problem

(P_0) $\min f_0(x)$
subject to $x\in R^n$

where $f_0(x)=\max\{\sum_{i=1}^{p}\lambda_i(f_i(.)-\ \overline{v_i}g_i(.))(x)-\alpha,h_1(x),h_2(x),\ldots,h_m(x)\}$and $\alpha=\sum_{i=1}^{p}\lambda_i(f_i(.)-\overline{v_i}g_i(.))(\bar{x})$.

Since $\bar{x}$ is an $\sum_{i=1}^{p}\lambda_i g_i(\bar{x})\varepsilon_i$-optimal solution of (P_0); therefore, we have $f_0(x)\geq f_0(\bar{x})-\sum_{i=1}^{p}\lambda_i g_i(\bar{x})\varepsilon_i$, for all $x\in E$. that is, $f_0(x)\geq-\sum_{i=1}^{p}\lambda_i g_i(\bar{x})\varepsilon_i$, for all $x\in E$,

$$\Rightarrow 0\in\partial_{\sum_{i=1}^{p}\lambda_i g_i(\bar{x})\varepsilon_i}f_0(\bar{x})$$

Hence

$$\left(0,\sum_{i=1}^{p}\lambda_i g_i(\bar{x})\varepsilon_i\right)\in epif_0^* \tag{8}$$

By Proposition 2.3, we have

$$\mathrm{epi}f_0^*=\mathrm{clco}\left\{epi\left(\sum_{i=1}^{p}\lambda_i(f_i(.)-\overline{v_i}g_i(.))(x)-\alpha\right)^*\bigcup\bigcup_{j\in J}\mathrm{epi}h_j^*\right\}. \tag{9}$$

Using (9) in (8) we get

$$\left(0,\sum_{i=1}^{p}\lambda_i g_i(\bar{x})\varepsilon_i\right)\in\mathrm{clco}\left\{\mathrm{epi}\left(\sum_{i=1}^{p}\lambda_i(f_i(.)-\overline{v_i}g_i(.))(x)-\alpha\right)^*\right.$$
$$\left.\bigcup\bigcup_{j\in J}\mathrm{epi}h_j^*\right\}.$$

By Remark 2.1, we have

$$\left(0,\sum_{i=1}^{p}\lambda_i g_i(\bar{x})\varepsilon_i\right) \in \text{clco}\left\{\text{epi}\left(\sum_{i=1}^{p}\lambda_i(f_i(.) - \bar{v}_i g_i(.))(x)\right)^{*} + (0,\alpha)\bigcup\bigcup_{j\in J}\text{epi}h_j^{*}\right\}.$$

Thus, there exist

$$\sum_{i=1}^{p}\xi_{ni} \in \partial_{\varepsilon_n}\left(\sum_{i=1}^{p}\lambda_i(f_i(.) - \bar{v}_i g_i(.))\right)(\bar{x}),\ \xi''_{nj} \in \partial_{\varepsilon''_{nj}} h_j(\bar{x}),$$
$$j = 1,2,\dots,m$$

and $\{\varepsilon_n\},\{\varepsilon''_{nj}\},\{\lambda'_{ni}\},\{\mu_{nj}\} \subset R_+,\quad i = 1,2,\dots,p,\ j = 1,2,\dots,m$ with $\sum_{i=1}^{p}\lambda'_{ni} + \sum_{j=1}^{m}\mu_{nj} = 1$ as $n \to \infty$ such that

$$\lim_{n\to\infty}\left[\sum_{i=1}^{p}\lambda'_{ni}\lambda_i(f_i(.) - \bar{v}_i g_i(.))(x) - \sum_{i=1}^{p}\lambda'_{ni}\lambda_i(f_i(.) - \bar{v}_i g_i(.))(\bar{x})\right] \geq$$
$$\lim_{n\to\infty}\left(-\sum_{j=1}^{m}\mu_{nj}h_j(x)\right) + \lim_{n\to\infty}\left(\sum_{j=1}^{m}\mu_{nj}h_j(\bar{x}) - \left(\sum_{i=1}^{p}\lambda'_{ni}\varepsilon_n + \sum_{j=1}^{m}\mu_{nj}\varepsilon''_{nj}\right)\right)$$

Using the conditions $\sum_{j=1}^{m}\mu_{nj}h_j(\bar{x}) \to 0,\ \{\varepsilon_n\}\downarrow 0,\ \{\mu_{nj}\}\downarrow 0,\ j = 1,2,\dots,m$ as $n \to \infty$ and the fact that x is feasible for (P), we get

$$\lim_{n\to\infty}\left[\sum_{i=1}^{p}\lambda'_{ni}\lambda_i(f_i(.) - \bar{v}_i g_i(.))(x) - \sum_{i=1}^{p}\lambda'_{ni}\lambda_i(f_i(.) - \bar{v}_i g_i(.))(\bar{x})\right] \geq 0$$

Since $\lambda_i \geq 0, \lambda'_{ni} \geq 0, i = 1,2,\dots,p$ for all $n \in N$, we have

$$f_i(x) - \bar{v}_i g_i(x) \geq f_i(\bar{x}) - \bar{v}_i g_i(\bar{x}),\quad i = 1,2,\dots,p$$

$$\left(0,\sum_{i=1}^{p}\lambda_i g_i(\bar{x})\varepsilon_i\right) = \lim_{n\to\infty}\left[\left(\sum_{i=1}^{p}\lambda'_{ni}\xi_{ni},\left\langle\sum_{i=1}^{p}\lambda'_{ni}\xi_{ni},\bar{x}\right\rangle + \sum_{i=1}^{p}\lambda'_{ni}\varepsilon_n - \sum_{i=1}^{p}\lambda'_{ni}\alpha\right) + \left(0,\sum_{i=1}^{p}\lambda'_{ni}\alpha\right) + \left(\sum_{j=1}^{m}\mu_{nj}\xi''_{nj},\left\langle\sum_{j=1}^{m}\mu_{nj}\xi''_{nj},\bar{x}\right\rangle + \sum_{j=1}^{m}\mu_{nj}\varepsilon''_{nj} - \sum_{j=1}^{m}\mu_{nj}h_j(\bar{x})\right)\right]$$

Now, comparing both the sides we get $\sum_{i=1}^{p}\lambda'_{ni}\xi_{ni} + \sum_{j=1}^{m}\mu_{nj}\xi''_{nj} \to 0$, as $n \to \infty$ and

$$\sum_{i=1}^{p}\lambda_i g_i(\bar{x})\varepsilon_i = \lim_{n\to\infty}\left[\left\langle\sum_{i=1}^{p}\lambda'_{ni}\xi_{ni},\bar{x}\right\rangle + \sum_{i=1}^{p}\lambda'_{ni}\varepsilon_n + \left(\left\langle\sum_{j=1}^{m}\mu_{nj}\xi''_{nj},\bar{x}\right\rangle + \sum_{j=1}^{m}\mu_{nj}\varepsilon''_{nj} - \sum_{j=1}^{m}\mu_{nj}h_j(\bar{x})\right)\right].$$

Using (7) we get

$$\sum_{i=1}^{p}\lambda_i g_i(\bar{x})\varepsilon_i = \lim_{n\to\infty}\left[\sum_{i=1}^{p}\lambda'_{ni}\varepsilon_n + \sum_{j=1}^{m}\mu_{nj}\varepsilon''_{nj} - \left(\sum_{j=1}^{m}\mu_{nj}h_j(\bar{x})\right)\right].$$

This equation along with the conditions $\varepsilon_i \geq 0, \lambda_i \geq 0, g_i(\bar{x}) > 0, \varepsilon_n \geq 0,\ \varepsilon''_{nj} \geq 0, \lambda'_{ni} \geq 0, i = 1,2,\dots,p,$ $\mu_{nj} \geq 0,\ j = 1,2,\dots,m$ and the fact that $\bar{x}$ is feasible for (P) implies $\{\varepsilon_n\}\downarrow 0,\quad \left\{\varepsilon''_{nj}\right\}\downarrow 0, j = 1,2,\dots,m,$ $\sum_{j=1}^{m}\mu_{nj}h_j(\bar{x}) \to 0$ as $n \to \infty$.

Conversely proceeding on the similar lines of Theorem 3.1, we arrive at the following condition

That is, $f_i(x) - \bar{v}_i g_i(x) \geq f_i(\bar{x}) - \bar{v}_i g_i(\bar{x}) \geq f_i(\bar{x}) - \bar{v}_i g_i(\bar{x}) - g_i(\bar{x})\varepsilon_i,\ i = 1, 2, \dots, p$ as $\varepsilon_i \geq 0,\ g_i(\bar{x}) > 0,$ $i = 1, 2, \dots, p$.

Since $\lambda_i \geq 0, i = 1,2,\dots,p$ we get

$$\sum_{i=1}^{p}\lambda_i(f_i(.) - \bar{v}_i g_i(.))(x) - \sum_{i=1}^{p}\lambda_i(f_i(.) - \bar{v}_i g_i(.))(\bar{x}) \geq -\sum_{i=1}^{p}\lambda_i g_i(\bar{x})\varepsilon_i$$

which gives contradiction to (5).

Hence the result.

We now give an example to illustrate the above theorem.

Remark 3.1 Consider Example 3.1 with $h(x)$ replaced by $h(x) = \begin{cases} x, & x \geq -1 \\ -x, & x < -1 \end{cases}$. Set of feasible solutions is given by

$$E = \{x \in R : -1 \leq x \leq 0\}.$$

It can be seen that $\bar{x} = 0$ is not an optimal solution but it is an ε_2-optimal solution.

Then, there exist $\varepsilon_n = \frac{1}{n}, \varepsilon''_{n1} = \frac{1}{n},\ \lambda_1 = \frac{1}{2}, \lambda_2 = \frac{1}{2},$ $\{\xi_{ni}\},\left\{\xi''_{n1}\right\} \in R,\quad i = 1,2,$ with $\xi_{n1} + \xi_{n2} = 1 \in$

$\partial_{\varepsilon_n}(\lambda_1(f_1-\overline{v_1}g_1)+\lambda_2(f_2-\overline{v_2}g_2))(\bar{x})$ as $x\leq \frac{x^2}{2}+x+\varepsilon_n$, for all $x\in R$ and

$$\xi''_{n1}=1\in\partial_{\varepsilon''_{n1}}h(\bar{x}) \text{ as} \begin{cases} x\leq x+\varepsilon''_{n1}, & x\geq -1 \\ x\leq -x+\varepsilon''_{n1}, & x<-1 \end{cases}$$

and $\lambda'_{n1}=1+\frac{1}{n},\lambda'_{n2}=\frac{1}{n},\mu_{n1}=\frac{1}{n}$ such that $\lambda'_{n1}\xi_{n1}+\lambda'_{n2}\xi_{n2}+\mu_{n1}\xi''_{n1}\to 0$ with $\lambda'_{n1}+\lambda'_{n2}+\mu_{n1}=1$ as $n\to\infty$

and $\mu_{n1}h(\bar{x})\to 0$, $\varepsilon_n\downarrow 0,\varepsilon''_{n1}\downarrow 0$ as $n\to\infty$.

Sequential duality results

In this section, we prove sequential duality results for (P).

For (P), the sequential Lagrange function $L: R^n\times R^m_+\to R\cup\{\pm\infty\}$ is defined as

$$L(x,\mu_n)=\sum_{i=1}^{p}\lambda_i(f_i(.)-\overline{v_i}g_i(.))(x)+\sum_{j=1}^{m}\mu_{nj}h_j(x)$$

The sequential dual for (P) is given by

$$\max_{\mu_n\in R^m_+}\min_{x\in E}\lim_{n\to\infty}\inf L(x,\mu_n)$$

In the following theorem, we establish sequential duality result.

Theorem 4.1 *Let $\bar{x}$ be an ε-weak efficient solution of (P) with optimal value $\alpha-\sum_{i=1}^{p}\lambda_i g_i(\bar{x})\varepsilon_i$ where $\alpha=\sum_{i=1}^{p}\lambda_i(f_i(.)-\overline{v_i}g_i(.))(\bar{x})$. Then*

$$\max_{\mu_n\in R^m_+}\min_{x\in E}\lim_{n\to\infty}\inf L(x,\mu_n)=\alpha-\sum_{i=1}^{p}\lambda_i g_i(\bar{x})\varepsilon_i.$$

Assume that $f(x)<+\infty, g(x)<+\infty, h(x)<+\infty$, for all $x\in R^n$.

Proof Since $\bar{x}$ is an ε-weak efficient solution of (P), therefore, proceeding on the lines of Theorem 3.1, there exist $\sum_{i=1}^{p}\xi_i\in\partial_{\varepsilon'}\left(\sum_{i=1}^{p}\lambda_i(f_i(.)-\overline{v_i}g_i(.))\right)(\bar{x})$, $\xi''_{nj}\in\partial_{\varepsilon''_{nj}}h_j(\bar{x})$, $j=1,2,\ldots,m$, $\varepsilon'\geq 0$ and sequences $\{\xi_i\},\{\xi''_{nj}\}\subset R^n, i=1,2,\ldots,p$, $j=1,2,\ldots,m$ and $\{\varepsilon''_{nj}\},\{\mu_{nj}\}\subset R_+$, $j=1,2,\ldots,m$ such that

$$\left(0,-\sum_{i=1}^{p}\lambda_i(f_i(.)-\overline{v_i}g_i(.))(\bar{x})+\sum_{i=1}^{p}\lambda_i g_i(\bar{x})\varepsilon_i\right)=$$
$$\left(\sum_{i=1}^{p}\xi_i,\left\langle\sum_{i=1}^{p}\xi_i,\bar{x}\right\rangle+\varepsilon'-\sum_{i=1}^{p}\lambda_i(f_i(.)-\overline{v_i}g_i(.))(\bar{x})\right)$$
$$+\lim_{n\to\infty}\left[\left(\sum_{j=1}^{m}\mu_{nj}\xi''_{nj},\left\langle\sum_{j=1}^{m}\mu_{nj}\xi''_{nj},\bar{x}\right\rangle+\sum_{j=1}^{m}\mu_{nj}\varepsilon''_{nj}-\sum_{j=1}^{m}\mu_{nj}h_j(\bar{x})\right)\right]$$

Using definitions of conjugates of $f_i(.)-\overline{v_i}g_i(.), i=1,2,\ldots,p, h_j(.)$, $j=1,2,\ldots,m$ and comparing both the sides we get

$$\sum_{i=1}^{p}\xi_i+\sum_{j=1}^{m}\mu_{nj}\xi''_{nj}\to 0 \quad\text{as}\quad n\to\infty$$

and

$$-\alpha+\sum_{i=1}^{p}\lambda_i g_i(\bar{x})\varepsilon_i=\left(\sum_{i=1}^{p}\lambda_i(f_i(.)-\overline{v_i}g_i(.))\right)^*\left(\sum_{i=1}^{p}\xi_i\right)$$
$$+\varepsilon'+\lim_{n\to\infty}\left[\left(\left(\sum_{j=1}^{m}\mu_{nj}h_j(.)\right)^*\left(\sum_{j=1}^{m}\mu_{nj}\xi''_{nj}\right)+\sum_{j=1}^{m}\mu_{nj}\varepsilon''_{nj}\right)\right]$$

Using definition of $\left(\sum_{i=1}^{p}\lambda_i(f_i(.)-\overline{v_i}g_i(.))\right)^*\left(\sum_{i=1}^{p}\xi_i\right)$ we get $\langle\sum_{i=1}^{p}\xi_i,x\rangle-\left(\sum_{i=1}^{p}\lambda_i(f_i(.)-\overline{v_i}g_i(.))\right)(x)\leq-\lim_{n\to\infty}\left[\left(\sum_{j=1}^{m}\mu_{nj}h_j(.)\right)^*\left(\sum_{j=1}^{m}\mu_{nj}\xi''_{nj}\right)+\sum_{j=1}^{m}\mu_{nj}\varepsilon''_{nj}\right]-\alpha-\varepsilon'+\sum_{i=1}^{p}\lambda_i g_i(\bar{x})\varepsilon_i$, for all $x\in E$, which gives that

$$\left(\sum_{i=1}^{p}\lambda_i(f_i(.)-\overline{v_i}g_i(.))\right)(x)$$
$$\geq\lim_{n\to\infty}\left[\left(\sum_{j=1}^{m}\mu_{nj}h_j(.)\right)^*\left(\sum_{j=1}^{m}\mu_{nj}\xi''_{nj}\right)+\sum_{j=1}^{m}\mu_{nj}\varepsilon''_{nj}\right]$$
$$+\left\langle\sum_{i=1}^{p}\xi_i,x\right\rangle+\alpha+\varepsilon'-\sum_{i=1}^{p}\lambda_i g_i(\bar{x})\varepsilon_i,$$
$$\geq\lim_{n\to\infty}\left[\left(\sum_{j=1}^{m}\mu_{nj}h_j(.)\right)^*\left(\sum_{j=1}^{m}\mu_{nj}\xi''_{nj}\right)+\sum_{j=1}^{m}\mu_{nj}\varepsilon''_{nj}\right]$$
$$+\left\langle\sum_{i=1}^{p}\xi_i,x\right\rangle+\alpha-\sum_{i=1}^{p}\lambda_i g_i(\bar{x})\varepsilon_i \tag{9}$$

as $\varepsilon'\geq 0$.

Now since $\{\varepsilon''_{nj}\},\{\mu_{nj}\}\subset R_+, j=1,2,\ldots,m$

$$\left(\sum_{j=1}^{m}\mu_{nj}h_j(.)\right)^*\left(\sum_{j=1}^{m}\mu_{nj}\xi''_{nj}\right)+\sum_{j=1}^{m}\mu_{nj}\varepsilon''_{nj}\geq\left(\sum_{j=1}^{m}\mu_{nj}h_j(.)\right)^*\left(\sum_{j=1}^{m}\mu_{nj}\xi''_{nj}\right),$$

for all $n\in N$

and therefore

$$\lim_{n\to\infty}\left[\left(\sum_{j=1}^{m}\mu_{nj}h_j(.)\right)^*\left(\sum_{j=1}^{m}\mu_{nj}\xi''_{nj}\right)+\sum_{j=1}^{m}\mu_{nj}\varepsilon''_{nj}\right]$$
$$\geq\lim_{n\to\infty}\sup\left[\left(\sum_{j=1}^{m}\mu_{nj}h_j(.)\right)^*\left(\sum_{j=1}^{m}\mu_{nj}\xi''_{nj}\right)\right] \tag{10}$$

(9) and (10) give

$$\left(\sum_{i=1}^{p}\lambda_i(f_i(.)-\overline{v}_i g_i(.))\right)(x)\geq$$

$$\limsup_{n\to\infty}\left[\left(\sum_{j=1}^{m}\mu_{nj}h_j(.)\right)^{*}\left(\sum_{j=1}^{m}\mu_{nj}\xi''_{nj}\right)\right]+\left\langle\sum_{i=1}^{p}\xi_i,x\right\rangle$$

$$+\alpha-\sum_{i=1}^{p}\lambda_i g_i(\overline{x})\varepsilon_i \qquad (11)$$

Now

$$\limsup_{n\to\infty}\left[\left(\sum_{j=1}^{m}\mu_{nj}h_j(.)\right)^{*}\left(\sum_{j=1}^{m}\mu_{nj}\xi''_{nj}\right)\right]$$

$$=\limsup_{n\to\infty}\left[\sup_{x\in dom(\sum_{j=1}^{m}\mu_{nj}h_j(.))}\left\{\left\langle\sum_{j=1}^{m}\mu_{nj}\xi''_{nj},x\right\rangle-\sum_{j=1}^{m}\mu_{nj}h_j(x)\right\}\right]$$

$$\geq\limsup_{n\to\infty}\left[\left\{\left\langle\sum_{j=1}^{m}\mu_{nj}\xi''_{nj},x\right\rangle-\sum_{j=1}^{m}\mu_{nj}h_j(x)\right\}\right]$$

Using above in (11) we get

$$\left(\sum_{i=1}^{p}\lambda_i\,(f_i(.)-\overline{v}_i g_i(.))\right)(x)$$

$$\geq\limsup_{n\to\infty}\left[\left\langle\sum_{j=1}^{m}\mu_{nj}\xi''_{nj},x\right\rangle-\sum_{j=1}^{m}\mu_{nj}h_j(x)\right] \qquad (12)$$

$$+\left\langle\sum_{i=1}^{p}\zeta_i,x\right\rangle+\alpha-\sum_{i=1}^{p}\lambda_i g_i(\overline{x})\varepsilon_i$$

Now,

$$\limsup_{n\to\infty}\left[-\sum_{j=1}^{m}\mu_{nj}h_j(x)\right]=$$

$$\limsup_{n\to\infty}\left[\begin{array}{l}\left(\left\langle\sum_{j=1}^{m}\mu_{nj}\xi''_{nj},x\right\rangle-\sum_{j=1}^{m}\mu_{nj}h_j(x)\right)-\\ \left\langle\sum_{j=1}^{m}\mu_{nj}\xi''_{nj},x\right\rangle\end{array}\right]$$

$$\leq\limsup_{n\to\infty}\left[\left(\left\langle\sum_{j=1}^{m}\mu_{nj}\xi''_{nj},x\right\rangle-\sum_{j=1}^{m}\mu_{nj}h_j(x)\right)\right]$$

$$+\limsup_{n\to\infty}-\left\{\left\langle\sum_{j=1}^{m}\mu_{nj}\xi''_{nj},x\right\rangle\right\}$$

which implies

$$\limsup_{n\to\infty}\left[\left(\left\langle\sum_{j=1}^{m}\mu_{nj}\xi''_{nj},x\right\rangle-\sum_{j=1}^{m}\mu_{nj}h_j(x)\right)\right]$$

$$\geq\limsup_{n\to\infty}\left[-\sum_{j=1}^{m}\mu_{nj}h_j(x)\right]+\liminf_{n\to\infty}\left[\left\langle\sum_{j=1}^{m}\mu_{nj}\xi''_{nj},x\right\rangle\right]$$

Using above in (12) and then using (1) we get

$$\left(\sum_{i=1}^{p}\lambda_i(f_i(.)-\overline{v}_i g_i(.))\right)(x)\geq\limsup_{n\to\infty}\left[-\sum_{j=1}^{m}\mu_{nj}h_j(x)\right]$$

$$+\alpha-\sum_{i=1}^{p}\lambda_i g_i(\overline{x})\varepsilon_i$$

which implies

$$\left(\sum_{i=1}^{p}\lambda_i(f_i(.)-\overline{v}_i g_i(.))\right)(x)+\liminf_{n\to\infty}\left[\sum_{j=1}^{m}\mu_{nj}h_j(x)\right]\geq\alpha$$

$$-\sum_{i=1}^{p}\lambda_i g_i(\overline{x})\varepsilon_i.$$

Hence $\liminf_{n\to\infty} L(x,\mu_n)\geq\alpha-\sum_{i=1}^{p}\lambda_i g_i(\overline{x})\varepsilon_i$.
Thus,

$$\max_{\mu_n\in R_+^m}\min_{x\in E}\liminf_{n\to\infty} L(x,\mu_n)\geq\alpha-\sum_{i=1}^{p}\lambda_i g_i(\overline{x})\varepsilon_i. \qquad (13)$$

To show

$$\max_{\mu_n\in R_+^m}\min_{x\in E}\liminf_{n\to\infty} L(x,\mu_n)\leq\alpha-\sum_{i=1}^{p}\lambda_i g_i(\overline{x})\varepsilon_i.$$

Since x is feasible for (P) and $\{\mu_{nj}\},\ j=1,2,\ldots,m\subset R_+$, we have

$$\left(\sum_{i=1}^{p}\lambda_i\,(f_i(.)-\overline{v}_i g_i(.))\right)(x)\geq\left(\sum_{i=1}^{p}\lambda_i\,(f_i(.)-\overline{v}_i g_i(.))\right)(x)$$

$$+\sum_{j=1}^{m}\mu_{nj}h_j(x)\quad\text{for all }\ x\in E$$

which implies

$$\inf_{x\in E}\left(\sum_{i=1}^{p}\lambda_i(f_i(.)-\overline{v}_i g_i(.))\right)(x)\geq L(x,\mu_n).$$

That is,

$$\max_{\mu_n\in R_+^m}\min_{x\in E}\liminf_{n\to\infty} L(x,\mu_n)\leq\alpha-\sum_{i=1}^{p}\lambda_i g_i(\overline{x})\varepsilon_i. \qquad (14)$$

(13) and (14) imply the required result.
Hence proved.

Corollary 4.1 *Let $\bar{x}$ be an ε-weak efficient solution of (P) with optimal value $\alpha - \sum_{i=1}^{p} \lambda_i g_i(\bar{x})\varepsilon_i$ where $\alpha = \sum_{i=1}^{p} \lambda_i (f_i(.) - \overline{v_i} g_i(.))(\bar{x})$. If, in the above theorem we impose a constraint qualification, that is, the set* $\left[\bigcup_{\mu_j \geq 0} \sum_{j=1}^{m} \mu_j epih_j^*\right]$ *is closed, then*

$\max_{\mu \in R_+^m} \min_{x \in E} L(x, \mu) = \alpha - \sum_{i=1}^{p} \lambda_i g_i(\bar{x})\varepsilon_i.$

Application [26]

The most important and common application of multiobjective fractional programming problem is transportation problem. Multiobjective linear fractional transportation problem is the problem with several criteria such as the maximization of the transport profitability like profit/cost or profit/time, and its two properties are source and destination. The problem is as follows:

Let there be m sources and n destinations. At each source, let $a_i, i = 1, 2, \ldots, m$ be the amount of homogenous products which are transported to n destinations to satisfy the demand for b_j, $j = 1, 2, \ldots, n$ units of the product there. Let x_{ij} be units of goods shipped from source i to destination j. For the objective function $Z_q(x), q = 1, 2, \ldots, Q$, profit matrix $p_q = [p_{ij}^q]_{m \times n}$ which determines the profit p_{ij}^q gained from shipment i to j, cost matrix $d_q = [d_{ij}^q]_{m \times n}$ which determines the cost d_{ij}^q per unit of shipment from i to j, scalars p_0^q, d_0^q which determine some constant profit and cost, respectively, the problem is

$$\text{maximize } Z_q(x) = \frac{p_q(x)}{d_q(x)} = \frac{\sum_{i=1}^{m} \sum_{j=1}^{n} p_{ij}^q x_{ij} + p_0^q}{\sum_{i=1}^{m} \sum_{j=1}^{n} d_{ij}^q x_{ij} + d_0^q}$$
$$q = 1, 2, \ldots, Q$$

subject to

$$\sum_{j=1}^{n} x_{ij} \leq a_i, \quad i = 1, 2, \ldots, m \tag{5.1}$$

$$\sum_{i=1}^{m} x_{ij} \geq b_j, \quad j = 1, 2, \ldots, n \tag{5.2}$$

$$x_{ij} \geq 0, i = 1, 2, \ldots, m, \quad j = 1, 2, \ldots, n \tag{5.3}$$

where $Z_q(x) = (Z_1(x), Z_2(x), \ldots, \quad Z_Q(x))$ is a vector of objective functions.

We suppose that $d_q(x) > 0, q = 1, 2, \ldots, Q$ and for all $x = (x_{ij}) \in S$, where $S \neq \phi$ denotes a convex and compact feasible set defined by (5.1), (5.2), (5.3). $p_q(x)$ and $d_q(x)$ are continuous on S.

Further, $a_i > 0$, for all i, $b_j > 0$, for all j, $p_{ij}^q > 0$, $d_{ij}^q > 0$, $p_0^q > 0$, $d_0^q > 0$, for all i, j and

$$\sum_{i=1}^{m} a_i \geq \sum_{j=1}^{n} b_j.$$

Conclusion

We know that constraint qualifications are required to obtain necessary optimality conditions but sometimes these constraint qualifications become very difficult to compute. In this paper, we develop sequential optimality conditions in the absence of any constraint qualification for multiobjective fractional programming problem (P) via scalarization and using the concept of epigraph of conjugate function in terms of ε-subdifferential computed at ε-weak efficient solution. Also, we derive sequential duality results for the problem (P).

Acknowledgments The author wishes to thank the unknown referees of this paper for valuable suggestions which have improved the final presentation of the paper.

References

1. Bai, F., Wu, Z., Zhu, D.: Sequential Lagrange multiplier condition for ε-optimal solution in convex programming. Optimization **57**(5), 669–680 (2008)
2. Bector, C.R., Chandra, S., Dutta, J.: Principles of Optimization Theory. Narosa, New Delhi (2005)
3. Choo, E.U., Atkins, D.R.: Proper efficiency in nonconvex programming. Math. Oper. Res. **8**, 467–470 (1983)
4. Coladas, L., Li, Z., Wang, S.: Optimality conditions for multiobjective and nonsmooth minimization in abstract spaces. Bull. Aust. Math. Soc. **50**, 205–218 (1994)
5. Gajek, L., Zagrodny, D.: Approximate necessary conditions for locally weak pareto optimality. J. Optim. Theory Appl. **82**, 49–58 (1994)
6. Geoffrion, A.M.: Proper efficiency and the theory of vector maximization. J. Optim. Theory Appl. **22**, 618–630 (1968)
7. Gerth, Chr: Nonconvex separation theorems and some applications in vector optimization. J. Optim. Theory Appl. **67**, 297–320 (1990)
8. Hiriart-Urruty, J.B.: ε-subdifferential. In: Aubin, J.B., Vinter, R. (eds.) Convex Analysis and Optimization, pp. 43–92. Pitman, London (1982)
9. Jeyakumar, V.: Asymptotic dual conditions characterizing optimality for convex programs. J. Optim. Theory Appl. **93**, 153–165 (1997)
10. Jeyakumar, V., Lee, G.M., Dinh, N.: New sequential lagrange multiplier conditions characterizing optimality without constraint qualifications for convex programs. SIAM J. Optim. **14**(2), 534–547 (2003)
11. Jeyakumar, V., Rubinov, A.M., Glover, B.M., Ishizuka, Y.: Inequality systems and global optimization. J. Math. Anal. Appl. **202**, 900–919 (1996)
12. Jeyakumar, V., Wu, Z.Y., Lee, G.M., Dinh, N.: Liberating the subgradient optimality conditions from constraint qualifications. J. Global Optim. **36**, 127–137 (2006)

13. Kaliszewski, I.: A theorem on nonconvex functions and its application to vector optimization. Eur. J. Oper. Res. **80**, 439–449 (1995)
14. Li, Z., Wang, S.: Lagrangian multipliers and saddle points in multiobjective programming. J. Optim. Theory Appl. **83**, 63–81 (1994)
15. Loridan, P.: Necessary conditions for ε-optimality. Math. Program. Study **19**, 140–152 (1982)
16. Loridan, P., Morgan, J.: Penalty functions in ε-programming and ε-minimax problems. Math. Program. **26**, 213–231 (1983)
17. Lui, J.C.: ε-Duality theorem of nondifferentiable nonconvex multiobjective programming. J. Optim. Theory Appl. **69**, 152–167 (1991)
18. Rockafellar, R.T., Wets, R.J.-B.: Variational Analysis. Springer, Berlin (1998)
19. Strodiot, J.J., Nguyen, V.H., Heukemes, N.: ε-Optimal solutions in nondifferentiable convex programming and some related questions. Math. Program. **25**, 307–328 (1983)
20. Tanaka, T.: A new approach to approximation of solutions in vector optimization problems. In: Proceedings of APROS, 1994, World Scientific, Singapore, pp 497–504 (1995)
21. Thibault, L.: Sequential convex subdifferential calculus and sequential lagrange multipliers. SIAM J. Control Optim. **35**, 1434–1444 (1997)
22. Yokoyama, K.: ε-Optimality criteria for convex programming problems via exact penalty functions. Math. Program. **56**, 233–243 (1992)
23. Yokoyama, K.: ε-Optimality criteria for vector minimization problems via exact penalty functions. J. Math. Anal. Appl. **187**, 296–305 (1994)
24. Yokoyama, K.: Epsilon approximate solutions for multiobjective programming problems. J. Math. Anal. Appl. **203**, 142–149 (1996)
25. Li, Z., Wang, S.: ε-Approximate solutions in multiobjective optimization. Optimization **44**, 161–174 (1998)

Additional Reference

26. Cetin N., Tiryaki F.: A Fuzzy Approach Using Generalized Dinkelbach's Algorithm for Multiobjective Linear Fractional Transportation Problem. Math. Probl. Eng. Article ID 702319, 1–10 (2014)

2

An analytical treatment to fractional Fornberg–Whitham equation

Mohamed S. Al-luhaibi[1]

Abstract In this paper, an analytical technique, namely the new iterative method (NIM), is applied to obtain an approximate analytical solution of the fractional Fornberg–Whitham equation. The obtained approximate solutions are compared with the exact or existing numerical results in the literature to verify the applicability, efficiency, and accuracy of the method.

Keywords New iterative method · Fractional Fornberg–Whitham equation · Approximate solution · Caputo's derivative · Partial differential equation

Introduction

In recent years, the fractional calculus used in many phenomena in engineering, physics, biology, fluid mechanics, and other sciences [1–8] can be described very successfully by models using mathematical tools from fractional calculus. Fractional derivatives provide an excellent instrument for the description of memory and hereditary properties of various materials and processes [9]. The fractional derivative has been occurring in many physical and engineering problems such as frequency-dependent damping behavior materials, signal processing and system identification, diffusion and reaction processes, creeping and relaxation for viscoelastic materials.

✉ Mohamed S. Al-luhaibi
alluhaibi.mohammed@yahoo.com

1 Department of Mathematics, Faculty of Science, Kirkuk University, Kirkuk, Iraq

The new iterative method (NIM), proposed first by Gejji and Jafari [10], has proven useful for solving a variety of nonlinear equations such as algebraic equations, integral equations, ordinary and partial differential equations of integer, and fractional order and systems of equations as well. The NIM is simple to understand and easy to implement using computer packages and yield better results than the existing Adomain decomposition method [11], homotopy perturbation method [12], and variational iteration method [13].

In the present paper, we have to solve the nonlinear time-fractional Fornberg–Whitham equation by the NIM. This equation can be written in operator form as:

$$u_t^{\alpha} u - u_{xxt} + u_x = uu_{xxx} - uu_x + 3u_x u_{xx}, t > 0, 0 < \alpha \leq 1 \quad (1)$$

with the initial condition

$$u(x, 0) = e^{\frac{x}{2}}, \quad (2)$$

where $u(x, t)$ is the fluid velocity, α is constant and lies in the interval (0, 1), t is the time and x is the spatial coordinate. Subscripts denote the partial differentiation unless stated otherwise. Fornberg and Whitham obtained a peaked solution of the form $u(x, t) = Ae^{-1/2(|x-4t/3|)}$ where A is an arbitrary constant.

Preliminaries and notations

In this section, we set up notation and review some basic definitions from fractional calculus [14, 15].

Definition 2.1 A real function $f(x)$, $x > 0$ is said to be in space C_{α}, $\alpha \in R$ if there exists a real number $p(> \alpha)$, such that $f(x) = x^p f_1(x)$ where $f_1(x) \in C[0, \infty)$.

Definition 2.2 A real function $f(x)$, $x > 0$ is said to be in space C_α^m, $m \in IN \cup \{0\}$ if $f^{(m)} \in C_\alpha$.

Definition 2.3 Let $f \in C_\alpha$and $\alpha \geq -1$, then the (left-sided) Riemann–Liouville integral of order μ, $\mu > 0$ is given by:

$$I_t^\mu f(x,t) = \frac{1}{\Gamma(\mu)} \int_0^t (t-\tau)^{\mu-1} f(x,\tau) d\tau, \qquad t > 0.$$

Definition 2.4 The (left-sided) Caputo's fractional derivative of f, $f \in C_{-1}^m$, $m \in IN \cup \{0\}$, is defined as:

$$D_t^\mu f(x,t) = \frac{\partial^m}{\partial t^m} f(x,t), \quad m = \mu$$

$$= I_t^{m-\mu} \frac{\partial^m f(x,t)}{\partial t^m}, \quad m-1 < \mu < m, m \in IN.$$

Note that

$$I_t^\mu D_t^\mu f(x,t) = f(x,t) - \sum_{k=0}^{m-1} \frac{\partial^k}{\partial t^k} f(x,0) \frac{t^k}{k!},$$
$$m-1 < \mu < m, m \in IN.$$

$$I_t^\mu t^\nu = \frac{\Gamma(\nu+1)}{\Gamma(\mu+\nu+1)} t^{\nu+\mu}.$$

Basic idea of new iterative method (NIM)

To describe the idea of the NIM, consider the following general functional equation [10, 16–20]:

$$u(x) = f(x) + N(u(x)), \tag{3}$$

where N is a nonlinear operator from a Banach space $B \to B$ and f is a known function. We are looking for a solution u of (3) having the series form:

$$u(x) = \sum_{i=0}^{\infty} u_i(x). \tag{4}$$

The nonlinear operator N can be decomposed as follows:

$$N\left(\sum_{i=0}^{\infty} u_i\right) = N(u_0) + \sum_{i=1}^{\infty} \left\{ N\left(\sum_{j=0}^{i} u_j\right) - N\left(\sum_{j=0}^{i-1} u_j\right) \right\}. \tag{5}$$

From Eqs. (4) and (5), Eq. (3) is equivalent to:

$$\sum_{i=0}^{\infty} u_i = f + N(u_0) + \sum_{i=1}^{\infty} \left\{ N\left(\sum_{j=0}^{i} u_j\right) - N\left(\sum_{j=0}^{i-1} u_j\right) \right\}. \tag{6}$$

We define the recurrence relation:

$$u_0 = f, \tag{7a}$$

$$u_1 = N(u_0), \tag{7b}$$

$$u_{n+1} = N(u_0 + u_1 + \cdots + u_n) - N(u_0 + u_1 + \cdots + u_{n-1}), \quad n = 1, 2, 3, \ldots. \tag{7c}$$

Then:

$$(u_1 + \cdots + u_{n+1}) = N(u_0 + u_1 + \cdots + u_n), \qquad n = 1, 2, 3, \ldots,$$

$$u = \sum_{i=0}^{\infty} u_i = f + N\left(\sum_{i=0}^{\infty} u_i\right) \tag{8}$$

If N is a contraction, i.e.,

$$\|N(x) - N(y)\| \leq k\|x-y\|, \quad 0 < k < 1,$$

then:

$$\|u_{n+1}\| = \|N(u_0 + u_1 + \cdots + u_n) - N(u_0 + u_1 + \cdots + u_{n-1})\|$$
$$\leq k\|u_n\| \leq \cdots \leq k^n \|u_0\| \; n = 0, 1, 2, \ldots, \tag{9}$$

and the series $\sum_{i=0}^{\infty} u_i$ absolutely and uniformly converges to a solution of (3) [21], which is unique, in view of the Banach fixed point theorem [22]. The k term approximate solution of (3) and (4) is given by $\sum_{i=0}^{k-1} u_i$.

Convergence analysis of the new iterative method (NIM)

Now, we introduce the condition of convergence of the NIM, which is proposed by Daftardar-Gejji and Jafari in (2006) [10], also called (DJM) [23].

From (5), the nonlinear operator N is decomposed as follows:

$N(u) = N(u_0) + [N(u_0 + u_1) - N(u_0)] + [N(u_0 + u_1 + u_2) - N(u_0 + u_1)] + \ldots$.

Let $G_0 = N(u_0)$ and

$$G_n = N\left(\sum_{i=0}^{n} u_i\right) - N\left(\sum_{i=0}^{n-1} u_i\right), \qquad n = 1, 2, 3, \ldots. \tag{10}$$

Then $N(u) = \sum_{i=0}^{\infty} G_i$.

Set:

$$u_0 = f, \tag{11}$$

$$u_n = G_{n-1}, \quad n = 1, 2, 3, \ldots. \tag{12}$$

Then:

$$u = \sum_{i=0}^{\infty} u_i \tag{13}$$

is a solution of the general functional Eq. (3). Also, the recurrence relation (7) becomes

$$u_0 = f,\qquad u_n = G_{n-1},\quad n = 1,2,\ldots. \tag{14}$$

Using Taylor series expansion for G_i, i = 1, 2, ..., n, we have

$$\begin{aligned} G_1 &= N(u_0+u_1) - N(u_0) \\ &= N(u_0) + N'(u_0)u_1 + N''(u_0)\frac{u_1^2}{2!} + \cdots - N(u_0) \\ &= \sum_{k=1}^{\infty} N^k(u_0)\frac{u_1^k}{k!}, \end{aligned} \tag{15}$$

$$\begin{aligned} G_2 &= N(u_0+u_1+u_2) - N(u_0+u_1) \\ &= N'(u_0+u_1)u_2 + N''(u_0+u_1)\frac{u_2^2}{2!} + \cdots \\ &= \sum_{j=1}^{\infty}\left[\sum_{i=0}^{\infty} N^{(i+j)}(u_0)\frac{u_1^i}{i!}\right]\frac{u_2^j}{j!} \end{aligned} \tag{16}$$

$$G_3 = \sum_{i_3=1}^{\infty}\sum_{i_2=0}^{\infty}\sum_{i_1=0}^{\infty} N^{(i_1+i_2+i_3)}(u_0)\frac{u_3^{i_3}\,u_2^{i_2}\,u_1^{i_1}}{i_3!\,i_2!\,i_1!}. \tag{17}$$

In general:

$$G_n = \sum_{i_n=1}^{\infty}\sum_{i_{n-1}=0}^{\infty}\cdots\sum_{i_1=0}^{\infty}\left[N^{(\sum_{k=1}^{n} i_k)}(u_0)\left(\prod_{j=1}^{n}\frac{u_j^{i_j}}{i_j!}\right)\right]. \tag{18}$$

In the following theorem, we state and prove the condition of convergence of the method.

Theorem 3.1 *If N is $C^{(\infty)}$ in a neighborhood of u_0 and*

$$\|N^{(n)}(u_0)\| = \sup\left\{N^{(n)}(u_0)(h_1,\ldots,h_n) : \|h_i\| \le 1,\ 1\le i\le n\right\} \le L, \tag{19}$$

for any n and for some real $L > 0$ and $\|u_i\| \le M < \frac{1}{e}$, $i = 1,2,\ldots$, then the series $\sum_{n=0}^{\infty} G_n$ is absolutely convergent, and moreover,

$$\|G_n\| \le LM^n e^{n-1}(e-1),\ n = 1,2,\ldots. \tag{20}$$

Proof In view of (18)

$$\begin{aligned} \|G_n\| &\le LM^n \sum_{i_n=1}^{\infty}\sum_{i_{n-1}=0}^{\infty}\cdots\sum_{i_1=0}^{\infty}\left[\left(\prod_{j=1}^{n}\frac{u_j^{i_j}}{i_j!}\right)\right] \\ &= LM^n e^{n-1}(e-1). \end{aligned} \tag{21}$$

□

Thus, the series $\sum_{n=1}^{\infty}\|G_n\|$ is dominated by the convergent series $LM(e-1)\sum_{n=1}^{\infty}(Me)^{n-1}$, where $M < \frac{1}{e}$. Hence, $\sum_{n=0}^{\infty} G_n$ is absolutely convergent, due to the comparison test. For more details, see [23].

Reliable algorithm of new iterative method (NIM) for solving the Linear and Nonlinear fractional partial differential equations

After the above presentation of the NIM, we introduce a reliable algorithm for solving nonlinear fractional PDEs using the NIM. Consider the following nonlinear fractional PDE of arbitrary order:

$$D_t^{\alpha}u(x,t) = A(u,\partial u) + B(x,t),\qquad m-1 < \alpha \le m,\ m \in IN \tag{22}$$

with the initial conditions

$$\frac{\partial^k}{\partial t^k}u(x,0) = h_k(x),\quad k = 0,1,2,\ldots,m-1, \tag{23}$$

where A is a nonlinear function of u and ∂u (partial derivatives of u with respect to x and t) and B is the source function. In view of the integral operators, the initial value problem (22) is equivalent to the following integral equation

$$u(x,t) = \sum_{k=0}^{m-1} h_k(x)\frac{t^k}{k!} + I_t^{\alpha}B(x,t) + I_t^{\alpha}A = f + N(u), \tag{24}$$

where

$$f = \sum_{k=0}^{m-1} h_k(x)\frac{t^k}{k!} + I_t^{\alpha}B(x,t), \tag{25}$$

and

$$N(u) = I_t^{\alpha}A, \tag{26}$$

where I_t^n is an integral operator of n fold. We get the solution of (24) by employing the algorithm (7).

Solution of the problem

We first consider the following time-fractional Fornberg–Whitham equation [24, 25]:

$$D_t^{\alpha}u - D_{xxt}u + D_x u = uD_{xxx}u - uD_x u + 3D_x u D_{xx}u, \qquad (27)$$

with the initial condition:

$$u(x,0) = e^{\frac{x}{2}}. \qquad (28)$$

Then, the exact solution is given by:

$$u(x,t) = e^{\frac{x}{2}-\frac{2t}{3}}. \qquad (29)$$

Note that Eq. (27) is equivalent to the integral equation

$$u(x,t) = e^{\frac{x}{2}} + I_t^{\alpha}\left[D_t^{\alpha}u - D_{xxt}u + D_x u + uD_{xxx}u - uD_x u + 3D_x u D_{xx}u\right],$$

where $f = e^{\frac{x}{2}}$ and $N(u) = I_t^{\alpha}\left[D_t^{\alpha}u - D_{xxt}u + D_x u + uD_{xxx}u - uD_x u + 3D_x u D_{xx}u\right]$, using (7) we get

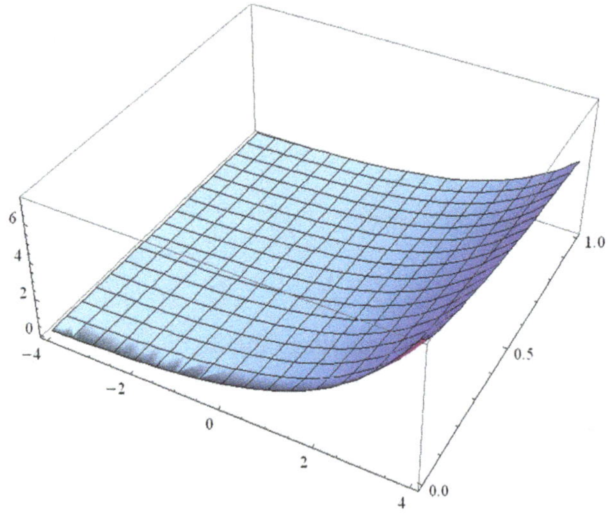

Fig. 1 The surfaces show the approximate solutions $u_5(x, t)$ for $\alpha = 2/3$

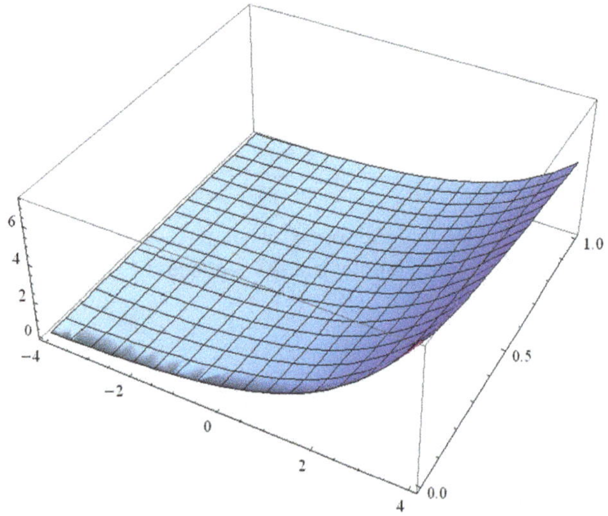

Fig. 2 The surfaces show the approximate solutions $u_5(x, t)$ for $\alpha = 3/4$

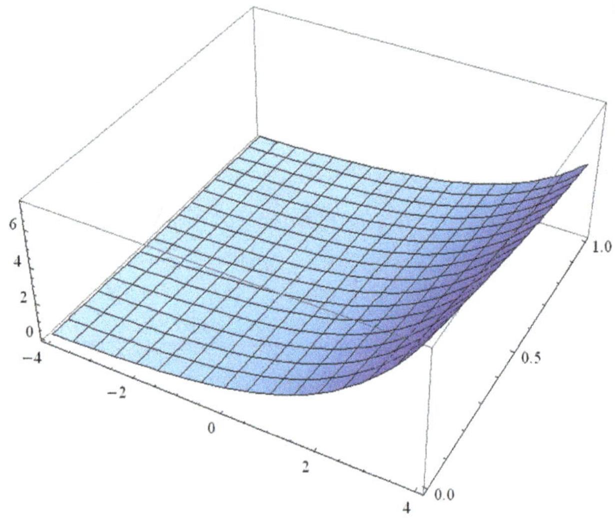

Fig. 3 The surfaces show the approximate solutions $u_5(x, t)$ for $\alpha = 1$

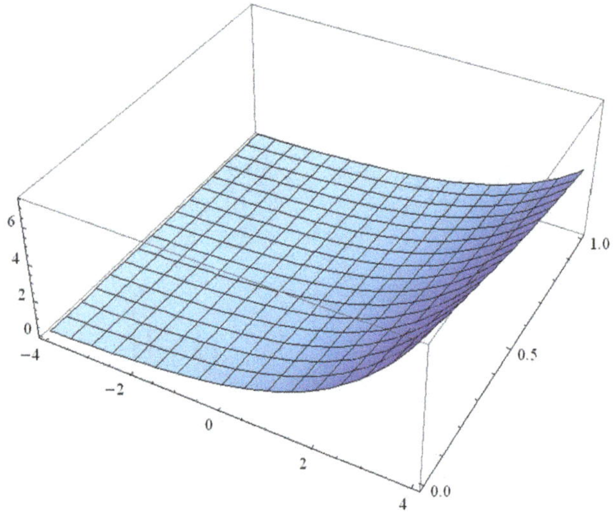

Fig. 4 The surfaces show the exact solution of $u(x, t)$

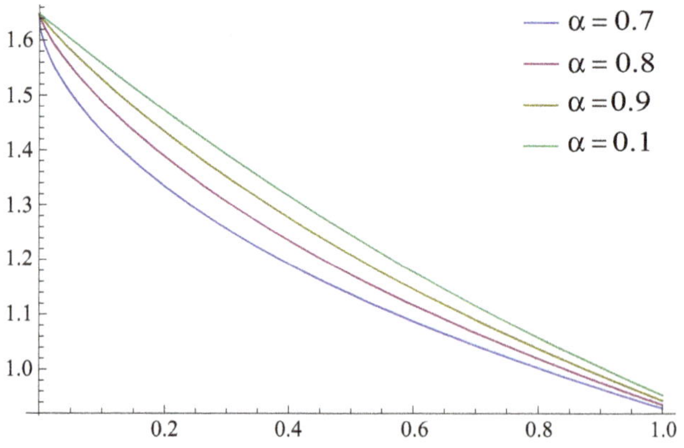

Fig. 5 Plots of $u(x, t)$ at $x = 1$ for different values of α

Table 1 The absolute errors for differences between the exact solution and 5th-order NIM when $\alpha = 1$

x_j/t_j	0.2	0.4	0.6	0.8	1.0
−4	2.22193×10^{-5}	9.47416×10^{-6}	4.83886×10^{-5}	6.71560×10^{-5}	5.36314×10^{-5}
−2	6.03987×10^{-5}	2.57532×10^{-5}	1.31533×10^{-4}	1.82549×10^{-4}	1.45785×10^{-4}
0	1.64180×10^{-5}	7.00049×10^{-5}	3.57546×10^{-4}	4.96219×10^{-4}	3.96285×10^{-4}
2	4.46289×10^{-4}	1.90293×10^{-4}	9.71910×10^{-4}	1.34886×10^{-3}	1.07721×10^{-3}
4	1.21314×10^{-4}	5.17269×10^{-4}	2.64192×10^{-3}	3.66659×10^{-3}	2.92817×10^{-3}

$$
\begin{aligned}
u_0(x,t) &= e^{\frac{x}{2}},\\
u_1(x,t) &= -\frac{1}{2}e^{\frac{x}{2}}\frac{t^{\alpha}}{\Gamma(\alpha+1)},\\
u_2(x,t) &= -\frac{1}{8}e^{\frac{x}{2}}\frac{t^{2\alpha-1}}{\Gamma(2\alpha)}+\frac{1}{2}e^{\frac{x}{2}}\frac{t^{2\alpha}}{\Gamma(2\alpha+1)},\\
u_3(x,t) &= -\frac{1}{32}e^{\frac{x}{2}}\frac{t^{3\alpha-2}}{\Gamma(3\alpha-1)}+\frac{1}{8}e^{\frac{x}{2}}\frac{t^{3\alpha-1}}{\Gamma(3\alpha)}\\
&\quad -\frac{1}{8}e^{\frac{x}{2}}\frac{t^{3\alpha}}{\Gamma(3\alpha+1)},\\
&\vdots
\end{aligned}
$$

Therefore, the NIM series solution is:

$$
\begin{aligned}
u(x,t) = e^{\frac{x}{2}}\bigg(1 &- \frac{t^{\alpha}}{2\Gamma(\alpha+1)} - \frac{t^{2\alpha-1}}{8\Gamma(2\alpha)} + \frac{t^{2\alpha}}{2\Gamma(2\alpha+1)}\\
&- \frac{t^{3\alpha-2}}{32\Gamma(3\alpha-1)} + \frac{t^{3\alpha-1}}{8\Gamma(3\alpha)} - \frac{t^{3\alpha}}{8\Gamma(3\alpha+1)} + \cdots\bigg)
\end{aligned} \tag{30}
$$

Numerical results and discussion

In this section, we calculate numerical results of the displacement $u(x, t)$ for different time-fractional Brownian motions $\alpha = 2/3$, 3/4, 1 and for various values of t and x. The numerical results for the approximate solution (30) obtained using NIM and the exact solution for various values of t, x, and a are shown by Figs. 1, 2, 3, 4 and those for different values of t and α at $x = 1$ are depicted in Fig. 5. It is observed from Figs. 1, 2, 3, 4 that $u(x, t)$ increases with the increase in both x and t for $\alpha = 2/3$, 3/4 and $\alpha = 1$. Figure 4 clearly show that, when $\alpha = 1$, the approximate solution (30) obtained by the present method is very near to the exact solution. It is also seen from Fig. 5 that as the value of a increases, the displacement $u(x, t)$ increases but afterward its nature is opposite. Finally, we remark that the approximate solution (30) is in full agreement with the results obtained homotopy perturbation method [24] and homotopy perturbation transform method [25]. In Table 1, we compute the absolute errors for differences between the exact solution (29) and the approximate solution (30) obtained by the NIM at some points.

Conclusion

In this paper, the new iterative method (NIM) has been applied for approximating the solution for the nonlinear fractional Fornberg–Whitham equation. The accuracy of the NIM for solving nonlinear fractional Fornberg–Whitham equation is good compared to the literature; however, it has the advantage of reducing the computations complexity presented in other perturbation techniques. In fact, in NIM, nonlinear problems are solved without using Adomian's polynomials or He's polynomials that appear in the decomposition methods. The numerical results show that the proposed method is reliable and efficient technique in finding approximate solutions for nonlinear differential equations.

Acknowledgements I am grateful to the prof. Mohamed A. Ramadan (Department of Mathematics, Faculty of Science, Menoufia University, Egypt) and referees for their helpful comments and suggestions that enhanced the paper.

References

1. Oldham, K.B., Spanier, J.: The fractional calculus. Academic Press, New York (1974)
2. Podlubny, I.: Fractional differential equations. Academic Press, New York (1999)
3. Kumar, S.: A numerical study for solution of time fractional nonlinear shallow-water equation in oceans. Zeitschrift fur Naturforschung A **68a**, 1–7 (2013)
4. Kumar, S.: A new analytical modelling for telegraph equation via Laplace transform. Appl. Math. Model. **38**(13), 3154–3163 (2014)
5. Kumar, S., Rashidi, M.M.: New analytical method for gas dynamics equation arising in shock fronts. Comput. Phys. Commun. **185**(7), 1947–1954 (2014)
6. Kumar, S.: A fractional model to describing the Brownian motion of particles and its analytical solution. Adv. Mech. Eng. **7**(12), 1–11 (2015)
7. Kumar, S.: A modified homotopy analysis method for solution of fractional wave equations. Adv. Mech. Eng. **7**(12), 1–8 (2015)
8. Kumar, S.: An analytical algorithm for nonlinear fractional Fornberg–Whitham equation arising in wave breaking based on a new iterative method. Alex. Eng. J. **53**(1), 225–231 (2014)
9. Kilbas, A.A., Srivastava, H.M., Trujillo, J.J.: Theory and applications of fractional differential equations. Elsevier, Amsterdam (2006)
10. Gejji, V.D., Jafari, H.: An iterative method for solving nonlinear functional equations. J. Math. Anal. Appl. **316**(2), 753–763 (2006)

11. Adomian, G.: Solving Frontier problems of physics. The decomposition method. Kluwer, Boston (1994)
12. He, J.H.: Homotopy perturbation technique. Comput. Methods Appl. Mech. Eng. **178**, 257–262 (1999)
13. He, J.H.: Variational iteration method-akind of nonlinear analytical technique: some examples. Int. J. Nonlinear Mech. **34**, 699–708 (1999)
14. Podlubny, I.: Fractional differential equations. Academic Press, San Diego (1999)
15. Luchko, Yu., Gorenflo, R.: An operational method for solving fractional differential equations with the Caputo derivatives. Acta Math. Vietnamica **24**, 207–233 (1999)
16. Al-luhaibi, M.S.: New iterative method for fractional gas dynamics and coupled burger's equations, The Scientific World Journal, Volume 2015, Article ID 153124, 8 pages
17. Ramadan, M.A., Al-luhaibi, M.S.: New iterative method for Cauchy problems. J. Math. Comput. Sci. **5**(6), 826–835 (2015)
18. Ramadan, M.A., Al-luhaibi, M.S.: New iterative method for solving the Fornberg–Whitham equation and comparison with homotopy perturbation transform method. Br. J. Math. Comput. Sci. **4**(9), 1213–1227 (2014)
19. Hemeda, A.A., Al-luhaibi, M.S.: New iterative method for solving gas dynamic equation. Int. J. Appl. Math. Res. **3**(2), 190–195 (2014)
20. Hemeda, A.A.: New iterative method: an application for solving fractional physical differential equations, Abstract and Applied Analysis Volume 2013, Article ID 617010, 9 pages
21. Cherruault, Y.: Convergence of Adomian's method. Kybernetes **18**(2), 31–38 (1989)
22. Jerri, A.J.: Introduction to integral equations with applications, 2nd edn. Wiley-Interscience, New York (1999)
23. Bhalekar, S., Gejji, V.D.: Convergence of the new iterative method. Int. J. Differ. Equ. 2011, Article ID 989065, 10 pages
24. Guptaa, P.K., Singh, M.: Homotopy perturbation method for fractional Fornberg–Whitham equation. Comput. Math Appl. **61**, 250–254 (2011)
25. Singh, J., Kumar, D., Kumar, S.: New treatment of fractional Fornberg–Whitha equation via Laplace transform. Ain Shams Eng. J. **4**, 557–562 (2013)

3

Optimal performance of heterogeneous networks based on the bit rate

Déthié Dione[1] · Salimata Gueye Diagne [1] · Bakary Koné [1] · Youssou Gningue[2]

Abstract Networks LTE(4G) and Wi-Fi complementarity establish a heterogeneous system of wireless and mobile networks. We study and analyze the optimal performances of this heterogeneous system based on the bit rate, the blocking probability and user connection loss. Random Waypoint is the user mobility model. User's provided mobile terminal equipped with multiple accesses interfaces. We have developed a Markov chain to estimate the performances obtained from the heterogeneous networks system, which allowed us to propose an average bit rate value in a sub-zone of this system then the average blocking probability user connection in this zone. We have also proposed a sensitivity factor of maximal decrease of these selection network parameters. This factor informs about the heterogeneous networks congestion and dis-congestion system.

Keywords LTE · Wi-Fi · Random waypoint · Handover · Markov chain · BLER · Bit rate · Blocking probability

Introduction

The integration of wireless and mobile networks such as longterm evolution (LTE) and Wi-Fi is nowadays a necessity for the satisfaction of user request which is stronger and stronger. The global services and user mobility makes this task difficult. However these networks are provided with characteristics to support services and user mobility.

The Random waypoint (RWP) is model chosen for users mobility in this heterogeneous networks system. The RWP model corresponds to the ideal behavior of a user in an urban area because according to the model of mobility RWP, every user chooses randomly a place of destination and goes to this one in a constant velocity. The user movement starting up from a point to a destination is named "one movement epoch". The users velocity in every epoch is random variable and is chosen from a uniform distribution of the velocity [0, *V*max], where *V*max is the eligible maximal velocity for user. In the RWP, the user can wait for a period of time called "time of reflection" before his departure for another point. The browsed path of a user is independent from its previous path and from other users path. Thus, at the end of every time of movement, the user stops a duration of the time and then chooses another destination place and, possibly, new velocity, and moves towards this destination in a constant velocity, and so on. RWP is one of the mobility models widely used in the performances analysis of the wireless and mobile networks. It represents well the individual movements which include the stop, the starting up and the other actions bound to the individual movements in cities.

So new methods for saving, transmission and sharing of bandwidth are imperative. Among these we can mention the selection technique of best network based on the bit rate which the selection parameters of which are the blocking probability and connections losses.

We noted in the literature the most used selection strategies of network. In their works [10], we were able to analyze the signal power received (RSNS) and then the available bandwidth (TBNS) of a heterogeneous networks system. They emphasized the parameters of this system such as the blocking probability and connections losses but they did not take into account the interference in the selection techniques

✉ Déthié Dione
dethiedione79@gmail.com

1 Department of Mathematic and Computer Science, University of Cheikh Anta Diop, Dakar, Senegal

2 University of Laurentia, Sudbury, Canada

that they developed which made less successful results obtained from the blocking probability and connections losses. Besides the authors [1, 2, 12, 13] considered the interference in the selection strategies they adopted which is the one based on the SINR, which allowed them to decrease the probability connections losses during a vertical handover. But through their studies, they did not approach the parameters connections blocking. On the other hand, these are analyzed by [7] who in their works obtained results more satisfactory than those who are preceded them. Besides, [11] took into account the users mobility(terminal-controlled mobility management) and other aspects such as the cost, the battery life cycle and the handover frequency.

However, through their studies, none of these authors took into account the constraint related to the block error rate (BLER) in a sub-cell given by the cluster. The contribution in our works bases on our model users connected and disconnected. When he is connected, the system quits a well-defined state and moves to an other state before returning to this first one in a given time interval. Besides, unlike the other works, we took the bit rate as an important selection technique of the best network which consists in choosing the biggest bit rate value.

The paper is organized as follows: "Model of heterogeneous networks system" section introduces the model of the studied heterogeneous networks where all the parameters of the system are defined. The selection method algorithm based on the bit rate is established at the level of "Selection method" section. In "User mobility model" section, we have developed the users mobility model, it is Random Waypoint (RWP) which we consider more adequate to the individual users movements. The average access demand rate for a service is given in "Average new access demand rate $\lambda_{C_i}^{C(k)}$ for a service" section. The average rate demand of vertical and horizontal handover is calculated in "Average demand rate of handovers" section. In "Modeling approach based on a Markov chain" section, we have used a Markov chain to analyze the number busy bandwidth units. The system studied performances such as the bit rate and the blocking probability and connection losses are estimated in "Evaluations of optimal performances" section. The results obtained are simulated in "Numerical tests" section. We finished our study in "Conclusion" section by conclusion and future work.

Model of heterogeneous networks system

The model of heterogeneous networks system which we study is represented by Fig. 1.

Indeed, we have an hexagonal area of service C_1 covered entirely by the mobile network LTE (4G). In this service area are present several homogeneous circular subcells $(C_j)_{2\leq j\leq m}$ of radius r_i among which each is also covered by a Wi-Fi wireless. So both mobile network LTE and Wi-Fi wireless overlap in cells C_i and the Wi-Fi wireless are separated between them. We denote by C_0 the part of the service area not covered by a Wi-Fi wireless. Where from we have:

$$C_0 = C_1 - \bigcup_{j=2}^{m} C_j \tag{1}$$

As the users are equipped with devices of multiple accesses, they have the possibility to connect or disconnect from a network in the cells where networks overlap by choosing automatically the network which has the best bit rate. In our study, we suppose that the LTE network supplies two types of services: those Multicasts or Unicasts whose numbers of units of bandwidth are respectively B_1^{mc} et B_1^{uc}. Besides, the number of units of bandwidth for every Wi-Fi wireless is B_i.

In the service area C_1, we suppose to have Q interferences sources distributed following a normal random distribution: $Q = \{I(q), q = 1, \ldots, Q\}$.

The selection technique is based on the bit rate then these interferences play an important role at the level of

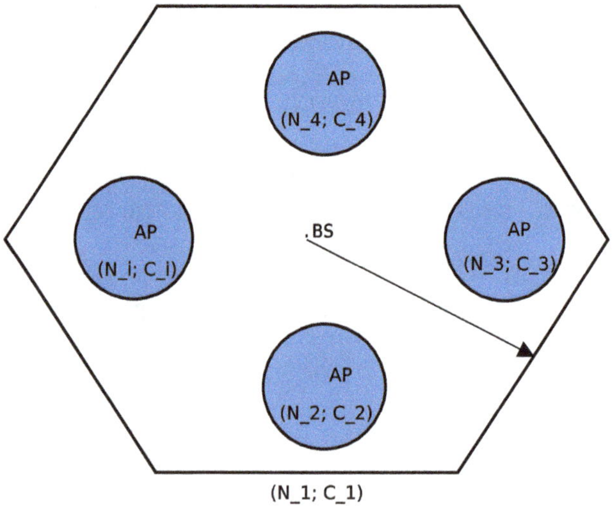

Fig. 1 Model cluster service zone

Table 1 LTE network parameters

Parameters	LTE networks N_1	
Interferences	Q sources	
Covered zone	C_1	
Service quality	Low bandwidth	
Types of services	Multicast	Unicast
Units of bandwidth	B_1^{mc}	B_1^{uc}

Table 2 Wi-Fi wireless parameters

Parameters	Wi-Fi wireless $N_{i=2,\ldots,m}$
Interferences	Q sources
Covered zones	$C_{i=2,\ldots,m}$
Service quality	Large bandwidth
Types of services	Unicast
Units of bandwidth	B_i

this strategy allowing to select a network. The different parameters of this heterogeneous networks system are given in the following Tables 1 and 2:

Selection method

When a user is in a cell C_i where both LTE mobile network and Wi-Fi wireless overlap then he has the possibility of connecting recently or by handover to the LTE network or Wi-Fi. If the LTE network has more free bandwidth units then user connects there otherwise he is blocked to connect to the Wi-Fi wireless as indicates Fig. 2. Besides, when a user wishes to connect recently in one sub-zone C_i then his

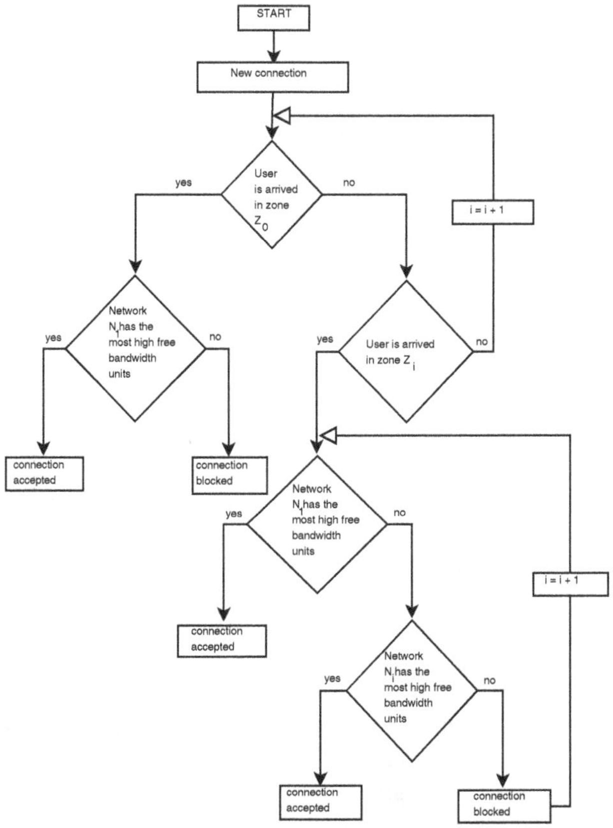

Fig. 2 Selection technique

terminal makes the same technique to select that previously.

Beyond choices connections accepted and blocked, we can add the third choice which we would nickname: "the selection technique of a network is striggered". However, we supposed indirectly this choice without quoting it in particular and we do not consider necessary to add this choice in the plan of selection technique.

As the selection technique is based on the network which has free bandwidth units then the mobile terminal detects all the accesses points and all the base stations in its neighborhood then estimates automatically the bit rate of each of them. This one depends on the sub-carrier bandwidth and modulation.

User mobility model

In the Random WayPoint (RWP) model, we suppose that a user moves in a convex space $C_1 \subset \mathbb{R}^2$ along a straight segment coming from any point towards another point. The various points which were occupied by the user are denoted by P_i. These points are uniformly distributed in C_1, $P_i \sim U(C_1)$. The transition of P_{i-1} to P_i corresponds to the i_{th} straight segment, and the velocity of the user on this segment is defined as a random variable v_i, $v_i \sim v$. Particularly, the RWP model assures that variables P_i and v_i are independent. By basing itself of this notation, the process of the RWP for a user is defined by an infinite sequence of triplets,

$$\{(P_0, P_1, v_1), (P_1, P_2, v_2), \ldots, (P_{n-1}, P_n, v_n)\}$$

We can note that in the process of the RWP model, the consecutive straight segments of movement are independent because they share a point in common. However, let us note that this process is reversible in the time as far as, any path $P_0, P_1, \ldots, P_n$ is equivalent at time to the inverse path $P_n, P_{n-1}, \ldots, P_0$.

Probability density

The probability density of finding a user situated at a distance x of the convex cluster center C_1 and in a circular sub-cell of radius r_i placed at a distance d_i of the service area center is defined by [5]:

$$h(x) = (1 - x^2) \int_0^{\pi} \sqrt{1 - x^2 \cos(\Phi + \alpha - \beta)} \mathrm{d}\Phi \qquad (2)$$

with

$$x = \sqrt{d_i^2 + 2 d_i r_i \cos\alpha + r_i^2}$$

$$\beta = \arctan(d_i + r_i \cos\alpha; r_i \sin\alpha)$$

The arrival rate $\tau_a(C_i)$ of a user in a sub-cell C_i

By means Random WayPoint model and results obtained by [4], we have proved that the average arrived rate of a user in a sub-cell C_i of radius r_i and situated at a distance d_i of the cluster center is established by equation:

$$\tau_a(Z_i) = \frac{2}{C_v} \int_0^{\pi} \int_0^{\pi} r_i.h(x) \sin \Phi \mathrm{d}\Phi \mathrm{d}\alpha \tag{3}$$

Probability $\mathbb{P}(C_i)$ of finding users in a sub-cell C_i

The probability of finding users in a cell C_i of radius r_i situated at a distance d_i of the service area center depends directly on the users mobility. Thus we are based on the Random WayPoint and works results of the authors [3] to show this probability by equation:

$$\mathbb{P}(C_i) = \int_0^{R} \int_0^{\pi} \int_0^{2\pi} r_i.h(x) \mathrm{d}\Phi \mathrm{d}\alpha \mathrm{d}r_i \tag{4}$$

Average new access demand rate $\lambda_{C_i}^{C(k)}$ for a service

Let us denote by $\lambda_{C_1}^{C(k)}$ the average demand access rate for a service k in a cell C_1. So the average demand access rate for a service k, $\lambda_{C_i}^{C(k)}$, being in one sub-cell C_i is defined by formula:

$$\lambda_{C_i}^{C(k)} = \mathbb{P}(C_i) \cdot \lambda_{C_1}^{C(k)} \tag{5}$$

Average demand rate of handovers

A connected user and then in mobility makes indirectly handover in the heterogeneous networks system. This handover is horizontal or vertical.

Horizontal

Let $u_{C_0}^{k}$ the average user number having access to the service k in the cell C_0. The average demands rate of horizontal handover $\tau_{C_0}^{H(k)}$ to the network N_1 for a service k is given by relation 6 referring to the author (Jabban et al. [6]):

$$\tau_{C_1}^{H(k)} = u_{C_0}^{k} \cdot \eta_{C_0}^{C_1} \tag{6}$$

where $\eta_{C_0}^{C_1}$ is the users exit flow of the zone C_0 outside cell C_1 and is defined by:

$$\eta_{C_0}^{C_1} = \frac{\mathbb{P}(C_0)}{\Delta_{C(k)}}$$

with $\Delta_{C(k)}$ the average residence time of users in a zone C_i.

Vertical

We denote by $u_{C_0}^{k}$ the average number of mobile users who have accessed to the service k in the cell C_0. The average demands rate of vertical handover $\tau_{C_0}^{V(k)}$ to the network N_i of users who have accessed to the service k in the zone C_0 and moving towards the cell C_i without finishing their connections is defined by Eq. 7 refer to the author (Jabban et al. [6]):

$$\tau_{C_i}^{V(k)} = u_{C_0}^{k} \cdot \eta_{C_0}^{C_i} \tag{7}$$

where $\eta_{C_0}^{C_i}$ is the users exit flow of the cell C_0 towards cell C_i and is given by:

$$\eta_{C_0}^{C_i} = \frac{\mathbb{P}(C_0)}{\Delta_{C(k)}}$$

with $\Delta_{C(k)}$ the average residence time of the users in a zone C_i.

Modeling approach based on a Markov chain

We have developed a Markov chain for modeling the dynamic fluctuations and define all the stages and states of the heterogeneous networks system.

Supposing that the set cells and services which are present in the service area, are respectively denoted M and S so the Markov chain size is given by : $s.(2m+1)$ with $\mid M \mid = m$ and $\mid S \mid = s$.

Different stages and states of the system

When we take the system at the given moment then we define it as being a stage of dynamic change.

Besides, user connections and disconnections from a network give the various states of the system which define the states space by (see Table 3):

$$\mathcal{E} = \{(b_{1,1}^{k}; b_{1,2}^{k}; \ldots; b_{1,i}^{k}; \ldots; b_{1,m}^{k}; ; b_{2}^{k}; \ldots; b_{i}^{k}; \ldots; b_{m}^{k})\}$$

$$s.q/\begin{cases} \sum_{j=1}^{m} \sum_{k=1}^{s} (b_{1j}^{k}) \leq B_1^{uc} \\ \sum_{k=1}^{s} (b_i^{k}) \leq B_i \end{cases}$$

States are differentiated by the possible variation of the units of busy bandwidth in a given zone. For example when the system is in a given state, it changes state if a user connects recently or to make a handover either disconnects by freeing units of busy bandwidth.

Table 3 Definition of system parameters

Parameters	Definitions
$b_{1,1}^k$	Number of units of busy bandwidth of LTE network to the service k in the cell C_0
$b_{1,i}^k$	Number of units of busy bandwidth of LTE network to the service k in the cell C_i
b_i^k	Number of units of busy bandwidth of LTE network to the service k in the cell C_i
N_{PRB}^k	Number of resources blocks asked to supply a service k by the LTE network in the cell C_i
$\sum_{k=1}^{s} b_{1,1}^k = b_{1,1}$	Number of units of busy bandwidth of LTE network in the cell C_0
$\sum_{k=1}^{s} b_{1,i}^k = b_{1,i}$	Number of units of busy bandwidth of LTE network in the cell C_i
$\sum_{k=1}^{s} b_i^k = b_i$	Number of units of busy bandwidth of Wifi wireless in the cell C_i

$$b_{11} \longrightarrow b_{11} + N_{\mathrm{PRB}}^k$$

Where b_{11} is the number of units of busy bandwidth in the zone C_1 by the network N_1 and the number of busy blocks resources by a user for a service k in the zone C_1 to the network N_1.

- Stage 0: $E_0 = (b_{1,1}^k;\ldots;b_{1,i}^k;\ldots;b_{1,m}^k;b_2^k;\ldots;b_i^k;\ldots;b_m^k)$
- Stage 1: $E_1 = (b_{1,1}^k + \mu N_{prb}^k;b_{1,2}^k;\ldots;b_{1,m}^k;b_2^k;\ldots;b_m^k)$

$$\mu = \begin{cases} 1 & \text{if a user } \mathrm{u_0} \text{ connects to the network LTE in the cell } \mathrm{C_0} \quad \mathrm{State(E_{1,1})} \\ -1 & \text{if a user } \mathrm{u_0} \text{ disconnects of the LTE from the cell } \mathrm{C_0} \quad \mathrm{State\ (E_{1,2})} \\ 0 & \text{Otherwise} \quad \mathrm{State(E_{1,3})} \end{cases}$$

- Stage 2: $E_2 = (b_{1,1}^k;\ldots;b_{1,i}^k + \mu N_{prb}^k;\ldots;b_{1,m}^k;b_2^k;\ldots;b_m^k)$

$$\mu = \begin{cases} 1 & \text{if a user } \mathrm{u_0} \text{ connects to the network LTE in the cell } \mathrm{C_i} \quad \mathrm{State(E_{2,1})} \\ -1 & \text{if a user } \mathrm{u_0} \text{ disconnects of the network LTE from the cell } \mathrm{C_i} \quad \mathrm{State(E_{2,2})} \\ 0 & \text{Otherwise} \quad \mathrm{State(E_{2,3})} \end{cases}$$

- Stage 3: $E_3 = (b_{1,1}^k;\ldots;b_{1,m}^k;b_2^k;\ldots;b_i^k + \mu;\ldots;b_m^k)$

$$\mu = \begin{cases} 1 & \text{if a user } \mathrm{u_0} \text{ connects to the wireless Wi} - \text{Fi in the cell } \mathrm{C_i} \quad \mathrm{State(E_{3,1})} \\ -1 & \text{ifauser } \mathrm{u_0} \text{ disconnects of the wireless Wi} - \text{Fi from the cell } \mathrm{C_i} \quad \mathrm{State(E_{3,1})} \\ 0 & \text{Otherwise} \quad \mathrm{State(E_{3,1})} \end{cases}$$

- Stage 4: $E_4 = (b^k_{1,1} + \mu N^k_{PRB}; \ldots; b^k_{1,i} + \mu' N^k_{PRB}; \ldots; b^k_{1,m}; b^k_2; \ldots; b^k_m)$

$$(\mu, \mu') = \begin{cases} (1,-1) & \text{if a user } \mathrm{u_0} \text{ connects to the network LTE in the cell } \mathrm{C_0} \\ & \text{by disconnecting of network LTE from the cell } \mathrm{C_i} \qquad \mathrm{State}(\mathrm{E_{4,1}}) \\ (-1,1) & \text{if a user } \mathrm{u_0} \text{ disconnects of the network LTE from the cell } \mathrm{C_0} \\ & \text{by connecting to the network LTE in the cell } \mathrm{C_i} \qquad \mathrm{State}(\mathrm{E_{4,2}}) \\ (0,0) & \text{Otherwise} \qquad \mathrm{State}(\mathrm{E_{4,3}}) \end{cases}$$

- Stage 5: $E_4 = (b^k_{1,1} + \mu N^k_{PRB}; b^k_{1,2}; \ldots; b^k_{1,m}; b^k_2; \ldots; b^k_i + \mu'; \ldots; b^k_m)$

$$(\mu, \mu') = \begin{cases} (1,-1) & \text{if a user } \mathrm{u_0} \text{ connects to network LTE in the cell } \mathrm{C_0}\text{by} \\ & \text{disconnecting of the wireless Wi} - \text{Fi from the cell } \mathrm{C_i} \qquad \mathrm{State}(\mathrm{E_{5,1}}) \\ (-1,1) & \text{if a user } \mathrm{u_0} \text{ disconnects of the network LTE from the cell } \mathrm{C_0} \\ & \text{by connecting to the wireless Wi} - \text{Fi in the cell } \mathrm{C_i} \qquad \mathrm{State}(\mathrm{E_{5,2}}) \\ (0,0) & \text{Otherwise} \qquad \mathrm{State}(\mathrm{E_{5,3}}) \end{cases}$$

- Stage 6: $E_5 = (b^k_{1,1}; b^k_{1,2}; \ldots; b^k_{1,i} + \mu N^k_{PRB}; \ldots; b^k_{1,m}; b^k_2; \ldots; b^k_i + \mu'; \ldots; b^k_m)$

$$(\mu, \mu') = \begin{cases} (1,-1) & \text{if a user } \mathrm{u_0} \text{ connects to the network LTE in the cell } \mathrm{C_i} \text{ by} \\ & \text{disconnecting of the wireless Wi} - \text{Fi from the cell } \mathrm{C_i} \qquad \mathrm{State}(\mathrm{E_{6,1}}) \\ (-1,1) & \text{if a user } \mathrm{u_0} \text{ disconnects of the network LTE from the cell } \mathrm{C_i} \\ & \text{by connecting to the wireless Wi} - \text{Fi in the cell } \mathrm{C_i} \qquad \mathrm{State}(\mathrm{E_{6,2}}) \\ (0,0) & \text{Otherwise} \qquad \mathrm{State}(\mathrm{E_{6,3}}) \end{cases}$$

Transition rate

Proposition 1 Let us consider the heterogeneous system of networks in a stage $(E_p)_{1 \le p \le 6}$. The transition rate towards the state $E_{p,t}$ of the stage E_p is established by:

(1) In the zone C_0 of the cluster:

$$\tau_{(E_0 \rightleftarrows E_{1,1})} = \tau_{1,1} = (\lambda^{C(k)}_{C_0} + \lambda^{H(k)}_{N_1}) \cdot \left(\frac{b^k_{11}}{N^k_{PRB}} + 1\right) \cdot \left(\frac{1}{\Delta_{C(k)}} + \eta^{\overline{C_1}}_{C_1}\right) \tag{8}$$

(2) In the cell C_i of the cluster:

$$\tau_{(E_1 \rightleftarrows E_{2,1})} = \tau_{2,1} = (\lambda_{C_i}^{c(k)} + \lambda_{C_i}^{c(k)} \cdot \mathbb{P}(N_1 \to N_i)) \cdot \left(\frac{b_{1i}^k}{N_{PRB}^k} + 1\right) \cdot \frac{1}{\Delta_{c(k)}} \quad (9)$$

(3) From the zone C_0 to the cell C_i of cluster:

✔ By horizontal handover:

$$\tau_{(E_3 \rightleftarrows E_{4,1})} = \tau_{4,1} = \left(\frac{b_{11}^k}{N_{PRB}^k} + 1\right) \cdot \left(\frac{b_{1i}^k}{N_{PRB}^k}\right) \cdot \tau_{(C_0 \leftarrow C_i)}^{H(k)} \quad (10)$$

✔ By vertical handover:

$$\tau_{(E_4 \rightleftarrows E_{5,1})} = \tau_{5,1} = (b_i^k) \cdot \left(\frac{b_{11}^k}{N_{PRB}^k} + 1\right) \cdot \tau_{(C_0 \leftarrow C_i)}^{V(k)} \quad (11)$$

Proof

(1) When the number of units of busy bandwidth in one under zone C_1 of the network N_1 (LTE) crosses of $b_(11)$ to $b_{11} + N_{\mathrm{PRB}}^k$ then the variation rate of the units of busy bandwidth is defined by:

$$\frac{b_{11} + N_{\mathrm{PRB}}^k}{N_{\mathrm{PRB}}^k} = \frac{b_{11}}{N_{\mathrm{PRB}}^k} + 1$$

If there is variation of units of busy bandwidth, it is because a user who is recently connected in it in sub-zone C_1 with a access demand rate $\lambda_{C_0}^{C(k)}$ or to make a horizontal handover $\lambda_{N_1}^{H(k)}$ and the residence time in this sub-zone C_1 is $\Delta_{C(k)}$ or the time of passage in this cell C_0 towards C_1. The residence time is inversely proportional to the transition rate. The transition rate of the state E_0 to the state E_1 is equal to the product of these various rates calculated previously. So the found rate is:

$$(\lambda_{C_0}^{C(k)} + \lambda_{N_1}^{H(k)}) \cdot \left(\frac{b_{11}^k}{N_{\mathrm{PRB}}^k} + 1\right) \cdot \left(\frac{1}{\Delta_{C(k)}} + \eta_{C_0}^{C_1}\right)$$

(2) If a user u connects to the network N_1 (LTE) in a sub-zone C_i then the number of bandwidth units occupied in this sub-zone passes to b_{1i} $b_{1i} + N_{PRB}^k$. So bandwidth units rate variation occupied in this sub-zone is calculated by:

$$\frac{b_{1i} + N_{\mathrm{PRB}}^k}{N_{\mathrm{PRB}}^k} = \frac{b_{1i}}{N_{\mathrm{PRB}}^k} + 1$$

This variation of the bandwidth units occupation rate is due to the fact that when a user is recently connected in this sub-zone C_i with an access request rate $\lambda_{C_i}^{c(k)}$ rr disconnects from the network N_i by connecting to the network N_1 with a probability $\mathbb{P}(N_i \to N_1)$ applied to an access request rate $\lambda_{C_i}^{c(k)}$ in this sub-zone. The transition rate to the state $E_{2,1}$ of the stage E_2 being inversely proportional of stay time $\Delta_{c(k)}$ in this su-zone C_i then it's found by making the product of the various rates calculated previously. So the rate $\tau_{(E_1 \rightleftarrows E_{2,1})}$ found is equal to:

$$(\lambda_{C_i}^{c(k)} + \lambda_{C_i}^{c(k)} \cdot \mathbb{P}(N_1 \to N_i)) \cdot \left(\frac{b_{1i}^k}{N_{PRB}^k} + 1\right) \cdot \frac{1}{\Delta_{c(k)}}$$

(3) Besides, if a user u connects to the network N_1 (LTE) in sub-zone C_0 by disconnecting from the same network N_1 but in a zone C_i then the number of bandwidth units occupied in this sub-zone C_0 of the network N_1 (LTE) passes from b_{11} to $b_{11} + N_{\mathrm{PRB}}^k$ and b_{1i} to $b_{1,i}^k - N_{\mathrm{PRB}}^k$ in sub-zone C_i. So the variation rate of the bandwidth units occupied in zones C_0 and C_i are respectively defined by:

$$\frac{b_{11} + N_{\mathrm{PRB}}^k}{N_{\mathrm{PRB}}^k} = \frac{b_{11}}{N_{\mathrm{PRB}}^k} + 1$$

and

$$\frac{b_{1i}^k}{N_{\mathrm{PRB}}^k}$$

As the user makes a horizontal handover(vertical resp.) of the zone C_i towards C_1 then the rate of transition is directly proportional at the horizontal handover rate (vertical resp.) $\tau_{(C_0 \leftarrow C_i)}^{H(k)}$(resp. $\tau_{(C_0 \leftarrow C_i)}^{V(k)}$). So the transition rate of the state $E(4,1)$ in the stage E_3 is equal to the product of these rates calculated previously. Thus the found rate is:

$$\left(\frac{b_{11}^k}{N_{\mathrm{PRB}}^k} + 1\right) \cdot \left(\frac{b_{1i}^k}{N_{\mathrm{PRB}}^k}\right) \cdot \tau_{(C_0 \leftarrow C_i)}^{H(k)}$$

Respectively:

$$((b_i^k) \cdot \left(\frac{b_{11}^k}{N_{\mathrm{PRB}}^k} + 1\right) \cdot \tau_{(C_0 \leftarrow C_i)}^{V(k)})$$

□

The various transitions:

✔ $\tau_{(E_0 \rightleftarrows E_{1,1})} = (\lambda_{C_0}^{C(k)} + \lambda_{N_1}^{H(k)}) \cdot \left(\frac{b_{11}^k}{N_{\mathrm{PRB}}^k} + 1\right) \cdot \left(\frac{1}{\Delta_{C(k)}} + \eta_{C_0}^{C_1}\right)$ $s.q/ \sum_{j=1}^{m} (b_{1j} + N_{\mathrm{PRB}}^k) \leq B_1^{\mathrm{uc}}$

✔ $\tau_{(E_0 \rightleftarrows E_{1,2})} = (\lambda_{C_0}^{C(k)} + \lambda_{N_1}^{H(k)}) \cdot \left(\frac{b_{11}^k}{N_{\mathrm{PRB}}^k}\right) \cdot \left(\frac{1}{\Delta_{C(k)}} + \eta_{C_0}^{C_1}\right)$ $s.q/$ $b_{11}^k \geq N_{\mathrm{PRB}}^k$

✔ $\tau_{(E_1 \rightleftarrows E_{2,1})} = (\lambda_{C_i}^{C(k)} + \lambda_{C_i}^{C(k)} \cdot \mathbb{P}(N_1 > N_i)) \cdot \left(\frac{b_{1i}^k}{N_{\text{PRB}}^k} + 1\right) \cdot \frac{1}{\Delta_{C(k)}} \quad s.q/\begin{cases} b_i = B_i \\ \sum_{j=1}^m (b_{1j} + N_{\text{PRB}}^k) \leq B_1^{\text{uc}} \end{cases}$

✔ $\tau_{(E_1 \rightleftarrows E_{2,2})} = (\lambda_{C_i}^{C(k)} + \lambda_{C_i}^{C(k)} \cdot \mathbb{P}(N_1 > N_i)) \cdot \frac{b_{1i}^k}{N_{\text{PRB}}^k} \cdot \frac{1}{\Delta_{C(k)}} \quad s.q/\begin{cases} b_i = B_i \\ b_{1i}^k \geq N_{\text{PRB}}^k \end{cases}$

✔ $\tau_{(E_2 \rightleftarrows E_{3,1})} = (\lambda_{C_i}^{C(k)} + \lambda_{C_i}^{C(k)} \cdot \mathbb{P}(N_1 > N_i)) \cdot (b_i^k + 1) \cdot \frac{1}{\Delta_{C(k)}} \quad s.q/\begin{cases} b_i \leq B_i \\ \sum_{j=1}^m (b_{1j} + N_{\text{PRB}}^k) = B_1^{\text{uc}} \end{cases}$

✔ $\tau_{(E_2 \rightleftarrows E_{3,2})} = (\lambda_{C_i}^{C(k)} + \lambda_{C_i}^{C(k)} \cdot \mathbb{P}(N_1 > N_i)) \cdot (b_i^k) \cdot \frac{1}{\Delta_{C(k)}} \quad s.q/\begin{cases} b_i \geq 1 \\ \sum_{j=1}^m (b_{1j} + N_{\text{PRB}}^k) = B_1^{\text{uc}} \end{cases}$

✔ $\tau_{(E_3 \rightleftarrows E_{4,1})} = \left(\frac{b_{11}^k}{N_{\text{PRB}}^k} + 1\right) \cdot \left(\frac{b_{1i}^k}{N_{\text{PRB}}^k}\right) \cdot \tau_{(C_> C_i)}^{H(k)} \quad s.q/\begin{cases} b_i = B_i \\ \sum_{j=1}^m (b_{1j} + N_{\text{PRB}}^k) \leq B_1^{\text{uc}} \\ b_{1i}^k \geq N_{\text{PRB}}^k \end{cases}$

✔ $\tau_{(E_3 \rightleftarrows E_{4,2})} = \left(\frac{b_{11}^k}{N_{\text{PRB}}^k}\right) \cdot \left(\frac{b_{1i}^k}{N_{\text{PRB}}^k} + 1\right) \cdot \tau_{(C_i > C_0)}^{H(k)} \quad s.q/\begin{cases} b_i = B_i \\ \sum_{j=1}^m (b_{1j} + N_{\text{PRB}}^k) \leq B_1^{\text{uc}} \\ b_{11}^k \geq N_{\text{PRB}}^k \end{cases}$

✔ $\tau_{(E_4 \rightleftarrows E_{5,1})} = (b_i^k) \cdot \left(\frac{b_{11}^k}{N_{\text{PRB}}^k} + 1\right) \cdot \tau_{(C_i > C_0)}^{V(k)} \quad s.q/\begin{cases} b_i \geq 1 \\ \sum_{j=1}^m (b_{1j} + N_{\text{PRB}}^k) \leq B_1^{\text{uc}} \end{cases}$

✔ $\tau_{(E_4 \rightleftarrows E_{5,2})} = (b_i^k + 1) \cdot \left(\frac{b_{11}^k}{N_{\text{PRB}}^k}\right) \cdot \tau_{(C_0 > C_i)}^{V(k)} \quad s.q/\begin{cases} b_i \leq B_i \\ b_{1i}^k \geq N_{PRB}^k \end{cases}$

✔ $\tau_{(E_5 \rightleftarrows E_{6,1})} = \left(\frac{b_{1i}^k}{N_{\text{PRB}}^k} + 1\right) \cdot (b_i^k) \quad s.q/\begin{cases} b_i \geq 1 \\ \sum_{j=1}^m (b_{1j} + N_{\text{PRB}}^k) \leq B_1^{\text{uc}} \end{cases}$

✔ $\tau_{(E_5 \rightleftarrows E_{6,2})} = \left(\frac{b_{1i}^k}{N_{\text{PRB}}^k}\right) \cdot (b_i^k + 1) \quad s.q/\begin{cases} b_{1i}^k \geq N_{\text{PRB}}^k \\ b_i \leq B_i \end{cases}$

Evaluations of optimal performances

We estimate the selection strategy performances of a network based on the bit rate related to the parameters such as the blocking probability and the connections quality.

Average bit rate of the system in a sub-zone C_i

The bit rate received in a zone Z_i from the network N_1 is in function of the number of present units of bandwidth in networks N_1 and N_i. By denoting $D_1^{\text{avg}}(E)$ the average bit rate value received from the network and $\mathbb{P}(E)$ the probability of system balance state so the total average bit rate value received in a zone Z_i from the network N_1 is given by:

$$D_{N_1}^{\text{tot}} = \sum_{k=1}^{s} (\lambda_{C_i}^{c(k)} + \mathbb{P}(C_i > C_1)) \cdot \mathbb{P}(E) \cdot D_1^{\text{avg}}(E) \tag{12}$$

$$s.q/ \quad \sum_{j=1}^{m} (b_{1j} + N_{\text{PRB}}^k) \leq B_1^{\text{uc}}$$

with:

$$D_1^{\text{avg}}(E) = D_1^{\text{avg}} \cdot \left(1 - \Lambda \cdot \sqrt{\frac{\sum_{j=1}^m (b_{1j} + b_j)}{B_1^{\text{uc}} + B_i}}\right)$$

where the average instantaneous bit rate $D_1^{\text{avg}}(E)$ is defined by the product of a sub-carrier bandwidth and modulation(numbers of modulated sub-carriers):

$$D_1^{\text{avg}} = B_{\text{sp}} \times N_{\text{sub}} \tag{13}$$

where $N_{\text{sub}} = K \times B \times E_i \times (1 - \text{BLER}_i)$ such as:

- K is the frequencies number;
- B is the numbers of symbols per second;
- E_i is the modulation efficiency;
- BLER_i is the Block error rate in a sub-cell C_i;

As result we have:

$$\overline{D}_{C_{1i}} = \frac{D_{N_1}^{\text{tot}}}{\eta_{C_i}^{c(k)}} \tag{14}$$

Blocking probability et connections losses in sub-cell C_i

We have calculated the mean blocking probability of connections in a sub-zone C_i in function of an equilibrium sate probability $\mathbb{P}(E)$ of the system. Indeed, we added the system states probability where the numbers of units of busy bandwidth is higher than these available in the network N_1 by formula:

$$\mathbb{P}_{N_1}^B = \sum_{k=1}^{s} (\lambda_{C_i}^{C(k)} + \mathbb{P}(C_i > C_1)) \cdot \mathbb{P}(E) \cdot \mathbb{P}_1^B(E) \tag{15}$$

$$s.q/ \quad \sum_{j=1}^{m} (b_{1j} + N_{\text{PRB}}^k) > B_1^{\text{uc}}$$

with

$$\mathbb{P}_1^B(E) = \mathbb{P}_1^B \cdot \left(1 - \Theta \cdot \sqrt{\frac{\sum_{j=1}^m (b_{1j} + b_j)}{B_1^{\text{uc}} + B_i}}\right)$$

Table 4 LTE network parameters test

Test	Value
Modulation	16 QAM
Symbols	6
Efficiency	1.4766
Sub-carriers number	72
Bandwidth	1.4 MHz

Table 5 System network parameters test

Parameters test	Value
Average access rate to a service	70 %
Average handover rate	60 %
Average state balance rate	80 %
Bandwidth units in LTE	60
Noise power	−174 dBm/Hz
Signal power	400 dBm
Services number	Two services unicast
Cell radius	600 m Z_1, 200 m Z_2
Distance between area centers	300 m

$$\mathbb{P}_1^B = \frac{\frac{\rho_{C_i}^s}{s!}}{\sum_{k=1}^{s} \frac{\rho_{C_i}^k}{k!}} \tag{16}$$

where s is the number of available services in the cluster and

$$\rho_{C_i} = \sum_{k=1}^{s} \left(\frac{\lambda_{C_i}^{c(k)}}{\lambda_{C_1}^{c}} \right)$$

is the probability that a user is blocked in the sub-zone C_i.

Numerical tests

To test the theoretical results that we obtained, we simplified our field of study in a service cell C_1 covered by the network N_1 (LTE) in which we implanted a wireless N_2 (Wi-Fi) in sub-cell C_2 of C_1. By means of the simulator NS3 [9] and working with the parameters below we managed to obtain satisfactory results as show by the obtained curves.

We have supposed that the data used for the test are the ones relative to the characteristics of the network LTE. So, we chose the modulation 16QAM of the network LTE the number of symbols which is six (06), Efficiency is equal 1.4766, sub-carriers number is 72 and as the bandwidth of the network LTE varies between 1.4 and 20 MHz then we have worked with the minimal value (1.4MHz) as indicated in Table 4; refer to Jabban [8].

However, the parameters chosen for the heterogeneous mobile and wireless system of networks are supposed to allow us to have satisfactory results for the numerical results. These parameters are given in Table 5.

The main parameters of both available networks are illustrated in the Table 5. We suppose that the coverage radius of networks N_1 and N_2 are respectively equal to 600 M and 200 m. We also suppose that the power transmitted by N_1 is equal to 400 dBM. The noise power is supposed equal to −174 dBm/Hm for the network LTE. We also suppose that the average access requests rates for a service, handover and the system balance state are, respectively, 70, 60 and 80 %. The number of units of bandwidth is supposed equal to 60. Finally the number of unicasts services is fixed to 2.

Fig. 3 Average bit rate of network N_1 in the cell C_2 in function of the BLER

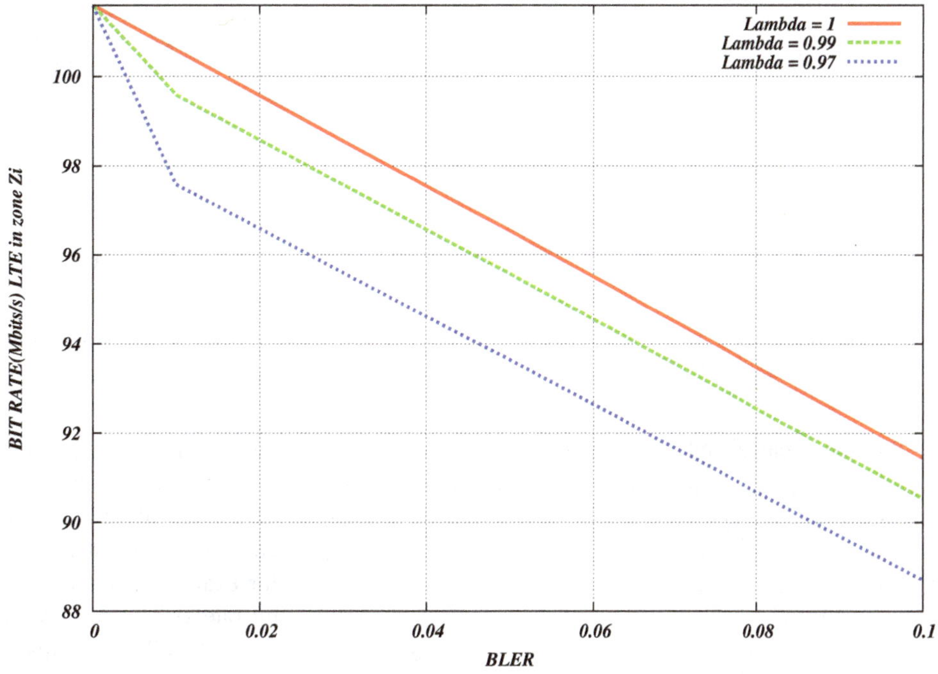

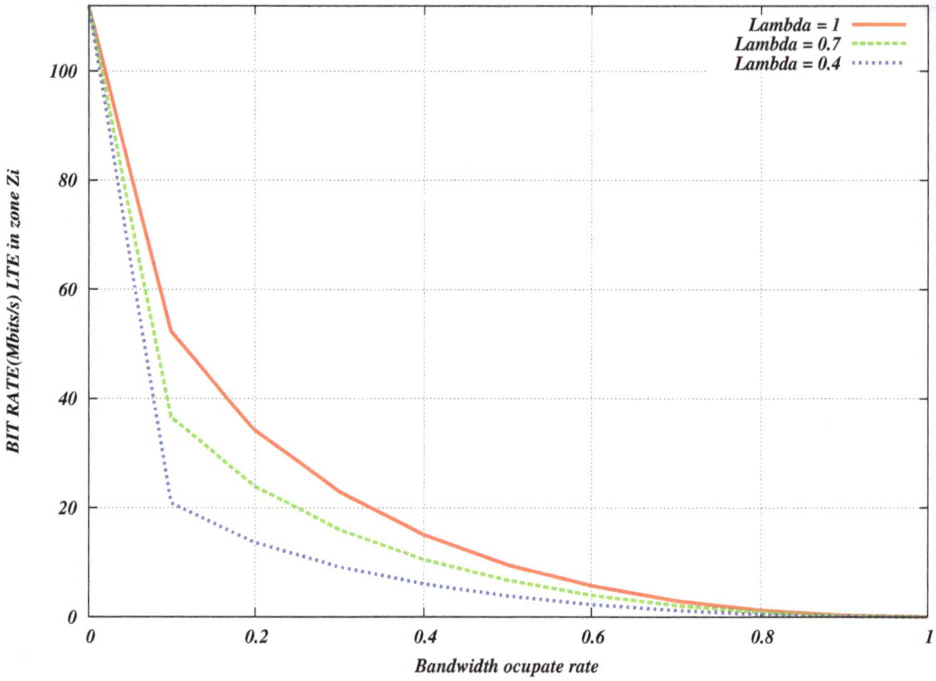

Fig. 4 Average bit rate of the network N_1 in the zone C_2 in function of the busy bandwidth rate

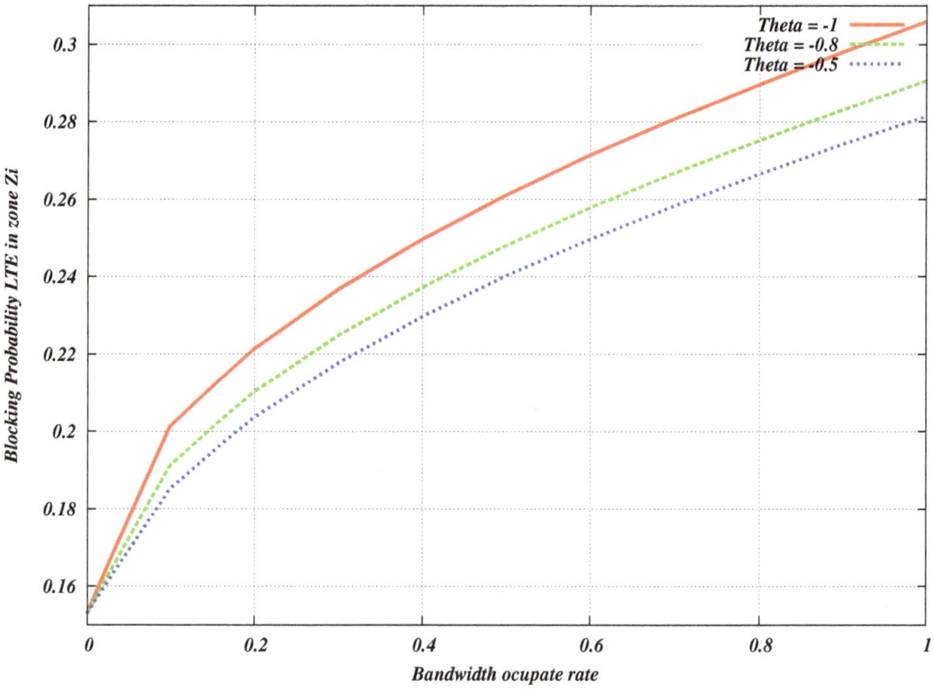

Fig. 5 Blocking probability in the zone C_2 in function of the busy bandwidth rate

We have calculated the average bit rate value $D_{N_1}^{\text{tot}}$ received by a user u_0 from the network N_1 in the cell C_2. The selection strategy is based on the most high bit rate, in a sub-cell C_2, the user chooses the unicasts services of the networks N_1 or N_2 according to the selection technique established. Through the obtained results, we have noticed that a small modification of the bit rate occurs due to the network saturation. This modification is represented by a parameter Λ which indicates a very high sensitivity of the bit rate because of the network congestion. For a low modification of the sensitivity factor Λ for example from 1 to 0.99 we obtained a net fall of the bit rate as illustrate by Fig. 2 in function of the BLER in the sub-cell C_2. This factor also informs about the

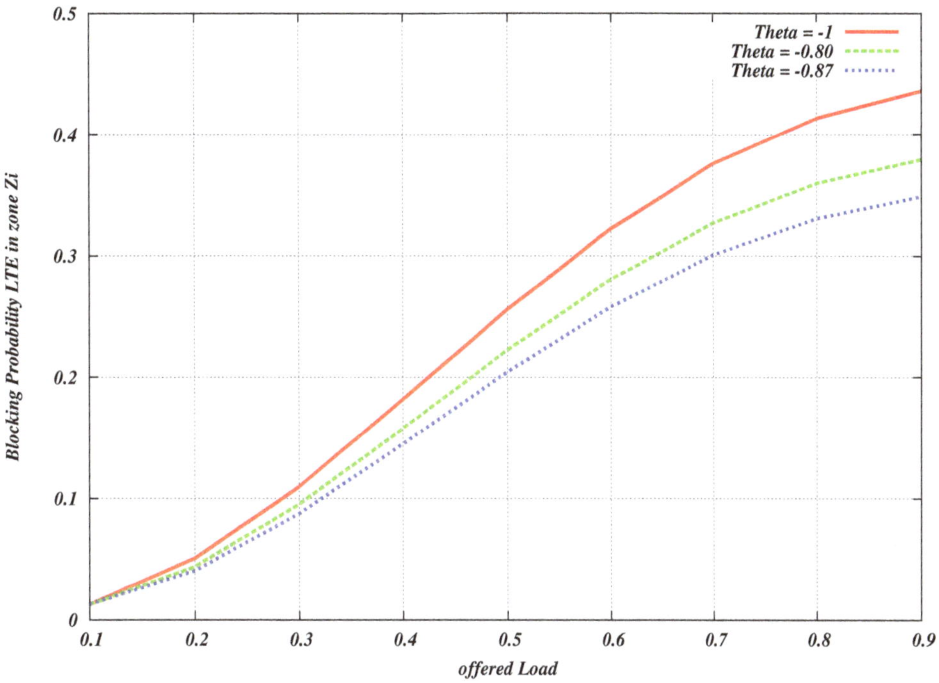

Fig. 6 Blocking probability in the zone C_2 in function of the offered load rate

maximal decrease of the bit rate because of the network congestion.

However the change of the factor Λ acts less on the bit rate when it is estimated in function of the occupation bandwidth rate by fixing the BLER as indicated by Fig. 3; BLER is fixed to 50 %.

Besides, we have estimated the network performances related to the average blocking probability and connections losses to the services in the sub-cell C_2. The obtained results depend on a sensitivity factor Θ as represented by Figs. 4 and 5. For sensitivity parameters $\Theta \in \{1; 0.8; 0.5\}$ we analyzed the blocking probability in the sub-cell C_2 in function of the busy bandwidth rate (Fig. 4) then to the offered load by the traffic (Fig. 5). Indeed, seen the satisfactory obtained results, we realize that the blocking probability do not pass 40 % when they are determined with the occupation bandwidth rate. They do not reach either the level 50 % when they are estimated in function of the offered load by the traffic whatever is the given sensitivity factor (Fig. 6).

Conclusion

At the end of our analysis on the system integration performances of new generation wireless and mobile networks, we found a factor which remains very sensitive to the variations of the bit rate received in a sub-cell C_i when it is calculated in function of the BLER in this sub-zone. Besides, this sensitivity factor remains so determining for the blocking probability theory in a sub-zone C_i by means of the busy bandwidth rate or the offered load traffic rate. The satisfactory results obtained on the system performances of wireless and mobile networks such the LTE and the Wi-Fi based on the bit rate allowed us to discover the dynamic fluctuations system. The parameters related to the bit rate such as the blocking probability are estimated with lower rates the bar 40 %.

In our future works, we intend to calculate the same sensitivity factor when we consider the performances of the system related to the SINR. We planned also to take into account the number of users having consumed these numbers of busy bandwidth units. This will allow us to encircle better the congestion and dis-congestion rates of the heterogeneous networks.

Acknowledgments We are solemnly anxious to thank all the members of research team of mathematical decision of Cheikh Anta Diop University for their collaboration, their critic and their support to realize this project. We also thank all the mathematics teachers of the department of mathematics and computing at Cheikh Anta Diop university of Dakar for the realization of these scientific works.

References

1. Al-Ghadi, M., Ababneh, I., Mardini, W.: Performance study of SINR scheme for vertical handoff in wireless networks. In: Proceeding of International Conference on Information and Communication Systems (ICICS), pp. 1–3 (2011)

2. Ayyappan, K., Narasimman, K., Dananjavan, P.: SINR based vertical handoff scheme for Qos in heterogenous wireless networks. In: Proceeding of International Conference on Future Computer and Communication (ICFCC), pp. 117–121 (2009)
3. Bettstetter, C., Wagner, C.: The spacial node distribution of the random waypoint mobility model. In: Proceedings of German Workshop on mobile ad hoc networks (WMAN), Ulm, Germany (2002)
4. Hyytia, E., Virtamo, J.: Random waypoint mobility model in cellular networks, Springer, Wireless Networks, vol. 13, no. 2 (2007)
5. Hyytia, E., Lassila, P., Virtamo, J.: Spacial node distribution of the random waypoint mobility model with applications. IEEE Transaction on Mobile Computing **5**(6), 680–694 (2006)
6. Jabban, A., Nasser, Y., Helard, M.: Performance analysis of heterogenous networks based on SINR selection strategy. In: Proceeding of International Conference on Telecommunication (ICT2013) (2013)
7. Jabban, A., Nasser, Y., Helard, M.: SINR based network selection strategy in integrated heterogenous networks. In: Proceeding of International Conference on Telecommunication (ICT2012), pp. 1–6 (2012)
8. Jabban, A.: Optimisation et analyse des réseaux intelligents et des réseaux hétérogènes. THESE INSA Rennes en Electronique et Télécommunications, Aout (2013)
9. NS3 CONSORTIUM (Georgia Institute of Technology, ICSI Center for Internet Research and 'Planète' INRIA Sophia-Antipolis), Website of the research educational network on development and application (2006). http://www.nsnam.org/docs/doxygen
10. Shen, W., Zeng, Q.A.: Cost-function-based network selection strategy in integrated wireless and mobile networks. IEEE Trans. Veh. Technol. **57**(6), 3778–3788 (2008)
11. Vuong, Q.N., Agoulmine, N., Doudane, Y.G.: Terminal-controlled mobility management in heterogeneous wireless networks. IEEE Commun. Mag. **45**, 122–129 (2007)
12. Yang, K., Gondal, I., Qiu, B., Dooley, L.S.: Combined SINR based vertical handoff algorithm for next generation heterogeneous wireless networks. In: Proceeding of IEEE Global Telecommunication Conferences (GLOBECOM '07), pp. 4483–4487 (2007)
13. Yang, K., Gondal, I., Qiu, B., Dooley, L.S.: Multi-dimensional adaptive SINR based vertical handoff for heterogeneous wireless networks. IEEE Commun. Lett. **12**(6), 4483–4487 (2007)

4

Solvability of impulsive (n, n − p) boundary value problems for higher order fractional differential equations

Yuji Liu[1]

Abstract We present a new general method for converting an impulsive fractional differential equation to an equivalent integral equation. Using this method and employing a fixed point theorem in Banach space, we establish existence results of solutions for a boundary value problem of impulsive singular higher order fractional differential equation. An example is presented to illustrate the efficiency of the results obtained. A conclusion section is given at the end of the paper.

Keywords Solvability · Singular fractional differential system · Impulse effect · Caputo fractional derivative · Fixed point theorem

Mathematics Subject Classification 92D25 · 34A37 · 34K15

Introduction

Fractional differential equation is a generalization of ordinary differential equation to arbitrary non-integer orders. Fractional differential equations, therefore, find numerous applications in different branches of physics, chemistry and biological sciences such as visco-elasticity, feed back amplifiers, electrical circuits, electro-analytical chemistry, fractional multipoles and neuron modelling. The reader may refer to the books and monographs [1–3] for fractional calculus and developments on fractional differential and fractional integro-differential equations with applications.

On the other hand, the theory of impulsive differential equations describes processes which experience a sudden change of their state at certain moments. Processes with such characteristics arise naturally and often; for example, phenomena studied in physics, chemical technology, population dynamics, biotechnology and economics. For an introduction of the basic theory of impulsive differential equation, we refer the reader to [4].

Solvability of boundary value problems for higher order ordinary differential equations were investigated by many authors. For example, in [5–16], the following $(n, n-k)$ type problems were studied:

$$\begin{cases} (-1)^{n-k}y^{(n)} = f(t,y), & t\in(0,1),\\ y^{(i)}(0)=0, & i\in\mathbb{N}_0^{k-1},\\ y^{(j)}(1)=0, & j\in\mathbb{N}_0^{n-k-1}. \end{cases} \tag{1.1}$$

In [17, 18], the following more general boundary value problems were studied:

$$\begin{cases} (-1)^{n-k}y^{(n)} = f(t,y), & t\in(0,1),\\ y^{(i)}(0)=0, & i\in\mathbb{N}_0^{k-1},\\ y^{(j)}(1)=0, & j\in\mathbb{N}_q^{n+q-k-1} \end{cases} \tag{1.2}$$

where $k\in\mathbb{N}_1^{n-1}, q\in\mathbb{N}_0^{k}$. In [6, 19, 20], authors studied existence of solutions of the following problems:

$$\begin{cases} (-1)^{n-p}y^{(n)} = f(t,y,y',\ldots,y^{(p-1)}), & t\in(0,1),\\ y^{(i)}(0)=0, & i\in\mathbb{N}_0^{p-1},\\ y^{(j)}(1)=0, & j\in\mathbb{N}_p^{n-1}. \end{cases} \tag{1.3}$$

Supported by the Natural Science Foundation of Guangdong province (No: S2011010001900) and Natural science research project for colleges and universities of Guangdong Province (No: 2014KTSCX126)

✉ Yuji Liu
liuyuji888@sohu.com

1 Department of Mathematics, Guangdong University of Finance and Economics, Guangzhou 510320, People's Republic of China

On the one hand, it is interesting to generalize results on boundary value problems for higher order ordinary differential equations; in mentioned papers, in [21], authors studied existence of solutions of the following boundary value problem for higher order fractional differential equation

$$\begin{cases} D_{0^+}^\alpha u(t) + \lambda f(t, u(t)) = 0, & 0 < t < b, \lambda > 0, \alpha \in [n, n+1), \\ u^{(j)}(0) = 0, & j \in \mathbb{N}_0^{n-1}, u^{(n-1)}(b) = 0. \end{cases}$$

(1.4)

In [22], solutions of the following problem were presented:

$$\begin{cases} D_{0^+}^\alpha u(t) + p(t)u(t) = 0, & 0 < t < 1, \lambda > 0, \\ u^{(j)}(0) = 0, & j \in \mathbb{N}_0^{n-2}, u(1) = 0. \end{cases}$$

(1.5)

On the other hand, higher order fractional differential equations have applications such as the fractional order elastic beam equations see [23], the fractional order viscoelastic material model see [24], the fractional viscoelastic model see [25–27] and so on.

There has been no papers concerned with the solvability of boundary value problems for higher order impulsive fractional differential equations since it is difficult to convert an impulsive fractional differential equation to an equivalent integral equation.

To fill this gap, in this paper, we discuss the following two boundary value problems for nonlinear impulsive singular fractional differential equation

$$\begin{cases} {}^cD_{0^+}^\alpha u(t) = f(t, u(t)), & t \in (t_s, t_{s+1}], s \in \mathbb{N}_0^m, \\ u^{(i)}(0) = 0, & i \in \mathbb{N}_0^{k-1}, \\ u^{(j)}(1) = 0, & j \in \mathbb{N}_l^{n+l-k-1}, \\ \Delta u^{(j)}(t_s) = I_j(t_s, u(t_s)), & j \in \mathbb{N}_0^{n-1}, s \in \mathbb{N}_1^m, \end{cases}$$

(1.6)

and

$$\begin{cases} {}^cD_{0^+}^\alpha u(t) = f(t, u(t)), & t \in (t_s, t_{s+1}], s \in \mathbb{N}_0^m, \\ u^{(i)}(0) = -u^{(j)}(1), & j \in \mathbb{N}_0^{n-1}, \\ \Delta u^{(j)}(t_s) = I_j(t_s, u(t_s)), & j \in \mathbb{N}_0^{n-1}, s \in \mathbb{N}_1^m, \end{cases}$$

(1.7)

where

(a) $n - 1 < \alpha < n$, n is a positive integer, ${}^cD_{0^+}^*$ is the Caputo fractional derivative of orders $*$ with starting point 0,
(b) $0 = t_0 < t_1 < \cdots < t_s < \cdots < t_m < t_{m+1} = 1$ with m being a positive integer, $\mathbb{N}_a^b = \{a, a+1, a+2, \ldots,\}$ for nonnegative integers $a < b$,
(c) $k \in \mathbb{N}_1^{n-1}$, and $l \in \mathbb{N}_0^k$,
(d) $f : (0, 1) \times \mathbb{R} \to \mathbb{R}$, $I_j : \{t_s : s \in \mathbb{N}_1^m\} \times \mathbb{R} \to \mathbb{R}$, f is a Carathéodory function, $I_j (j \in \mathbb{N}_1^m)$ are discrete Carathéodory functions.

A function $x : (0, 1] \to \mathbb{R}$ is said to be a solution of (1.6) or (1.7) if

$$x|_{(t_s, t_{s+1}]} \in C^0(t_s, t_{s+1}], s \in N_0^m, \quad \lim_{t \to t_s^+} x(t) \text{ exist for all } s \in \mathbb{N}_0^m,$$
$${}^cD_{0^+}^\alpha x \text{ is measurable on each } (t_i, t_{i+1}], \quad i \in \mathbb{N}_0^m$$

and x satisfies all equations in (1.6) or (1.7), respectively.

In [28], a general method for converting an impulsive fractional differential equation to an equivalent integral equation was presented. We present a new method (Lemma 2.2) for converting BVP (1.6) to an equivalent integral equation in this paper. We shall construct a weighted Banach space and apply the Leray–Schauder nonlinear alternative to obtain the existence of at least one solution of (1.6) and (1.7), respectively. Our results are new and naturally complement the literature on fractional differential equations.

The paper is outlined as follows. "Preliminaries" contains some preliminary results. Main results are presented in "Main results". In "Examples", we give an example to illustrate the efficiency of the results obtained. A conclusion section is given at the end of the paper.

Preliminaries

For the convenience of the readers, we shall state the necessary definitions from fractional calculus theory.

For $\phi \in L^1(0, \infty)$, denote $\|\phi\|_1 = \int_0^\infty |\phi(s)| ds$. Let the Gamma and beta functions $\Gamma(\alpha)$ and $\mathbf{B}(p, q)$ be defined by

$$\Gamma(\alpha) = \int_0^{+\infty} x^{\alpha-1} e^{-x} dx, \quad \mathbf{B}(p, q) = \int_0^1 x^{p-1}(1-x)^{q-1} dx.$$

Definition 2.1 [3] The Riemann–Liouville fractional integral of order $\alpha > 0$ of a function $g : (0, \infty) \to \mathbb{R}$ is given by

$$I_{0^+}^\alpha g(t) = \frac{1}{\Gamma(\alpha)} \int_0^t (t-s)^{\alpha-1} g(s) ds,$$

provided that the right-hand side exists.

Definition 2.2 [3] The Caputo fractional derivative of order $\alpha > 0$ of a continuous function $g : (0, \infty) \to \mathbb{R}$ is given by

$${}^cD_{0^+}^\alpha g(t) = \frac{1}{\Gamma(n-\alpha)} \int_0^t \frac{g^{(n)}(s)}{(t-s)^{\alpha-n+1}} ds,$$

where $n - 1 \leq \alpha < n$, provided that the right-hand side exists.

Definition 2.3 Let $p > -1, q \in (-1, 0]$. We say $K : (0,1) \times \mathbb{R} \to \mathbb{R}$ is a Carathéodory function if it satisfies the following:

(i) $t \to K(t,x)$ is integral on $(t_s, t_{s+1}]$ for every $x \in \mathbb{R}$, $s \in \mathbb{N}_0^m$,

(ii) $x \to K(t,x)$ is continuous on $\mathbb{R}$ for all $t \in (t_s, t_{s+1}]$ $(s \in \mathbb{N}_0^m)$,

(iii) for each $r > 0$ there exists a constant $A_{r,f} \geq 0$ satisfying

$$|K(t,x)| \leq A_{r,K} t^p (1-t)^q$$

holds for $t \in (t_s, t_{s+1}], s \in \mathbb{N}_0^m, |x| \leq r$.

Definition 2.4 $G : \{t_s : s \in \mathbb{N}_1^m\} \times \mathbb{R} \to \mathbb{R}$ is called a discrete Carathéodory function if

(i) $x \to G(t_s, x)$ is continuous on $\mathbb{R}$ for each $s \in \mathbb{N}_1^m$,

(ii) for each $r > 0$ there exists $A_{r,G,s} \geq 0$ such that $|G(t_s, x)| \leq A_{r,G,s}$

holds for $|x| \leq r, s \in \mathbb{N}_1^m$.

Lemma 2.1 (Lemma 2.22 in [29]) *Suppose that* $h \in L^1(0, t_1) \bigcap C^0(0, t_1)$. *Then x is a solution of* ${}^cD_{0^+}^\alpha x(t) = h(t), a.e., t \in (0, t_1]$ *if and only if there exist constants* $c_{v0} \in \mathbb{R}$ *such that*

$$x(t) = \sum_{v=0}^{n-1} \frac{c_{v0}}{\Gamma(v+1)} t^v + \int_0^t \frac{(t-s)^{\alpha-1}}{\Gamma(\alpha)} h(s) \mathrm{d}s, \quad t \in (0, t_1].$$

Lemma 2.2 *Suppose that h is integral on each subinterval of (0, 1). Then x satisfying*

$$\begin{aligned} &x|_{(t_s,t_{s+1}]} \in C^0(t_s, t_{s+1}], s \in \mathbb{N}_0, \quad j \in \mathbb{N}_0^{n-1}, \\ &\lim_{t \to t_s^+} x(t) \text{ exists for all } s \in \mathbb{N}_0, \quad j \in \mathbb{N}_0^{n-1} \end{aligned} \tag{2.1}$$

is a solution of

$${}^cD_{0^+}^\alpha x(t) = h(t), a.e., \quad t \in (t_i, t_{i+1}] (i \in \mathbb{N}_0^m) \tag{2.2}$$

if and only if there exist constants $c_{v0} \in \mathbb{R}$ *such that*

$$\begin{aligned} x(t) = &\sum_{j=0}^{i} \sum_{v=0}^{n-1} \frac{c_{vj}}{\Gamma(v+1)} (t - t_j)^v \\ &+ \int_0^t \frac{(t-s)^{\alpha-1}}{\Gamma(\alpha)} h(s) \mathrm{d}s, \quad t \in (t_i, t_{i+1}], i \in \mathbb{N}_0^m. \end{aligned} \tag{23}$$

Proof By Lemma 2.1, we know that x satisfying (2.1) is a solution of (2.2) if and only if x satisfies (2.3) on $(t_0, t_1]$. To complete the proof, we consider two steps:

Step 1. We prove that x satisfies (2.1) and (2.2) if x satisfies (2.3). From (2.3), we know obviously that (2.1) holds. We need to prove that (2.2) holds on all $(t_i, t_{i+1}] (i \in \mathbb{N}_0^m)$. In fact, for $t \in (t_0, t_1]$, by Lemma 2.1, we know $D_{0^+}^\alpha x(t) = h(t)$. For $t \in (t_i, t_{i+1}]$, we have

$$\begin{aligned}
{}^cD_{0^+}^\alpha x(t) &= \frac{1}{\Gamma(n-\alpha)} \int_0^t (t-s)^{n-\alpha-1} x^{(n)}(s) \mathrm{d}s \\
&= \frac{\sum_{j=0}^{i-1} \int_{t_j}^{t_{j+1}} (t-s)^{n-\alpha-1} x^{(n)}(s) \mathrm{d}s + \int_{t_i}^t (t-s)^{n-\alpha-1} x^{(n)}(s) \mathrm{d}s}{\Gamma(n-\alpha)} \\
&= \frac{\sum_{j=0}^{i-1} \int_{t_j}^{t_{j+1}} (t-s)^{n-\alpha-1} \left(\sum_{v=0}^{j} \sum_{\kappa=0}^{n-1} \frac{c_{\kappa v}}{\Gamma(\kappa+1)} (s-t_v)^\kappa + \int_0^s \frac{(s-u)^{\alpha-1}}{\Gamma(\alpha)} h(u) \mathrm{d}u \right)^{(n)} \mathrm{d}s}{\Gamma(n-\alpha)} \\
&\quad + \frac{\int_{t_i}^t (t-s)^{n-\alpha-1} \left(\sum_{v=0}^{i} \sum_{\kappa=0}^{n-1} \frac{c_{\kappa v}}{\Gamma(\kappa+1)} (s-t_v)^\kappa + \int_0^s \frac{(s-u)^{\alpha-1}}{\Gamma(\alpha)} f(u) \mathrm{d}u \right)^{(n)} \mathrm{d}s}{\Gamma(n-\alpha)} \\
&= \frac{\sum_{j=0}^{i-1} \int_{t_j}^{t_{j+1}} (t-s)^{n-\alpha-1} \left(\int_0^s \frac{(s-u)^{\alpha-1}}{\Gamma(\alpha)} h(u) \mathrm{d}u \right)^{(n)} \mathrm{d}s}{\Gamma(n-\alpha)} \\
&\quad + \frac{\int_{t_i}^t (t-s)^{n-\alpha-1} \left(\int_0^s \frac{(s-u)^{\alpha-1}}{\Gamma(\alpha)} f(u) \mathrm{d}u \right)^{(n)} \mathrm{d}s}{\Gamma(n-\alpha)} \\
&= \frac{\int_0^t (t-s)^{n-\alpha-1} \left(\int_0^s \frac{(s-u)^{\alpha-n}}{\Gamma(\alpha-n+1)} h(u) \mathrm{d}u \right)' \mathrm{d}s}{\Gamma(n-\alpha)} \\
&= \left[\frac{\int_0^t (t-s)^{n-\alpha} \left(\int_0^s \frac{(s-u)^{\alpha-n}}{\Gamma(\alpha-n+1)} h(u) \mathrm{d}u \right)' \mathrm{d}s}{(n-\alpha)\Gamma(n-\alpha)} \right]' \\
&= \left[\frac{(t-s)^{n-\alpha} \int_0^s \frac{(s-u)^{\alpha-n}}{\Gamma(\alpha-n+1)} h(u) \mathrm{d}u \Big|_0^t + (n-\alpha) \int_0^t (t-s)^{n-\alpha-1} \int_0^s \frac{(s-u)^{\alpha-n}}{\Gamma(\alpha-n+1)} h(u) \mathrm{d}u \mathrm{d}s}{(n-\alpha)\Gamma(n-\alpha)} \right]' \\
&= \left[\frac{\int_0^t (t-s)^{n-\alpha-1} \int_0^s \frac{(s-u)^{\alpha-n}}{\Gamma(\alpha-n+1)} h(u) \mathrm{d}u \mathrm{d}s}{\Gamma(n-\alpha)} \right]' \\
&= \left[\frac{\int_0^t \int_s^t (t-s)^{n-\alpha-1} \frac{(s-u)^{\alpha-n}}{\Gamma(\alpha-n+1)} \mathrm{d}s h(u) \mathrm{d}u}{\Gamma(n-\alpha)} \right]' \\
&= \left[\frac{\int_0^t \int_0^1 (1-w)^{n-\alpha-1} \frac{w^{\alpha-n}}{\Gamma(\alpha-n+1)} \mathrm{d}w h(u) \mathrm{d}u}{\Gamma(n-\alpha)} \right]' \\
&= h(t), t \in (t_i, t_{i+1}], i \in \mathbb{N}_1.
\end{aligned}$$

It follows that x satisfies (2.2).

Step 2. We prove that x satisfies (2.3) if x satisfies (2.1) and (2.2). By Lemma 2.1, from (2.1) and (2.2), we know that (2.3) holds for $i = 0$. We suppose that (2.3) holds for $0, 1, 2, \ldots, i$. We will prove that (2.3) holds for $i + 1$. Then by mathematical induction method, we see that (2.3) holds for all $i \in \mathbb{N}_0^m$.

In fact, we suppose that

$$x(t) = \Phi(t) + \sum_{j=0}^{i} \sum_{v=0}^{n-1} \frac{c_{vj}}{\Gamma(v+1)} (t-t_j)^v + \int_0^t \frac{(t-s)^{\alpha-1}}{\Gamma(\alpha)} h(s) \mathrm{d}s, \quad t \in (t_{i+1}, t_{i+2}]. \tag{2.4}$$

Then for $t \in (t_{i+1}, t_{i+2}]$ we have

$$\begin{aligned}
h(t) &= {}^cD_{0^+}^\alpha x(t) \\
&= \frac{1}{(n-\alpha)} \left[\sum_{j=0}^{i} \int_{t_j}^{t_{j+1}} (t-s)^{n-\alpha-1} x^{(n)}(s) \mathrm{d}s + \int_{t_{i+1}}^t (t-s)^{n-\alpha-1} x^{(n)}(s) \mathrm{d}s \right] \\
&= \frac{\sum_{j=0}^{i} \int_{t_j}^{t_{j+1}} (t-s)^{n-\alpha-1} \left(\sum_{\kappa=0}^{j} \sum_{v=0}^{n-1} \frac{c_{v\kappa}}{\Gamma(v+1)} (s-t_\kappa)^v + \int_0^s \frac{(s-v)^{\alpha-1}}{\Gamma(\alpha-1)} h(v) \mathrm{d}v \right)^{(n)} \mathrm{d}s}{\Gamma(n-\alpha)} \\
&\quad + \frac{\int_{t_{i+1}}^t (t-s)^{n-\alpha-1} \left(\Phi(s) + \sum_{\kappa=0}^{k} \sum_{v=0}^{n-1} \frac{c_{v\kappa}}{\Gamma(\alpha-v+1)} (s-t_\kappa)^v + \int_0^s \frac{(s-v)^{\alpha-1}}{\Gamma(\alpha-1)} h(v) \mathrm{d}v \right)^{(n)} \mathrm{d}s}{\Gamma(n-\alpha)} \\
&= {}^cD_{t_{i+1}^+}^\alpha \Phi(t) + \frac{\int_0^t (t-s)^{n-\alpha-1} \left(\int_0^s \frac{(s-v)^{\alpha-1}}{\Gamma(\alpha)} h(v) \mathrm{d}v \right)^{(n)} \mathrm{d}s}{\Gamma(n-\alpha)}.
\end{aligned}$$

Similarly to Step 1 we can get that

$$h(t) = {}^cD^{\alpha}_{0^+}x(t) = h(t) + {}^cD^{\alpha}_{t^+_{i+1}}\Phi(t).$$

So ${}^cD^{\alpha}_{t^+_{i+1}}\Phi(t) = 0$ on $(t_{i+1}, t_{i+2}]$. Then there exist constants $c_{vi+1} \in R(v \in \mathbb{N}_0^{n-1})$ such that $\Phi(t) = \sum_{v=0}^{n} c_{vi+1}\frac{(t-t_{i+1})^v}{\Gamma(v+1)}$ on $(t_{i+1}, t_{i+2}]$. Substituting Φ in (2.4), we get that (2.3) holds for $i+1$. By mathematical induction method, we know that (2.3) holds for all $i \in \mathbb{N}_0^m$. So x satisfies (2.3) if x satisfies (2.1) and (2.2). The proof is complete. □

(i) $\{t \to x(t) : x \in M\}$ *is uniformly bounded,*

(ii) $\{t \to x(t) : x \in M\}$ *is equicontinuous in any interval* $(t_s, t_{s+1}](s \in \mathbb{N}_0)$.

Proof The proof is standard and omitted.

For $x \in X$, denote $f_x(t) = f(t, x(t))$ and $I_{jx}(t_s) = I_j(t_s, x(t_s))$. Denote

$$M = (m_{ij})_{(n-k)\times(n-k)}$$

$$= \begin{pmatrix}
\frac{1}{\Gamma(k-l+1)} & \frac{1}{\Gamma(k-l)} & \cdots & 1 & 0 & 0 & 0 & \cdots & 0 \\
\frac{1}{\Gamma(k-l+2)} & \frac{1}{\Gamma(k-l+1)} & \cdots & \frac{1}{\Gamma(2)} & 1 & 0 & 0 & \cdots & 0 \\
\frac{1}{\Gamma(k-l+3)} & \frac{1}{\Gamma(k-l+2)} & \cdots & \frac{1}{\Gamma(3)} & \frac{1}{\Gamma(2)} & 1 & 0 & \cdots & 0 \\
\vdots & \vdots & \vdots & \vdots & \vdots & \vdots & \vdots & \cdots & \vdots \\
\frac{1}{\Gamma(n-k)} & \frac{1}{\Gamma(n-k-1)} & \cdots & \cdots & \cdots & \cdots & \cdots & \cdots & 1 \\
\frac{1}{\Gamma(n-k+1)} & \frac{1}{\Gamma(n-k)} & \cdots & \cdots & \cdots & \cdots & \cdots & \cdots & \frac{1}{\Gamma(2)} \\
\vdots & \vdots & \vdots & \vdots & \vdots & \vdots & \vdots & \cdots & \vdots \\
\frac{1}{\Gamma(n-l)} & \frac{1}{\Gamma(n-l-1)} & \cdots & \frac{1}{\Gamma(n-k)} & \frac{1}{\Gamma(n-k-1)} & \frac{1}{\Gamma(n-k-2)} & \frac{1}{\Gamma(n-k-3)} & \cdots & \frac{1}{\Gamma(k-l+1)}
\end{pmatrix},$$

$$N = (n_{ij})_{(n)\times(n)} = \begin{pmatrix}
1 & 0 & 0 & 0 & \cdots & 0 \\
\frac{1}{2\Gamma(2)} & 1 & 0 & 0 & \cdots & 0 \\
\frac{1}{2\Gamma(3)} & \frac{1}{2\Gamma(2)} & 1 & 0 & \cdots & 0 \\
\cdots & \cdots & \cdots & \cdots & \cdots & \cdots \\
\frac{1}{2\Gamma(n)} & \frac{1}{2\Gamma(n-1)} & \frac{1}{2\Gamma(n-2)} & \frac{1}{2\Gamma(n-3)} & \cdots & 1
\end{pmatrix}.$$

Define

$$X = \left\{x : (0,1] \to \mathbb{R} : x|_{(t_s,t_{s+1}]} \in C^0(t_s, t_{s+1}](s \in \mathbb{N}_0),\ \lim_{t\to t_s^+} x(t) \text{ exist}(s \in \mathbb{N}_0^m)\right\}.$$

For $x \in X$, define the norms by $\|x\| = \|x\|_X = \sup_{t\in(0,1]} |x(t)|$.

Lemma 2.3 X is a Banach space.

Proof The proof is standard and omitted. □

Lemma 2.4 *Let M be a subset of X. Then M is relatively compact if and only if the following conditions are satisfied:*

Then $|M| \neq 0$ and $|N| = 1$. One has for a determinant $|a_{i,j}|_{(n-k)\times(n-k)}$ that

$$|a_{i,j}|_{(n-k)\times(n-k)} = \sum_{i=1}^{n-k} a_{i,n-j}A_{i,n-j}, j \in \mathbb{N}_k^{n-1}, \tag{2.5}$$

where $A_{i,n-j}$ is the algebraic cofactor of $a_{i,n-j}$.

Suppose that $|a_{i,j}| \leq 1$. It is easy to show that

$$|A_{i,n-j}| \leq (n-k-1)! = \Gamma(n-k), \quad i \in \mathbb{N}_1^{n-k}, j \in \mathbb{N}_k^{n-1}. \tag{2.6}$$

Then

$$M^{-1}=M^{*}=\begin{pmatrix} M_{11} & M_{21} & M_{31} & M_{41} & \cdots & M_{n-k1}\\ M_{12} & M_{22} & M_{32} & M_{42} & \cdots & M_{n-k2}\\ M_{13} & M_{23} & M_{33} & M_{43} & \cdots & M_{n-k3}\\ \cdot & \cdot & \cdot & \cdot & \cdots & \cdot\\ M_{1n-k} & M_{2n-k} & M_{3n-k} & M_{4n-k} & \cdots & M_{n-kn-k}\end{pmatrix},$$

$$N^{-1}=N^{*}=\begin{pmatrix} N_{11} & N_{21} & N_{31} & N_{41} & \cdots & N_{n1}\\ N_{12} & N_{22} & N_{32} & N_{42} & \cdots & N_{n2}\\ N_{13} & N_{23} & N_{33} & N_{43} & \cdots & N_{n3}\\ \cdot & \cdot & \cdot & \cdot & \cdots & \cdot\\ N_{1n} & N_{2n} & N_{3n} & N_{4n} & \cdots & N_{nn}\end{pmatrix},$$

where M_{ij} and N_{ij} are the algebraic cofactors of m_{ij} and n_{ij}, respectively. M^* and N^* are the adjoint matrix of M and N, respectively. From (2.5) and (2.6), we know that $|M_{ij}|\le\Gamma(n-k)$ and $|N_{ij}|\le\Gamma(n)$. □

Lemma 2.5 *Suppose that $u\in X$. Then $x\in X$ is a solution of*

$$\begin{cases} {}^cD_{0^+}^{\alpha}x(t)=f_u(t), & t\in(t_s,t_{s+1}],\, s\in\mathbb{N}_0^m,\\ x^{(i)}(0)=0, & i\in\mathbb{N}_0^{k-1},\\ x^{(j)}(1)=0, & j\in\mathbb{N}_l^{n+l-k-1},\\ \Delta x^{(j)}(t_s)=I_{jx}(t_s), & j\in\mathbb{N}_0^{n-1},\, s\in\mathbb{N}_1^m \end{cases}\tag{2.7}$$

if and only if

$$\begin{aligned} x(t)=&-\sum_{i=k}^{n-1}\sum_{j=1}^{n-k}\frac{M_{jn-i}}{\Gamma(i+1)|M|}\sum_{w=1}^{m}\sum_{v=n+l-k-j}^{n-1}\\ &\times\frac{(1-t_w)^{v-(n+l-k-j)}}{\Gamma(v-(n+l-k-j)+1)}I_{vu}(t_w)t^i\\ &-\sum_{i=k}^{n-1}\sum_{j=1}^{n-k}\frac{M_{jn-i}}{\Gamma(i+1)|M|}\int_0^1\frac{(1-s)^{\alpha-(n+l-k-j)-1}}{\Gamma(\alpha-(n+l-k-j))}f_u(s)\mathrm{d}st^i\\ &+\sum_{w=1}^{s}\sum_{v=0}^{n-1}\frac{(t-t_w)^v}{\Gamma(v+1)}I_{vu}(t_w)\\ &+\int_0^t\frac{(t-s)^{\alpha-1}}{\Gamma(\alpha)}f_u(s)\mathrm{d}s, t\in(t_s,t_{s+1}],\quad s\in\mathbb{N}_1^m. \end{aligned}\tag{2.8}$$

Proof First, we prove that x satisfies (2.8) if $x\in X$ and x is a solution of (2.7). Since $u\in X$, there exists $r>0$ such that

$$\|u\|=\max\left\{\sup_{t\in(0,1]}|u(t)|,\ \sup_{t\in(0,1]}|{}^cD_{0^+}^{\beta}u(t)|\right\}\le r.\tag{2.9}$$

Since f is a Carathéodory function, there exist constants $A_{r,f}\ge 0$ such that

$$|f(t,u(t),{}^cD_{0^+}^{\beta}u(t))|\le A_{r,f}t^p(1-t)^q,\quad t\in(t_s,t_{s+1}],\quad s\in\mathbb{N}_0^m.\tag{2.10}$$

Similarly, since I_j is a discrete Carathéodory function, there exist positive constants $A_{r,I_j,s}\ge 0(s\in\mathbb{N}_1^m, j\in\mathbb{N}_0^{n-1})$ such that

$$|I_j(t_s,u(t_s){}^cD_{0^+}^{\beta}u(t_s))|\le A_{r,I_j,s}.\tag{2.11}$$

Suppose that $x\in X$ and x is a solution of (2.7). By Lemma 2.2, we know that there exist constants $c_{vj}\in\mathbb{R}$ such that

$$\begin{aligned} x(t)=&\sum_{w=0}^{s}\sum_{v=0}^{n-1}\frac{c_{vw}}{\Gamma(v+1)}(t-t_w)^v\\ &+\int_0^t\frac{(t-s)^{\alpha-1}}{\Gamma(\alpha)}f_u(s)\mathrm{d}s,\quad t\in(t_s,t_{s+1}],\quad s\in\mathbb{N}_0. \end{aligned}\tag{2.12}$$

By Definition 2, we have

$$\begin{aligned} {}^cD_{0^+}^{\beta}x(t)=&\sum_{w=0}^{s}\sum_{v=1}^{n-1}\frac{c_{vw}}{\Gamma(v-\beta+1)}(t-t_w)^{v-\beta}\\ &+\int_0^t\frac{(t-s)^{\alpha-\beta-1}}{\Gamma(\alpha-\beta)}f_u(s)\mathrm{d}s,\quad t\in(t_s,t_{s+1}],s\in\mathbb{N}_0^m,\\ x^{(j)}(t)=&\sum_{w=0}^{s}\sum_{v=j}^{n-1}\frac{c_{vw}}{\Gamma(v-j+1)}(t-t_w)^{v-j}\\ &+\int_0^t\frac{(t-s)^{\alpha-j-1}}{\Gamma(\alpha-j)}f_u(s)\mathrm{d}s,\quad t\in(t_s,t_{s+1}],s\in\mathbb{N}_0^m,j\in\mathbb{N}_0^{n-1}. \end{aligned}\tag{2.13}$$

We have

$$\begin{aligned} \left|\int_0^t\frac{(t-s)^{\alpha-1}}{\Gamma(\alpha)}f_u(s)(s)\mathrm{d}s\right|&\le A_{r,f}\int_0^t\frac{(t-s)^{\alpha-1}}{\Gamma(\alpha)}s^p(1-s)^q\mathrm{d}s\\ &\le A_{r,f}\int_0^t\frac{(t-s)^{\alpha+q-1}}{\Gamma(\alpha)}s^p\mathrm{d}s\\ &=A_{r,f}t^{\alpha+p+q}\frac{\mathbf{B}(\alpha+q,p+1)}{\Gamma(\alpha)}, \end{aligned}$$

$$\begin{aligned} \left|\int_0^t\frac{(t-s)^{\alpha-j-1}}{\Gamma(\alpha-j)}f_u(s)(s)\mathrm{d}s\right|&\le A_{r,f}\int_0^t\frac{(t-s)^{\alpha-j-1}}{\Gamma(\alpha-j)}s^p(1-s)^q\mathrm{d}s\\ &\le A_{r,f}\int_0^t\frac{(t-s)^{\alpha+q-j-1}}{\Gamma(\alpha-j)}s^p\mathrm{d}s\\ &=A_{r,f}t^{\alpha-j+p+q}\frac{\mathbf{B}(\alpha-j+q,p+1)}{\Gamma(\alpha-j)},\\ &j\in\mathbb{N}_0^{n-1}. \end{aligned}$$

(i) It follows from $x^{(j)}(0)=0$ that $c_{j0}=0(j\in\mathbb{N}_0^{k-1})$.
(ii) From $\Delta x^{(j)}(t_s)=I_{ju}(t_s)$ and (2.13), we get $c_{js}=I_{ju}(t_s)(j\in\mathbb{N}_0^{n-1}, s\in\mathbb{N}_1^m)$.
(iii) From $x^{(j)}(1)=0$, $j\in\mathbb{N}_l^{n+l-k-1}$, we get

$$\sum_{w=0}^{m}\sum_{v=j}^{n-1}\frac{c_{vw}}{\Gamma(v-j+1)}(1-t_w)^{v-j}+\int_0^1\frac{(1-s)^{\alpha-j-1}}{\Gamma(\alpha-j)}f_u(s)\mathrm{d}s=0.$$

Use (i) and (ii), we get

$$\sum_{v=j}^{n-1}\frac{c_{v0}}{\Gamma(v-j+1)}+\sum_{w=1}^{m}\sum_{v=j}^{n-1}\frac{(1-t_w)^{v-j}}{\Gamma(v-j+1)}I_{vu}(t_w)$$
$$+\int_0^1\frac{(1-s)^{\alpha-j-1}}{\Gamma(\alpha-j)}f_u(s)\mathrm{d}s=0,\quad j\in\mathbb{N}_l^{n+l-k-1}.$$

Then

$$\begin{pmatrix}
\frac{1}{\Gamma(k-l+1)} & \frac{1}{\Gamma(k-l)} & \cdots & 1 & 0 & 0 & 0 & \cdots & 0\\
\frac{1}{\Gamma(k-l+2)} & \frac{1}{\Gamma(k-l+1)} & \cdots & \frac{1}{\Gamma(2)} & 1 & 0 & 0 & \cdots & 0\\
\frac{1}{\Gamma(k-l+3)} & \frac{1}{\Gamma(k-l+2)} & \cdots & \frac{1}{\Gamma(3)} & \frac{1}{\Gamma(2)} & 1 & 0 & \cdots & 0\\
\cdots & \cdots & \cdots & \cdots & \cdots & \cdots & \cdots & \cdots & \cdots\\
\frac{1}{\Gamma(n-k)} & \frac{1}{\Gamma(n-k-1)} & \cdots & \cdots & \cdots & \cdots & \cdots & \cdots & 1\\
\frac{1}{\Gamma(n-k+1)} & \frac{1}{\Gamma(n-k)} & \cdots & \cdots & \cdots & \cdots & \cdots & \cdots & \frac{1}{\Gamma(2)}\\
\cdots & \cdots & \cdots & \cdots & \cdots & \cdots & \cdots & \cdots & \cdots\\
\frac{1}{\Gamma(n-l)} & \frac{1}{\Gamma(n-l-1)} & \cdots & \frac{1}{\Gamma(n-k)} & \frac{1}{\Gamma(n-k-1)} & \frac{1}{\Gamma(n-k-2)} & \frac{1}{\Gamma(n-k-3)} & \cdots & \frac{1}{\Gamma(k-l+1)}
\end{pmatrix}$$

$$\times\begin{pmatrix} c_{n-10}\\ c_{n-20}\\ c_{n-30}\\ \cdots\\ \\ c_{k0}\end{pmatrix}=-\begin{pmatrix}
\sum_{w=1}^{m}\sum_{v=n+l-k-1}^{n-1}\frac{(1-t_w)^{v-(n+l-k-1)}}{\Gamma(v-(n+l-k-1)+1)}I_{vu}(t_w)+\int_0^1\frac{(1-s)^{\alpha-(n+l-k-1)-1}}{\Gamma(\alpha-(n+l-k-1))}f_u(s)\mathrm{d}s\\
\sum_{w=1}^{m}\sum_{v=n+l-k-2}^{n-1}\frac{(1-t_w)^{v-(n+l-k-2)}}{\Gamma(v-(n+l-k-2)+1)}I_{vu}(t_w)+\int_0^1\frac{(1-s)^{\alpha-(n+l-k-2)-1}}{\Gamma(\alpha-(n+l-k-2))}f_u(s)\mathrm{d}s\\
\sum_{w=1}^{m}\sum_{v=n+l-k-3}^{n-1}\frac{(1-t_w)^{v-(n+l-k-3)}}{\Gamma(v-(n+l-k-3)+1)}I_{vu}(t_w)+\int_0^1\frac{(1-s)^{\alpha-(n+l-k-3)-1}}{\Gamma(\alpha-(n+l-k-3))}f_u(s)\mathrm{d}s\\
\cdots\\
\sum_{w=1}^{m}\sum_{v=l}^{n-1}\frac{(1-t_w)^{v-l}}{\Gamma(v-l+1)}I_{vu}(t_w)+\int_0^1\frac{(1-s)^{\alpha-l-1}}{\Gamma(\alpha-l)}f_u(s)\mathrm{d}s
\end{pmatrix}.$$

Hence,

$$\begin{pmatrix} c_{n-10} \\ c_{n-20} \\ c_{n-30} \\ \cdots \\ c_{k0} \end{pmatrix} = -M^{-1} \begin{pmatrix} \sum_{w=1}^{m}\sum_{v=n+l-k-1}^{n-1} \frac{(1-t_w)^{v-(n+l-k-1)}}{\Gamma(v-(n+l-k-1)+1)} I_{vu}(t_w) + \int_0^1 \frac{(1-s)^{\alpha-(n+l-k-1)-1}}{\Gamma(\alpha-(n+l-k-1))} f_u(s)\mathrm{d}s \\ \sum_{w=1}^{m}\sum_{v=n+l-k-2}^{n-1} \frac{(1-t_w)^{v-(n+l-k-2)}}{\Gamma(v-(n+l-k-2)+1)} I_{vu}(t_w) + \int_0^1 \frac{(1-s)^{\alpha-(n+l-k-2)-1}}{\Gamma(\alpha-(n+l-k-2))} f_u(s)\mathrm{d}s \\ \sum_{w=1}^{m}\sum_{v=n+l-k-3}^{n-1} \frac{(1-t_w)^{v-(n+l-k-3)}}{\Gamma(v-(n+l-k-3)+1)} I_{vu}(t_w) + \int_0^1 \frac{(1-s)^{\alpha-(n+l-k-3)-1}}{\Gamma(\alpha-(n+l-k-3))} f_u(s)\mathrm{d}s \\ \cdots \\ \sum_{w=1}^{m}\sum_{v=l}^{n-1} \frac{(1-t_w)^{v-l}}{\Gamma(v-l+1)} I_{vu}(t_w) + \int_0^1 \frac{(1-s)^{\alpha-l-1}}{\Gamma(\alpha-l)} f_u(s)\mathrm{d}s \end{pmatrix}.$$

It follows for $i \in \mathbb{N}_k^{n-1}$ that

$$c_{i0} = -\sum_{j=1}^{n-k} \frac{M_{jn-i}}{|M|} \times \left(\sum_{w=1}^{m}\sum_{v=n+l-k-j}^{n-1} \frac{(1-t_w)^{v-(n+l-k-j)}}{\Gamma(v-(n+l-k-j)+1)} I_{vu}(t_w) + \int_0^1 \frac{(1-s)^{\alpha-(n+l-k-j)-1}}{\Gamma(\alpha-(n+l-k-j))} f_u(s)\mathrm{d}s \right). \tag{2.14}$$

From (i), (ii) and (iii), we have (2.14) and

$$c_{j0} = 0, j \in \mathbb{N}_0^{k-1}, \quad c_{js} = I_{ju}(t_s)(s \in \mathbb{N}_1^m, \quad j \in \mathbb{N}_0^{n-1}). \tag{2.15}$$

Substituting (2.14) and (2.15) in (2.12), we get (2.8).

Second, we prove that $x \in X$ and x satisfies (2.7) if x satisfies (2.8). It is easy to see that $x \in X$ and

$$\lim_{t\to 0} x^{(j)}(t) = u_j, \quad j \in \mathbb{N}_0^{k-1}, \quad x^{(j)}(1) = v_j, \quad j \in \mathbb{N}_k^{n-1},$$
$$\Delta x^{(j)}(t_s) = I_{ju}(t_s), \quad j \in \mathbb{N}_0^{n-1}, \quad s \in\in \mathbb{N}_1^m.$$

Now, we prove that x satisfies ${}^cD_{0^+}^{\alpha}x(t) = f_u(t)$ if (2.8) holds. We remember (2.15) and (2.14), then it suffices to prove ${}^cD_{0^+}^{\alpha}x(t) = f_u(t)$ on (0, 1) if x satisfies (2.8).

In fact, for $t \in (t_i, t_{i+1}](i \in \mathbb{N}_0^m$, by Definition 2.2, we have

$$\begin{aligned} D_{0^+}^{\alpha}x(t) &= \frac{1}{\Gamma(n-\alpha)}\int_0^t (t-s)^{n-\alpha-1}x^{(n)}(s)\mathrm{d}s \\ &= \frac{1}{\Gamma(n-\alpha)}\left[\sum_{j=0}^{i-1}\int_{t_j}^{t_{j+1}}(t-s)^{n-\alpha-1}x^{(n)}(s)\mathrm{d}s + \int_{t_i}^t (t-s)^{n-\alpha-1}x^{(n)}(s)\mathrm{d}s\right] \\ &= \frac{\sum_{j=0}^{i-1}\int_{t_j}^{t_{j+1}}(t-s)^{n-\alpha-1}\left(\sum_{w=0}^{j}\sum_{\theta=0}^{n-1}\frac{c_{\theta w}}{\Gamma(\theta+1)}(s-t_w)^{\theta} + \int_0^s \frac{(s-\sigma)^{\alpha-1}}{\Gamma(\alpha)}f_u(\sigma)d\sigma\right)^{(n)}\mathrm{d}s}{\Gamma(n-\alpha)} \\ &\quad + \frac{\int_{t_i}^t (t-s)^{n-\alpha-1}\left(\sum_{w=0}^{i}\sum_{\theta=0}^{n-1}\frac{c_{\theta w}}{\Gamma(\theta+1)}(s-t_w)^{\theta} + \int_0^s \frac{(s-\sigma)^{\alpha-1}}{\Gamma(\alpha)}f_u(\sigma)d\sigma\right)^{(n)}\mathrm{d}s}{\Gamma(n-\alpha)} \\ &= \frac{\int_0^t (t-s)^{n-\alpha-1}\left(\int_0^s \frac{(s-\sigma)^{\alpha-1}}{\Gamma(\alpha)}f_u(\sigma)d\sigma\right)^{(n)}\mathrm{d}s}{\Gamma(n-\alpha)} \\ &= \left[\frac{\int_0^t (t-s)^{n-\alpha}\left(\int_0^s \frac{(s-\sigma)^{\alpha-n}}{\Gamma(\alpha-n+1)}f_u(\sigma)d\sigma\right)'\mathrm{d}s}{(n-\alpha)\Gamma(n-\alpha)}\right]' \\ &= \left[\frac{(t-s)^{n-\alpha}\left(\int_0^s \frac{(s-\sigma)^{\alpha-n}}{\Gamma(\alpha-n+1)}f_u(\sigma)d\sigma\right)\Big|_0^t + (n-\alpha)\int_0^t (t-s)^{n-\alpha-1}\left(\int_0^s \frac{(s-\sigma)^{\alpha-n}}{\Gamma(\alpha-n+1)}f_u(\sigma)d\sigma\right)\mathrm{d}s}{(n-\alpha)\Gamma(n-\alpha)}\right]' \\ &= \left[\frac{\int_0^t\int_s^t (t-s)^{n-\alpha-1}\frac{(s-\sigma)^{\alpha-n}}{\Gamma(\alpha-n+1)}\mathrm{d}s f_u(\sigma)d\sigma}{\Gamma(n-\alpha)}\right]' \\ &= \left[\frac{\int_0^t\int_0^1 (1-w)^{n-\alpha-1}\frac{w^{\alpha-n}}{\Gamma(\alpha-n+1)}\mathrm{d}w f_u(\sigma)d\sigma}{\Gamma(n-\alpha)}\right]' \\ &= f_u(t), \quad t \in (t_i, t_{i+1}], i \in \mathbb{N}_0. \end{aligned}$$

From above discussion, we know that $x \in X$ and x satisfies (2.7) if (2.8) holds. The proof is completed. □

Remark 2.1 It is easy to see from Lemma 2.6 that $x \in X$ is a solution of (2.10) if and only if x satisfies that there exists constants $d_{vs} \in \mathbb{R}$ such that

$$x(t) = \sum_{v=0}^{n-1} d_{vs}t^v + \int_0^t \frac{(t-w)^{\alpha-1}}{\Gamma(\alpha)} f_{uv}(w)\mathrm{d}w, t \in (t_s, t_{s+1}], \quad s \in \mathbb{N}_0.$$

In [28], authors have proved this result but our proof of Lemma 2.6 is different from that in [28].

Now, we define the operator T_1 on X by

$$\begin{aligned}(T_1x)(t) = & -\sum_{i=k}^{n-1}\sum_{j=1}^{n-k}\frac{M_{jn-i}}{\Gamma(i+1)|M|}\sum_{w=1}^{m}\sum_{v=n+l-k-j}^{n-1}\frac{(1-t_w)^{v-(n+l-k-j)}}{\Gamma(v-(n+l-k-j)+1)}I_{vx}(t_w)t^i \\ & -\sum_{i=k}^{n-1}\sum_{j=1}^{n-k}\frac{M_{jn-i}}{\Gamma(i+1)|M|}\int_0^1\frac{(1-s)^{\alpha-(n+l-k-j)-1}}{\Gamma(\alpha-(n+l-k-j))}f_x(s)\mathrm{d}st^i \\ & +\sum_{w=1}^{s}\sum_{v=0}^{n-1}\frac{(t-t_w)^v}{\Gamma(v+1)}I_{vx}(t_w)+\int_0^t\frac{(t-s)^{\alpha-1}}{\Gamma(\alpha)}f_x(s)\mathrm{d}s, \\ & t\in(t_s,t_{s+1}],\quad s\in\mathbb{N}_1^m.\end{aligned} \tag{2.16}$$

Remark 2.2 By Lemma 2.5, we know that $T_1 : X \to X$ is well defined and $x \in X$ is a solution of system (1.6) if and only if $x \in X$ is a fixed point of the operator T_1.

Lemma 2.6 *The operator* $T_1 : X \to X$ *is completely continuous.*

Proof The proof is standard and is omitted, one may see [21]. □

Lemma 2.7 *Suppose that* $u \in X$. *Then* $x \in X$ *is a solution of*

$$\begin{cases} {}^cD_{0^+}^\alpha x(t) = f_u(t), & t\in(t_s,t_{s+1}], s\in\mathbb{N}_0^m, \\ x^{(i)}(0) = -x^{(i)}(1), & j\in\mathbb{N}_0^{n-1}, \\ \Delta x^{(j)}(t_s) = I_{jx}(t_s), & j\in\mathbb{N}_0^{n-1}, s\in\mathbb{N}_1^m \end{cases} \tag{2.17}$$

if and only if

$$\begin{aligned}x(t) = & -\sum_{i=0}^{n-1}\sum_{j=0}^{n-1}\frac{N_{jn-i}}{\Gamma(i+1)}\sum_{w=1}^{m}\sum_{v=n-j}^{n-1}\frac{(1-t_w)^{v-(n-j)}}{\Gamma(v-(n-j)+1)}I_{vu}(t_w)t^i \\ & -\sum_{i=0}^{n-1}\sum_{j=0}^{n-1}\frac{N_{jn-i}}{\Gamma(i+1)}\int_0^1\frac{(1-s)^{\alpha-(n-j)-1}}{\Gamma(\alpha-(n-j))}f_u(s)\mathrm{d}st^i \\ & +\sum_{w=1}^{s}\sum_{v=0}^{n-1}\frac{(t-t_w)^v}{\Gamma(v+1)}I_{vu}(t_w)+\int_0^t\frac{(t-s)^{\alpha-1}}{\Gamma(\alpha)}f_u(s)\mathrm{d}s, \\ & t\in(t_s,t_{s+1}],\quad s\in\mathbb{N}_1^m.\end{aligned} \tag{2.18}$$

Proof Similarly to the proof of Lemma 2.5, we get Lemma 2.7.

Now, we define the operator T_2 on X by

$$\begin{aligned}(T_2x)(t) = & -\sum_{i=0}^{n-1}\sum_{j=0}^{n-1}\frac{N_{jn-i}}{\Gamma(i+1)}\sum_{w=1}^{m}\sum_{v=n-j}^{n-1}\frac{(1-t_w)^{v-(n-j)}}{\Gamma(v-(n-j)+1)}I_{vx}(t_w)t^i \\ & -\sum_{i=0}^{n-1}\sum_{j=0}^{n-1}\frac{N_{jn-i}}{\Gamma(i+1)}\int_0^1\frac{(1-s)^{\alpha-(n-j)-1}}{\Gamma(\alpha-(n-j))}f_x(s)\mathrm{d}st^i \\ & +\sum_{w=1}^{s}\sum_{v=0}^{n-1}\frac{(t-t_w)^v}{\Gamma(v+1)}I_{vx}(t_w)+\int_0^t\frac{(t-s)^{\alpha-1}}{\Gamma(\alpha)}f_x(s)\mathrm{d}s, \\ & t\in(t_s,t_{s+1}],\quad s\in\mathbb{N}_1^m.\end{aligned} \tag{21.9}$$

□

Remark 2.3 By Lemma 2.7, we know that $T_2 : X \to X$ is well defined, $x \in X$ is a solution of system (1.7) if and only if $x \in X$ is a fixed point of the operator T_2.

Lemma 2.8 *The operator* $T_2 : X \to X$ *is completely continuous.*

Proof The proof is standard and is omitted, one may see [21]. □

Main results

In this section, we are ready to present the main theorems. We need the following assumptions:

(H1) there exist nonnegative numbers σ_i, $a_i, A_i (i \in \mathbb{N}_0^n)$ such that

$$|f(t,x)| \le [a_0 + \sum_{i=1}^{\omega} a_i|x|^{\sigma_i}]t^p(1-t)^q, \quad t\in(0,1), x\in\mathbb{R},$$

$$|I_j(t_s,x)| \le A_0 + \sum_{i=1}^{\omega} A_i|x|^{\sigma_i}, \quad s\in\mathbb{N}_1^m, x\in\mathbb{R}.$$

Denote

$$\sigma = \max\{\sigma_i : i\in\mathbb{N}_1^n\},$$

$$\begin{aligned}M_0 = & A_0\sum_{i=k}^{n-1}\sum_{j=1}^{n-k}\frac{\Gamma(n-k)}{\Gamma(i+1)||M||}\sum_{w=1}^{m}\sum_{v=n+l-k-j}^{n-1}\frac{(1-t_w)^{v-(n+l-k-j)}}{\Gamma(v-(n+l-k-j)+1)} \\ & +A_0\sum_{v=0}^{n-1}\frac{m}{\Gamma(v+1)} \\ & +a_0\sum_{i=k}^{n-1}\sum_{j=1}^{n-k}\frac{\Gamma(n-k)}{\Gamma(i+1)||M||}\frac{\mathbf{B}(\alpha-(n+l-k-j)+q,p+1)}{\Gamma(\alpha-(n+l-k-j))} \\ & +a_0\frac{\mathbf{B}(\alpha+q,p+1)}{\Gamma(\alpha)},\end{aligned}$$

$$\begin{aligned}M_u = & \sum_{i=k}^{n-1}\sum_{j=1}^{n-k}\frac{\Gamma(n-k)}{\Gamma(i+1)||M||}\sum_{w=1}^{m}\sum_{v=n+l-k-j}^{n-1}\frac{(1-t_w)^{v-(n+l-k-j)}}{\Gamma(v-(n+l-k-j)+1)}A_u \\ & +\sum_{v=0}^{n-1}\frac{m}{\Gamma(v+1)}A_u \\ & +\sum_{i=k}^{n-1}\sum_{j=1}^{n-k}\frac{\Gamma(n-k)}{\Gamma(i+1)||M||}\frac{\mathbf{B}(\alpha-(n+l-k-j)+q,p+1)}{\Gamma(\alpha-(n+l-k-j))}a_u \\ & +\frac{\mathbf{B}(\alpha+q,p+1)}{\Gamma(\alpha)}a_u, \quad u\in\mathbb{N}_1^{\omega}.\end{aligned}$$

Theorem 3.1 *Suppose that (a)–(d) (defined in "Introduction") and (H1) hold. Then, the system* (1.6) *has at least one solution in* $X \times Y$ *if*

(i) $\sigma < 1$ or
(ii) $\sigma = 1$ with

$$\sum_{u=1}^{\omega} M_u < 1 \tag{3.1}$$

or

(iii) $\sigma > 1$ *with*

$$\sigma_{u_0} > 1,\ M_0 + \sum_{u=1}^{\omega} M_u \left(\frac{M_0}{M_{u_0}(\sigma_{u_0}-1)}\right)^{\sigma_u/\sigma_{u_0}} \le \left(\frac{M_0}{M_{u_0}(\sigma_{u_0}-1)}\right)^{1/\sigma_{u_0}}. \tag{3.2}$$

Proof We shall apply the Schauder's fixed point theorem. From Lemma 2.6 and Remark 2.2 we note that T_1 is completely continuous. If x is a fixed point of T_1, the system (1.6) has a solution x.

Let $\Omega_r = \{x \in X : ||x|| \le r\}$. For $x \in \Omega_r$. Then $||x|| \le r$, i.e., $|x(t)| \le r$ for all $t \in (0,1]$. So

(H1) implies

$$|f(t,x(t))| \le [a_0 + \sum_{i=1}^{\omega} a_i|x(t)|^{\sigma_i}]t^p(1-t)^q \le [a_0 + \sum_{i=1}^{\omega} a_i r^{\sigma_i}]t^p(1-t)^q,$$

$$|I_j(t_s,x(t_s))| \le A_0 + \sum_{i=1}^{\omega} A_i|x(t_s)|^{\sigma_i} \le A_0 + \sum_{i=1}^{\omega} A_i r^{\sigma_i}.$$

We know $|M_{ij}| \le \Gamma(n-k)$. By (2.16), we have

$$\begin{aligned}
|(T_1x)(t)| \le{}& \sum_{i=k}^{n-1}\sum_{j=1}^{n-k} \frac{|M_{jn-i}|}{\Gamma(i+1)||M||}\sum_{w=1}^{m}\sum_{v=n+l-k-j}^{n-1}\frac{(1-t_w)^{v-(n+l-k-j)}}{\Gamma(v-(n+l-k-j)+1)}|I_{vx}(t_w)|t^i \\
&+ \sum_{i=k}^{n-1}\sum_{j=1}^{n-k} \frac{|M_{jn-i}|}{\Gamma(i+1)||M||}\int_0^1 \frac{(1-s)^{\alpha-(n+l-k-j)-1}}{\Gamma(\alpha-(n+l-k-j))}|f_x(s)|\mathrm{d}s t^i \\
&+ \sum_{w=1}^{s}\sum_{v=0}^{n-1}\frac{(t-t_w)^v}{\Gamma(v+1)}|I_{vx}(t_w)| + \int_0^t \frac{(t-s)^{\alpha-1}}{\Gamma(\alpha)}|f_x(s)|\mathrm{d}s \\
\le{}& \sum_{i=k}^{n-1}\sum_{j=1}^{n-k} \frac{\Gamma(n-k)}{\Gamma(i+1)||M||}\sum_{w=1}^{m}\sum_{v=n+l-k-j}^{n-1}\frac{(1-t_w)^{v-(n+l-k-j)}}{\Gamma(v-(n+l-k-j)+1)}\left[A_0 + \sum_{i=1}^{\omega} A_i r^{\sigma_i}\right] \\
&+ \sum_{i=k}^{n-1}\sum_{j=1}^{n-k} \frac{\Gamma(n-k)}{\Gamma(i+1)||M||}\int_0^1 \frac{(1-s)^{\alpha-(n+l-k-j)-1}}{\Gamma(\alpha-(n+l-k-j))}[a_0 + \sum_{i=1}^{\omega} a_i r^{\sigma_i}]s^p(1-s)^q\mathrm{d}s \\
&+ \sum_{w=1}^{s}\sum_{v=0}^{n-1}\frac{(t-t_w)^v}{\Gamma(v+1)}\left[A_0 + \sum_{i=1}^{\omega} A_i r^{\sigma_i}\right] + \int_0^t \frac{(t-s)^{\alpha-1}}{\Gamma(\alpha)}[a_0 + \sum_{i=1}^{\omega} a_i r^{\sigma_i}]s^p(1-s)^q\mathrm{d}s \\
\le{}& \sum_{i=k}^{n-1}\sum_{j=1}^{n-k} \frac{\Gamma(n-k)}{\Gamma(i+1)||M||}\sum_{w=1}^{m}\sum_{v=n+l-k-j}^{n-1}\frac{(1-t_w)^{v-(n+l-k-j)}}{\Gamma(v-(n+l-k-j)+1)}\left[A_0 + \sum_{i=1}^{\omega} A_i r^{\sigma_i}\right] \\
&+ \sum_{i=k}^{n-1}\sum_{j=1}^{n-k} \frac{\Gamma(n-k)}{\Gamma(i+1)||M||}\frac{\mathbf{B}(\alpha-(n+l-k-j)+q,p+1)}{\Gamma(\alpha-(n+l-k-j))}[a_0 + \sum_{i=1}^{\omega} a_i r^{\sigma_i}] \\
&+ \sum_{v=0}^{n-1}\frac{m}{\Gamma(v+1)}\left[A_0 + \sum_{i=1}^{\omega} A_i r^{\sigma_i}\right] + \frac{\mathbf{B}(\alpha+q,p+1)}{\Gamma(\alpha)}[a_0 + \sum_{i=1}^{\omega} a_i r^{\sigma_i}] \\
={}& A_0\sum_{i=k}^{n-1}\sum_{j=1}^{n-k} \frac{\Gamma(n-k)}{\Gamma(i+1)||M||}\sum_{w=1}^{m}\sum_{v=n+l-k-j}^{n-1}\frac{(1-t_w)^{v-(n+l-k-j)}}{\Gamma(v-(n+l-k-j)+1)} + A_0\sum_{v=0}^{n-1}\frac{m}{\Gamma(v+1)} \\
&+ a_0\sum_{i=k}^{n-1}\sum_{j=1}^{n-k} \frac{\Gamma(n-k)}{\Gamma(i+1)||M||}\frac{\mathbf{B}(\alpha-(n+l-k-j)+q,p+1)}{\Gamma(\alpha-(n+l-k-j))} + a_0\frac{\mathbf{B}(\alpha+q,p+1)}{\Gamma(\alpha)} \\
&+ \sum_{u=1}^{\omega}\left(\sum_{i=k}^{n-1}\sum_{j=1}^{n-k} \frac{\Gamma(n-k)}{\Gamma(i+1)||M||}\sum_{w=1}^{m}\sum_{v=n+l-k-j}^{n-1}\frac{(1-t_w)^{v-(n+l-k-j)}}{\Gamma(v-(n+l-k-j)+1)}A_u + \sum_{v=0}^{n-1}\frac{m}{\Gamma(v+1)}A_u\right. \\
&\left.+ \sum_{i=k}^{n-1}\sum_{j=1}^{n-k} \frac{\Gamma(n-k)}{\Gamma(i+1)||M||}\frac{\mathbf{B}(\alpha-(n+l-k-j)+q,p+1)}{\Gamma(\alpha-(n+l-k-j))}a_u + \frac{\mathbf{B}(\alpha+q,p+1)}{\Gamma(\alpha)}a_u\right)r^{\sigma_u}.
\end{aligned}$$

It follows that

$$||T_1x|| \le M_0 + \sum_{u=1}^{\omega} M_u r^{\sigma_u}. \tag{3.3}$$

To use Schauder's fixed point theorem, from (3.4), we should choose $r > 0$ such that

$$M_0 + \sum_{u=1}^{\omega} M_u r^{\sigma_u} \le r. \tag{3.4}$$

Then $T_1\Omega_r \subseteq \Omega_r$. So T_1 has a fixed point in Ω_r. Then BVP (1.6) has a solution. We consider the following three cases:

Case 1 $\sigma < 1$. Since $\lim\limits_{r\to\infty} \frac{M_0+\sum_{u=1}^{\omega} M_u r^{\sigma_u}}{r} = 0$, we can choose $r > 0$ sufficiently small such that (3.4) holds. Then $T_1\Omega_r \subseteq \Omega_r$. So T_1 has a fixed point in Ω_r. Then BVP(1.6) has a solution.

Case 2 $\sigma = 1$. Since $\lim\limits_{r\to\infty} \frac{M_0+\sum_{u=1}^{\omega} M_u r^{\sigma_u}}{r} < \sum_{u=1}^{n} M_u < 1$, we can choose $r > 0$ sufficiently small such that (3.4) holds. Then $T_1\Omega_r \subseteq \Omega_r$. So T_1 has a fixed point in Ω_r. Then BVP (1.6) has a solution.

Case 3 $\sigma > 1$. Choose $r = \left(\frac{M_0}{M_{u_0}(\sigma_{u_0}-1)}\right)^{1/\sigma_{u_0}}$. Then we have by the inequality in (iii) that

$$||T_1x|| \le M_0 + \sum_{u=1}^{\omega} M_u r^{\sigma_u} \le r.$$

Then $T_1\Omega_r \subseteq \Omega_r$. So T_1 has a fixed point in Ω_r. Then BVP (1.6) has a solution. The proof of Theorem 3.1 is completed.

(H2) there exist constants $M_f, M_I \ge 0$ such that $|f(t,x,y)| \le M_f$, $|I_j(t_s,x,y)| \le M_I$ hold for all $t \in (0,1), s \in \mathbb{N}_1^m, j \in \mathbb{N}_0^{n-1}, (x,y) \in \mathbb{R}^2$.

□

Theorem 3.2 *Suppose that (a)–(d) and (H2) hold. Then BVP* (1.6) *has at least one solution.*

Proof Choose $p = q = 0$, $a_0 = M_f, A_0 = M_I$ and $a_i = 0$, $A_i = 0$, $\sigma_i = 0$. One sees by (H2) that (H1) holds. By Theorem 3.1 (i), we get its proof. □

Remark 3.1 BVP (1.7) can be called a anti-periodic boundary value problem. By similar method, we can establish existence results for BVP (1.7). We omit the details, readers should try it.

Examples

To illustrate the usefulness of our main result, we present an example that Theorem 3.1 can readily apply.

Example 4.1 Consider the following impulsive boundary value problem

$$\begin{cases} {}^cD_{0^+}^{\frac{18}{5}}u(t) = t^{-\frac{1}{5}}(1-t)^{-\frac{1}{5}}[a_0 + a_1[u(t)]^\sigma], & t \in (s,s+1], s \in \mathbb{N}_0^m, \\ u^{(i)}(0) = 0, i \in \mathbb{N}_0^1 \ u^{(j)}(1) = 0, & j \in \mathbb{N}_0^1, \\ \Delta u^{(i)}(1/2) = A_0 + A_1[u(1/2)]^\sigma, \end{cases} \tag{4.1}$$

where $a_i, A_i (i = 0,1)$ are nonnegative constants.

Corresponding to system (1.6) we have $\alpha = \frac{18}{5}$ with $n = 4$. So equation in BVP (4.1) is a fractional elastic beam equation. We also have $p = q = -\frac{1}{5}$, $k = 2$ and $l = 0$, $0 = t_0 < t_1 = \frac{1}{2} < t_2 = 1$ with $m = 1$ and

$$f(t,x) = t^{-\frac{1}{5}}(1-t)^{-\frac{1}{5}}[a_0 + a_1x^\sigma], \quad I_j(s,x) = A_0 + A_1x^\sigma.$$

It is easy to know that (a)–(d) and (H1) hold with $\omega = 1$. By direct computation, we have

$$M = (m_{ij})_{2\times 2} = \begin{pmatrix} \frac{1}{\Gamma(3)} & 1 \\ \frac{1}{\Gamma(4)} & \frac{1}{\Gamma(3)} \end{pmatrix}, \quad |M| = -2.$$

Now, $n = 4, k = 2, l = 0, |M| = -2, \quad m = 1, \quad t_1 = 1/2$, $\omega = 1$, $\alpha = \frac{18}{5}$, $p = q = -\frac{1}{5}$, we have

$$M_0 = A_0 \sum_{i=2}^{3}\sum_{j=1}^{2} \frac{1}{2\Gamma(i+1)} \sum_{v=2-j}^{3} \frac{(1/2)^{v+j-2}}{\Gamma(v+j-1)} + \frac{13A_0}{6}$$
$$+ a_0 \sum_{i=2}^{3}\sum_{j=1}^{2} \frac{1}{2\Gamma(i+1)} \frac{\mathbf{B}(7/5+j,4/5)}{\Gamma(8/5+j)} + a_0 \frac{\mathbf{B}(17/5,4/5)}{\Gamma(18/5)},$$

$$M_1 = \sum_{i=2}^{3}\sum_{j=1}^{2} \frac{1}{2\Gamma(i+1)} \sum_{v=2-j}^{3} \frac{(1/2)^{v+j-2}}{\Gamma(v+j-1)} A_1 + \frac{13A_1}{6}$$
$$+ \sum_{i=2}^{3}\sum_{j=1}^{2} \frac{1}{2\Gamma(i+1)} \frac{\mathbf{B}(7/5+j,4/5)}{\Gamma(8/5+j)} a_1 + \frac{\mathbf{B}(17/5,4/5)}{\Gamma(18/5)} a_1.$$

By Theorem 3.1, BVP (4.1) has at least one solution if one of the following items holds:

(i) $\sigma < 1$.
(ii) $\sigma = 1$ with $M_1 < 1$.
(iii) $\sigma > 1$ with $M_0^{1-1/\sigma} M_1^{1/\sigma} \frac{\sigma}{\sigma-1} \le \frac{1}{(\sigma-1)^{1/\sigma}}$.

Conclusion

In this paper, we discuss the solvability of two classes of boundary value problems or higher order fractional differential equations involving the Caputo fractional derivatives. Using some fixed point theorems in Banach spaces, we establish sufficient conditions for the existence of solutions of these kinds of problems.

In recent years, there have been several kinds of fractional derivatives proposed such as the Riemann–Liouville fractional derivative, the Hadamard fractional derivative, etc., see [29, 30]. Hence, it is interesting to study the existence and uniqueness of solutions of boundary value problems for other kinds of fractional differential equations. It is also interesting to find the similar properties and the difference properties between these different kinds of fractional differential equations.

The fixed point theorems in Banach spaces [31] are main tools for investigating the solvability of boundary value problems for fractional differential equations. It needs to find other methods for finding solutions for these kinds of problems.

References

1. Hilfer, R.: Applications of Fractional Calculus in Physics. World Scientific, River Edge (2000)
2. Miller, K.S., Ross, B.: An Introduction to the Fractional Calculus and Fractional Differential Equations. John Wiley, New York (1993)
3. Podlubny, I.: Fractional Differential Equations. Mathmatics in Science and Engineering, vol 198, Academic Press, San Diego, California, USA (1999)
4. Lakshmikantham, V.V., Bainov, D.D., Simeonov, P.S.: Theory of Impulsive Differential Equations. World Scientific, Singapore (1989)
5. Agarwal, R.P., Bohner, M., Wong, P.J.Y.: Positive solutions and eigenvalues of conjugate boundary value problems. Proceedings of the Edinburgh Mathematical Society (Series 2). **42**(02), 349–374 (1999)
6. Agarwal, R.P., O'Regan, D.: Positive solutions for $(p, n-p)$ conjugate boundary value problems. J. Differ. Equ. **150**(2), 462–473 (1998)
7. Davis, J.M., Henderson, J.: Triple positive solutions for (k, nk) conjugate boundary value problems. Math. Slovaca **51**(3), 313–320 (2001)
8. Eloe, P.W., Henderson, J.: Singular nonlinear $(k, n-k)$ conjugate boundary value problems. J. Differ. Equ. **133**(1), 136–151 (1997)
9. Il'in, V.A., Moiseev, E.I.: An a priori bound for a solution of the problem conjugate to a nonlocal boundary-value problem of the first kind. Differ. Equ. **24**(5), 795–804 (1988)
10. Jiang, D., Liu, H.: Existence of positive solutions to $(k, n-k)$ conjugate boundary value problems. Kyushu J. Math. **53**(1), 115–125 (1999)
11. Kosmatov, N.: On a singular conjugate boundary value problem with infinitely many solutions. Math. Sci. Res. Hot-Line **4**, 9–17 (2000)
12. Kong, L., Wang, J.: The Green's function for $(k, n-k)$ conjugate boundary value problems and its applications. J. Math. Anal. Appl. **255**(2), 404–422 (2001)
13. Lin, X., Jiang, D., Li, X.: Existence and uniqueness of solutions for singular $(k, n-k)$ conjugate boundary value problems. Comput. Math. Appl. **52**(3), 375–382 (2006)
14. Ma, R.: Positive solutions for semipositone $(k, n-k)$ conjugate boundary value problems. J. Math. Anal. Appl. **252**(1), 220–229 (2000)
15. Tian, S., Gao, W.: Positive solutions of singular $(k, n-k)$ conjugate eigenvalue problem. J. Appl. Math. Bioinf. **5**(2), 85–97 (2015)
16. Wong, P.J.Y.: Triple positive solutions of conjugate boundary value problems. Comput. Math. Appl. **36**(9), 19–35 (1998)
17. Henderson, J., Yin, W.: Singular $(k, n-k)$ boundary value problems between conjugate and right focal. J. Comput. Appl. Math. **88**(1), 57–69 (1998)
18. Kong, L., Lu, T.: Positive solutions of singular $(n, n-k)$ conjugate boundary value problem. J. Appl. Math. Bioinf. **5**(1), 13–24 (2015)
19. Davis, J.M., Henderson, J., Rajendra, P.K.: Eigenvalue intervals for nonlinear right focal problems. Appl. Anal. **74**(1–2), 215–231 (2000)
20. Agarwal, R.P., O'Regan, D., Wong, P.J.Y.: Positive solutions of Differential, Difference and Integral Equations. Kluwer Academic, Dordrecht (1999)
21. Yuan, C.: Multiple positive solutions for $(n-1, 1)$-type semipositone conjugate boundary value problems of nonlinear fractional differential equations. Electr. J. Qual. Theor. Differ. Equ. **36**, 1–12 (2010)
22. Yang, A., Henderson Jr., J., Nelms, C.: Extremal points for a higher-order fractional boundary-value problem. Electr. J. Differ. Equ. **2015**(161), 1–12 (2015)
23. Chen, S., Liu, Y.: Solvability of boundary value problems for fractional order elastic beam equations. Adv. Differ. Equ. **2014**, 204 (2014)
24. Freundlich, J.: Vibrations of a simply supported beam with a fractional viscoelastic material model-supports movement excitation. Shock Vibration. **20**, 1103C1112 (2013)
25. Bagley, R.L., Torvik, P.J.: A theoretical basis for the application of fractional calculus to viscoelasticity. J. Rheol. **27**(3), 201C210 (1983)
26. Bagley, R.L., Torvik, P.J.: On the fractional calculus model of viscoelastic behavior. J. Rheol. **30**(1), 133C155 (1986)
27. Adolfsson, K., Enelund, M., Olsson, P.: On the fractional order model of viscoelasticity. Mech. Time Depend. Mater. **9**, 15C34 (2005)
28. ur Rehman, M., Eloe, P.W.: Existence and uniqueness of solutions for impulsive fractional differential equations. Appl. Math. Comput. **224**, 422–431 (2013)
29. Kilbas, A.A., Srivastava, H.M., Trujillo, J.J.: Theory and Applications of Fractional Differential Equations. Elsevier (2006)
30. Lashmikanthan, V., Vatsala, A.S.: Basic theory of fractional differential equations. Nonlinear Anal. **69**, 2715–2682 (2008)
31. Mawhin, J.: Topological degree methods in nonlinear boundary value problems. In: CBMS Regional Conference Series in Mathematics 40, American Math. Soc. Providence, R.I., (1979)

5

Generalized parameter-free duality models in discrete minmax fractional programming based on second-order optimality conditions

G. J. Zalmai[1] · Ram U. Verma[2]

Abstract In this paper, we construct six generalized second-order parameter-free duality models, and prove several weak, strong, and strict converse duality theorems for a discrete minmax fractional programming problem using two partitioning schemes and various types of generalized second-order $(\mathcal{F}, \beta, \phi, \rho, \theta, m)$-univexity (more compactly, 'second-order univexity' is referred to as 'sounivexity') assumptions. The obtained results are new and generalize most of results on discrete minmax fractional programming involving the second-order invexity as well as on second-order univexity in the literature.

Keywords Discrete minmax fractional programming · Generalized second-order ($\mathcal{F}$, β, ϕ, ρ, θ, m)-univex · Generalized parameter-free duality models · Duality theorems

Mathematics Subject Classification 90C26 · 90C30 · 90C32 · 90C46 · 90C47

✉ Ram U. Verma
verma99@msn.com

G. J. Zalmai
gzalmai@nmu.edu

1 Department of Mathematics and Computer Science, Northern Michigan University, Marquette, MI 49855, USA

2 Department of Mathematics, University of North Texas, Denton, TX 76201, USA

Introduction and preliminaries

Based on a close observation on second-order necessary and sufficient optimality conditions for minmax fractional programming problems, which have not received much attention in the literature of mathematical programming, that is in sharp contrast to the case of minmax programming problems, numerous second-order necessary and sufficient optimality conditions for various classes of nonlinear programming problems with single and multiple objective functions have been investigated in the literature, including [1, 8–11, 13, 15, 19–21]. However, none of the sufficient optimality conditions discussed in these publications involve developing any kind of second-order duality theory for any type of optimization problems. The notion of duality for generalized linear fractional programming was initially considered by von Neumann [14] to the context of an economic equilibrium problem. However, a significant number of optimality criteria, duality results, and computational algorithms for several classes of generalized linear and nonlinear fractional programming problems have appeared in the related literature, for example in the publications [2–7, 12, 14, 16–18, 22–25]. Verma and Zalmai [10] dealt with some details on discrete minmax fractional programming, a fairly extensive list of currently available publications dealing with various second-order necessary and sufficient optimality conditions for several types of optimization problems, some modifications of the concepts of second-order invexity, pseudoinvexity, and quasiinvexity originally defined by Hanson [3], a set of second-order necessary optimality conditions, and making use of the new classes of generalized second-order invex functions, a fairly large number of sets of second-order sufficient optimality criteria. The sufficient optimality conditions established in [10] are further generalized in [12] using various generalized second-order $(\phi, \eta, \rho, \theta, m)$-invexity assumptions. For more details on generalized linear and nonlinear fractional programming problems, we refer the reader [1–25].

In this paper, it is our intention to lay the theoretical foundation which will enable us to fully investigate the second-order optimality and duality aspects of our following principal problem (P) as well as its semiinfinite counterpart in a series of papers. We begin our investigation here by establishing a set of second-order parametric necessary optimality conditions and several sets of sufficient optimality conditions for principal problem (P). Furthermore, we utilize two partitioning schemes due to Mond and Weir [7] and Yang [18], in conjunction with the generalized versions of the new classes of second-order invex functions introduced in (Verma and Zalmai [10]) to formulate six generalized parameter-free duality models for principal problem (P) and prove appropriate duality theorems. The duality models and the related duality theory established in this paper generalize most of results available in the literature, including those results published in [2–7, 12, 16, 18, 22–25].

To the best of our knowledge, all of these duality results established in this paper are new in the area of discrete minmax fractional programming. In fact, it seems that results of this type, which are based on second-order necessary and sufficient optimality conditions, have not yet appeared in any shape or form for any type of mathematical programming problems in the literature.

Now, we formulate six generalized second-order parameter-free duality models and prove a variety of weak, strong, and strict converse duality theorems for the following discrete minmax fractional programming problem:

$$(P) \qquad \text{Minimize} \max_{1 \leq i \leq p} \frac{f_i(x)}{g_i(x)}$$

subject to $G_j(x) \leq 0, j \in \underline{q}, H_k(x) = 0, k \in \underline{r}, x \in X$,

where x is an open convex subset of $\mathbb{R}^n$ (n-dimensional Euclidean space), $f_i, g_i, i \in \underline{p} \equiv \{1, 2, \ldots, p\}$, $G_j, j \in \underline{q}$, and $H_k, k \in \underline{r}$, are real-valued functions defined on X, and for each $i \in \underline{p}$, $g_i(x) > 0$ for all x satisfying the constraints of (P).

Evidently, all the duality results established in this paper can be modified and restated for each one of the following three classes of nonlinear programming problems, which are special cases of (P):

$$(P1) \qquad \text{Minimize}_{x \in \mathbb{F}} \frac{f_1(x)}{g_1(x)};$$

$$(P2) \qquad \text{Minimize}_{x \in \mathbb{F}} \max_{1 \leq i \leq p} f_i(x);$$

$$(P3) \qquad \text{Minimize}_{x \in \mathbb{F}} f_1(x),$$

where $\mathbb{F}$ (assumed to be nonempty) is the feasible set of (P), that is,

$$\mathbb{F} = \{x \in X : G_j(x) \leq 0, j \in \underline{q}, H_k(x) = 0, k \in \underline{r}\}.$$

Since, in most cases, these results can easily be altered and rephrased for each one of the above three problems, we shall not state them explicitly.

The rest of this paper is organized as follows: In Sect. 1, we present the historical development and introduce/recall a few basic definitions and auxiliary results that will be used in the sequel. In Sect. 2, we utilize a partitioning scheme due to Mond and Weir [7], and formulate two general second-order parameter-free duality models for (P) and prove weak, strong, and strict converse duality theorems using various generalized $(\mathcal{F}, \beta, \phi, \rho, \theta, m)$-sounivexity assumptions. We continue our discussion of duality in Sects. 3 and 4 where we construct four additional general second-order parameter-free duality models with different constraint structures and prove several second-order duality results under a variety of generalized $(\mathcal{F}, \beta, \phi, \rho, \theta, m)$-sounivexity conditions. Finally, in Sect. 5, we summarize our main results and also point out some research opportunities arising from certain modifications of the principal minmax model investigated in this study.

We next introduce the new classes of 'second-order univex' functions (referred to as "sounivex" functions). The notion of 'sounivexity' generalizes the notion of 'second-order invexity,' which is referred to as "sonvexity" in the literature. Let $f : X \to \mathbb{R}$ be a twice differentiable function.

Definition 1.1 The function f is said to be *(strictly)* $(\mathcal{F}, \beta, \phi, \rho, \theta, m)$*-sounivex* at x^* if there exist functions $\beta : X \times X \to \mathbb{R}_+ \equiv (0, \infty)$, $\phi : \mathbb{R} \to \mathbb{R}$, $\rho : X \times X \to \mathbb{R}$, $\theta : X \times X \to \mathbb{R}^n$, a sublinear function $\mathcal{F}(x, x^*; \cdot) : \mathbb{R}^n \to \mathbb{R}$, and a positive integer m, such that for each $x \in X (x \neq x^*)$ and $z \in \mathbb{R}^n$,

$$\phi\big(f(x) - f(x^*)\big)(>) \geq \mathcal{F}\big(x, x^*; \beta(x, x^*)\nabla f(x^*)\big) + \frac{1}{2}\langle z, \nabla^2 f(x^*) z\rangle + \rho(x, x^*)\|\theta(x, x^*)\|^m,$$

where $\|\cdot\|$ is a norm on $\mathbb{R}^n$ and $\langle a, b\rangle$ is the inner product of the vectors a and b.

The function f is said to be *(strictly)* $(\mathcal{F}, \beta, \phi, \rho, \theta, m)$*-sounivex* on X if it is *(strictly)* $(\mathcal{F}, \beta, \phi, \rho, \theta, m)$ *-sounivex* at each $x^* \in X$.

Definition 1.2 The function f is said to be *(strictly)* $(\mathcal{F}, \beta, \phi, \rho, \theta, m)$*-pseudosounivex* at x^* if there exist functions $\beta : X \times X \to \mathbb{R}_+$, $\phi : \mathbb{R} \to \mathbb{R}$, $\rho : X \times X \to \mathbb{R}$, $\theta : X \times X \to \mathbb{R}^n$, a sublinear function $\mathcal{F}(x, x^*; \cdot) : \mathbb{R}^n \to \mathbb{R}$, and a positive integer m, such that for each $x \in X (x \neq x^*)$ and $z \in \mathbb{R}^n$,

$$\mathcal{F}\big(x,x^*;\beta(x,x^*)\nabla f(x^*)\big)+\frac{1}{2}\langle z,\nabla^2 f(x^*)z\rangle$$
$$\geq -\rho(x,x^*)\|\theta(x,x^*)\|^m$$
$$\Rightarrow \phi\big(f(x)-f(x^*)\big)(>)\geq 0.$$

The function f is said to be *(strictly)* $(\mathcal{F},\beta,\phi,\rho,\theta,m)$-*pseudosounivex* on X if it is *(strictly)* $(\mathcal{F},\beta,\phi,\rho,\theta,m)$ - *pseudosounivex* at each $x^*\in X$.

Definition 1.3 The function f is said to be *(prestrictly)* $(\mathcal{F},\beta,\phi,\rho,\theta,m)$-*quasisounivex* at x^* if there exist functions $\beta: X\times X\to\mathbb{R}_+,\phi:\mathbb{R}\to\mathbb{R},\rho: X\times X\to\mathbb{R},\theta: X\times X\to\mathbb{R}^n$, a sublinear function $\mathcal{F}(x,x^*;\cdot):\mathbb{R}^n\to\mathbb{R}$, and a positive integer m, such that for each $x\in X$ and $z\in\mathbb{R}^n$,

$$\phi\big(f(x)-f(x^*)\big)(<)\leq 0$$
$$\Rightarrow \mathcal{F}\big(x,x^*;\beta(x,x^*)\nabla f(x^*)\big)+\frac{1}{2}\langle z,\nabla^2 f(x^*)z\rangle$$
$$\leq -\rho(x,x^*)\|\theta(x,x^*)\|^m.$$

The function f is said to be *(prestrictly)* $(\mathcal{F},\beta,\phi,\rho,\theta,m)$-*quasisounivex* on X if it is *(prestrictly)* $(\mathcal{F},\beta,\phi,\rho,\theta,m)$-*quasisounivex* at each $x^*\in X$.

From the above definitions, it is clear that if f is $(\mathcal{F},\beta,\phi,\rho,\theta,m)$-sounivex at x^*, then it is both $(\mathcal{F},\beta,\phi,\rho,\theta,m)$-pseudosounivex and $(\mathcal{F},\beta,\phi,\rho,\theta,m)$-quasisounivex at x^*, if f is $(\mathcal{F},\beta,\phi,\rho,\theta,m)$-quasisounivex at x^*, then it is prestrictly $(\mathcal{F},\beta,\phi,\rho,\theta,m)$-quasisounivex at x^*, and if f is strictly $(\mathcal{F},\beta,\phi,\rho,\theta,m)$-pseudosounivex at x^*, then it is $(\mathcal{F},\beta,\phi,\rho,\theta,m)$-quasisounivex at x^*.

In the proofs of the duality theorems, sometimes, it may be more convenient to use certain alternative but equivalent forms of the above definitions. These are obtained by considering the contrapositive statements. For example, $(\mathcal{F},\beta,\phi,\rho,\theta,m)$-quasisounivexity can be defined in the following equivalent way:

The function f is said to be $(\mathcal{F},\beta,\phi,\rho,\theta,m)$-quasisounivex at x^* if there exist functions $\beta: X\times X\to\mathbb{R}_+,\phi:\mathbb{R}\to\mathbb{R},\rho: X\times X\to\mathbb{R},\theta: X\times X\to\mathbb{R}^n$, a sublinear function $\mathcal{F}(x,x^*;\cdot):\mathbb{R}^n\to\mathbb{R}$, and a positive integer m, such that for each $x\in X$ and $z\in\mathbb{R}^n$,

$$\mathcal{F}\big(x,x^*;\beta(x,x^*)\nabla f(x^*)\big)+\frac{1}{2}\langle z,\nabla^2 f(x^*)z\rangle$$
$$> -\rho(x,x^*)\|\theta(x,x^*)\|^m$$
$$\Rightarrow \phi\big(f(x)-f(x^*)\big)>0.$$

Needless to say that the new classes of generalized convex functions specified in Definitions 1.1–1.3 contain a variety of special cases that can easily be identified by appropriate choices of $\mathcal{F},\beta,\phi,\rho,\theta$, and m. For example, if let $\mathcal{F}\big(x,x^*;\nabla f(x^*)\big)=\langle\nabla f(x^*),\eta(x,x^*)\rangle$ and $\beta(x,x^*)\equiv 1$, then we obtain the definitions of (strictly) $(\phi,\eta,\rho,\theta,m)$-sonvex, (strictly) $(\phi,\eta,\rho,\theta,m)$-pseudosonvex, and (prestrictly) $(\phi,\eta,\rho,\theta,m)$-quasisonvex functions introduced recently in [10], where the "second-order invexity" is compactly abbreviated as "*sonvexity*." The notion of the sonvexity/generalized sonvexity has been applied in developing a new optimality-duality theory in nonlinear programming based on second-order necessary and sufficient optimality conditions [1, 8–10, 12, 22].

Definition 1.4 The function f is said to be *(strictly)* $(\phi,\eta,\rho,\theta,m)$-*sonvex* at x^* if there exist functions $\phi:\mathbb{R}\to\mathbb{R},\eta: X\times X\to\mathbb{R}^n,\rho: X\times X\to\mathbb{R}$, and $\theta: X\times X\to\mathbb{R}^n$, and a positive integer m, such that for each $x\in X(x\neq x^*)$ and $z\in\mathbb{R}^n$,

$$\phi\big(f(x)-f(x^*)\big)(>)\geq\langle\nabla f(x^*),\eta(x,x^*)\rangle+\frac{1}{2}\langle z,\nabla^2 f(x^*)z\rangle$$
$$+\rho(x,x^*)\|\theta(x,x^*)\|^m.$$

The function f is said to be (strictly) $(\phi,\eta,\rho,\theta,m)$-sonvex on X if it is (strictly) $(\phi,\eta,\rho,\theta,m)$-sonvex at each $x^*\in X$.

Definition 1.5 The function f is said to be *(strictly)* $(\phi,\eta,\rho,\theta,m)$-*pseudosonvex* at x^* if there exist functions $\phi:\mathbb{R}\to\mathbb{R},\eta: X\times X\to\mathbb{R}^n,\rho: X\times X\to\mathbb{R}$, and $\theta: X\times X\to\mathbb{R}^n$, and a positive integer m, such that for each $x\in X(x\neq x^*)$ and $z\in\mathbb{R}^n$,

$$\langle\nabla f(x^*),\eta(x,x^*)\rangle+\frac{1}{2}\langle z,\nabla^2 f(x^*)z\rangle\geq -\rho(x,x^*)\|\theta(x,x^*)\|^m$$
$$\Rightarrow \phi\big(f(x)-f(x^*)\big)(>)\geq 0,$$

equivalently,

$$\phi\big(f(x)-f(x^*)\big)(\leq)<0\Rightarrow\langle\nabla f(x^*),\eta(x,x^*)\rangle+\frac{1}{2}\langle z,\nabla^2 f(x^*)z\rangle$$
$$<-\rho(x,x^*)\|\theta(x,x^*)\|^m.$$

The function f is said to be (strictly) $(\phi,\eta,\rho,\theta,m)$-pseudosonvex on X if it is (strictly) $(\phi,\eta,\rho,\theta,m)$-pseudosonvex at each $x^*\in X$.

Definition 1.6 The function f is said to be *(prestrictly)* $(\phi,\eta,\rho,\theta,m)$-*quasisonvex* at x^* if there exist functions $\phi:\mathbb{R}\to\mathbb{R},\eta: X\times X\to\mathbb{R}^n,\rho: X\times X\to\mathbb{R}$, and $\theta: X\times X\to\mathbb{R}^n$, and a positive integer m, such that for each $x\in X$ and $z\in\mathbb{R}^n$,

$$\phi\big(f(x)-f(x^*)\big)(<)\leq 0\Rightarrow\langle\nabla f(x^*),\eta(x,x^*)\rangle$$
$$+\frac{1}{2}\langle z,\nabla^2 f(x^*)z\rangle\leq -\rho(x,x^*)\|\theta(x,x^*)\|^m,$$

equivalently

$$\langle\nabla f(x^*),\eta(x,x^*)\rangle+\frac{1}{2}\langle z,\nabla^2 f(x^*)z\rangle>-\rho(x,x^*)\|\theta(x,x^*)\|^m$$
$$\Rightarrow \phi\big(f(x)-f(x^*)\big)(\geq)>0.$$

The function f is said to be (prestrictly) $(\phi,\eta,\rho,\theta,m)$-quasisonvex on X if it is (prestrictly) $(\phi,\eta,\rho,\theta,m)$-quasisonvex at each $x^* \in X$.

Duality model I and duality theorems

We begin this section by recalling a set of second-order parameter-free necessary optimality conditions for (*P*). This result, which is obtained from Theorem 3.1 of [10] by eliminating the parameter λ^* and redefining the Lagrange multipliers, will be needed for proving strong and strict converse duality theorems.

Theorem 2.1 [10] *Let x^* be a normal optimal solution of (P) and assume that the functions f_i, $g_i, i \in \underline{p}, G_j, j \in \underline{q}$, and H_k, $k \in \underline{r}$, are twice continuously differentiable at x^*. Then, for each $z^* \in C(x^*)$, there exist $u^* \in U \equiv \{u \in \mathbb{R}^p : u \geq 0, \sum_{i=1}^p u_i = 1\}, v^* \in \mathbb{R}_+^q \equiv \{v \in \mathbb{R}^q : v \geq 0\}$, and $w^* \in \mathbb{R}^r$, such that*

$$\sum_{i=1}^p u_i^*[D(x^*,u^*)\nabla f_i(x^*) - N(x^*,u^*)\nabla g_i(x^*)] + \sum_{j=1}^q v_j^*\nabla G_j(x^*) + \sum_{k=1}^r w_k^*\nabla H_k(x^*) = 0,$$

$$\left\langle z^*, \left\{ \sum_{i=1}^p u_i^*[D(x^*,u^*)\nabla^2 f_i(x^*) - N(x^*,u^*)\nabla^2 g_i(x^*)] + \sum_{j=1}^q v_j^*\nabla^2 G_j(x^*) + \sum_{k=1}^r w_k^*\nabla^2 H_k(x^*)\right\} z^* \right\rangle \geq 0,$$

$$u_i^*[D(x^*,u^*)f_i(x^*) - N(x^*,u^*)g_i(x^*)] = 0, \ i \in \underline{p},$$

$$\max_{1\leq i\leq p} \frac{f_i(x^*)}{g_i(x^*)} = \frac{N(x^*,u^*)}{D(x^*,u^*)},$$

$$v_j^* G_j(x^*) = 0, j \in \underline{q},$$

where $C(x^*)$ is the set of all critical directions of (P) at x^*, that is

$$C(x^*) = \{z^* \in \mathbb{R}^n : \langle D(x^*,u^*)\nabla f_i(x^*) - N(x^*,u^*)g_i(x^*), z^*\rangle = 0, \ i \in A(x^*), \langle \nabla G_j(x^*), z^*\rangle \leq 0, j \in B(x^*), \langle \nabla H_k(x^*), z^*\rangle = 0, k \in \underline{r}\},$$

$$A(x^*) = \{j \in \underline{p} : f_j(x^*)/g_j(x^*) = \max_{1\leq i\leq p} f_i(x^*)/g_i(x^*)\}, \ B(x^*) = \{j \in \underline{q} : G_j(x^*) = 0\}, N(x^*,u^*) = \sum_{i=1}^p u_i^* f_i(x^*), \text{ and } D(x^*,u^*) = \sum_{i=1}^p u_i^* g_i(x^*).$$

In the above theorem, a normal optimal solution refers to an optimal solution at which an appropriate second-order constraint qualification is satisfied.

In the remainder of this paper, we shall assume that the functions f_i, g_i, $i \in \underline{p}$, G_j, $j \in \underline{q}$, and H_k, $k \in \underline{r}$, are twice continuously differentiable on the open set X. Moreover, we shall assume, without loss of generality, that for each $i \in \underline{p}, f_i(x) \geq 0$ and $g_i(x) > 0$ for all $x \in X$.

Duality model I

In this section, we discuss several families of duality results under various generalized $(\mathcal{F},\beta,\phi,\rho,\theta,m)$-sounivexity hypotheses imposed on certain combinations of the problem functions. This is accomplished by employing a certain partitioning scheme which was originally proposed in [7] for the purpose of constructing generalized dual problems for nonlinear programming problems. For this, we need some additional notation.

Let $\{J_0, J_1, \ldots, J_M\}$ and $\{K_0, K_1, \ldots, K_M\}$ be partitions of the index sets $\underline{q}$ and $\underline{r}$, respectively; thus, $J_\mu \subseteq \underline{q}$ for each $\mu \in \underline{M} \cup \{0\}, J_\mu \cap J_\nu = \emptyset$ for each $\mu, \nu \in \underline{M} \cup \{0\}$ with $\mu \neq \nu$, and $\cup_{\mu=0}^M J_\mu = \underline{q}$. Obviously, similar properties hold for $\{K_0, K_1, \ldots, K_M\}$. Moreover, if m_1 and m_2 are the numbers of the partitioning sets of $\underline{q}$ and $\underline{r}$, respectively, then $M = \max\{m_1, m_2\}$ and $J_\mu = \emptyset$ or $K_\mu = \emptyset$ for $\mu > \min\{m_1, m_2\}$.

In addition, we use the real-valued functions $\xi \to \Phi(\xi, u, v, w, \lambda)$, and $\xi \to \Lambda_t(\xi, v, w)$ defined, for fixed λ, u, v, and w, on X as follows:

$$\Phi(\xi,y,u,v,w) = \sum_{i=1}^p u_i \Big\{ D(y,u)\Big[f_i(\xi) + \sum_{j\in J_0} v_j G_j(\xi) + \sum_{k\in K_0} w_k H_k(\xi)\Big] - [N(y,u) + \Lambda_0(y,v,w)]g_i(\xi)\Big\},$$

$$\Lambda_t(\xi,v,w) = \sum_{j\in J_t} v_j G_j(\xi) + \sum_{k\in K_t} w_k H_k(\xi), \quad t \in \underline{M} \cup \{0\}.$$

Making use of the sets and functions defined above, we can now formulate our first pair of second-order parameter-free duality models for (*P*).

Consider the following two problems:

(*DI*) Maximize

$$\frac{\sum_{i=1}^p u_i f_i(y) + \sum_{j\in J_0} v_j G_j(y) + \sum_{k\in K_0} w_k H_k(y)}{\sum_{i=1}^p u_i g_i(y)}$$

subject to

$$\sum_{i=1}^p u_i \Big\{ D(y,u)\Big[\nabla f_i(y) + \sum_{j\in J_0} v_j \nabla G_j(y) + \sum_{k\in K_0} w_k \nabla H_k(y)\Big] - [N(y,u) + \Lambda_0(y,v,w)]\nabla g_i(y)\Big\} + \sum_{j\in \underline{q}\setminus J_0} v_j \nabla G_j(y) + \sum_{k\in \underline{r}\setminus K_0} w_k \nabla H_k(y) = 0, \tag{2.1}$$

$$\left\langle z, \sum_{i=1}^{p} u_i \left\{ D(y,u) \left[\nabla^2 f_i(y) + \sum_{j\in J_0} v_j \nabla^2 G_j(y) + \sum_{k\in K_0} w_k \nabla^2 H_k(y) \right] - [N(y,u) + \Lambda_0(y,v,w)] \nabla^2 g_i(y) \right\} + \sum_{j\in \underline{q}\setminus J_0} v_j \nabla^2 G_j(y) + \sum_{k\in \underline{r}\setminus K_0} w_k \nabla^2 H_k(y) \Big\} z \right\rangle \geq 0, \tag{2.2}$$

$$\sum_{j\in J_t} v_j G_j(y) + \sum_{k\in K_t} w_k H_k(y) \geq 0, t \in \underline{M}, \tag{2.3}$$

$$y \in X, z \in C(y), u \in U, v \in \mathbb{R}^q_+, w \in \mathbb{R}^r; \tag{2.4}$$

$(\tilde{DI})$ Maximize

$$\frac{\sum_{i=1}^{p} u_i f_i(y) + \sum_{j\in J_0} v_j G_j(y) + \sum_{k\in K_0} w_k H_k(y)}{\sum_{i=1}^{p} u_i g_i(y)}$$

subject to (2.2)–(2.4) and

$$\mathcal{F}\left(x, y; \sum_{i=1}^{p} u_i \left\{ D(y,u) \left[\nabla f_i(y) + \sum_{j\in J_0} v_j \nabla G_j(y) + \sum_{k\in K_0} w_k \nabla H_k(y) \right] - [N(y,u) + \Lambda_0(y,v,w)] \nabla g_i(y) \right\} + \sum_{j\in \underline{q}\setminus J_0} v_j \nabla G_j(y) + \sum_{k\in \underline{r}\setminus K_0} w_k \nabla H_k(y) \right) \geq 0 \text{ for all } x \in \mathbb{F}, \tag{2.5}$$

where $\mathcal{F}(x, y; \cdot)$ is a sublinear function from $\mathbb{R}^n$ to $\mathbb{R}$.

Comparing (DI) and $(\tilde{DI})$, we see that $(\tilde{DI})$ is relatively more general than (DI) in the sense that any feasible solution of (DI) is also feasible for $(\tilde{DI})$, but the converse is not necessarily true. Furthermore, we observe that (2.1) is a system of n equations, whereas (2.5) is a single inequality. Clearly, from a computational point of view, *(DI)* is preferable to $(\tilde{DI})$ because of the dependence of (2.5) on the feasible set of *(P)*.

Despite these apparent differences, it turns out that the statements and proofs of all the duality theorems for $(P)-(DI)$ and $(P)-(\tilde{DI})$ are almost identical and, therefore, we shall consider only the pair $(P)-(DI)$.

In the proofs of our duality theorems, we shall make frequent use of the following auxiliary result which provides an alternative expression for the objective function of *(P)*.

Lemma 2.1 [8] *For each* $x \in X$,

$$\varphi(x) = \max_{1\leq i\leq p} \frac{f_i(x)}{g_i(x)} = \max_{u\in U} \frac{\sum_{i=1}^{p} u_i f_i(x)}{\sum_{i=1}^{p} u_i g_i(x)}.$$

The next two theorems show that (DI) is a dual problem for (P).

Theorem 2.2 (Weak duality) *Let x and* $\mathcal{S} \equiv (y, z, u, v, w)$ *be arbitrary feasible solutions of (P) and (DI), respectively, and assume that any one of the following four sets of hypotheses is satisfied:*

1.
   - (a) $\xi \to \Phi(\xi, y, u, v, w)$ is $(\mathcal{F}, \beta, \bar{\phi}, \bar{\rho}, \theta, m)$-pseudosounivex *at y and* $\bar{\phi}(a) \geq 0 \Rightarrow a \geq 0$;
   - (b) *For each* $t \in \underline{M}, \xi \to \Lambda_t(z, v, w)$ is $(\mathcal{F}, \beta, \tilde{\phi}_t, \tilde{\rho}_t, \theta, m)$-quasisounivex at y, $\tilde{\phi}_t$ *is increasing, and* $\tilde{\phi}_t(0) = 0$;
   - (c) $\bar{\rho}(x, y) + \sum_{t=1}^{M} \tilde{\rho}_t(x, y) \geq 0$;
2.
   - (a) $\xi \to \Phi(\xi, y, u, v, w)$ is prestrictly $(\mathcal{F}, \beta, \bar{\phi}, \bar{\rho}, \theta, m)$-quasisounivex *at y and* $\bar{\phi}(a) \geq 0 \Rightarrow a \geq 0$;
   - (b) *for each* $t \in \underline{M}, \xi \to \Lambda_t(\xi, v, w)$ *is* $(\mathcal{F}, \beta, \tilde{\rho}_t, \rho_t, \theta, m)$-quasisounivex *at y,* $\tilde{\phi}_t$*is increasing, and* $\tilde{\phi}_t(0) = 0$;
   - (c) $\bar{\rho}(x, y) + \sum_{t=1}^{M} \tilde{\rho}_t(x, y) > 0$;
3.
   - (a) $\xi \to \Phi(\xi, y, u, v, w)$ *is prestrictly* $(\mathcal{F}, \beta, \bar{\phi}, \bar{\rho}, \theta, m)$*-quasisounivex at y,* $\bar{\phi}$ *is strictly increasing, and* $\bar{\phi}(0) = 0$;
   - (b) *For each* $t \in \underline{M}, \xi \to \Lambda_t(\xi, v, w)$ *is strictly* $(\mathcal{F}, \beta, \tilde{\phi}_t, \tilde{\rho}_t, \theta, m)$-pseudosounivex *at y,* $\tilde{\phi}_t$ *is increasing, and* $\tilde{\phi}_t(0) = 0$;
   - (c) $\bar{\rho}(x, y) + \sum_{t=1}^{M} \tilde{\rho}_t(x, y) \geq 0$;
4.
   - (a) $\xi \to \Phi(\xi, y, u, v, w)$ is prestrictly $(\mathcal{F}, \beta, \bar{\phi}, \bar{\rho}, \theta, m)$-quasisounivex *at y,* $\bar{\phi}$ *is strictly increasing, and* $\bar{\phi}(0) = 0$;
   - (b) *For each* $t \in \underline{M_1}, \xi \to \Lambda_t(\xi, v, w)$ is $(\mathcal{F}, \beta, \tilde{\phi}_t, \tilde{\rho}_t, \theta, m)$-quasisounivex at *y, for each* $t \in \underline{M_2} \neq \emptyset, \xi \to \Lambda_t(\xi, v, w)$ is strictly $(\mathcal{F}, \beta, \tilde{\phi}_t, \tilde{\rho}_t, \theta, m)$-pseudosounivex *at y, and for each* $t \in \underline{M}, \tilde{\phi}_t$ is *increasing and* $\tilde{\phi}_t(0) = 0$, *where* $\{\underline{M_1}, \underline{M_2}\}$ *is a partition of* $\underline{M}$;
   - (c) $\bar{\rho}(x, y) + \sum_{t=1}^{M} \tilde{\rho}_t(x, y) \geq 0$.

Then, $\varphi(x) \geq \psi_I(y, u, v, w)$, *where* ψ_I *is the objective function of (DI).*

Proof (a) : Since $\mathcal{F}(x, y; \cdot)$ is sublinear and $\beta(x, y) > 0$, it is clear that (2.1) and (2.2) can be expressed as follows:

$$\mathcal{F}\Bigg(x,y;\beta(x,y)\sum_{i=1}^{p}u_i\Bigg\{D(y,u)\Bigg[\nabla f_i(y)+\sum_{j\in J_0}v_j\nabla G_j(y)+\sum_{k\in K_0}w_k\nabla H_k(y)\Bigg]-[N(y,u)+\Lambda_0(y,v,w)]\nabla g_i(y)\Bigg\}\Bigg)+\mathcal{F}\Bigg(x,y;\beta(x,y)\sum_{t=1}^{M}\Bigg[\sum_{j\in J_t}v_j\nabla G_j(y)+\sum_{k\in K_t}w_k\nabla H_k(y)\Bigg]\Bigg)\geq 0. \tag{2.6}$$

$$\Bigg\langle z,\sum_{i=1}^{p}u_i\Bigg\{D(y,u)\Bigg[\nabla^2 f_i(y)+\sum_{j\in J_0}v_j\nabla^2 G_j(y)+\sum_{k\in K_0}w_k\nabla^2 H_k(y)\Bigg]-[N(y,u)+\Lambda_0(y,v,w)]\nabla^2 g_i(y)\Bigg\}z\Bigg\rangle+\Bigg\langle z,\sum_{t=1}^{M}\Bigg[\sum_{j\in J_t}v_j\nabla^2 G_j(y)+\sum_{k\in K_t}w_k\nabla^2 H_k(y)\Bigg]z\Bigg\rangle\geq 0. \tag{2.7}$$

Since for each $t\in \underline{M}$,

$$\begin{aligned}\Lambda_t(x,v,w)&=\sum_{j\in J_t}v_jG_j(x)+\sum_{k\in K_t}w_kH_k(x)\\&\leq 0\ (\text{by the primal feasibility of } x)\\&\leq\sum_{j\in J_t}v_jG_j(y)+\sum_{k\in K_t}w_kH_k(y)\\&\quad(\text{by (2.3) and the dual feasibility of } \mathcal{S})\\&=\Lambda_t(y,v,w),\end{aligned}$$

and hence, $\tilde{\phi}_t(\Lambda_t(x,v,w)-\Lambda_t(y,v,w))\leq 0$, it follows from (ii) that

$$\mathcal{F}\Bigg(x,y;\beta(x,y)\Bigg[\sum_{j\in J_t}v_j\nabla G_j(y)+\sum_{k\in K_t}w_k\nabla H_k(y)\Bigg]\Bigg)+\frac{1}{2}\Bigg\langle z,\Bigg[\sum_{j\in J_t}v_j\nabla^2 G_j(y)+\sum_{k\in K_t}w_k\nabla^2 H_k(y)\Bigg]z\Bigg\rangle\leq-\tilde{\rho}_t(x,y)\|\theta(x,y)\|^m.$$

Summing over $t\in\underline{M}$ and using the sublinearity of $\mathcal{F}(x,y;\cdot)$, we obtain

$$\mathcal{F}\Bigg(x,y;\beta(x,y)\sum_{t=1}^{M}\Bigg[\sum_{j\in J_t}v_j\nabla G_j(y)+\sum_{k\in K_t}w_k\nabla H_k(y)\Bigg]\Bigg)+\frac{1}{2}\Bigg\langle z,\sum_{t=1}^{M}\Bigg[\sum_{j\in J_t}v_j\nabla^2 G_j(y)+\sum_{k\in K_t}w_k\nabla^2 H_k(y)\Bigg]z\Bigg\rangle\leq-\sum_{t=1}^{M}\tilde{\rho}_t(x,y)\|\theta(x,y)\|^m. \tag{2.8}$$

Combining (2.6)–(2.8) and using (iii), we get

$$\mathcal{F}\Bigg(x,y;\beta(x,y)\sum_{i=1}^{p}u_i\Bigg\{D(y,u)\Bigg[\nabla f_i(y)+\sum_{j\in J_0}v_j\nabla G_j(y)+\sum_{k\in K_0}w_k\nabla H_k(y)\Bigg]-[N(y,u)+\Lambda_0(y,v,w)]\nabla g_i(y)\Bigg\}\Bigg)+\frac{1}{2}\Bigg\langle z,\sum_{i=1}^{p}u_i\Bigg\{D(y,u)\Bigg[\nabla^2 f_i(y)+\sum_{j\in J_0}v_j\nabla^2 G_j(y)+\sum_{k\in K_0}w_k\nabla^2 H_k(y)\Bigg]-[N(y,u)+\Lambda_0(y,v,w)]\nabla^2 g_i(y)\Bigg\}z\Bigg\rangle\geq\sum_{t=1}^{M}\tilde{\rho}_t(x,y)\|\theta(x,y)\|^m\geq-\bar{\rho}(x,y)\|\theta(x,y)\|^m, \tag{2.9}$$

which by virtue of (i) implies that

$\bar{\phi}(\Phi(x,y,u,v,w)-\Phi(y,y,u,v,w))\geq 0.$

However, $\bar{\phi}(a)\geq 0\Rightarrow a\geq 0$, and hence, we get

$\Phi(x,y,u,v,w)\geq\Phi(y,y,u,v,w)=0,$

where the equality follows from the definitions of $D(y,u)$, $N(y,u)$, and $\Lambda_0(y,v,w)$. Since $x\in\mathbb{F}$, the above inequality reduces to

$$\sum_{i=1}^{p}u_i\{D(y,u)f_i(x)-[N(y,u)+\Lambda_0(y,v,w)]g_i(x)\}\geq 0. \tag{2.10}$$

Now, using (2.10) and Lemma 2.1, we obtain the weak duality inequality as follows:

$$\varphi(x)=\max_{a\in U}\frac{\sum_{i=1}^{p}a_if_i(x)}{\sum_{i=1}^{p}a_ig_i(x)}\geq\frac{\sum_{i=1}^{p}u_if_i(x)}{\sum_{i=1}^{p}u_ig_i(x)}\geq\frac{N(y,u)+\Lambda_0(y,v,w)}{D(y,u)}=\psi_I(y,z,u,v,w).$$

(b) The proof is similar to that of part (a).

(c) Suppose to the contrary that $\varphi(x)<\psi_I(y,z,u,v,w)$. This implies that for each $i\in\underline{p}$,

$$D(y,u)f_i(x)-[N(y,u)+\Lambda_0(y,v,w)]g_i(x)<0. \tag{2.11}$$

Using these inequalities, we see that

$$\begin{aligned}\Phi(x,y,u,v,w)&=\sum_{i=1}^{p}u_i\Bigg\{D(y,u)\Bigg[f_i(x)+\sum_{j\in J_0}v_jG_j(x)+\sum_{k\in K_0}w_kH_k(x)\Bigg]-[N(y,u)+\Lambda_0(y,v,w)]g_i(x)\Bigg\},\\&\leq\sum_{i=1}^{p}u_i\{D(y,u)f_i(x)-[N(y,u)+\Lambda_0(y,v,w)]g_i(x)\}\\&\quad(\text{by the primal of feasibility of } x)\\&<0\ (\text{by (2.11)})\\&=\Phi(y,y,u,v,w)(\text{by the definitions of } D(y,u),N(y,u),\text{ and }\Lambda_0(y,v,w)),\end{aligned}$$

and hence, $\bar{\phi}(\Phi(x,y,u,v,w)-\Phi(y,y,u,v,w))<0$ which by virtue of (i) implies that

$$\mathcal{F}\Big(x,y;\beta(x,y)\sum_{i=1}^{p}u_i\Big\{D(y,u)\Big[\nabla f_i(y)+\sum_{j\in J_0}v_j\nabla G_j(y)+\sum_{k\in K_0}w_k\nabla H_k(y)\Big] - [N(y,u)+\Lambda_0(y,v,w)]\nabla g_i(y)\Big\}\Big)+\frac{1}{2}\Big\langle z,\sum_{i=1}^{p}u_i\Big\{D(y,u)\Big[\nabla^2 f_i(y) + \sum_{j\in J_0}v_j\nabla^2 G_j(y)+\sum_{k\in K_0}w_k\nabla^2 H_k(y)\Big]-[N(y,u) + \Lambda_0(y,v,w)]\nabla^2 g_i(y)\Big\}z\Big\rangle\leq -\bar{\rho}(x,y)\|\theta(x,y)\|^m. \tag{2.12}$$

Proceeding as in the proof of part (a), we obtain $\tilde{\phi}_t\big(\Lambda_t(x,v,w)-\Lambda_t(y,v,w)\big)\leq 0$, which, because of (ii), implies that

$$\mathcal{F}\Big(x,y;\beta(x,y)\Big[\sum_{j\in J_t}v_j\nabla G_j(y)+\sum_{k\in K_t}w_k\nabla H_k(y)\Big]\Big)+\frac{1}{2}\Big\langle z,\Big[\sum_{j\in J_t}v_j\nabla^2 G_j(y) + \sum_{k\in K_t}w_k\nabla^2 H_k(y)\Big]z\Big\rangle < -\tilde{\rho}_t(x,y)\|\theta(x,y)\|^m.$$

Summing over $t\in \underline{M}$ and using the sublinearity of $\mathcal{F}(x,y;\cdot)$, we obtain

$$\mathcal{F}\Big(x,y;\beta(x,y)\sum_{t=1}^{M}\Big[\sum_{j\in J_t}v_j\nabla G_j(y)+\sum_{k\in K_t}w_k\nabla H_k(y)\Big]\Big) +\frac{1}{2}\Big\langle z,\sum_{t=1}^{M}\Big[\sum_{j\in J_t}v_j\nabla^2 G_j(y)+\sum_{k\in K_t}w_k\nabla^2 H_k(y)\Big]z\Big\rangle < -\sum_{t=1}^{M}\tilde{\rho}_t(x,y)\|\theta(x,y)\|^m.$$

Combining this inequality with (2.6) and (2.7) and using (iii), we get

$$\mathcal{F}\Big(x,y;\beta(x,y)\sum_{i=1}^{p}u_i\Big\{D(y,u)\Big[\nabla f_i(y)+\sum_{j\in J_0}v_j\nabla G_j(y)+\sum_{k\in K_0}w_k\nabla H_k(y)\Big] - [N(y,u)+\Lambda_0(y,v,w)]\nabla g_i(y)\Big\}\Big)+\frac{1}{2}\Big\langle z,\sum_{i=1}^{p}u_i\Big\{D(y,u)\Big[\nabla^2 f_i(y) + \sum_{j\in J_0}v_j\nabla^2 G_j(y)+\sum_{k\in K_0}w_k\nabla^2 H_k(y)\Big]-[N(y,u)+\Lambda_0(y,v,w)]\nabla^2 g_i(y)\Big\}z\Big\rangle > \sum_{t=1}^{M}\tilde{\rho}_t(x,y)\|\theta(x,y)\|^m\geq -\bar{\rho}(x,y)\|\theta(x,y)\|^m,$$

which contradicts (2.12). Therefore, we conclude that $\bar{\phi}(x)\geq \psi_I(y,z,u,v,w)$.

(d) The proof is similar to that of part (c). □

Theorem 2.3 (Strong duality) *Let x^* be a normal optimal solution of (P) and assume that any one of the four sets of conditions specified in Theorem 2.2 is satisfied for all feasible solutions of (DI). Then, for each $z^*\in C(x^*)$, there exist $u^*\in U, v^*\in \mathbb{R}^q_+$, and $w^*\in\mathbb{R}^r$, such that $\mathcal{S}^*\equiv (x^*,z^*,u^*,v^*,w^*)$ is an optimal solution of (DI) and $\varphi(x^*)=\psi_I(\mathcal{S}^*)$.*

Proof Since x^* is a normal optimal solution of (P), by Theorem 2.1, for each $z^*\in C(x^*)$, there exist $u^*\in U, \bar{v}\in\mathbb{R}^q_+$, and $\bar{w}\in\mathbb{R}^r$, such that

$$\sum_{i=1}^{p}u_i^*[D(x^*,u^*)\nabla f_i(x^*)-N(x^*,u^*)\nabla g_i(x^*)] + \sum_{j=1}^{q}\bar{v}_j\nabla G_j(x^*)+\sum_{k=1}^{r}\bar{w}_k\nabla H_k(x^*)=0, \tag{2.13}$$

$$\Big\langle z^*,\Big\{\sum_{i=1}^{p}u_i^*[D(x^*,u^*)\nabla^2 f_i(x^*)-N(x^*,u^*)\nabla^2 g_i(x^*)] + \sum_{j=1}^{q}\bar{v}_j\nabla^2 G_j(x^*)+\sum_{k=1}^{r}\bar{w}_k\nabla^2 H_k(x^*)\Big\}z^*\Big\rangle\geq 0, \tag{2.14}$$

$$\max_{1\leq i\leq p}\frac{f_i(x^*)}{g_i(x^*)}=\frac{N(x^*,u^*)}{D(x^*,u^*)}, \tag{2.15}$$

$$\bar{v}_jG_j(x^*)=0,\ j\in\underline{q}. \tag{2.16}$$

Now, choosing $v_j^*=\bar{v}_j/D(x^*,u^*)$ for each $j\in J_0, v_j^*=\bar{v}_j$ for each $j\in\underline{q}\backslash J_0, w_k^*=\bar{w}_k/D(x^*,u^*)$ for each $k\in K_0$, and $w_k^*=\bar{w}_k$ for each $k\in\underline{r}\backslash K_0$, and noticing that $x^*\in\mathbb{F}$, we deduce the following relations from (2.13) to (2.16):

$$\sum_{i=1}^{p}u_i^*\Big\{D(x^*,u^*)\Big[\nabla f_i(x^*)+\sum_{j\in J_0}v_j^*\nabla G_j(x^*)+\sum_{k\in K_0}w_k^*\nabla H_k(x^*)\Big] - [N(x^*,u^*)+\Lambda_0(x^*,v^*,w^*)]\nabla g_i(x^*)\Big\}+\sum_{j\in\underline{q}\backslash J_0}v_j^*\nabla G_j(x^*) + \sum_{k\in\underline{r}\backslash K_0}w_k^*\nabla H_k(x^*)=0, \tag{2.17}$$

$$\Big\langle z^*,\sum_{i=1}^{p}u_i^*\Big\{D(x^*,u^*)\Big[\nabla^2 f_i(x^*)+\sum_{j\in J_0}v_j^*\nabla^2 G_j(x^*)+\sum_{k\in K_0}w_k^*\nabla^2 H_k(x^*)\Big] - [N(x^*,u^*)+\Lambda_0(x^*,v^*,w^*)]\nabla^2 g_i(x^*)\Big\}+\sum_{j\in\underline{q}\backslash J_0}v_j^*\nabla^2 G_j(x^*) + \sum_{k\in\underline{r}\backslash K_0}w_k^*\nabla^2 H_k(x^*)\Big\}z^*\Big\rangle\geq 0, \tag{2.18}$$

$$\sum_{j\in J_t}v_j^*G_j(x^*)+\sum_{k\in K_t}w_k^*H_k(x^*)=0, t\in\underline{M}\cup\{0\}, \tag{2.19}$$

$$\varphi(x^*)=\frac{N(x^*,u^*)+\Lambda_0(x^*,v^*,w^*)}{D(x^*,u^*)}. \tag{2.20}$$

From (2.17) to (2.19), it is clear that $\mathcal{S}^*$ is a feasible solution of (DI), and from (2.20), we see that $\varphi(x^*)=\psi_I(\mathcal{S}^*)$. If $\mathcal{S}^*$ were not an optimal solution of (DI), then there would exist a feasible solution $\mathcal{S}^\circ\equiv(x^\circ,z^\circ,u^\circ,v^\circ,w^\circ)$ of (DI), such that $\psi_I(\mathcal{S}^\circ)>\psi_I(\mathcal{S}^*)=\varphi(x^*)$, which contradicts Theorem 2.2. Therefore, we conclude that $\mathcal{S}^*$ is an optimal solution of (DI). □

Theorem 2.4 (Strict Converse Duality) *Let x^* be a normal optimal solution of (P), let $\tilde{S} \equiv (\tilde{x}, \tilde{z}, \tilde{u}, \tilde{v}, \tilde{w})$ be an optimal solution of (DI), and assume that any one of the following four sets of conditions holds:*

(a) *The assumptions specified in part (a) of Theorem* 2.1 *are satisfied for the feasible solution $\tilde{S}$* of (DI), *$\bar{\phi}(a) > 0 \Rightarrow a > 0$, and the function $\xi \to \Phi(\xi, \tilde{x}, \tilde{u}, \tilde{v}, \tilde{w})$ is strictly $(\mathcal{F}, \beta, \bar{\phi}, \bar{\rho}, \theta, m)$-pseudosounivex at $\tilde{x}$.*

(b) *The assumptions specified in part (b) of Theorem* 2.1 *are satisfied for the feasible solution $\tilde{S}$* of (DI), *$\bar{\phi}(a) > 0 \Rightarrow a > 0$, and the function $\xi \to \Phi(\xi, \tilde{x}, \tilde{u}, \tilde{v}, \tilde{w})$ is $(\mathcal{F}, \beta, \bar{\phi}, \bar{\rho}, \theta, m)$-quasisounivex at $\tilde{x}$.*

(c) *The assumptions specified in part (c) of Theorem* 2.1 *are satisfied for the feasible solution $\tilde{S}$ of (DI), $\bar{\phi}(a) > 0 \Rightarrow a > 0$, and the function $\xi \to \Phi(\xi, \tilde{x}, \tilde{u}, \tilde{v}, \tilde{w})$ is $(\mathcal{F}, \beta, \bar{\phi}, \bar{\rho}, \theta, m)$-quasisounivex at $\tilde{x}$.*

(d) *The assumptions specified in part (d) of Theorem* 2.1 *are satisfied for the feasible solution $\tilde{S}$ of (DI), $\bar{\phi}(a) > 0 \Rightarrow a > 0$, and the function $\xi \to \Phi(\xi, \tilde{x}, \tilde{u}, \tilde{v}, \tilde{w})$ is $(\mathcal{F}, \beta, \bar{\phi}, \bar{\rho}, \theta, m)$-quasisounivex at $\tilde{x}$.*

Then, $\tilde{x} = x^*$ and $\varphi(x^*) = \psi_I(\tilde{S})$.

Proof Since x^* is a normal optimal solution of (*P*), by Theorem 2.3, there exist $z^* \in C(x^*), u^*, v^*$, and w^*, such that $S^* \equiv (x^*, z^*, u^*, v^*, w^*)$ is a feasible solution of (*DI*) and $\varphi(x^*) = \psi_I(S^*)$. (a): Suppose to the contrary that $\tilde{x} \neq x^*$. Now, proceeding as in the proof of part (a) of Theorem 2.2 (with x replaced by x^* and $\mathcal{S}$ by $\tilde{S}$), we arrive at the strict inequality

$$\sum_{i=1}^{p} \tilde{u}_i\{D(\tilde{x}, \tilde{u})f_i(x) - [N(\tilde{x}, \tilde{u}) + \Lambda_0(\tilde{x}, \tilde{v}, \tilde{w})]g_i(x)\} > 0.$$

Using this inequality along with Lemma 2.1, as in the proof of Theorem 2.2, we get $\varphi(x^*) > \psi_I(\tilde{S})$, which contradicts the fact that $\varphi(x^*) = \psi_I(S^*) \leq \psi_I(\tilde{S})$. (b)–(d): The proofs are similar to that of part (a). □

As pointed out earlier, the duality models (*DI*) and ($\tilde{D}I$) are two families of dual problems whose members can easily be identified by appropriate choices of the partitioning sets $J_0, J_1, \ldots, J_M, K_0, K_1, \ldots, K_M$. To illustrate this possibility, we shall next briefly discuss some special cases of (DI) and ($\tilde{D}I$).

If we choose $J_0 = \underline{q}$ and $K_0 = \underline{r}$ in (DI) and ($\tilde{D}I$), then we obtain the following dual problems for (*P*):

(*DIa*) Maximize
$$\frac{\sum_{i=1}^{p} u_i f_i(y) + \sum_{j=1}^{q} v_j G_j(y) + \sum_{k=1}^{r} w_k H_k(y)}{\sum_{i=1}^{p} u_i g_i(y)}$$

subject to

$$D(y, u)\Big[\sum_{i=1}^{p} u_i \nabla f_i(y) + \sum_{j=1}^{q} v_j \nabla G_j(y) + \sum_{k=1}^{r} w_k \nabla H_k(y)\Big] - [N(y, u) + \Lambda(y, v, w)]\sum_{i=1}^{p} u_i \nabla g_i(y) = 0,$$

$$\Big\langle z, \Big\{D(y, u)\Big[\sum_{i=1}^{p} u_i \nabla^2 f_i(y) + \sum_{j=1}^{q} v_j \nabla^2 G_j(y) + \sum_{k=1}^{r} w_k \nabla^2 H_k(y)\Big] - [N(y, u) + \Lambda(y, v, w)]\sum_{i=1}^{p} u_i \nabla^2 g_i(y)\Big]\Big\}z\Big\rangle \geq 0,$$

$$y \in X, z \in C(y), u \in U, v \in \mathbb{R}^q_+, w \in \mathbb{R}^r,$$

where

$$\Lambda(y, v, w) = \sum_{j=1}^{q} v_j G_j(y) + \sum_{k=1}^{r} w_k H_k(y);$$

($\tilde{D}Ia$) Maximize
$$\frac{\sum_{i=1}^{p} u_i f_i(y) + \sum_{j=1}^{q} v_j G_j(y) + \sum_{k=1}^{r} w_k H_k(y)}{\sum_{i=1}^{p} u_i g_i(y)}$$

subject to

$$\mathcal{F}\Big(x, y; D(y, u)\Big[\sum_{i=1}^{p} u_i \nabla f_i(y) + \sum_{j=1}^{q} v_j \nabla G_j(y) + \sum_{k=1}^{r} w_k \nabla H_k(y)\Big] - [N(y, u) + \Lambda(y, v, w)]\sum_{i=1}^{p} u_i \nabla g_i(y)\Big) \geq 0 \text{ for all } x \in \mathbb{F},$$

$$\Big\langle z, \Big\{D(y, u)\Big[\sum_{i=1}^{p} u_i \nabla^2 f_i(y) + \sum_{j=1}^{q} v_j \nabla^2 G_j(y) + \sum_{k=1}^{r} w_k \nabla^2 H_k(y)\Big] - [N(y, u) + \Lambda(y, v, w)]\sum_{i=1}^{p} u_i \nabla^2 g_i(y)\Big]\Big\}z\Big\rangle \geq 0,$$

$$y \in X, z \in C(y), u \in U, v \in \mathbb{R}^q_+, w \in \mathbb{R}^r,$$

where $\mathcal{F}(x, y; \cdot)$ is a sublinear function from $\mathbb{R}^n$ to $\mathbb{R}$.

If we choose $M = q + r, J_0 = \emptyset, K_0 = \emptyset, J_t = \{t\}, K_t = \emptyset, t \in \underline{q}$, and $J_t = \emptyset, K_t = \{t\}, t \in \underline{r}$, then (*DI*) and ($\tilde{D}I$) reduce to the following dual problems for (*P*):

(*DIb*) Maximize $\dfrac{\sum_{i=1}^{p} u_i f_i(y)}{\sum_{i=1}^{p} u_i g_i(y)}$

subject to

$$\sum_{i=1}^{p} u_i[D(y, u)\nabla f_i(y) - N(y, u)\nabla g_i(y)] + \sum_{j=1}^{q} v_j \nabla G_j(y) + \sum_{k=1}^{r} w_k \nabla H_k(y) = 0,$$

$$\Big\langle z, \Big\{ \sum_{i=1}^{p} u_i[D(y,u)\nabla^2 f_i(y) - N(y,u)\nabla^2 g_i(y)] + \sum_{j=1}^{q} v_j \nabla^2 G_j(y) + \sum_{k=1}^{r} w_k \nabla^2 H_k(y) \Big\} z \Big\rangle \geq 0,$$

$$v_j G_j(y) \geq 0,\ j \in \underline{q},$$

$$w_k H_k(y) \geq 0,\ k \in \underline{r},$$

$$y \in X, z \in C(y), u \in U, v \in \mathbb{R}_+^q,\ w \in \mathbb{R}^r;$$

$(\tilde{D}Ib)$ Maximize $\dfrac{\sum_{i=1}^{p} u_i f_i(y)}{\sum_{i=1}^{p} u_i g_i(y)}$

subject to

$$\mathcal{F}\Big(x, y; D(y,u)\Big[\sum_{i=1}^{p} u_i \nabla f_i(y) + \sum_{j=1}^{q} v_j \nabla G_j(y) + \sum_{k=1}^{r} w_k \nabla H_k(y)\Big] - [N(y,u) + \Lambda(y,v,w)] \sum_{i=1}^{p} u_i \nabla g_i(y)\Big) \geq 0 \text{ for all } x \in \mathbb{F},$$

$$\Big\langle z, \Big\{ \sum_{i=1}^{p} u_i[D(y,u)\nabla^2 f_i(y) - N(y,u)\nabla^2 g_i(y)] + \sum_{j=1}^{q} v_j \nabla^2 G_j(y) + \sum_{k=1}^{r} w_k \nabla^2 H_k(y) \Big\} z \Big\rangle \geq 0,$$

$$v_j G_j(y) \geq 0, j \in \underline{q},$$

$$w_k H_k(y) \geq 0, k \in \underline{r},$$

$$y \in X, z \in C(y), u \in U, v \in \mathbb{R}_+^q, w \in \mathbb{R}^r.$$

In a similar manner, we can identify many other special cases of *(DI)* and $(\tilde{D}I)$. Evidently, Theorems 2.1–2.3 can be specialized for $(DIa), (\tilde{D}Ia), (DIb)$, and $(\tilde{D}Ib)$ in a straightforward fashion.

The dual problems $(DIa), (\tilde{D}Ia), (DIb)$, and $(\tilde{D}Ib)$ were investigated previously in [10] with $\mathcal{F}(x, x^*; \nabla f(x^*)) = \langle \nabla f(x^*), \eta(x, x^*) \rangle$, where η is a function from $X \times X$ to $\mathbb{R}^n$, and a great variety of duality results were established under various (strict) $(\phi, \eta, \rho, \theta, m)$-sonvexity, (strict) $(\phi, \eta, \rho, \theta, m)$-pseudosonvexity, and (prestrict) $(\phi, \eta, \rho, \theta, m)$-quasisonvexity hypotheses.

Duality model II and duality theorems

In Theorems 2.2–2.4, various generalized $(\mathcal{F}, \beta, \phi, \rho, \theta, m)$-sounivexity conditions were imposed on the function $\xi \to \Phi(\xi, y, u, v, w)$, which is the weighted sum of the functions

$$\Phi_i(\xi, y, v, w) = D(y,u)\Big[f_i(\xi) + \sum_{j \in J_0} v_j G_j(\xi) + \sum_{k \in K_0} w_k H_k(\xi)\Big] - [N(y,u) + \Lambda_0(y,v,w)] g_i(\xi), i \in \underline{p}.$$

In this section, we consider some generalized versions of *(DI)* and $(\tilde{D}I)$, and prove weak and strong duality theorems in which we assume that the individual functions $\xi \to \Phi_i(\xi, y, v, w), i \in \underline{p}$, satisfy appropriate generalized $(\mathcal{F}, \beta, \phi, \rho, \theta, m)$-sounivexity hypotheses. This can be accomplished by appending an additional system of inequality constraints to *(DI)* and $(\tilde{D}I)$.

Consider the following two problems:

(DII) Maximize $\dfrac{\sum_{i=1}^{p} u_i f_i(y) + \sum_{j \in J_0} v_j G_j(y) + \sum_{k \in K_0} w_k H_k(y)}{\sum_{i=1}^{p} u_i g_i(y)}$

subject to

$$\sum_{i=1}^{p} u_i \Big\{ D(y,u)\Big[\nabla f_i(y) + \sum_{j \in J_0} v_j \nabla G_j(y) + \sum_{k \in K_0} w_k \nabla H_k(y)\Big] - [N(y,u) + \Lambda_0(y,v,w)] \nabla g_i(y) \Big\} + \sum_{j \in \underline{q} \backslash J_0} v_j \nabla G_j(y) + \sum_{k \in \underline{r} \backslash K_0} w_k \nabla H_k(y) = 0, \tag{3.1}$$

$$\Big\langle z, \sum_{i=1}^{p} u_i \Big\{ D(y,u)\Big[\nabla^2 f_i(y) + \sum_{j \in J_0} v_j \nabla^2 G_j(y) + \sum_{k \in K_0} w_k \nabla^2 H_k(y)\Big] - [N(y,u) + \Lambda_0(y,v,w)] \nabla^2 g_i(y) \Big\} + \sum_{j \in \underline{q} \backslash J_0} v_j \nabla^2 G_j(y) + \sum_{k \in \underline{r} \backslash K_0} w_k \nabla^2 H_k(y) \Big\} z \Big\rangle \geq 0, \tag{3.2}$$

$$D(y,u)\Big[f_i(y) + \sum_{j \in J_0} v_j G_j(y) + \sum_{k \in K_0} w_k H_k(y)\Big] - [N(y,u) + \Lambda_0(y,v,w)] g_i(y) \geq 0, i \in \underline{p}, \tag{3.3}$$

$$\sum_{j \in J_t} v_j G_j(y) + \sum_{k \in K_t} w_k H_k(y) \geq 0, t \in \underline{M}, \tag{3.4}$$

$$y \in X, z \in C(y), u \in U, v \in \mathbb{R}_+^q, w \in \mathbb{R}^r; \tag{3.5}$$

$(\tilde{D}II)$ Maximize $\dfrac{\sum_{i=1}^{p} u_i f_i(y) + \sum_{j \in J_0} v_j G_j(y) + \sum_{k \in K_0} w_k H_k(y)}{\sum_{i=1}^{p} u_i g_i(y)}$

subject to (3.2)–(3.5) and

$$\mathcal{F}\Big(x, y; \sum_{i=1}^{p} u_i \Big\{ D(y,u)\Big[\nabla f_i(y) + \sum_{j \in J_0} v_j \nabla G_j(y) + \sum_{k \in K_0} w_k \nabla H_k(y)\Big] - [N(y,u) + \Lambda_0(y,v,w)] \nabla g_i(y) \Big\} + \sum_{j \in \underline{q} \backslash J_0} v_j \nabla G_j(y) + \sum_{k \in \underline{r} \backslash K_0} w_k \nabla H_k(y)\Big) \geq 0 \text{ for all } x \in \mathbb{F},$$

where $\mathcal{F}(x, y; \cdot)$ is a sublinear function from $\mathbb{R}^n$ to $\mathbb{R}$.

The comments and observations made earlier about the relationship between *(DI)* and $(\tilde{D}I)$ are, of course, also valid for *(DII)* and $(\tilde{D}II)$.

The following two theorems show that *(DII)* is a dual problem for *(P)*.

Theorem 3.1 (Weak duality) *Let x and* $\mathcal{S} \equiv (y, z, u, v, w)$ *be arbitrary feasible solutions of (P) and (DII), respectively, and assume that any one of the following seven sets of hypotheses is satisfied:*

(a)

- (i) For each $i \in I_+ \equiv \{i \in \underline{p} : u_i > 0\}, \xi \to \Phi_i(\xi, y, v, w)$ is $(\mathcal{F}, \beta, \bar{\phi}_i, \bar{\rho}_i, \theta, m)$-pseudosounivex at $y, \bar{\phi}_i$ is strictly increasing, and $\bar{\phi}_i(0) = 0$;
- (ii) For each $t \in \underline{M}, \xi \to \Lambda_t(\xi, v, w)$ is $(\mathcal{F}, \beta, \tilde{\phi}_t, \tilde{\rho}_t, \theta, m)$-quasisounivex at $y, \tilde{\phi}_t$ is increasing, and $\tilde{\phi}_t(0) = 0$;
- (iii) $\sum_{i \in I_+} u_i \bar{\rho}_i(x, y) + \sum_{t=1}^{M} \tilde{\rho}_t(x, y) \geq 0$;

(b)

- (i) For each $i \in I_+, \xi \to \Phi_i(\xi, y, v, w)$ is prestrictly $(\mathcal{F}, \beta, \bar{\phi}_i, \bar{\rho}_i, \theta, m)$-quasisounivex at $y, \bar{\phi}_i$ is strictly increasing, and $\bar{\phi}_i(0) = 0$;
- (ii) For each $t \in \underline{m}, \xi \to \Lambda_t(\xi, v, w)$ is strictly $(\mathcal{F}, \beta, \tilde{\phi}_t, \tilde{\rho}_t, \theta, m)$-pseudosounivex at $y, \tilde{\phi}_t$ is increasing, and $\tilde{\phi}_t(0) = 0$;
- (iii) $\sum_{i \in I_+} u_i \bar{\rho}_i(x, y) + \sum_{t=1}^{M} \tilde{\rho}_t(x, y) \geq 0$;

(c)

- (i) For each $i \in I_+, \xi \to \Phi_i(\xi, y, v, w)$ is prestrictly $(\mathcal{F}, \beta, \bar{\phi}_i, \bar{\rho}_i, \theta, m)$-quasisounivex at $y, \bar{\phi}_i$ is strictly increasing, and $\bar{\phi}_i(0) = 0$;
- (ii) For each $t \in \underline{M}, \xi \to \Lambda_t(\xi, v, w)$ is $(\mathcal{F}, \beta, \tilde{\phi}_t, \tilde{\rho}_t, \theta, m)$-quasisounivex at $y, \tilde{\phi}_t$ is increasing, and $\tilde{\phi}_t(0) = 0$;
- (iii) $\sum_{i \in I_+} u_i \bar{\rho}_i(x, y) + \sum_{t=1}^{M} \tilde{\rho}_t(x, y) > 0$;

(d)

- (i) For each $i \in I_{1+}, \xi \to \Phi_i(\xi, y, v, w)$ is $(\mathcal{F}, \beta, \bar{\phi}_i, \bar{\rho}_i, \theta, m)$-pseudosounivex at y , for each $i \in I_{2+}, \xi \to \Phi_i(\xi, y, v, w)$ is prestrictly $(\mathcal{F}, \beta, \bar{\phi}_i, \bar{\rho}_i, \theta, m)$-quasisounivex at y, and for each $i \in I_+, \bar{\phi}_i$ is strictly increasing and $\bar{\phi}_i(0) = 0$, where $\{I_{1+}, I_{2+}\}$ is a partition of I_+;
- (ii) For each $t \in \underline{M}, \xi \to \Lambda_t(\xi, v, w)$ is strictly $(\mathcal{F}, \beta, \tilde{\phi}_t, \tilde{\rho}_t, \theta, m)$-pseudosounivex at $y, \tilde{\phi}_t$ is increasing, and $\tilde{\phi}_t(0) = 0$;
- (iii) $\sum_{i \in I_+} u_i \bar{\rho}_i(x, y) + \sum_{t=1}^{M} \tilde{\rho}_t(x, y) \geq 0$;

(e)

- (i) For each $i \in I_{1+} \neq \emptyset, \xi \to \Phi_i(\xi, y, v, w)$ is $(\mathcal{F}, \beta, \bar{\phi}_i, \bar{\rho}_i, \theta, m)$-pseudosounivex at y, for each $i \in I_{2+}, \xi \to \Phi_i(\xi, y, v, w)$ is prestrictly $(\mathcal{F}, \beta, \bar{\phi}_i, \bar{\rho}_i, \theta)$-quasisounivex at y, and for each $i \in I_+, \bar{\phi}_i$ is strictly increasing and $\bar{\phi}_i(0) = 0$, where $\{I_{1+}, I_{2+}\}$ is a partition of I_+;
- (ii) For each $t \in \underline{M}, \xi \to \Lambda_t(\xi, v, w)$ is $(\mathcal{F}, \beta, \tilde{\phi}_t, \tilde{\rho}_t, \theta, m)$-quasisounivex at $y, \tilde{\phi}_t$ is increasing, and $\tilde{\phi}_t(0) = 0$;
- (iii) $\sum_{i \in I_+} u_i \bar{\rho}_i(x, y) + \sum_{t=1}^{M} \tilde{\rho}_t(x, y) \geq 0$;

(f)

- (i) For each $i \in I_+, \xi \to \Phi_i(\xi, y, v, w)$ is prestrictly $(\mathcal{F}, \beta, \bar{\phi}_i, \bar{\rho}_i, \theta)$-quasisounivex at $y, \bar{\phi}_i$ is strictly increasing, and $\bar{\phi}_i(0) = 0$;
- (ii) For each $t \in \underline{M_1} \neq \emptyset, \xi \to \Lambda_t(\xi, v, w)$ is strictly $(\mathcal{F}, \beta, \tilde{\phi}_t, \tilde{\rho}_t, \theta, m)$-pseudosounivex at y, for each $t \in \underline{M_2}, \xi \to \Lambda_t(\xi, v, w)$ is $(\mathcal{F}, \beta, \tilde{\phi}_t, \tilde{\rho}_t, \theta, m)$-quasisounivex at y, and for each $t \in \underline{M}, \tilde{\phi}_t$ is increasing and $\tilde{\phi}_t(0) = 0$, where $\{\underline{M_1}, \underline{M_2}\}$ is a partition of $\underline{M}$;
- (iii) $\sum_{i \in I_+} u_i \bar{\rho}_i(x, y) + \sum_{t=1}^{M} \tilde{\rho}_t \geq 0$;

(g)

- (i) For each $i \in I_{1+}, \xi \to \Phi_i(\xi, y, v, w)$ is $(\mathcal{F}, \beta, \bar{\phi}_i, \bar{\rho}_i, \theta, m)$-pseudosounivex at y, for each $i \in I_{2+}, \xi \to \Phi_i(\xi, y, v, w)$ is prestrictly $(\mathcal{F}, \beta, \bar{\phi}_i, \bar{\rho}_i, \theta, m)$-quasisounivex at y, and for each $i \in I_+, \bar{\phi}_i$ is strictly increasing and $\bar{\phi}_i(0) = 0$, where $\{I_{1+}, I_{2+}\}$ is a partition of I_+;
- (ii) For each $t \in \underline{M_1}, \xi \to \Lambda_t(\xi, v, w)$ is strictly $(\mathcal{F}, \beta, \tilde{\phi}_t, \tilde{\rho}_t, \theta, m)$-pseudosounivex at y,for each $t \in \underline{M_2}, \xi \to \Lambda_t(\xi, v, w)$ is $(\mathcal{F}, \beta, \tilde{\phi}_t, \tilde{\rho}_t, \theta, m)$-quasisounivex at y, and for $t \in \underline{M}, \tilde{\phi}_t$ is increasing and $\tilde{\phi}_t(0) = 0$, where $\{\underline{M_1}, \underline{M_2}\}$ is a partition of $\underline{M}$;
- (iii) $\sum_{i \in I_+} u_i \bar{\rho}_i(x, y) + \sum_{t=1}^{M} \tilde{\rho}_t(x, y) \geq 0$;
- (iv) $I_{1+} \neq \emptyset$, $\underline{M_1} \neq \emptyset$, or $\sum_{i \in I_+} u_i \bar{\rho}_i(x, y) + \sum_{t=1}^{M} \tilde{\rho}_t(x, y) > 0$.

Then, $\varphi(x) \geq \psi_{II}(\mathcal{S})$, where ψ_{II} is *the objective function of (DII).*

Proof (a): Suppose to the contrary that $\varphi(x) < \psi_{II}(\mathcal{S})$. This implies that

$$D(y, u) f_i(x) - [N(y, u) + \Lambda_0(y, v, w)] g_i(x) < 0, i \in \underline{p}. \tag{3.6}$$

Keeping in mind that $v \geq 0$, we see that for each $i \in I_+$,

$$\Phi_i(x,y,v,w)=D(y,u)\Big[f_i(x)+\sum_{j\in J_0}v_jG_j(x)+\sum_{k\in K_0}w_kH_k(x)\Big]$$
$$-[N(y,u)+\Lambda_0(y,u,v)]g_i(x)$$
$$\leq D(y,u)f_i(x)-[N(y,u)+\Lambda_0(y,v,w)]g_i(x)$$
$$\text{(by the primal feasibility of } x)$$
$$<0 \text{ (by (3.6))} \leq \Phi_i(y,y,v,w) \text{ (by (3.3))},$$

and so it follows from the properties of $\bar{\phi}_i$ that $\bar{\phi}_i\big(\Phi_i(x,y,v,w)-\Phi_i(y,y,v,w)\big)<0,$

which in view of (i) implies that

$$\mathcal{F}\Big(x,y;\beta(x,y)\Big\{D(y,u)\Big[\nabla f_i(y)+\sum_{j\in J_0}u_j\nabla G_j(y)+\sum_{k\in K_0}w_k\nabla H_k(y)\Big]$$
$$-[N(y,u)+\Lambda_0(y,u,v)]\nabla g_i(x)\Big\}\Big)+\frac{1}{2}\Big\langle z,\Big\{D(y,u)\Big[\nabla^2 f_i(y)$$
$$+\sum_{j\in J_0}v_j\nabla^2G_j(y)+\sum_{k\in K_0}w_k\nabla^2H_k(y)\Big]$$
$$-[N(y,u)+\Lambda_0(y,v,w)]\nabla^2 g_i(x)\Big\}z\Big\rangle<-\bar{\rho}_i(x,y)\|\theta(x,y)\|^m.$$

Since $u\geq 0, u_i=0$ for each $i\in \underline{p}\backslash I_+, \sum_{i=1}^p u_i=1$, and $\mathcal{F}(x,y;\cdot)$ is sublinear, the above inequalities yield

$$\mathcal{F}\Big(x,y;\beta(x,y)\Big\{\sum_{i=1}^p u_i\Big\{D(y,u)\Big[\nabla f_i(y)+\sum_{j\in J_0}u_j\nabla G_j(y)+\sum_{k\in K_0}w_k\nabla H_k(y)\Big]$$
$$-[N(y,u)+\Lambda_0(y,u,v)]\nabla g_i(x)\Big\}\Big\}\Big)+\frac{1}{2}\Big\langle z,\Big\{\sum_{i=1}^p u_i\Big\{D(y,u)\Big[\nabla^2 f_i(y)$$
$$+\sum_{j\in J_0}v_j\nabla^2G_j(y)+\sum_{k\in K_0}w_k\nabla^2H_k(y)\Big]-[N(y,u)+\Lambda_0(y,u,v)]\nabla^2 g_i(x)\Big\}z\Big\rangle$$
$$<-\sum_{i\in I_+}u_i\bar{\rho}_i(x,y)\|\theta(x,y)\|^m. \tag{3.7}$$

As seen in the proof of Theorem 2.2, our assumptions in (ii) lead to

$$\mathcal{F}\Big(x,y;\beta(x,y)\sum_{t=1}^M\Big[\sum_{j\in J_t}u_j\nabla G_j(y)+\sum_{k\in K_t}w_k\nabla H_k(y)\Big]\Big)$$
$$+\frac{1}{2}\Big\langle z,\Big[\sum_{j\in J_t}u_j\nabla^2G_j(y)+\sum_{k\in K_t}w_k\nabla^2H_k(y)\Big]z\Big\rangle$$
$$\leq-\sum_{t=1}^M\tilde{\rho}_t(x,y)\|\theta(x,y)\|^m,$$

which when combined with (3.6) and (3.7) results in

$$\mathcal{F}\Big(x,y;\beta(x,y)\Big\{\sum_{i=1}^p u_i\Big\{D(y,u)\Big[\nabla f_i(y)+\sum_{j\in J_0}u_j\nabla G_j(y)+\sum_{k\in K_0}w_k\nabla H_k(y)\Big]$$
$$-[N(y,u)+\Lambda_0(y,u,v)]\nabla g_i(x)\Big\}\Big\}\Big)+\frac{1}{2}\Big\langle z,\Big\{\sum_{i=1}^p u_i\Big\{D(y,u)\Big[\nabla^2 f_i(y)$$
$$+\sum_{j\in J_0}v_j\nabla^2G_j(y)+\sum_{k\in K_0}w_k\nabla^2H_k(y)\Big]$$
$$-[N(y,u)+\Lambda_0(y,v,w)]\nabla^2 g_i(x)\Big\}z\Big\rangle\geq\sum_{t=1}^M\tilde{\rho}_t(x,y)\|\theta(x,y)\|^m.$$

In view of (iii), this inequality contradicts (3.7). Hence, $\varphi(x)\geq\psi_{II}(\mathcal{S})$. (b)–(g) : The proofs are similar to that of part (a). □

Theorem 3.2 (Strong duality) *Let x^* be a normal optimal solution of (P) and assume that any one of the seven sets of conditions set forth in Theorem 3.1 is satisfied for all feasible solutions of (DII). Then, for each $z^*\in C(x^*)$, there exist u^*, v^*, and w^*, such that (x^*,z^*,u^*,v^*,w^*) is an optimal solution of (DII) and $\varphi(x^*)=\psi_{II}(x^*,z^*,u^*,v^*,w^*)$.*

Proof The proof is similar to that of Theorem 2.2. □

The duality models (*DII*) and ($\tilde{D}II$) contain numerous special cases that can easily be identified by appropriate choices of the partitioning sets.

Duality model III and duality theorems

In this section, we discuss two additional duality models for (*P*). In these duality formulations, we utilize a partition of $\underline{p}$ in addition to those of $\underline{q}$ and $\underline{r}$. This partitioning scheme, which is an extended version of the one initially proposed by Mond and Weir [7], was used by Yang [18] for formulating a generalized duality model for a multiobjective fractional programming problem. In our duality theorems, we impose appropriate generalized $(\mathcal{F},\beta,\phi,\rho,\theta,m)$-sounivexity requirements on certain combinations of the problem functions.

Let $\{I_0,I_1,\ldots,I_\ell\}$ be a partition of $\underline{p}$, such that $\mathcal{L}=\{0,1,2,\ldots,\ell\}\subset\mathcal{M}=\{0,1,\ldots,M\}$, and let the real-valued function $\xi\to\Pi_t(\xi,y,u,v,w)$ be defined, for fixed u, v, and w, on X by

$$\Pi_t(\xi,y,u,v,w)=\sum_{i\in I_t}u_i[D(y,u)f_i(x)$$
$$-N(y,u)g_i(x)]+\sum_{j\in J_t}v_jG_j(x)+\sum_{k\in K_t}w_kH_k(x), t\in\mathcal{M}.$$

Consider the following two problems:

(*DIII*) Maximize $\dfrac{\sum_{i=1}^p u_if_i(y)}{\sum_{i=1}^p u_ig_i(y)}$

subject to

$$\sum_{i=1}^p u_i[D(y,u)\nabla f_i(y)-N(y,u)\nabla g_i(y)]+\sum_{j=1}^q v_j\nabla G_j(y)$$
$$+\sum_{k=1}^r w_k\nabla H_k(y)=0, \tag{4.1}$$

$$\Big\langle z,\Big\{\sum_{i=1}^p u_i[D(y,u)\nabla^2 f_i(y)-N(y,u)\nabla^2 g_i(y)]$$
$$+\sum_{j=1}^q v_j\nabla^2G_j(y)+\sum_{k=1}^r w_k\nabla^2H_k(y)\Big\}z\Big\rangle\geq 0, \tag{4.2}$$

$$\sum_{i\in I_t} u_i[D(y,u)f_i(y) - N(y,u)g_i(y)] + \sum_{j\in J_t} v_jG_j(y) + \sum_{k\in K_t} w_kH_k(y) \geq 0, t \in \mathcal{M}, \tag{4.3}$$

$$\sum_{j\in J_t} v_jG_j(y) + \sum_{k\in K_t} w_kH_k(y) \geq 0, t \in \mathcal{L}\backslash\mathcal{M}, \tag{4.4}$$

$$y \in X, z \in C(y), u \in U, v \in \mathbb{R}^q_+, w \in \mathbb{R}^r; \tag{4.5}$$

$(\tilde{D}III)$ Maximize $\dfrac{\sum_{i=1}^p u_if_i(y)}{\sum_{i=1}^p u_ig_i(y)}$

subject to (4.2)–(4.5) and

$$\mathcal{F}\Big(x,y;\sum_{i=1}^p u_i[D(y,u)\nabla f_i(y) - N(y,u)\nabla g_i(y)] + \sum_{j=1}^q v_j\nabla G_j(y) + \sum_{k=1}^r w_k\nabla H_k(y)\Big) \geq 0 \text{ for all } x \in \mathbb{F},$$

where $\mathcal{F}(x,y;\cdot)$ is a sublinear function from $\mathbb{R}^n$ to $\mathbb{R}$.

The comments and observations made earlier about the relationship between *(DI)* and $(\tilde{D}I)$ are, of course, also valid for *(DIII)* and $(\tilde{D}III)$.

The following two theorems show that *(DIII)* is a dual problem for *(P)*.

Theorem 4.1 (Weak duality) *Let x and $\mathcal{S} \equiv (y,z,u,v,w)$ be arbitrary feasible solutions of (P) and (DIII), respectively, and assume that any one of the following seven sets of hypotheses is satisfied:*

(a)

(i) for each $t \in \mathcal{L}, \xi \to \Pi_t(\xi,y,u,v,w)$ is strictly $(\mathcal{F},\beta,\phi_t,\rho_t,\theta,m)$-pseudosounivex at y, ϕ_t *is increasing, and* $\phi_t(0) = 0$;

(ii) *for each* $t \in \mathcal{M}\backslash\mathcal{L}, \xi \to \Lambda_t(\xi,v,w)$ is $(\mathcal{F},\beta,\phi_t,\rho_t,\theta,m)$-quasisounivex at y, ϕ_t *is increasing, and* $\phi_t(0) = 0$;

(iii) $\sum_{t\in\mathcal{M}} \rho_t(x,y) \geq 0$ *for all* $x \in \mathbb{F}$;

(b)

(i) *For each $t \in \mathcal{L}, \xi \to \Pi_t(\xi,y,u,v,w)$ is prestrictly $(\mathcal{F},\beta,\phi_t,\rho_t,\theta,m)$-quasisounivex at y, ϕ_t is increasing, and $\phi_t(0) = 0$;*

(ii) *For each $t \in \mathcal{M}\backslash\mathcal{L}, \xi \to \Lambda_t(\xi,v,w)$ is strictly $(\mathcal{F},\beta,\phi_t,\rho_t,\theta,m)$-pseudosounivex at y, ϕ_t is increasing, and $\phi_t(0) = 0$;*

(iii) $\sum_{t\in\mathcal{M}} \rho_t(x,y) \geq 0$ for all $x \in \mathbb{F}$;

(c)

(i) *For each $t \in \mathcal{L}, \xi \to \Pi_t(\xi,y,u,v,w)$ is prestrictly $(\mathcal{F},\beta,\phi_t,\rho_t,\theta,m)$-quasisounivex at y, ϕ_t is increasing, and $\phi_t(0) = 0$;*

(ii) *For each $t \in \mathcal{M}\backslash\mathcal{L}, \xi \to \Lambda_t(\xi,v,w)$is $(\mathcal{F},\beta,\phi_t,\rho_t,\theta,m)$-quasisounivex at y, ϕ_t is increasing, and $\phi_t(0) = 0$;*

(iii) $\sum_{t\in\mathcal{M}} \rho_t(x,y) > 0$ *for all* $x \in \mathbb{F}$;

(d)

(i) *For each $t \in \mathcal{L}_1, \xi \to \Pi_t(\xi,y,u,v,w)$ is strictly $(\mathcal{F},\beta,\phi_t,\rho_t,\theta,m)$-pseudosounivex at y, for each $t \in \mathcal{L}_2, \xi \to \Pi_t(\xi,y,u,v,w)$ is prestrictly $(\mathcal{F},\beta,\phi_t,\rho_t,\theta,m)$-quasisounivex at y, and for each $t \in \mathcal{L}, \phi_t$ is increasing and $\phi_t(0) = 0$, where $\{\mathcal{L}_1,\mathcal{L}_2\}$ is a partition of $\mathcal{L}$;*

(ii) *For each $t \in \mathcal{M}\backslash\mathcal{L}, \xi \to \Lambda_t(\xi,v,w)$ is strictly* $(\mathcal{F},\beta,\phi_t,\rho_t,\theta,m)$-pseudosounivex at y, ϕ_t *is increasing, and* $\phi_t(0) = 0$;

(iii) $\sum_{t\in\mathcal{M}} \rho_t(x,y) \geq 0$ for all $x \in \mathbb{F}$;

(e)

(i) *for each $t \in \mathcal{L}_1 \neq \emptyset, \xi \to \Pi_t(\xi,y,u,v,w)$ is strictly $(\mathcal{F},\beta,\phi_t,\rho_t,\theta,m)$-pseudosounivex at y, for each $t \in \mathcal{L}_2, \xi \to \Pi_t(\xi,y,u,v,w)$ is prestrictly $(\mathcal{F},\beta,\phi_t,\rho_t,\theta,m)$-quasisounivex at y, and for each $t \in \mathcal{L}, \phi_t$ is increasing and $\phi_t(0) = 0$, where* $\{\mathcal{L}_1,\mathcal{L}_2\}$ is a *partition of* $\mathcal{L}$;

(ii) For each $t \in \mathcal{M}\backslash\mathcal{L}, \xi \to \Lambda_t(\xi,v,w)$ is $(\mathcal{F},\beta,\phi_t,\rho_t,\theta,m)$-quasisounivex at y, ϕ_t is increasing, and $\phi_t(0) = 0$;

(iii) $\sum_{t\in\mathcal{M}} \rho_t(x,y) \geq 0$ for all $x \in \mathbb{F}$;

(f)

(i) for each $t \in \mathcal{L}, \xi \to \Pi_t(\xi,y,u,v,w)$ is prestrictly $(\mathcal{F},\beta,\phi_t,\rho_t,\theta,m)$-quasisounivex at y, ϕ_t is increasing, and $\phi_t(0) = 0$;

(ii) for each $t \in (\mathcal{M}\backslash\mathcal{L})_1 \neq \emptyset, \xi \to \Lambda_t(\xi,v,w)$ is strictly $(\mathcal{F},\beta,\phi_t,\rho_t,\theta,m)$-pseudosounivex at y, for each $t \in (\mathcal{M}\backslash\mathcal{L})_2, \xi \to \Lambda_t(\xi,v,w)$ is $(\mathcal{F},\beta,\phi_t,\rho_t,\theta,m)$-quasisounivex at y, and for each $t \in \mathcal{L}, \phi_t$ is increasing and $\phi_t(0) = 0$, where $\{(\mathcal{M}\backslash\mathcal{L})_1,(\mathcal{M}\backslash\mathcal{L})_2\}$ is a partition of $\mathcal{M}\backslash\mathcal{L})$;

(iii) $\sum_{t\in\mathcal{M}} \rho_t(x,y) \geq 0$ for all $x \in \mathbb{F}$;

(g)

(i) for each $t \in \mathcal{L}_1, \xi \to \Pi_t(\xi,y,u,v,w)$ is $(\mathcal{F},\beta,\phi_t,\rho_t,\theta,m)$-pseudosounivex at y, for each $t \in \mathcal{L}_2, \xi \to \Pi_t(\xi,y,u,v,w)$ is prestrictly $(\mathcal{F},\beta,\phi_t,\rho_t,\theta,m)$-quasisounivex at y, and for each $t \in \mathcal{L}, \phi_t$ is increasing and $\phi_t(0) = 0$, where $\{\mathcal{L}_1,\mathcal{L}_2\}$ is a partition of $\mathcal{L}$;

(ii) for each $t \in (\mathcal{M}\backslash\mathcal{L})_1, \xi \to \Lambda_t(\xi,v,w)$ is strictly $(\mathcal{F},\beta,\phi_t,\rho_t,\theta,m)$-pseudosounivex at y, for each $t \in (\mathcal{M}\backslash\mathcal{L})_2, \xi \to \Lambda_t(\xi,v,w)$ is

$(\mathcal{F}, \beta, \phi_t, \rho_t, \theta, m)$-quasisounivex at y, and for each $t \in \mathcal{M}\backslash\mathcal{L}$, ϕ_t is increasing and $\phi_t(0) = 0$, where $\{(\mathcal{M}\backslash\mathcal{L})_1, (\mathcal{M}\backslash\mathcal{L})_2\}$ is a partition of $\mathcal{M}\backslash\mathcal{L}$;

(iii) $\sum_{t\in\mathcal{M}} \rho_t(x, y) \geq 0$ for all $x \in \mathbb{F}$;

(iv) $\mathcal{L}_1 \neq \emptyset$, $(\mathcal{M}\backslash\mathcal{L})_1 \neq \emptyset$, or $\sum_{t\in\mathcal{M}} \rho_t(x, y) > 0$.

Then, $\varphi(x) \geq \psi_{III}(\mathcal{S})$, where ψ_{III} is the objective function of (*DIII*).

Proof (a): Suppose to the contrary that $\varphi(x) < \psi_{III}(\mathcal{S})$. This implies that

$$D(y, u)f_i(x) - N(y, u)g_i(x) < 0, ; i \in \underline{p}.$$

Since $u \geq 0$ and $u \neq 0$, we see that for each $t \in \mathcal{L}$,

$$\sum_{i\in I_t} u_i[f_i(x) - N(y, u)g_i(x)] \leq 0. \tag{4.6}$$

Now, using this inequality, we see that

$$\begin{aligned}\Pi_t(x, y, u, v, w) &= \sum_{i\in I_t} u_i[D(y, u)f_i(x) - N(y, u)g_i(x)] + \sum_{j\in J_t} v_j G_j(x)\\ &+ \sum_{k\in K_t} w_k H_k(x) \leq \sum_{i\in I_t} u_i[D(y, u)f_i(x) - N(y, u)g_i(x)]\\ &\text{(by the primal feasibility of } x) \leq 0 \text{ (by (4.6))}\\ &\leq \sum_{i\in I_t} u_i[D(y, u)f_i(y) - N(y, u)g_i(y)] + \sum_{j\in J_t} v_j G_j(y)\\ &+ \sum_{k\in K_t} w_k H_k(y) \text{ (by (4.3) and the dual feasibility of } \mathcal{S})\\ &= \Pi_t(y, y, u, v, w),\end{aligned}$$

and hence

$$\phi_t(\Pi_t(x, y, u, v, w) - \Pi_t(y, y, u, v, w)) \leq 0,$$

which in view of (i) implies that

$$\begin{aligned}&\mathcal{F}\Big(x, y; \beta(x, y)\Big\{\sum_{i\in I_t} u_i[D(y, u)\nabla f_i(y) - N(y, u)\nabla g_i(y)] + \sum_{j\in J_t} v_j\nabla G_j(y)\\ &+ \sum_{k\in K_t} w_k\nabla H_k(y)\Big\}\Big) + \frac{1}{2}\Big\langle z, \Big\{\sum_{i\in I_t} u_i[D(y, u)\nabla^2 f_i(y) - N(y, u)\nabla^2 g_i(y)]\\ &+ \sum_{j\in J_t} v_j\nabla^2 G_j(y) + \sum_{k\in K_t} w_k\nabla^2 H_k(y)\Big\}z\Big\rangle < -\rho_t(x, y)\|\theta(x, y)\|^m.\end{aligned}$$

Summing over $t \in \mathcal{L}$ and using the sublinearity of $\mathcal{F}(x, y; \cdot)$, we obtain

$$\begin{aligned}&\mathcal{F}\Big(x, y; \beta(x, y)\Big\{\sum_{i=1}^{p} u_i[D(y, u)\nabla f_i(y) - N(y, u)\nabla g_i(y)] + \sum_{t\in\mathcal{L}}\Big[\sum_{j\in J_t} v_j\nabla G_j(y)\\ &+ \sum_{k\in K_t} w_k\nabla H_k(y)\Big]\Big\}\Big) + \frac{1}{2}\Big\langle z, \Big\{\sum_{i=1}^{p} u_i[D(y, u)\nabla^2 f_i(y) - N(y, u)\nabla^2 g_i(y)]\\ &+ \sum_{t\in\mathcal{L}}\Big[\sum_{j\in J_t} v_j\nabla^2 G_j(y) + \sum_{k\in K_t} w_k\nabla^2 H_k(y)\Big]\Big\}z\Big\rangle < -\sum_{t\in\mathcal{L}} \rho_t(x, y)\|\theta(x, y)\|^m.\end{aligned} \tag{4.7}$$

Proceeding as in the proof of Theorem 2.2, we get for each $t \in \mathcal{M}\backslash\mathcal{L}$,

$$\Lambda_t(x, v, w) \leq \Lambda_t(y, v, w),$$

and so

$$\phi_t(\Lambda_t(x, v, w) - \Lambda_t(y, v, w)) \leq 0,$$

which in view of (ii) implies that

$$\begin{aligned}&\mathcal{F}\Big(x, y; \beta(x, y)\Big[\sum_{j\in J_t} v_j\nabla G_j(y) + \sum_{k\in K_t} w_k\nabla H_k(y)\Big]\Big) + \frac{1}{2}\Big\langle z, \Big[\sum_{j\in J_t} v_j\nabla^2 G_j(y)\\ &+ \sum_{k\in K_t} w_k\nabla^2 H_k(y)\Big]z\Big\rangle \leq -\rho_t(x, y)\|\theta(x, y)\|^m.\end{aligned}$$

Summing over $t \in \mathcal{M}\backslash\mathcal{L}$ and using the sublinearity of $\mathcal{F}(x, y; \cdot)$, we get

$$\begin{aligned}&\mathcal{F}\Big(x, y; \beta(x, y)\sum_{t\in\mathcal{M}\backslash\mathcal{L}}\Big[\sum_{j\in J_t} v_j\nabla G_j(y) + \sum_{k\in K_t} w_k\nabla H_k(y)\Big]\Big)\\ &+ \frac{1}{2}\Big\langle z, \sum_{t\in\mathcal{M}\backslash\mathcal{L}}\Big[\sum_{j\in J_t} v_j\nabla^2 G_j(y) + \sum_{k\in K_t} w_k\nabla^2 H_k(y)\Big]z\Big\rangle\\ &< -\sum_{t\in\mathcal{M}\backslash\mathcal{L}} \rho_t(x, y)\|\theta(x, y)\|^m.\end{aligned} \tag{4.8}$$

Now, combining (4.7) and (4.8) and using (iii), we obtain

$$\begin{aligned}&\mathcal{F}\Big(x, y; \beta(x, y)\Big\{\sum_{i=1}^{p} u_i[D(y, u)\nabla f_i(y) - N(y, u)\nabla g_i(y)] + \sum_{j=1}^{q} v_j\nabla G_j(y)\\ &+ \sum_{k=1}^{r} w_k\nabla H_k(y)\Big\}\Big) + \frac{1}{2}\Big\langle z, \Big\{\sum_{i=1}^{p} u_i[D(y, u)\nabla^2 f_i(y) - N(y, u)\nabla^2 g_i(y)]\\ &+ \sum_{j=1}^{q} v_j\nabla^2 G_j(y) + \sum_{k=1}^{r} w_k\nabla^2 H_k(y)\Big\}z\Big\rangle\\ &< -\sum_{t\in\mathcal{M}} \rho_t(x, y)\|\theta(x, y)\|^m \leq 0.\end{aligned} \tag{4.9}$$

Now, multiplying (4.1) by β, applying the sublinear function $\mathcal{F}(x, y; \cdot)$ to both sides of the resulting equation, and then adding the equation to (4.2), we get

$$\begin{aligned}&\mathcal{F}\Big(x, y; \beta(x, y)\Big\{\sum_{i=1}^{p} u_i[D(y, u)\nabla f_i(y) - N(y, u)\nabla g_i(y)] + \sum_{j=1}^{q} v_j\nabla G_j(y)\\ &+ \sum_{k=1}^{r} w_k\nabla H_k(y)\Big\}\Big) + \frac{1}{2}\Big\langle z, \Big\{\sum_{i=1}^{p} u_i[D(y, u)\nabla^2 f_i(y) - N(y, u)\nabla^2 g_i(y)]\\ &+ \sum_{j=1}^{q} v_j\nabla^2 G_j(y) + \sum_{k=1}^{r} w_k\nabla^2 H_k(y)\Big\}z\Big\rangle \geq 0,\end{aligned}$$

which contradicts (4.9). Therefore, we conclude that $\varphi(x) \geq \psi_{III}(\mathcal{S})$. (b)–(g): The proofs are similar to that of part (a). □

Theorem 4.2 (Strong Duality) *Let x^* be a normal optimal solution of (P) and assume that any one of the seven sets of conditions set forth in Theorem* 4.1 *is satisfied for all feasible solutions of (DIII). Then, for each $z^* \in C(x^*)$, there exist u^*, v^*, w^*, and λ^*, such that (x^*, z^*, u^*, v^*) is an optimal solution of (DIII) and $\varphi(x^*) = \psi_{III}(x^*, z^*, u^*, v^*)$.*

Proof The proof is similar to that of Theorem 2.2. □

The generalized duality models (*DIII*) and ($\tilde{D}III$) subsume a great variety of special cases which can be identified explicitly by appropriate choices of the partitioning sets $\{I_0, I_1, \ldots, I_\ell\}, \{J_0, J_1, \ldots, J_M\}$, and $\{K_0, K_1, \ldots, K_M\}$.

Concluding remarks

Remark 5.1 Using a direct nonparametric approach, in this paper, we have formulated six generalized second-order parameter-free duality models for a discrete minmax fractional programming problem and established numerous duality results using a variety of generalized $(\mathcal{F}, \beta, \phi, \rho, \theta, m)$-sounivexity assumptions. Each one of the six duality models considered in this paper is, in fact, a family of dual problems whose members can easily be identified by appropriate choices of certain sets and functions. The generalized duality models and the related duality theorems collectively provide a vast number of new second-order dual problems and duality theorems for the principal minmax problem (*P*) and its special cases designated as $(P1) - (P3)$ in Sect. 2. Furthermore, the style of presentation adopted in this paper as well as the main results derived here will prove useful in investigating other related classes of nonlinear programming problems and utilizing similar generalized convexity concepts. For example, employing similar techniques, one can investigate the second-order sufficient optimality and duality aspects of the following 'semiinfinite' minmax fractional programming problem:

$$\text{Minimize} \max_{1 \le i \le p} \frac{f_i(x)}{g_i(x)}$$

subject to

$$G_j(x,t) \le 0 \text{ for all } t \in T_j, j \in \underline{q}; H_k(x,s) = 0 \text{ for all } s \in S_k, k \in \underline{r} x \in X,$$

where X, f_i, and $g_i, i \in \underline{p}$, are as defined in the description of (*P*), for each $j \in \underline{q}$ and $k \in \underline{r}$, T_j and S_k are compact subsets of complete metric spaces, for each $j \in \underline{q}, \xi \to G_j(\xi, t)$ is a real-valued function defined on X for all $t \in T_j$, for each $k \in \underline{r}, \xi \to H_k(\xi, s)$ is a real-valued function defined on X for all $s \in S_k$, for each $j \in \underline{q}$ and $k \in \underline{r}, t \to G_j(x,t)$ and $s \to H_k(x,s)$ are continuous real-valued functions defined, respectively, on T_j and S_k for all $x \in X$.

Remark 5.2 The generalized parametric duality model results, established in this paper applying generalized $(\mathcal{F}, \beta, \phi, \rho, \theta, m)$-sounivexity assumptions, can be generalized to the case of the generalized $(\mathcal{F}, \beta, \phi, h(x^*, z), \kappa(x^*, z), \rho, \theta, m)$-sounivexity.

Definition 5.1 The function f is said to be *(strictly)* $(\mathcal{F}, \beta, \phi, h(x^*, z), \kappa(x^*, z), \rho, \theta, m)$*-sounivex* at x^* of higher order if there exist functions $\beta : X \times X \to \mathbb{R}_+ \backslash \{0\} \equiv (0, \infty)$, $\phi : \mathbb{R} \to \mathbb{R}, \rho : X \times X \to \mathbb{R}, \theta : X \times X \to \mathbb{R}^n$, and a sublinear function $\mathcal{F}(x, x^*; \cdot) : \mathbb{R}^n \to \mathbb{R}$, such that for each $x \in X (x \neq x^*)$ and $z \in \mathbb{R}^n$,

$$\begin{aligned}\phi\big(f(x) - f(x^*)\big)(>) \ge & \mathcal{F}\big(x, x^*; \beta(x, x^*)[\nabla_z \kappa(x^*, z)]\big) \\ & + \langle z, \nabla_z h(x^*, z)\rangle - h(x^*, z) \\ & + \rho(x, x^*)\|\theta(x, x^*)\|^m,\end{aligned}$$

where $h, \kappa : \mathbb{R}^n \times \mathbb{R}^n \to \mathbb{R}^n$ are differentiable.

Acknowledgments The author is greatly indebted to the reviewer for all valuable comments leading to the revised version.

References

1. Antczak, T.: Generalized fractional minimax programming with B-(p, r)-invexity. Comp. Math. Appl. **56**, 1505–1525 (2008)
2. Craven, B.D., Mond, B.: On converse duality in nonlinear programming. Operat. Res. **19**, 1075–1078 (1971)
3. Hanson, M.A.: Second order invexity and duality in mathematical programming. Opsearch **30**, 313–320 (1993)
4. Mangasarian, O.L.: Second and higher-order duality in nonlinear programming. J. Math. Anal. Appl. **51**, 607–620 (1975)
5. Mond, B.: Second order duality for nonlinear programs. Opsearch **11**, 90–99 (1974)
6. Mond, B., Zhang, J.: Duality for multiobjective programming involving second-order V-invex functions. In: Glover, M., Jeyakumar, V. (eds.) Proceedings of the Optimization Miniconference II (B), pp. 89–100. University of New South Wales, Australia (1995)
7. Mond, B., Weir, T.: Generalized concavity and duality, *Generalized in Optimization and Economics* (S. Schaible and W. T. Ziemba, eds.), Academic Press, New York (1981) pp. 263–279
8. Verma, R.U.: Generalized B-$(b, \rho, \theta, \tilde{p}, \tilde{r})$-invexities and efficiency conditions for multiobjective fractional programming. Tbilisi Math. J. **8**(2), 159–180 (2015)
9. Verma, R.U.: Generalized higher order (ϕ,η,ω,π,ρ,θ, m)-invexities in parametric optimality conditions for discrete minmax fractional programming. J. Basic Appl. Sci. **12**, 283–300 (2016)
10. Verma, R.U., Zalmai, G.J.: Second-order parametric optimality conditions in discrete minmax fractional programming. Commun. Appl. Nonlinear Anal. **23**(3), 1–32 (2016)
11. Verma, R.U., Zalmai, G.J.: Generalized second-order parametric optimality conditions in discrete minmax fractional programming. Trans. Math. Program. Appl. **2**(12), 1–20 (2014)
12. Verma, R.U., Zalmai, G.J.: Parameter-free duality in discrete minmax fractional programming based on second-order optimality conditions. Trans. Math. Program. Appl. **2**(11), 1–37 (2014)
13. Verma, R.U., Zalmai, G.J.: Generalized second-order parametric optimality conditions in semiinfinite discrete minmax fractional programming and second order (F,β,ϕ,ρ,θ, m)-univexity. Stat. Optim. Inform. Comp. **4**, 15–29 (2016)
14. von Neumann, J.: A model of general economic equilibrium. Review Econ. Stud. **13**, 1–9 (1945)
15. Wang, S.Y.: Second order necessary and sufficient conditions in

multiobjective programming. Numer. Funct. Anal. Optim. **12**, 237–252 (1991)
16. Werner, J.: Duality in generalized fractional programming. Intl. Ser. Numer. Anal. **84**, 341–351 (1988)
17. Wolfe, P.: A duality theorem for nonlinear programming. Quart. Appl. Math. **19**, 239–244 (1961)
18. Yang, X.: Generalized convex duality for multiobjective fractional programs. Opsearch **31**, 155–163 (1994)
19. Yang, X.Q.: Second-order conditions in $C^{1,1}$ optimization with applications. Numer. Funct. Anal. Optim. **14**, 621–632 (1993)
20. Yang, X.Q.: Second-order global optimality conditions of convex composite optimization. Math. Prog. **81**, 327–347 (1998)
21. Yang, X.Q.: Second-order global optimality conditions for optimization problems. J. Global Optim. **30**, 271–284 (2004)
22. Zalmai, G.J.: Hanson-Antczak type generalized $(\alpha,\beta,\gamma,\xi,\eta,\rho,\theta)$-V-invex functions in semiinfinite multiobjective fractional programming Part II: First-order parametric duality models. Adv Nonlinear Variat. Inequal. **16**(2), 61–90 (2013)
23. Zalmai, G.J.: Hanson-Antczak type generalized $(\alpha, \beta,\gamma,\xi,\eta,\rho,\theta)$-V-invex functions in semiinfinite multiobjective fractional programming Part III: Second-order parametric duality models. Adv. Nonlinear Variat. Inequal. **16**(2), 91–126 (2013)
24. Zalmai, G.J.: Optimality conditions and duality models for a class of nonsmooth constrained fractional variational problems. Optimization **30**, 15–51 (1994)
25. Zalmai, G.J.: Optimality principles and duality models for a class of continuous-time generalized fractional programming problems with operator constraints. J. Stat. Manag. Syst. **1**, 61–100 (1998)

6

Analytical solutions of fractional foam drainage equation by residual power series method

Marwan Alquran

Abstract The current work highlights the following issues: a brief survey of the development in the theory of fractional differential equations has been raised. A very recent technique based on the generalized Taylor series called—residual power series (RPS)—is introduced in detailed manner. The time-fractional foam drainage equation is considered as a target model to test the validity of the RPS method. Analysis of the obtained approximate solution of the fractional foam model reveals that RPS is an alternative method to be added for the fractional theory and computations and considered to be a significant method for exploring nonlinear fractional models.

Keywords Generalized Taylor series · Residual power series · Time-fractional foam drainage equation

Mathematics Subject Classification 26A33 · 35F25 · 35C10

Introduction

The last four decades witness fundamental works and developments on the fractional derivative and fractional differential equations. Oldham and Spanier [1], Miller and Ross [2], Samko et al. [3], Podlubny [4], Kilbas et al. [5] and others [6, 7] are the pioneer in this field; their works form an introduction to the theory of fractional differential equations and provide a systematic understanding of the fractional calculus such as the existence and the uniqueness of solutions. Hernandez et al. [8] published a paper on recent developments in the theory of abstract differential equations with fractional derivative. Finally, interested various applications of fractional calculus in the field of interdisciplinary sciences such as image processing and control theory have been studied by Magin et al. [9] and Mainardi [10].

In the literature, there exist no methods that produce an exact solution for nonlinear fractional differential equations. Only approximate solutions can be derived using linearization or successive or perturbation methods. Such methods are: variational iteration method and multivariate Pade approximations [11], Iterative Laplace transform method [12], Adomian decomposition method [13–17], Homotopy analysis method [18, 19] and Sumudu transform method [20].

The main objective of this paper is to conduct a new novel technique called residual power series method to study the solution of time-fractional Foam drainage model described by

$$D_t^{\alpha}u(x,t) = \frac{1}{2}u(x,t)u_{xx}(x,t) - 2u^2(x,t)u_x(x,t) + u_x^2(x,t), \tag{1.1}$$

subject to the initial conditions:

$$u(x,0) = f(x). \tag{1.2}$$

It is a simple model of the flow of liquid through channels (Plateau borders) and nodes (intersection of four channels) between the bubbles, driven by gravity and capillarity [21, 22]. Approximate solutions of Eqs. (1.1)–(1.2) have been obtained by different methods; Adomian decomposition

M. Alquran (✉)
Department of Mathematics and Statistics, Jordan University of Science and Technology, P.O.Box(3030), Irbid 22110, Jordan
e-mail: marwan04@just.edu.jo; marwan04@squ.edu.om

M. Alquran
Department of Mathematics and Statistics, Sultan Qaboos University, P.O.Box(36), PC 123, Al-Khod, Muscat, Oman

method [23], the Homotopy analysis method [24] and the variational iteration method [25]. It should be noted here that none of the previous studies addressed the accuracy of such methods.

The pattern of the current paper is as follows: In Sect. 2, some definitions and theorems regarding Caputo's derivative and fractional power series are given. In Sect. 3, we derive a residual power series solution to the time-fractional foam drainage equation. Graphical results regarding the foam drainage model are presented in Sect. 4.

Preliminaries

Many definitions and studies of fractional calculus have been proposed in the literature. These definitions include: Grunwald–Letnikov, Riemann–Liouville, Weyl, Riesz and Caputo sense. In the Caputo case, the derivative of a constant is zero and one can define, properly, the initial conditions for the fractional differential equations which can be handled using an analogy with the classical integer case. For these reasons, researchers prefer to use the Caputo fractional derivative [26] which is defined as

Definition 2.1 For m to be the smallest integer that exceeds α, the Caputo fractional derivatives of order $\alpha > 0$ are defined as

$$D^{\alpha}u(x,t) = \frac{\partial^{\alpha}u(x,t)}{\partial t^{\alpha}} = \begin{cases} \frac{1}{\Gamma(m-\alpha)}\int_0^t (t-\tau)^{m-\alpha-1}\frac{\partial^m u(x,\tau)}{\partial \tau^m}\mathrm{d}\tau, & m-1<\alpha<m \\ \frac{\partial^m u(x,t)}{\partial t^m}, & \alpha = m \in N \end{cases}$$

Now, we survey some needed definitions and theorems regarding the fractional power series (RPS) where there is much theory to be found in [27, 28].

Definition 2.2 A power series expansion of the form

$$\sum_{m=0}^{\infty} c_m(t-t_0)^{m\alpha} = c_0 + c_1(t-t_0)^{\alpha} + c_2(t-t_0)^{2\alpha} + \cdots \quad 0\leq n-1<\alpha\leq n, \quad t\leq t_0$$

is called fractional power series PS about $t = t_0$

Theorem 2.1 *Suppose that f has a fractional PS representation at $t = t_0$ of the form*

$$f(t) = \sum_{m=0}^{\infty} c_m(t-t_0)^{m\alpha}, \quad t_0\leq t<t_0+R.$$

If $D^{m\alpha}f(t)$, $m = 0,1,2,\ldots$ are continuous on (t_0, t_0+R), then $c_m = \frac{D^{m\alpha}f(t_0)}{\Gamma(1+m\alpha)}$.

Definition 2.3 A power series expansion of the form

$$\sum_{m=0}^{\infty} f_m(x)(t-t_0)^{m\alpha}$$

is called multiple fractional power series PS about $t = t_0$

Theorem 2.2 *Suppose that $u(x,t)$ has a multiple fractional PS representation at $t = t_0$ of the form*

$$u(x,t) = \sum_{m=0}^{\infty} f_m(x)(t-t_0)^{m\alpha}, \quad x\in I, \quad t_0\leq t<t_0+R.$$

If $D_t^{m\alpha}u(x,t)$, $m = 0,1,2,\ldots$ are continuous on $I\times(t_0,t_0+R)$, then $f_m(x) = \frac{D_t^{m\alpha}u(x,t_0)}{\Gamma(1+m\alpha)}$.

From the last theorem, it is clear that if $n+1$-dimensional function has a multiple fractional PS representation at $t = t_0$, then it can be derived in the same manner, i.e.

Corollary 2.3 *Suppose that $u(x,y,t)$ has a multiple fractional PS representation at $t = t_0$ of the form*

$$u(x,y,t) = \sum_{m=0}^{\infty} g_m(x,y)(t-t_0)^{m\alpha}, \quad (x,y)\in I_1\times I_2, \quad t_0\leq t<t_0+R.$$

If $D_t^{m\alpha}u(x,y,t)$, $m = 0,1,2,\ldots$ are continuous on $I_1\times I_2\times(t_0,t_0+R)$, then $g_m(x,y) = \frac{D_t^{m\alpha}u(x,y,t_0)}{\Gamma(1+m\alpha)}$.

Next, we present in details the derivation of the residual power series solution to the generalized fractional DSW system.

Residual power series (RPS) of the foam drainage model

The aim of this section is to construct power series solution to the time-fractional Foam drainage model by substituting its power series (PS) expansion among its truncated residual function [29, 30]. From the resulting equation, a recursion formula for the computation of the coefficients is derived, while the coefficients in the fractional PS expansion can be computed recursively by recurrent fractional differentiation of the truncated residual function.

The RPS method proposes the solution for Eqs. (1.1–1.2) as a fractional PS about the initial point $t = 0$

$$u(x,t) = \sum_{n=0}^{\infty} f_n(x)\frac{t^{n\alpha}}{\Gamma(1+n\alpha)}, \quad 0<\alpha\leq 1,\ x\in I, \quad 0\leq t<R. \tag{3.1}$$

Next, we let $u_k(x,t)$ to denote the k-th truncated series of $u(x,t)$, i.e.,

$$u_k(x,t) = \sum_{n=0}^{k} f_n(x)\frac{t^{n\alpha}}{\Gamma(1+n\alpha)}, \quad 0<\alpha\leq 1, \quad x\in I,\ 0\leq t<R. \tag{3.2}$$

It is clear that by condition (1.2) the 0-th RPS approximate solutions of $u(x,t)$ is

$$u_0(x,t) = f_0(x) = u(x,0) = f(x). \tag{3.3}$$

Also, Eq. (3.2) can be written as

$$u_k(x,t) = f(x) + \sum_{n=1}^{k} f_n(x)\frac{t^{n\alpha}}{\Gamma(1+n\alpha)}, \quad 0<\alpha\leq 1,\ x\in I,\ 0\leq t<R,\ k=1,2,3,\ldots \tag{3.4}$$

Now, we define the residual functions, *Res*, for Eq. (1.1)

$$Res_u(x,t) = D_t^{\alpha}u(x,t) - \frac{1}{2}u(x,t)u_{xx}(x,t) + 2u^2(x,t)u_x(x,t) - u_x^2(x,t), \tag{3.5}$$

and, therefore, the k-th residual function, $Res_{u,k}$, is

$$Res_{u,k}(x,t) = D_t^{\alpha}u_k(x,t) - \frac{1}{2}u_k(x,t)\frac{\partial^2}{\partial x^2}u_k(x,t) + 2u_k^2(x,t)\frac{\partial}{\partial x}u_k(x,t) - \left(\frac{\partial}{\partial x}u_k(x,t)\right)^2. \tag{3.6}$$

As in [29, 30], $Res(x,t) = 0$ and $\lim_{k\to\infty} Res_k(x,t) = Res(x,t)$ for all $x\in I$ and $t\geq 0$. Therefore, $D_t^{r\alpha}Res(x,t) = 0$ since the fractional derivative of a constant in the Caputo's sense is 0. Also, the fractional derivative $D_t^{r\alpha}$ of $Res(x,t)$ and $Res_k(x,t)$ is matching at $t=0$ for each $r = 0,1,2,\ldots,k$.

To clarify the RPS technique, we substitute the k-th truncated series of $u(x,t)$ into Eq. (3.6), find the fractional derivative formula $D_t^{(k-1)\alpha}$ of both $Res_{u,k}(x,t)$, $k = 1,2,3,\ldots$, and then solve the obtained algebraic system

$$D_t^{(k-1)\alpha}Res_{u,k}(x,0) = 0,\ 0<\alpha\leq 1,\ x\in I,\ k=1,2,3\ldots \tag{3.7}$$

to get the required coefficients $f_n(x)$, $n = 1,2,3,\ldots,k$ in Eq. (3.4). Now, we follow the following steps.

Step 1. To determine $f_1(x)$, we consider $(k=1)$ in (3.6)

$$Res_{u,1}(x,t) = D_t^{\alpha}u_1(x,t) - \frac{1}{2}u_1(x,t)\frac{\partial^2}{\partial x^2}u_1(x,t) + 2u_1^2(x,t)\frac{\partial}{\partial x}u_1(x,t) - \left(\frac{\partial}{\partial x}u_1(x,t)\right)^2. \tag{3.8}$$

But, $u_1(x,t) = f(x) + f_1(x)\frac{t^{\alpha}}{\Gamma(1+\alpha)}$. Therefore,

$$\begin{aligned} Res_{u,1}(x,t) &= f_1(x) - \frac{1}{2}\left(f(x) + f_1(x)\frac{t^{\alpha}}{\Gamma(1+\alpha)}\right) \\ &\quad\times\left(f''(x) + f_1''(x)\frac{t^{\alpha}}{\Gamma(1+\alpha)}\right) \\ &\quad+ 2\left(f(x) + f_1(x)\frac{t^{\alpha}}{\Gamma(1+\alpha)}\right)^2 \\ &\quad\times\left(f'(x) + f_1'(x)\frac{t^{\alpha}}{\Gamma(1+\alpha)}\right) \\ &\quad- \left(f'(x) + f_1'(x)\frac{t^{\alpha}}{\Gamma(1+\alpha)}\right)^2. \end{aligned} \tag{3.9}$$

From Eq. (3.7), we deduce that $Res_{u,1}(x,0) = 0$ and thus,

$$f_1(x) = \frac{1}{2}f(x)f''(x) - 2f^2(x)f'(x) + f'^2(x) \tag{3.10}$$

Therefore, the 1st RPS approximate solution is

$$u_1(x,t) = f(x) + \left(\frac{1}{2}f(x)f''(x) - 2f^2(x)f'(x) + f'^2(x)\right)\frac{t^{\alpha}}{\Gamma(1+\alpha)}. \tag{3.11}$$

Step 2. To obtain $f_2(x)$, we substitute the 2nd truncated series $u_2(x,t) = f(x) + f_1(x)\frac{t^{\alpha}}{\Gamma(1+\alpha)} + f_2(x)\frac{t^{2\alpha}}{\Gamma(1+2\alpha)}$ into the 2nd residual function $Res_{u,2}(x,t)$, i.e.,

$$\begin{aligned} Res_{u,2}(x,t) &= D_t^{\alpha}u_2(x,t) - \frac{1}{2}u_2(x,t)\frac{\partial^2}{\partial x^2}u_2(x,t) \\ &\quad+ 2u_2^2(x,t)\frac{\partial}{\partial x}u_2(x,t) - \left(\frac{\partial}{\partial x}u_2(x,t)\right)^2. \\ &= f_1(x) + f_2(x)\frac{t^{\alpha}}{\Gamma(1+\alpha)} \\ &\quad- \frac{1}{2}\left(f(x) + \cdots + f_2(x)\frac{t^{2\alpha}}{\Gamma(1+2\alpha)}\right) \\ &\quad\times\left(f''(x) + \cdots + f_2''(x)\frac{t^{2\alpha}}{\Gamma(1+2\alpha)}\right) \\ &\quad+ 2\left(f(x) + \cdots + f_2(x)\frac{t^{2\alpha}}{\Gamma(1+2\alpha)}\right)^2 \\ &\quad\times\left(f'(x) + \cdots + f_2'(x)\frac{t^{2\alpha}}{\Gamma(1+2\alpha)}\right) \\ &\quad- \left(f'(x) + \cdots + f_2'(x)\frac{t^{2\alpha}}{\Gamma(1+2\alpha)}\right)^2. \end{aligned} \tag{3.12}$$

Applying D_t^{α} on both sides of Eq. (3.12) gives

$$\begin{aligned} D_t^{\alpha} Res_{u,2}(x,t) &= f_2(x) - \frac{1}{2}\left(f_1(x) + f_2(x)\frac{t^{\alpha}}{\Gamma(1+\alpha)}\right) \\ &\times \left(f''(x) + \cdots + f_2''(x)\frac{t^{2\alpha}}{\Gamma(1+2\alpha)}\right) \\ &- \frac{1}{2}\left(f(x) + \cdots + f_2(x)\frac{t^{2\alpha}}{\Gamma(1+2\alpha)}\right) \\ &\times \left(f_1''(x) + f_2''(x)\frac{t^{\alpha}}{\Gamma(1+\alpha)}\right) \\ &+ 4\left(f(x) + \cdots + f_2(x)\frac{t^{2\alpha}}{\Gamma(1+2\alpha)}\right) \\ &\times \left(f_1(x) + f_2(x)\frac{t^{\alpha}}{\Gamma(1+\alpha)}\right) \\ &\times \left(f'(x) + \cdots + f_2'(x)\frac{t^{2\alpha}}{\Gamma(1+2\alpha)}\right) \\ &+ 2\left(f(x) + \cdots + f_2(x)\frac{t^{2\alpha}}{\Gamma(1+2\alpha)}\right)^2 \\ &\times \left(f_1'(x) + f_2'(x)\frac{t^{\alpha}}{\Gamma(1+\alpha)}\right) \\ &- 2\left(f'(x) + \cdots + f_2'(x)\frac{t^{2\alpha}}{\Gamma(1+2\alpha)}\right) \\ &\times \left(f_1'(x) + f_2'(x)\frac{t^{\alpha}}{\Gamma(1+\alpha)}\right). \end{aligned} \tag{3.13}$$

By the fact that $D_t^{\alpha} Res_{u,2}(x,0) = 0$ and solving the resulting system in (3.13) for the unknown coefficient function $f_2(x)$, we get

$$\begin{aligned} f_2(x) &= \frac{1}{2} f(x) f_1''(x) + \frac{1}{2} f_1(x) f''(x) - 4 f(x) f_1(x) f'(x) \\ &- 2 f^2(x) f_1'(x) + 2 f'(x) f_1'(x). \end{aligned} \tag{3.14}$$

Step 3. We derive the formula of the coefficient function $f_3(x)$. Substitute the 3rd truncated series $u_3(x,t) = f(x) + f_1(x)\frac{t^{\alpha}}{\Gamma(1+\alpha)} + f_2(x)\frac{t^{2\alpha}}{\Gamma(1+2\alpha)} + f_3(x)\frac{t^{3\alpha}}{\Gamma(1+3\alpha)}$ into the 3rd residual function $Res_{u,3}(x,t)$. i.e.,

$$\begin{aligned} Res_{u,3}(x,t) &= D_t^{\alpha} u_3(x,t) - \frac{1}{2} u_3(x,t) \frac{\partial^2}{\partial x^2} u_3(x,t) \\ &+ 2u_3^2(x,t)\frac{\partial}{\partial x} u_3(x,t) - \left(\frac{\partial}{\partial x} u_3(x,t)\right)^2. \\ &= f_1(x) + f_2(x)\frac{t^{\alpha}}{\Gamma(1+\alpha)} + f_3(x)\frac{t^{2\alpha}}{\Gamma(1+2\alpha)} \\ &- \frac{1}{2}\left(f(x) + \cdots + f_3(x)\frac{t^{3\alpha}}{\Gamma(1+2\alpha)}\right) \\ &\times \left(f''(x) + \cdots + f_3''(x)\frac{t^{3\alpha}}{\Gamma(1+3\alpha)}\right) \\ &+ 2\left(f(x) + \cdots + f_3(x)\frac{t^{3\alpha}}{\Gamma(1+3\alpha)}\right)^2 \\ &\times \left(f'(x) + \cdots + f_3'(x)\frac{t^{3\alpha}}{\Gamma(1+3\alpha)}\right) \\ &- \left(f'(x) + \cdots + f_3'(x)\frac{t^{3\alpha}}{\Gamma(1+3\alpha)}\right)^2. \end{aligned} \tag{3.15}$$

Now, we apply the operator $D_t^{2\alpha}$ on both sides of Eq. (3.15)

$$\begin{aligned} D_t^{2\alpha} Res_{u,3}(x,t) &= D_t^{\alpha}(D_t^{\alpha} Res_{x,3}(x,t)) \\ &= f_3(x) - \left(f_1(x) + \cdots + f_3(x)\frac{t^{2\alpha}}{\Gamma(1+2\alpha)}\right) \\ &\times \left(f_1''(x) + \cdots + f_3''(x)\frac{t^{2\alpha}}{\Gamma(1+2\alpha)}\right) \\ &- \frac{1}{2}\left(f_2(x) + f_3(x)\frac{t^{\alpha}}{\Gamma(1+\alpha)}\right) \\ &\times \left(f''(x) + \cdots + f_3''(x)\frac{t^{3\alpha}}{\Gamma(1+3\alpha)}\right) \\ &- \frac{1}{2}\left(f(x) + \cdots + f_3(x)\frac{t^{3\alpha}}{\Gamma(1+3\alpha)}\right) \\ &\times \left(f_2''(x) + f_3''(x)\frac{t^{\alpha}}{\Gamma(1+\alpha)}\right) \\ &+ 4\left(f_1(x) + \cdots + f_3(x)\frac{t^{2\alpha}}{\Gamma(1+2\alpha)}\right)^2 \\ &\times \left(f'(x) + \cdots + f_3'(x)\frac{t^{3\alpha}}{\Gamma(1+3\alpha)}\right) \\ &+ 4\left(f(x) + \cdots + f_3(x)\frac{t^{3\alpha}}{\Gamma(1+3\alpha)}\right) \\ &\times \left(f_2(x) + f_3(x)\frac{t^{\alpha}}{\Gamma(1+\alpha)}\right) \\ &\times \left(f'(x) + \cdots + f_3'(x)\frac{t^{3\alpha}}{\Gamma(1+3\alpha)}\right) \\ &+ 8\left(f(x) + \cdots + f_3(x)\frac{t^{3\alpha}}{\Gamma(1+3\alpha)}\right) \\ &\times \left(f_1(x) + \cdots + f_3(x)\frac{t^{2\alpha}}{\Gamma(1+2\alpha)}\right) \\ &\times \left(f_1'(x) + \cdots + f_3'(x)\frac{t^{2\alpha}}{\Gamma(1+2\alpha)}\right) \\ &+ 2\left(f(x) + \cdots + f_3(x)\frac{t^{3\alpha}}{\Gamma(1+3\alpha)}\right)^2 \\ &\times \left(f_2'(x) + f_3'(x)\frac{t^{\alpha}}{\Gamma(1+\alpha)}\right) \\ &- 2\left(f_1'(x) + \cdots + f_3'(x)\frac{t^{2\alpha}}{\Gamma(1+2\alpha)}\right)^2 \\ &- 2\left(f'(x) + \cdots + f_3'(x)\frac{t^{3\alpha}}{\Gamma(1+3\alpha)}\right) \\ &\times \left(f_2'(x) + f_3'(x)\frac{t^{\alpha}}{\Gamma(1+\alpha)}\right). \end{aligned} \tag{3.16}$$

Thus, solving the equation $D_t^{2\alpha} Res_{u,3}(x,0) = 0$ results in the following recurrence formula

$$\begin{aligned} f_3(x) &= f_1(x) f_1''(x) + \frac{1}{2} f(x) f_2''(x) + \frac{1}{2} f_2(x) f''(x) \\ &- 4 f_1^2(x) f'(x) - 4 f(x) f_2(x) f'(x) \\ &- 8 f(x) f_1(x) f_1'(x) - 2 f^2(x) f_2'(x) \\ &+ 2(f_1'(x))^2 + 2 f'(x) f_2'(x). \end{aligned} \tag{3.17}$$

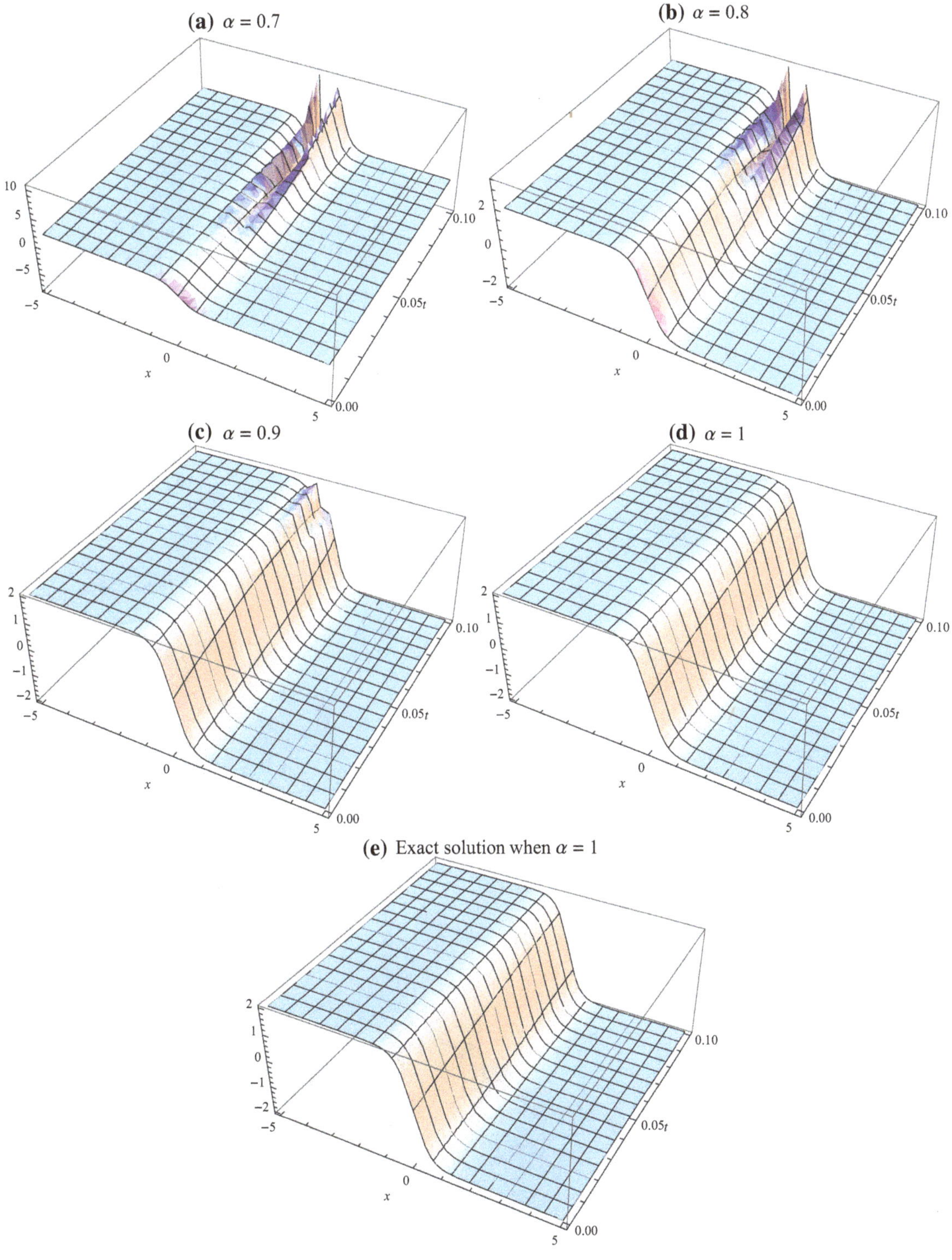

Fig. 1 The 4th RPS approximate solution of the Foam model when $f(x) = -\sqrt{c}\tanh(\sqrt{c}x)$: **a** $u_4(x,t,,\alpha = 0.7)$, **b** $u_4(x,t,\alpha = 0.8)$, **c** $u_4(x,t,\alpha = 0.9)$, **d** $u_4(x,t,\alpha = 1)$, **e** $u(x,t)$ for $\alpha = 1$, $-5 < x < 5$, $0 < t < 0.1$

Table 1 The obtained errors $|u_{exact} - u^4_{approx}|$ for the RPS method

x	t	Numerical scheme
1.0	0.02	6.6627E−06
	0.04	9.0798E−05
	0.06	3.8406E−04
	0.08	1.0024E−03
3.0	0.02	7.2440E−10
	0.04	2.4547E−08
	0.06	1.9773E−07
	0.08	8.8589E−07
5.0	0.02	2.4314E−13
	0.04	8.2420E−12
	0.06	6.6377E−11
	0.08	2.9735E−10

Finally, we follow the same manner as above and solve the equation $D_t^{3\alpha} Res_{u,4}(x,0) = 0$ to deduce the following result

$$\begin{aligned} f_4(x) = &\frac{3}{2} f_2(x) f_1''(x) + \frac{3}{2} f_1(x) f_2''(x) + \frac{1}{2} f_3(x) f''(x) \\ &+ \frac{1}{2} f(x) f_3''(x) \\ &- 12 f_1(x) f_2(x) f'(x) - 12 f_1^2(x) f_1'(x) \\ &- 4 f(x) f_3(x) f'(x) - 8 f(x) f_2(x) f_1'(x) \\ &- 8 f(x) f_1(x) f_2'(x) - 2 f^2(x) f_3'(x) + 6 f_1'(x) f_2'(x) \\ &+ 2 f'(x) f_3'(x). \end{aligned} \tag{3.18}$$

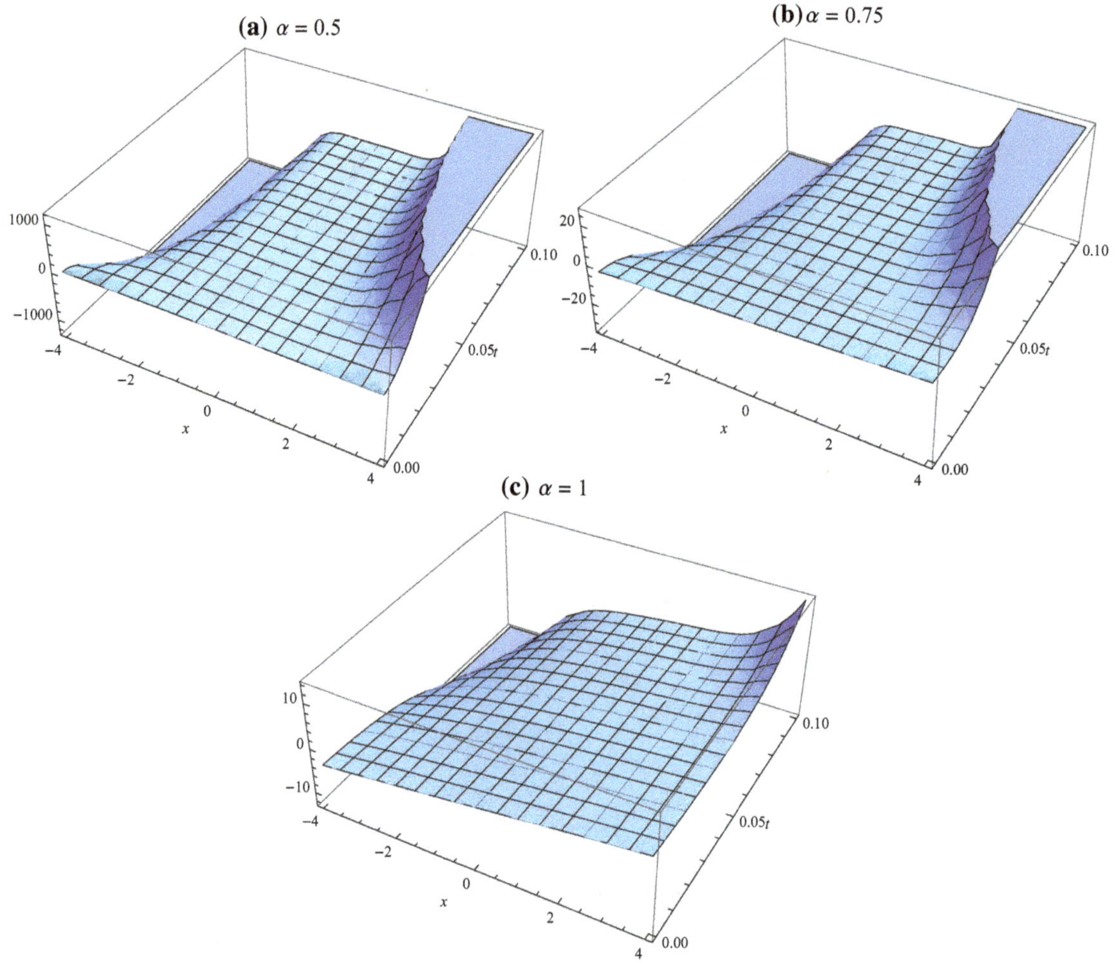

Fig. 2 The 4th RPS approximate solution of the Foam model when $f(x) = x$: **a** $u_4(x,t,\alpha = 0.5)$, **b** $u_4(x,t,\alpha = 0.75)$, **c** $u_4(x,t,\alpha = 1)$, $-4 < x < 4$, $0 < t < 0.1$

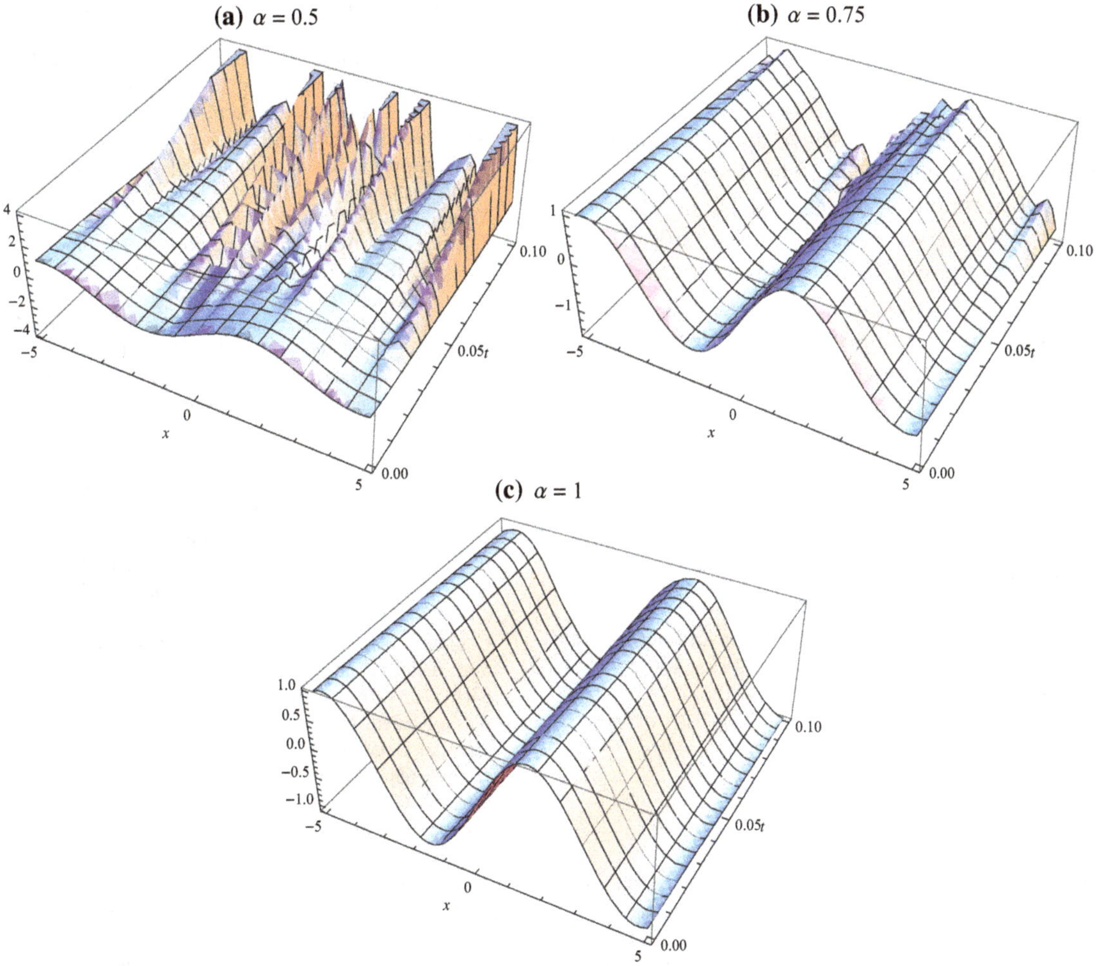

Fig. 3 The 4th RPS approximate solution of the Foam model when $f(x) = \sin(x)$: **a** $u_4(x, t, \alpha = 0.5)$, **b** $u_4(x, t, \alpha = 0.75)$, **c** $u_4(x, t, \alpha = 1)$, $-5 < x < 5,\ \ 0 < t < 0.1$

By the above recurrence relations, we are ready to present some graphical results regarding the time-fractional Foam drainage model.

Applications and results

The purpose of this section is to test the derivation of residual power series solutions of the following time-fractional foam drainage equation.

$$D_t^{\alpha} u(x,t) = \frac{1}{2} u(x,t) u_{xx}(x,t) - 2u^2(x,t) u_x(x,t) + u_x^2(x,t), \tag{4.1}$$

subject to the initial conditions:

$$u(x,0) = -\sqrt{c}\tanh(\sqrt{c}x). \tag{4.2}$$

Where the exact solution of this model when $\alpha = 1$ is $u(x,t) = -\sqrt{c}\tanh(\sqrt{c}(x - ct))$. Based on the obtained results from the previous section, we consider only the 4*th* RPS approximate solution. Figure 1 represents the 4th RPS approximate solution of the function $u(x,t)$ for different values of the fractional derivative α. To validate the efficiency and accuracy of the analytical scheme, we give explicit values of x and t and compute the absolute error compared with the exact solution when $\alpha = 1$ (See Table 1). Figure 2 represents solutions to the fractional foam drainage when the initial condition is $f(x) = x$ for $\alpha = 0.5,\ 1$. Finally, Fig. 3 represents solutions to the fractional foam drainage when the initial condition is $f(x) = \sin(x)$ for $\alpha = 0.5,\ 0.75,\ 1$.

In the theory of fractional calculus, it is obvious that when the fractional derivative $n - 1 < \alpha < n$ tends to positive integer n, the approximate solution continuously tends to the exact solution of the problem with derivative $n = 1$. It is also clear from Figs. 1, 2 and 3 that when α is close to 0, the solutions bifurcate and provide wave-like pattern. But, when α is close to 1, there is no pattern.

Conclusions

A new analytical iterative technique based on the residual power series is proposed to obtain an approximate solution to a nonlinear time-fractional foam drainage equation. Three different initial conditions of the Foam model are considered and accordingly different physical behaviors are produced. We observed when α is close to 0, the solutions bifurcate and provide wave-like pattern. But, when α is close to 1, there is no pattern (Figs. 1 and 3). This bifurcation phenomenon will help the researchers to open many new avenues to better understanding of time-fractional derivatives and its relation to real-life phenomena. The accuracy of the proposed method has been tested by studying the absolute errors of the obtained approximate of the Foam model (Table 1). The RPS method is a promising technique based on its simplicity and accuracy. It is to be considered an additive tool for the field of fractional theory and computations. As future work, we will extend the RPS method to handle (2 + 1)-dimensional linear and nonlinear space- and time-fractional physical models.

References

1. Oldham, K.B., Spanier, J.: The Fractional Calculus: Theory and Application of Differentiation and Integration to Arbitrary order. Academic Press, California (1974)
2. Miller, K., Ross, B.: An Introduction to the Fractional Calculus and Fractional Differential Equations. Wiley, New York (1993)
3. Ray, S.S.: Analytical solution for the space fractional diffusion equation by two-step Adomian decomposition method. Commun. Nonlinear Sci. Numer. Simul. **14**, 1130–1295 (2009)
4. Podlubny, I.: Fractional Differential Equations. Academic Press, California (1999)
5. Kilbas, A.A., Srivastava, H.M., Trujillo, J.J.: Theory and Applications of Fractional Differential Equations. Elsevier, Amstrdam (2006)
6. Diethelm, K.: The Analysis of Fractional Differential Equations. Springer, New York (2010)
7. Caponetto, R., Dongola, G., Fortuna, L., Petras, I.: Fractional Order Systems: Modeling and Control Applications. World Scientific Publishing Company, Singapore (2010)
8. Hernandez, E., O'Regan, D., Balachandran, K.: On recent developments in the theory of abstract differential equations with fractional derivatives. Nonlinear Anal.: Theory Methods Appl. **73**, 3462–3471 (2010)
9. Magin, R., Ortigueira, M.D., Podlubny, I., Trujillo, J.J.: On the fractional signals and systems. Signal Process. **91**, 350–371 (2011)
10. Mainardi, F.: Fractional Calculus and Waves in Linear Viscoelasticity: An Introduction to Mathematical Models. Imperial College Press, London (2010)
11. Turut, V., Guzel, N.: On solving partial differential equations of fractional order by using the variational iteration method and multivariate Pad approximations. Eur. J. Pure Appl. Math. **6**(2) 147–171 (2013)
12. Jafaria, J., Nazarib, M., Baleanuc, D., Khalique, C.M.: A new approach for solving a system of fractional partial differential equations. Comput. Math. Appl. **66**(5), 838–843 (2013)
13. Al-khaled, K., Momani, S.: An approximate solution for a fractional diffusion-wave equation using the decomposition method. Appl. Math. Comput. **165**(2), 473–483 (2005)
14. Wang, Q.: Numerical solutions for fractional KdV-Burgers equation by Adomian decomposition method. Appl. Math. Comput. **182**, 1048–1055 (2006)
15. Hu, Y., Luo, Y., Lu, Z.: Analytical solution of the linear fractional differential equation by Adomian decomposition method. J. Comput. Appl. Math. **215**, 220–229 (2008)
16. Parthiban, V., Balachandran, K.: Solutions of system of fractional partial differential equations. Appl. Appl. Math. **8**(1), 289–304 (2013)
17. Al-khaled, K., Alquran, M.: An approximate solution for a fractional model of generalized Harry Dym equation. Math Sci (2015). doi:10.1007/s40096-015-0137-x
18. Dehghan, M., Manafian, J., Saadatmandi, A.: Solving nonlinear fractional partial differential equations using the homotopy analysis method. Numer. Methods Partial Differ. Equ. **26**(2), 448–479 (2010)
19. Ganjiani, M.: Solution of nonlinear fractional differential equations using Homotopy analysis method. Appl. Math. Model. **34**(6), 1634–1641 (2010)
20. K. Al-Khaled, Numerical solution of time-fractional partial differential equations using Sumudu Decomposition method. Rom. J. Phys. **60**(1–2) (2015)
21. Verbist, G., Weuire, D., Kraynik, A.M.: The foam drainage equation. J. Phys. Condens. Matter **8**, 3715–3731 (1996)
22. Weaire, D., Hutzler, S., Cox, S., Alonso, M.D., Drenckhan, D.: The fluid dynamics of foams. J. Phys. Condens. Matter **15**, S65–S73 (2003)
23. Dahmani, Z., Mesmoudi, M.M., Bebbouch, R.: The Foam Drainage equation with time- and space-fractional derivatives solved by the Adomian method. Electron. J. Qual. Theory Differ. Equ. **30**, 1–10 (2008)
24. Fadravi, H.H., Nik, H.S., Buzhabadi, R.: Homotopy analysis method for solving Foam Drainage equation with space- and time-fractional derivatives. Int. J. Differ. Equ. (2011) (Article ID 237045)
25. Dahmani, Zoubir, Anber, Ahmed: The variational iteration method for solving the fractional foam drainage equation. Int. J. Nonlinear Sci. **10**(1), 39–45 (2010)
26. Gomez-Aguilar, J.F., Razo-Hernandez, R., Granados-Lieberman, D.: A physical interpretation of fractional calculus in observables terms: analysis of the fractional time constant and the transitory response. Revista Mexicana de Fisica **60**, 32–38 (2014)
27. El-Ajou, A., Arqub, O.A., Momani, S.: Approximate analytical solution of the nonlinear fractional KdV-Burgers equation: a new iterative algorithm. J. Comput. Phys. (2014) (In press)
28. El-Ajou, A., Abu, O., Al Zhour, Z.A., Momani, S.: New results on fractional power series: theories and applications. Entropy **15**, 5305–5323 (2013)
29. Arqub, O.A.: Series solution of fuzzy differential equations under strongly generalized differentiability. J. Adv. Res. Appl. Math. **5**, 31–52 (2013)
30. Arqub, O.A., El-Ajou, A., Bataineh, A., Hashim, I.: A representation of the exact solution of generalized Lane Emden equations using a new analytical method. Abstr. Appl. Anal. (2013) (Article ID 378593)

7

Some new contractive mappings on S-metric spaces and their relationships with the mapping (S25)

Nİhal Yilmaz Özgür[1] · Nİhal Taş[1]

Abstract Recently, S-metric spaces are introduced as a generalization of metric spaces. In this paper, we consider the relationships between of an S-metric space and a metric space, and give an example of an S-metric which does not generate a metric. Then, we introduce new contractive mappings on S-metric spaces and investigate relationships among them by counterexamples. In addition, we obtain new fixed point theorems on S-metric spaces.

Keywords S-metric space · Fixed point theorem · Periodic point · Diameter

Mathematics Subject Classification 54E35 · 54E40 · 54E45 · 54E50

Introduction

Recently, Sedghi, Shobe, and Aliouche have defined the concept of an S-metric space as a generalization of a metric space in [14] as follows:

Definition 1 [14] Let X be a nonempty set, and $S : X^3 \to [0, \infty)$ be a function satisfying the following conditions for all $x, y, z, a \in X$:

1. $S(x,y,z) = 0$ if and only if $x = y = z$,
2. $S(x,y,z) \leq S(x,x,a) + S(y,y,a) + S(z,z,a)$.

Then, S is called an S-metric on X and the pair (X, S) is called an S -metric space.

The fixed point theory on various metric spaces was studied by many authors. For example, A. Aghajani, M. Abbas, and J. R. Roshan proved some common fixed point results for four mappings satisfying generalized weak contractive condition on partially ordered complete b-metric spaces [1]; T. V. An, N. V. Dung, and V. T. L. Hang studied some fixed point theorems on G-metric spaces [2]; N. V. Dung, N. T. Hieu, and S. Radojevic proved some fixed point theorems on partially ordered S-metric spaces [6]. Gupta and Deep studied some fixed point results using mixed weakly monotone property and altering distance function in the setting of S-metric space [9]. The present authors investigated some generalized fixed point theorems on a complete S-metric space [11].

Motivated by the above studies, our aim is to obtain new fixed point theorems on S-metric spaces related to Rhoades' conditions.

We recall Rhoades' conditions in (X, d) and (X, S), respectively.

Let (X, d) be a complete metric space and T be a self-mapping of X. In [13], T is called a Rhoades' mapping (**RN**), ($N = 25, 50, 75, 100, 125$) if the following condition is satisfied, respectively:

(**R25**) $$d(Tx,Ty) < \max\{d(x,y), d(x,Tx), d(y,Ty), d(x,Ty), d(y,Tx)\},$$

for each $x, y \in X$, $x \neq y$.

(**R50**) There exists a positive integer p, such that

$$d(T^p x, T^p y) < \max\{d(x,y), d(x,T^p x), d(y,T^p y), d(x,T^p y), d(y,T^p x)\},$$

for each $x, y \in X$, $x \neq y$.

✉ Nİhal Taş
nihaltas@balikesir.edu.tr

Nİhal Yilmaz Özgür
nihal@balikesir.edu.tr

[1] Department of Mathematics, Balıkesir University, 10145 Balıkesir, Turkey

(**R75**) There exist positive integers p, q, such that

$$d(T^p x, T^q y) < \max\{d(x,y), d(x,T^p x), d(y,T^q y), d(x,T^q y), d(y,T^p x)\},$$

for each $x, y \in X$, $x \neq y$.

(**R100**) There exists a positive integer $p(x)$, such that

$$d(T^{p(x)}x, T^{p(x)}y) < \max\{d(x,y), d(x,T^{p(x)}x), d(y,T^{p(x)}y), d(x,T^{p(x)}y), d(y,T^{p(x)}x)\},$$

for any given x, every $y \in X$, $x \neq y$.

(**R125**) There exists a positive integer $p(x, y)$, such that

$$d(T^{p(x,y)}x, T^{p(x,y)}y) < \max\{d(x,y), d(x,T^{p(x,y)}x), d(y,T^{p(x,y)}y), d(x,T^{p(x,y)}y), d(y,T^{p(x,y)}x)\},$$

for any given $x, y \in X$, $x \neq y$.

Let (X, S) be an S-metric space and T be a self-mapping of X. In [12], the present authors defined Rhoades' condition (**S25**) on (X, S) as follows:

(**S25**) $S(Tx,Tx,Ty) < max\{S(x,x,y), S(Tx,Tx,x), S(Ty,Ty,y), S(Ty,Ty,x), S(Tx,Tx,y)\}$,

for each $x, y \in X$, $x \neq y$.

In this paper, we consider some forms of Rhoades' conditions and give some fixed point theorems on S-metric spaces. In Sect. 2, we investigate relationships between metric spaces and S-metric spaces. It is known that every metric generates an S-metric, and in [10], it was given an example of an S-metric which is not generated by a metric. Here, we give a new example of an S-metric which is not generated by a metric and use this new S-metric in the next sections. In [8], it is mentioned that every S-metric defines a metric. However, we give a counterexample to this result. We obtain an example of an S-metric which does not generate a metric. We introduce new contractive mappings, such as (**S50**), (**S75**), (**S100**), and (**S125**), and also study relations among them by counterexamples. In Sect. 3, we investigate some new fixed point theorems using periodic index on S-metric spaces for the contractive mappings defined in Sect. 2. In Sect. 4, we define the condition (**Q25**) and give new fixed point theorems on S-metric spaces.

New contractive mappings on S-metric spaces

In this section, we introduce new types of Rhoades' conditions on S-metric spaces, such as (**S50**), (**S75**), (**S100**), and (**S125**). At first, we recall some definitions and theorems.

Definition 2 [14] Let (X, S) be an S-metric space and $A \subset X$.

1. A sequence $\{x_n\}$ in X converges to x if and only if $S(x_n, x_n, x) \to 0$ as $n \to \infty$. That is, there exists $n_0 \in \mathbb{N}$ such that for all $n \geq n_0$, $S(x_n, x_n, x) < \varepsilon$ for each $\varepsilon > 0$. We denote this by $\lim_{n\to\infty} x_n = x$ or $\lim_{n\to\infty} S(x_n, x_n, x) = 0$.
2. A sequence $\{x_n\}$ in X is called a Cauchy sequence if $S(x_n, x_n, x_m) \to 0$ as $n, m \to \infty$. That is, there exists $n_0 \in \mathbb{N}$, such that for all $n, m \geq n_0$, $S(x_n, x_n, x_m) < \varepsilon$ for each $\varepsilon > 0$.
3. The S-metric space (X, S) is called complete if every Cauchy sequence is convergent.

Lemma 1 [14] Let (X, S) be an S-metric space. Then,

$$S(x,x,y) = S(y,y,x). \tag{2.1}$$

The relation between a metric and an S-metric is given in [10] as follows:

Lemma 2 [10] Let (X, d) be a metric space. Then, the following properties are satisfied:

1. $S_d(x,y,z) = d(x,z) + d(y,z)$ for all $x, y, z \in X$ is an S-metric on X.
2. $x_n \to x$ in (X, d) if and only if $x_n \to x$ in (X, S_d).
3. $\{x_n\}$ is Cauchy in (X, d) if and only if $\{x_n\}$ is Cauchy in (X, S_d).
4. (X, d) is complete if and only if (X, S_d) is complete.

We call the metric S_d as the S-metric generated by d.

Note that there exists an S-metric S satisfying $S \neq S_d$ for all metrics d [10]. Now, we give an another example which shows that there exists an S-metric S satisfying $S \neq S_d$ for all metrics d.

Example 1 Let $X = \mathbb{R}$ and define the function

$$S(x,y,z) = |x-z| + |x+z-2y|,$$

for all $x, y, z \in \mathbb{R}$. Then, (X, S) is an S-metric space. Now, we prove that there does not exist any metric d, such that $S = S_d$. Conversely, suppose that there exists a metric d, such that

$$S(x,y,z) = d(x,z) + d(y,z),$$

for all $x, y, z \in \mathbb{R}$. Then, we obtain

$$S(x,x,z) = 2d(x,z) = 2|x-z| \text{ and } d(x,z) = |x-z|$$

and

$$S(y,y,z) = 2d(y,z) = 2|y-z| \text{ and } d(y,z) = |y-z|,$$

for all $x, y, z \in \mathbb{R}$. Hence, we have

$$|x-z| + |x+z-2y| = |x-z| + |y-z|,$$

which is a contradiction. Therefore, $S \neq S_d$.

Now, we give the relationship between the Rhoades' condition (**R25**) and (**S25**).

Proposition 1 *Let (X, d) be a complete metric space, (X, S_d) be the S-metric space obtained by the S-metric generated by d, and T be a self-mapping of X. If T satisfies the inequality* (**R25**), *then T satisfies the inequality* (**S25**).

Proof Let the inequality (**R25**) be satisfied. Using the inequality (**R25**) and (2.1), we have

$$\begin{aligned}S_d(Tx,Tx,Ty) &= d(Tx,Ty)+d(Tx,Ty)=2d(Tx,Ty)\\ &< 2\max\{d(x,y),d(x,Tx),d(y,Ty),d(x,Ty),d(y,Tx)\}\\ &= \max\{2d(x,y),2d(x,Tx),2d(y,Ty),2d(x,Ty),2d(y,Tx)\}\\ &= \max\{S_d(x,x,y),S_d(x,x,Tx),S_d(y,y,Ty),S_d(x,x,Ty),\\ &\quad S_d(y,y,Tx)\}\\ &= \max\{S_d(x,x,y),S_d(Tx,Tx,x),S_d(Ty,Ty,y),S_d(Ty,Ty,x),\\ &\quad S_d(Tx,Tx,y)\},\end{aligned}$$

and so, the inequality (**S25**) is satisfied on (X, S_d). □

Let (X, S) be any S-metric space. In [8], it was shown that every S-metric on X defines a metric d_S on X as follows:

$$d_S(x,y) = S(x,x,y) + S(y,y,x), \qquad (2.2)$$

for all $x, y \in X$. However, the function $d_S(x,y)$ defined in (2.2) does not always define a metric because of the reason that the triangle inequality does not satisfied for all elements of X everywhen. If the S-metric is generated by a metric d on X, then it can be easily seen that the function d_S is a metric on X, especially we have $d_S(x,y) = 4d(x,y)$. However, if we consider an S-metric which is not generated by any metric, then d_S can or cannot be a metric on X. We call this metric d_S as the metric generated by S in the case d_S is a metric.

More precisely, we can give the following examples.

Example 2 Let $X = \{1,2,3\}$ and the function $S : X \times X \times X \to [0,\infty)$ be defined as:

$$\begin{aligned}&S(1,1,2) = S(2,2,1) = 5,\\ &S(2,2,3) = S(3,3,2) = S(1,1,3) = S(3,3,1) = 2,\\ &S(x,y,z) = 0 \text{ if } x = y = z,\\ &S(x,y,z) = 1 \text{ if otherwise},\end{aligned}$$

for all $x, y, z \in X$. Then, the function S is an S-metric which is not generated by any metric and the pair (X, S) is an S-metric space. However, the function d_S defined in (2.2) is not a metric on X. Indeed, for $x = 1$, $y = 2$, and $z = 3$, we get

$$d_S(1,2) = 10 \nleq d_S(1,3) + d_S(3,2) = 8.$$

Example 3 Let $X = \mathbb{R}$ and consider the S-metric defined in Example 1 which is not generated by any metric. Using the Eq. (2.2), we obtain

$$d_S(x,y) = 4|x-y|,$$

for all $x, y \in \mathbb{R}$. Then, $(\mathbb{R}, d_S)$ is a metric space on $\mathbb{R}$.

We give the following proposition.

Proposition 2 *Let (X, S) be a complete S-metric space, (X, d_S) be the metric space obtained by the metric generated by S , and T be a self-mapping of X. If T satisfies the inequality* (**S25**), *then T satisfies the inequality* (**R25**).

Proof Let the inequality (**S25**) be satisfied. Using the inequality (**S25**) and (2.1), we have

$$\begin{aligned}d_S(Tx,Ty) &= S(Tx,Tx,Ty)+S(Ty,Ty,Tx)\\ &= S(Tx,Tx,Ty)+S(Tx,Tx,Ty)=2S(Tx,Tx,Ty)\\ &< 2\max\{S(x,x,y),S(Tx,Tx,x),S(Ty,Ty,y),S(Ty,Ty,x),\\ &\quad S(Tx,Tx,y)\}\\ &= \max\{2S(x,x,y),2S(Tx,Tx,x),2S(Ty,Ty,y),2S(Ty,Ty,x),\\ &\quad 2S(Tx,Tx,y)\}\\ &= \max\{S(x,x,y)+S(y,y,x),S(Tx,Tx,x)+S(x,x,Tx),\\ &\quad S(Ty,Ty,y)+S(y,y,Ty),\\ &\quad S(Ty,Ty,x)+S(x,x,Ty),S(Tx,Tx,y)+S(y,y,Tx)\}\\ &= \max\{d_S(x,y),d_S(x,Tx),d_S(y,Ty),d_S(x,Ty),d_S(y,Tx)\},\end{aligned}$$

and so, the inequality (**R25**) is satisfied on (X, d_S). □

In [13], it was given another forms of (**R25**) as (**R50**), (**R75**), (**R100**), and (**R125**). Now, we extend the forms (**R50**) – (**R125**) for complete S-metric spaces. We can give the following definition.

Definition 3 Let (X, S) be an S-metric space and T be a self-mapping of X. We define (**S50**), (**S75**), (**S100**), and (**S125**), as follows :

(**S50**) There exists a positive integer p, such that

$$\begin{aligned}S(T^px,T^px,T^py) < \max\{&S(x,x,y),S(T^px,T^px,x),S(T^py,T^py,y),\\ &S(T^py,T^py,x),S(T^px,T^px,y)\},\end{aligned}$$

for any $x, y \in X$, $x \neq y$.

(**S75**) There exist positive integers p, q, such that

$$\begin{aligned}S(T^px,T^px,T^qy) < \max\{&S(x,x,y),S(T^px,T^px,x),S(T^qy,T^qy,y),\\ &S(T^qy,T^qy,x),S(T^px,T^px,y)\},\end{aligned}$$

for any $x, y \in X$, $x \neq y$.

(**S100**) For any given $x \in X$, there exists a positive integer p(x), such that

$$\begin{aligned}S(T^{p(x)}x,T^{p(x)}x,T^{p(x)}y) < \max\{&S(x,x,y),S(T^{p(x)}x,T^{p(x)}x,x),\\ &S(T^{p(x)}y,T^{p(x)}y,y),S(T^{p(x)}y,T^{p(x)}y,x),\\ &S(T^{p(x)}x,T^{p(x)}x,y)\},\end{aligned}$$

for any $y \in X$, $x \neq y$.

(**S125**) For any given $x, y \in X$, $x \neq y$, there exists a positive integer $p(x, y)$, such that

$$S(T^{p(x,y)}x, T^{p(x,y)}x, T^{p(x,y)}y) < \max\{S(x,x,y), S(T^{p(x,y)}x, T^{p(x,y)}x, x), S(T^{p(x,y)}y, T^{p(x,y)}y, y), S(T^{p(x,y)}y, T^{p(x,y)}y, x), S(T^{p(x,y)}x, T^{p(x,y)}x, y)\}.$$

Corollary 1 Let (X, d) be a complete metric space, (X, S_d) be the S-metric space obtained by the S-metric generated by d, and T be a self-mapping of X. If T satisfies the inequality (**R50**) [resp. (**R75**), (**R100**), and (**R125**)], then T satisfies the inequality (**S50**) [resp. (**S75**), (**S100**), and (**S125**)].

Corollary 2 Let (X, S) be a complete S-metric space, (X, d_S) be the metric space obtained by the metric generated by S, and T be a self-mapping of X. If Tsatisfies the inequality (**S50**) [resp. (**S75**), (**S100**), and (**S125**)], then T satisfies the inequality (**R50**) [resp. (**R75**), (**R100**), and (**R125**)].

The proof of following proposition is obvious, so it is omitted.

Proposition 3 *Let (X, S) be an S-metric space and T be a self-mapping of X. We obtain the following implications by the Definition* 3:

(**S25**) $\Longrightarrow$ (**S50**) $\Longrightarrow$ (**S75**)and(**S50**) $\Longrightarrow$ (**S100**) $\Longrightarrow$ (**S125**).

The converses of above implications in Proposition 3 are not always true as we have seen in the following examples.

Example 4 Let $\mathbb{R}$ be the real line. It can be easily seen that the following function defines an S-metric on $\mathbb{R}$ different from the usual S-metric defined in [15]:

$$S(x,y,z) = |x-z| + |x+z-2y|$$

for all $x, y, z \in \mathbb{R}$. Let

$$Tx = \begin{cases} 0 & \text{if } x \in [0,1], x \neq \frac{1}{4} \\ 1 & \text{if } x = \frac{1}{4} \end{cases}.$$

Then, T is a self-mapping on the S-metric space [0, 1].

For $x = \frac{1}{2}$, $y = \frac{1}{4}$, we have

$$S(Tx,Tx,Ty) = S(0,0,1) = 2,$$
$$S(x,x,y) = S\left(\frac{1}{2},\frac{1}{2},\frac{1}{4}\right) = \frac{1}{2},$$
$$S(Tx,Tx,x) = S\left(0,0,\frac{1}{2}\right) = 1,$$
$$S(Ty,Ty,y) = S\left(1,1,\frac{1}{4}\right) = \frac{3}{2}$$
$$S(Ty,Ty,x) = S\left(1,1,\frac{1}{2}\right) = 1,$$
$$S(Tx,Tx,y) = S\left(0,0,\frac{1}{4}\right) = \frac{1}{2}$$

and so

$$S(Tx,Tx,Ty) = 2 < \max\left\{\frac{1}{2},1,\frac{3}{2},1,\frac{1}{2}\right\} = \frac{3}{2},$$

which is a contradiction. Then, the inequality (**S25**) is not satisfied.

For each $x, y \in X$ $(x \neq y)$ and $p \geq 2$, T is satisfied the inequality (**S50**).

Example 5 We consider the self-mapping T in the example on page 105 in [3] and the usual S-metric defined in [15]. If we choose $x = (\frac{1}{n}+1, 0)$, $y = (\frac{1}{n}, 0)$ for each n, then the inequality (**S50**) is not satisfied. A positive integer $p(x)$ can be chosen for any given $x \in X$, such that the inequality (**S100**) is satisfied.

Example 6 Let $\mathbb{R}$ be the real line. Let us consider the S-metric defined in Example 4 on $\mathbb{R}$ and let

$$Tx = \begin{cases} 0 & \text{if } x \in \left[\frac{1}{2}, 1\right] \\ 1 & \text{if } x \in \left[0, \frac{1}{2}\right) \end{cases}.$$

Then, T is a self-mapping on the S-metric space [0, 1].

Let us choose $x = 0$ and $y = 1$.

For $p = 1$, we have

$$S(Tx,Tx,Ty) = S(1,1,0) = 2,$$
$$S(x,x,y) = S(0,0,1) = 2,$$
$$S(Tx,Tx,x) = S(1,1,0) = 2,$$
$$S(Ty,Ty,y) = S(0,0,1) = 2,$$
$$S(Ty,Ty,x) = S(0,0,0) = 0,$$
$$S(Tx,Tx,y) = S(1,1,1) = 0$$

and so

$$S(Tx,Tx,Ty) = 2 < \max\{2,2,2,0,0\} = 2,$$

which is a contradiction. Then, the inequality (**S50**) is not satisfied.

For $p = 2$, we have

$$S(T^2x,T^2x,T^2y) = S(0,0,1) = 2,$$
$$S(x,x,y) = S(0,0,1) = 2,$$
$$S(T^2x,T^2x,x) = S(0,0,0) = 0,$$
$$S(T^2y,T^2y,y) = S(1,1,1) = 0,$$
$$S(T^2y,T^2y,x) = S(1,1,0) = 2,$$
$$S(T^2x,T^2x,y) = S(0,0,1) = 2$$

and so

$$S(T^2x,T^2x,T^2y) = 2 < \max\{2,0,0,2,2\} = 2,$$

which is a contradiction. Then, the inequality (**S50**) is not satisfied.

For $p \geq 3$ using similar arguments, we can see that the inequality (**S50**) is not satisfied.

We now show that the inequality (**S75**) is satisfied under the following four cases:

Case 1 We take $x \in [0, \frac{1}{2})$, $y \in [\frac{1}{2}, 1]$, $p = 2$, and $q = 1$. Then, the inequality (**S75**) is satisfied, since

$$S(T^2x, T^2x, Ty) = 0,$$

for $x \in [0, \frac{1}{2})$, $y \in [\frac{1}{2}, 1]$, $x \neq y$.

Case 2 We take $y \in [0, \frac{1}{2})$, $x \in [\frac{1}{2}, 1]$, $p = 2$, and $q = 1$. Then, using similar arguments in Case 1, we can see that the inequality (**S75**) is satisfied.

Case 3 We take $x, y \in [0, \frac{1}{2})$, $p = 2$, and $q = 2$. Then, the inequality (**S75**) is satisfied, since

$$S(T^2x, T^2x, T^2y) = 0,$$

for $x, y \in [0, \frac{1}{2})$, $x \neq y$.

Case 4 We take $x, y \in [\frac{1}{2}, 1]$, $p = 2$, and $q = 2$. Then, using similar arguments in Case 3, we can see that the inequality (**S75**) is satisfied.

Example 7 Let $\mathbb{R}$ be the S-metric space with the S-metric defined in Example 4 and let

$$Tx = \begin{cases} \sqrt{x} & \text{if } x \in [0, 1], x \neq \frac{1}{2}, x \neq \frac{1}{3} \\ \frac{1}{3} & \text{if } x = \frac{1}{2} \\ 3 & \text{if } x = \frac{1}{3} \\ \frac{1}{2} & \text{if } x = 3 \end{cases}.$$

Then, T is a self-mapping on the S-metric space $[0, 1] \cup \{3\}$.

The inequality (**S100**) is not satisfied, since there is not a positive integer $p(x)$ for any given $x \in X$, such that T is satisfied the inequality (**S100**) for any $y \in X$, $x \neq y$. However, for any given $x, y \in X$, $x \neq y$, there exists a positive integer $p(x, y)$, such that the inequality (**S125**) is satisfied.

Remark 1 (**S75**) and (**S100**) are independent of each other by Examples 5 and 6.

Some fixed point theorems on *S*-metric spaces

In this section, we give some fixed point theorems by means of periodic points on S-metric spaces for the contractive mappings defined in Sect. 3.

Theorem 1 *Let (X, S) be an S-metric space* and *Tbe a self-mapping of X which satisfies the inequality* (**S125**). *If Thas a fixed point, then it is unique.*

Proof Suppose that x and y are fixed points of T, such that $x, y \in X$ $(x \neq y)$. Then, there exists a positive integer $p = p(x, y)$, such that

$$\begin{aligned} S(T^px, T^px, T^py) &< \max\{S(x,x,y), S(T^px, T^px, x), S(T^py, T^py, y), \\ &\quad S(T^py, T^py, x), S(T^px, T^px, y)\} \\ &= \max\{S(x,x,y), 0, 0, S(y,y,x), S(x,x,y)\} \\ &= S(x,x,y), \end{aligned}$$

by the inequality (**S125**). Then, using Lemma 1 and the fact that $T^px = x$, $T^py = y$, we obtain

$$S(T^px, T^px, T^py) = S(x,x,y) < S(x,x,y).$$

Thus, the assumption that x and y are fixed points of T has led to a contradiction. Consequently, the fixed point is unique. □

Corollary 3 Let (X, S) be an S-metric space, T be a self-mapping of X, and the inequality (**S25**) [resp. $T \in$ (**S50**), $T \in$ (**S100**)] be satisfied. If T has a fixed point, then it is unique.

Proof It can be seen from Proposition 3. □

Corollary 4 Let (X, S) be an S-metric space, T be a self-mapping of X, and the inequality (**S75**) be satisfied. If T has a fixed point, then it is unique.

Proof By a similar argument used in the proof of Theorem 1, the proof can be easily seen by the definition of (**S75**). □

Now, we recall the following definitions and corollary.

Definition 4 [14] Let (X, S) be an S-metric space and $A \subset X$. Then, A is called S-bounded if there exists $r > 0$, such that $S(x, x, y) < r$ for all $x, y \in A$.

Definition 5 [4] Let (X, S) be an S-metric space, T be a self-mapping of X, and $x \in X$. A point x is called a periodic point of T, if there exists a positive integer n, such that

$$T^nx = x. \tag{3.1}$$

The least positive integer satisfying the condition (3.1) is called the periodic index of x.

Definition 6 [10] Let (X, S) be an S-metric space, T, F be two self-mappings of X, and $A \subset X$, $x \in X$. Then

1. $\delta(A) = \sup\{S(x,x,y) : x, y \in A\}$.
2. $O_{T,F}(x, n) = \{Tx, TFx, TF^2x, \ldots, TF^nx\}$.
3. $O_{T,F}(x, \infty) = \{Tx, TFx, TF^2x, \ldots, TF^nx, \ldots\}$.
4. If T is identify, then $O_F(x, n) = O_{T,F}(x, n)$ and $O_F(x, \infty) = O_{T,F}(x, \infty)$.

Let A be a nonempty subset of X. In [12], it was called $\delta(A)$ as the diameter of A and we write

$$\delta(A) = diam\{A\} = sup\{S(x,x,y) : x,y \in A\}.$$

If A is S-bounded, then we will write $\delta(A) < \infty$.

The following corollary is a generalization of [6, Theorem1] into the structure of S-metric in [5].

Corollary 5 [10] Let (X, S) be an S-metric space and T be a self-mapping of X, such that

(1) Every Cauchy sequence of the form $\{T^n x\}$ is convergent in X for all $x \in X$;
(2) There exists $h \in [0, 1)$, such that

$$S(Tx, Tx, Ty) \leq h \max\{S(x, x, y), S(Tx, Tx, x), S(Tx, Tx, y), S(Ty, Ty, x), S(Ty, Ty, y)\},$$

for each $x, y \in X$.

Then

1. $\delta(T^i x, T^i x, T^j x) \leq h\delta[O_T(x, n)]$ for all $i,j \leq n$, $n \in \mathbb{N}$ and $x \in X$;
2. $\delta[O_T(x, \infty)] \leq \frac{2}{1-h} S(Tx, Tx, x)$ for all $x \in X$;
3. T has a unique fixed point x_0;
4. $\lim_{n\to\infty} T^n x = x_0$.

Theorem 2 *Let (X, S) be an S-metric space, T be a self-mapping of X, the inequality* (**S125**) *be satisfied, and $x \in X$. Assume that xis a periodic point of Twith periodic index m. Then, Thas a fixed point x in $\{T^n x\}(n \geq 0)$if and only if for any $T^{n_1}x$, $T^{n_2}x \in \{T^n x\}(n \geq 0)$, $T^{n_1}x \neq T^{n_2}x$, there exist $T^{n_3}x$, $T^{n_4}x \in \{T^n x\}$, such that*

$$T^{p(T^{n_3}x, T^{n_4}x)}(T^{n_3}x) = T^{n_1}x \mathrm{and} \mathrm{T}^{\mathrm{p}(\mathrm{T}^{n_3}\mathrm{x},\mathrm{T}^{n_4}\mathrm{x})}(\mathrm{T}^{n_4}\mathrm{x}) = \mathrm{T}^{n_2}\mathrm{x}.$$

Then, the point x is the unique fixed point of T in X.

Proof The proof of the if part of the theorem is obvious. Therefore, we prove the only if part. If x is a periodic point of T with periodic index m, then we have

$$\{T^n x\} = \{x, Tx, \ldots, T^{m-1}x\}.$$

If $x \neq Tx$, then there exist $T^{n_1}x$, $T^{n_2}x \in \{T^n x\}$, $T^{n_1}x \neq T^{n_2}x$, such that

$$\delta(\{T^n x\}) = \max_{0 \leq k,l \leq m-1, k \neq l} \{S(T^k x, T^k x, T^l x)\} = S(T^{n_1}x, T^{n_1}x, T^{n_2}x).$$

By the hypothesis, there exist $T^{n_3}x$, $T^{n_4}x \in \{T^n x\}$, such that

$$T^{p(T^n x, T^{n_4}x)}(T^{n_3}x) = T^{n_1}x \text{ and } T^{p(T^{n_3}x, T^{n_4}x)}(T^{n_4}x) = T^{n_2}x.$$

Since $T^{n_1}x \neq T^{n_2}x$, we obtain $T^{n_3}x \neq T^{n_4}x$. Hence, we have

$$\begin{aligned}\delta(\{T^n x\}) &= S(T^{n_1}x, T^{n_1}x, T^{n_2}x)\\ &= S(T^{p(T^{n_3}x,T^{n_4}x)}(T^{n_3}x), T^{p(T^{n_3}x,T^{n_4}x)}(T^{n_3}x), T^{p(T^{n_3}x,T^{n_4}x)}(T^{n_4}x))\\ &< \max\{S(T^{n_3}x, T^{n_3}x, T^{n_4}x), S(T^{n_1}x, T^{n_1}x, T^{n_3}x),\\ &\quad S(T^{n_2}x, T^{n_2}x, T^{n_4}x),\\ &\quad S(T^{n_2}x, T^{n_2}x, T^{n_3}x), S(T^{n_1}x, T^{n_1}x, T^{n_4}x)\}\\ &\leq \delta(\{T^n x\}),\end{aligned}$$

which is a contradiction, and so, we have $x = Tx$. It is obvious that x is unique fixed point of T in X by Theorem 1. □

Corollary 6 Let (X, S) be an S-metric space, T be a self-mapping of X, the inequality (**S100**) be satisfied, and $x \in X$ be a periodic point of T. Then, the following conditions are equivalent:

(1) T has a unique fixed point in $\{T^n x\}(n \geq 0)$,
(2) There exists $T^{n_0}x \in \{T^n x\}(n \geq 0)$, such that

$$T^{p(T^{n_0}x)}(T^{n_0}x) = T^{n_1}x,$$

for any $T^{n_1}x \in \{T^n x\}(n \geq 0)$, where $p(T^{n_0}x)$ is the positive integer.

Then, the point x is the unique fixed point of T in X.

Corollary 7 Let (X, S) be an S-metric space, T be a self-mapping of X, the inequality (**S75**) be satisfied, and $x \in X$ be a periodic point of T. Then, x is the unique fixed point of T if there exist $T^{n_3}x$, $T^{n_4}x \in \{T^n x\}(n \geq 0)$, and $T^{n_3}x \neq T^{n_4}x$, such that

$$T^p(T^{n_3}x) = T^{n_1}x \text{ and } T^q(T^{n_4}x) = T^{n_2}x,$$

for any $T^{n_1}x$, $T^{n_2}x \in \{T^n x\}(n \geq 0)$, $T^{n_1}x \neq T^{n_2}x$. Here, p and q are the positive integers.

Corollary 8 Let (X, S) be an S-metric space, T be a self-mapping of X, and the inequality (**S50**) be satisfied. Then, the following conditions are equivalent:

(1) T has a fixed point in X,
(2) There exists a periodic point $x \in X$ of T.

Then, the point x is the unique fixed point of T in X.

We give some sufficient conditions to guarantee the existence of fixed point for a self-mapping T satisfying the inequality (**S75**) in the following theorem.

Theorem 3 *Let (X, S) be an S-metric space, T be a self-mapping of X, the inequality* (**S75**) *be satisfied, and $x \in X$be a periodic point of T with periodic index m. Suppose that p and q are the positive integers and also the following conditions are satisfied:*

1. $p = p_1 m + p_2$, $q = q_1 m + q_2$, $0 \leq p_2, q_2 < m$, and p_1 and q_1 are non-negative integers.
2. $2|p_2 - q_2| \neq m$.

Then, the point x is the unique fixed point of T in X.

Proof We now show that x is the fixed point of T in X. On the contrary, assume that x is not the fixed point of T. Let

$A = \{T^n x\} = \{x, Tx, T^2x, \ldots, T^n x, \ldots\}.$

Since the periodic index of x is m, we have

$A = \{T^n x\} = \{x, Tx, T^2x, \ldots, T^{m-1}x\}$

and the elements in A are distinct. Therefore, there exist i, j, such that $0 \leq i < j < m$ and

$$\delta(A) = \max_{0 \leq k,l \leq m-1, k \neq l} S(T^k x, T^k x, T^l x) = S(T^i x, T^i x, T^j x).$$

We can assume that $p_2 \geq q_2$. In addition, we have $T^n(A) = A$ for any non-negative integer n. Therefore, there exist $T^{n_1}x$ and $T^{n_2}x \in A$, such that

$$T^i x = T^{p_2}(T^{n_1}x) \text{ and } T^j x = T^{q_2}(T^{n_2}x). \tag{3.2}$$

Similarly, there exist $T^{n_3}x$ and $T^{n_4}x \in A$, such that

$$T^i x = T^{q_2}(T^{n_3}x) \text{ and } T^j x = T^{p_2}(T^{n_4}x). \tag{3.3}$$

We prove that at least one of the statements $n_1 \neq n_2$ and $n_3 \neq n_4$ is true.

Suppose that $n_3 = n_4$. Since

$0 \leq i, j, p_2, q_2, n_1, n_2, n_3, n_4 < m,$

using (3.2) and (3.3), there exist $a, b, c, d \in \{0, 1\}$, such that

$$p_2 + n_1 = am + i, q_2 + n_2 = bm + j, \tag{3.4}$$

$$q_2 + n_3 = cm + i, p_2 + n_4 = dm + j. \tag{3.5}$$

If $n_1 = n_2$, we have $am + i \geq bm + j$, since $p_2 \geq q_2$. Since $i < j$, we have $a = 1$, $b = 0$. It follows from (3.4) that

$$(p_2 - q_2) + (j - i) = m. \tag{3.6}$$

Using the condition (3.5) and $n_3 = n_4$, we obtain

$$(p_2 - q_2) = (d - c)m + (j - i). \tag{3.7}$$

Since $0 \leq p_2 - q_2 \leq m - 1$, $0 \leq j - i < m$, we have $d - c = 0$ using the condition (3.7), and so, $p_2 - q_2 = j - i$.

By the condition (3.6), we have

$2(p_2 - q_2) = m,$

which is a contradiction. Hence, it should be $n_1 \neq n_2$. Then, $T^{n_1}x \neq T^{n_2}x$. Using $T^{p_2}(x) = T^p x$ and $T^{q_2}(x) = T^q x$, we obtain

$$\begin{aligned}\delta(A) &= S(T^i x, T^i x, T^j x) \\ &= S(T^{p_2}(T^{n_1}x), T^{p_2}(T^{n_1}x), T^{q_2}(T^{n_2}x)) = S(T^p(T^{n_1}x), \\ &\quad T^p(T^{n_1}x), T^q(T^{n_2}x)) \\ &< \max\{S(T^{n_1}x, T^{n_1}x, T^{n_2}x), S(T^p(T^{n_1}x), T^p(T^{n_1}x), T^{n_1}x), \\ &\quad S(T^q(T^{n_2}x), T^q(T^{n_2}x), T^{n_2}x), S(T^q(T^{n_2}x), T^q(T^{n_2}x), T^{n_1}x), \\ &\quad S(T^p(T^{n_1}x), T^p(T^{n_1}x), T^{n_2}x)\} \leq \delta(A),\end{aligned}$$

which is a contradiction. Consequently, $x = Tx$.

Similarly, it can be seen that if $n_1 = n_2$, then it should be $n_3 \neq n_4$, and hence, we get $x = Tx$.

It is obvious that x is the unique fixed point of T in X by Corollary 4. □

Some applications of contractive mappings on S-metric spaces

The following corollary was given in [15] on page 123 by Sedghi and Dung.

Corollary 9 [15] *Let (X, S) be a complete S -metric space, T be a self-mapping of X, and*

$$\begin{aligned}S(Tx, Tx, Ty) \leq h\max\{&S(x, x, y), S(Tx, Tx, x), S(Tx, Tx, y), \\ &S(Ty, Ty, x), S(Ty, Ty, y)\},\end{aligned} \tag{4.1}$$

for some $h \in [0, \frac{1}{3})$ and each $x, y \in X$. Then, T has a unique fixed point in X. In addition, T is continuous at this fixed point.

We call the inequality (4.1) as (**Q25**) in Corollary as follows:

There exists a number h with $h \in [0, \frac{1}{3})$, such that

(**Q25**) $S(Tx, Tx, Ty) \leq h\max\{S(x, x, y), S(Tx, Tx, x), S(Tx, Tx, y), S(Ty, Ty, x), S(Ty, Ty, y)\},$

for any $x, y \in X$.

In this section, we study fixed point theorems using the inequality (**Q25**) on S-metric spaces. Finally, we obtain a fixed point theorem for a self-mapping T of a compact S-metric space X satisfying the inequality (**S25**).

Now, we give the definition of T_S-orbitally complete space.

Definition 7 Let (X, S) be an S-metric space and T be a self-mapping of X. Then, an S-metric space X is said to be T_S-orbitally complete if and only if every Cauchy sequence which is contained in the sequence $\{x, Tx, \ldots, T^n x, \ldots\}$ for some $x \in X$ converges in X.

Theorem 4 *Let (X, S) be T_S-orbitally complete, T be a self-mapping of X, and the inequality* (**Q25**) *be satisfied. Then, T has a unique fixed point in X.*

Proof It is obvious from Corollary 5. □

Now, we will extend the definition (**Q25**) on an S-metric space as follows:

(**Q25a**) $S(T^p x, T^p x, T^q y) \leq h \max\{S(T^{r_1}x, T^{r_1}x, T^{s_1}y), S(T^{r_1}x, T^{r_1}x, T^{r_2}x), S(T^{s_1}y, T^{s_1}y, T^{s_2}y) : 0 \leq r_1, r_2 \leq p \text{ and } 0 \leq s_1, s_2 \leq q\},$

for each $x, y \in X$, some fixed positive integers p and q. Here, $h \in [0, \frac{1}{2})$.

The following theorems are the generalizations of the fixed point theorems given in [7] to an S-metric space (X, S).

Theorem 5 *Let (X, S) be a complete S-metric space, Tbe a continuous self-mapping of X, and the inequality* (**Q25a**) *be satisfied. Then, T has a unique fixed point in X.*

Proof Without loss of generality, we assume that $h \in [\frac{1}{3}, \frac{1}{2})$. Then, we have $\dfrac{h}{1-2h} \geq 1$. Suppose that $p \geq q$.

Let $x \in X$ and assume that the sequence $\{T^n x : n = 1, 2, \ldots\}$ is unbounded. Then, clearly, the sequence

$$\{S(T^n x, T^n x, T^q x) : n = 1, 2, \ldots\}$$

is unbounded. Hence, there exists an integer n, such that

$$S(T^n x, T^n x, T^q x) > \frac{h}{1-2h} \max\{S(T^i x, T^i x, T^q x) : 0 \leq i \leq p\}.$$

Suppose that m is the smallest such n. Clearly, we have $m > p \geq q$. Therefore

$$\begin{aligned} S(T^m x, T^m x, T^q x) &> \frac{h}{1-2h} \max\{S(T^i x, T^i x, T^q x) : 0 \leq i \leq p\} \\ &\geq \max\{S(T^{r_1}x, T^{r_1}x, T^q x) : 0 \leq r_1 < m\}. \end{aligned} \tag{4.2}$$

Using (4.2), we obtain

$$\begin{aligned} (1-2h)S(T^m x, T^m x, T^q x) &> h \max\{S(T^i x, T^i x, T^q x) : 0 \leq i \leq p\} \\ &\geq h \max\{S(T^i x, T^i x, T^{r_1}x) - 2S(T^{r_1}x, T^{r_1}x, T^q x) : 0 \leq i \leq p \text{ and } 0 \leq r_1 < m\} \\ &\geq h \max\{S(T^i x, T^i x, T^{r_1}x) - 2S(T^m x, T^m x, T^q x) : 0 \leq i \leq p \text{ and } 0 \leq r_1 < m\} \end{aligned}$$

and then

$$S(T^m x, T^m x, T^q x) > h \max\{S(T^i x, T^i x, T^{r_1}x) : 0 \leq i \leq p \text{ and } 0 \leq r_1 < m\}. \tag{4.3}$$

Now, we prove that

$$S(T^m x, T^m x, T^q x) > h \max\{S(T^i x, T^i x, T^{r_1}x) : 0 \leq i, r_1 < m\} \tag{4.4}$$

For if not

$$S(T^m x, T^m x, T^q x) \leq h \max\{S(T^i x, T^i x, T^{r_1}x) : 0 \leq i, r_1 < m\}$$

and so using (4.3)

$$S(T^m x, T^m x, T^q x) \leq h \max\{S(T^i x, T^i x, T^{r_1}x) : p < i, r_1 < m\}. \tag{4.5}$$

Using the inequality (**Q25a**), we can write

$$S(T^m x, T^m x, T^q x) \leq h^k \max\{S(T^i x, T^i x, T^{r_1}x) : p < i, r_1 < m\}$$

for $k = 1, 2, \ldots$, since we can omitted the terms of the form as $S(T^i x, T^i x, T^{r_1}x)$ with $0 \leq i \leq p$ by (4.3).

Now, we get $S(T^m x, T^m x, T^q x) = 0$ for $k \to \infty$, which is a contradiction by our assumption. Therefore, we obtain the inequality (4.4).

However, using the inequality (**Q25a**), we have

$$\begin{aligned} S(T^m x, T^m x, T^q x) &\leq h \max\{S(T^{r_1}x, T^{r_1}x, T^{s_1}x), S(T^{r_1}x, T^{r_1}x, T^{r_2}x), \\ &\quad S(T^{s_1}x, T^{s_1}x, T^{s_2}x) : m - p \leq r_1, r_2 \leq m \text{ and } 0 \leq s_1, s_2 \leq q\} \\ &\leq h \max\{S(T^{r_1}x, T^{r_1}x, T^{s_1}x) : 0 \leq r_1, s_1 \leq m\}, \end{aligned}$$

which is a contradiction from (4.4). Then, the sequence $\{T^n x : n = 1, 2, \ldots\}$ should be S-bounded.

Now, we put $N = \sup\{S(T^{r_1}x, T^{r_1}x, T^{s_1}x) : r_1, s_1 = 0, 1, 2, \ldots\} < \infty$. Therefore, for arbitrary $\varepsilon > 0$, choose M, so that $h^M N < \varepsilon$. For $m, n \geq M \max\{p, q\}$ and using the inequality (**Q25a**) M times, we have

$$S(T^m x, T^m x, T^n x) \leq h^M N < \varepsilon.$$

Hence, the sequence $\{T^n x : n = 1, 2, \ldots\}$ is a Cauchy sequence in the complete S-metric space (X, S) and so has a limit x_0 in X. Since T is continuous, we have $Tx_0 = x_0$ and then x_0 is a fixed point of T. It can be easily seen that the point x_0 is a unique fixed point of T. Then, the proof is completed. □

From the inequality (**Q25a**), for $q = 1$ (or $p = 1$), we define the following generalization of (**Q25**):

(**Q25b**) $S(T^p x, T^p x, Ty) \leq h \max\{S(T^{r_1}x, T^{r_1}x, T^s y), S(T^{r_1}x, T^{r_1}x, T^{r_2}x), S(Ty, Ty, y) : 0 \leq r_1, r_2 \leq p \text{ and } s = 0, 1\},$

for each $x, y \in X$, some fixed positive integer p. Here, $h \in [0, \frac{1}{2})$.

The condition that the self-mapping T be continuous is not necessary when the inequality (**Q25b**) is satisfied as we have seen in the following theorem.

Theorem 6 *Let (X, S) be a complete S-metric space and T be a self-mapping of X satisfying the inequality* (**Q25b**). *Then, T has a unique fixed point in X.*

Proof Let $x \in X$. Then, the sequence $\{T^n x : n = 1, 2, \ldots\}$ is a Cauchy sequence in the complete S-metric space X as we have seen in the proof of Theorem 5. Hence, the sequence has a limit x_0 in X. For $n \geq p$, we obtain

$$S(T^n x, T^n x, Tx_0) \leq h \max\{S(T^{r_1}x, T^{r_1}x, T^s x_0), S(T^{r_1}x, T^{r_1}x, T^{r_2}x), S(Tx_0, Tx_0, x_0) : n - p \leq r_1, r_2 \leq n \text{ and } s = 0, 1\}.$$

Then, by (2.1), we have

$$\begin{aligned} S(x_0, x_0, Tx_0) &= S(Tx_0, Tx_0, x_0) \\ &\leq h \max\{S(T^s x_0, T^s x_0, x_0) : s = 0, 1\} \\ &= hS(Tx_0, Tx_0, x_0), \end{aligned}$$

when n goes to infinity. Since $h < 1$, we have $Tx_0 = x_0$. Then, the proof is completed. □

Corollary 10 *Let (X, S) be a complete S-metric space and T be a self-mapping of X satisfying the inequality* (**Q25**). *Then, Thas a unique fixed point in X.*

Remark 2 The condition that T be continuous when $p, q \geq 2$ is necessary in Theorem 5. The following example shows that Theorem 5 cannot be always true when T is a discontinuous self-mapping of X.

Example 8 Let $\mathbb{R}$ be the real line. Let us consider the S-metric defined in Example on $\mathbb{R}$ and let

$$Tx = \begin{cases} 1 & \text{if } x = 0 \\ \frac{x}{4} & \text{if } x \neq 0 \end{cases}.$$

Then, T is a discontinuous self-mapping on the complete S-metric space [0, 1]. For each $x, y \in X$, we obtain

$$S(T^p x, T^p x, T^q y) = \frac{1}{4} S(T^{p-1}x, T^{p-1}x, T^{q-1}y)$$

and so the inequality (**Q25a**) is satisfied with $h = \frac{1}{4}$. However, T has not a fixed point.

Now, we consider compact S-metric spaces and prove the following theorem.

Theorem 7 *Let (X, S) be a compact S-metric space and T be a continuous self-mapping of X satisfying*

$$S(T^p x, T^p x, T^q y) < \max\{S(T^{r_1}x, T^{r_1}x, T^{s_1}y), S(T^{r_1}x, T^{r_1}x, T^{r_2}x), S(T^{s_1}y, T^{s_1}y, T^{s_2}y) : 0 \leq r_1, r_2 \leq p \text{ and } 0 \leq s_1, s_2 \leq q\} \tag{4.6}$$

for each $x, y \in X$. Here, the right-hand side of (4.6) *is positive. Then, T has a unique fixed point in X.*

Proof Let the inequality (**Q25a**) be satisfied. Then, T has a unique fixed point in X from Theorem 5.

Let the inequality (**Q25a**) be not satisfied. If $\{h_n : n = 1, 2, \ldots\}$ is a monotonically increasing sequence of numbers converging to 1, then there exist sequences $\{x_n : n = 1, 2, \ldots\}$ and $\{y_n : n = 1, 2, \ldots\}$ in X, such that

$$S(T^p x_n, T^p x_n, T^q y_n) > h_n \max\{S(T^{r_1}x_n, T^{r_1}x_n, T^{s_1}y_n), S(T^{r_1}x_n, T^{r_1}x_n, T^{r_2}x_n), S(T^{s_1}y_n, T^{s_1}y_n, T^{s_2}y_n) : 0 \leq r_1, r_2 \leq p \text{ and } 0 \leq s_1, s_2 \leq q\}$$

for $n = 1, 2, \ldots$. Using compactness of X, there exist subsequences $\{x_{n_k} : k = 1, 2, \ldots\}$ and $\{y_{n_k} : k = 1, 2, \ldots\}$ of $\{x_n\}$ and $\{y_n\}$ converging to x and y, respectively. Since T is continuous self-mapping, for $k \to \infty$, we have

$$S(T^p x, T^p x, T^q y) \geq \max\{S(T^{r_1}x, T^{r_1}x, T^{s_1}y), S(T^{r_1}x, T^{r_1}x, T^{r_2}x), S(T^{s_1}y, T^{s_1}y, T^{s_2}y) : 0 \leq r_1, r_2 \leq p \text{ and } 0 \leq s_1, s_2 \leq q\},$$

which is a contradiction unless $Tx = x = y$. Then, T has a fixed point x. It can be easily seen that x is the unique fixed point. □

We have the following corollary for $p = q = 1$.

Corollary 11 *Let (X, S) be a compact S-metric space and T be a continuous self-mapping of X satisfying the inequality* (**S25**). *Here, the right-hand side of the inequality* (**S25**) *is positive. Then, T has a unique fixed point in X.*

Acknowledgements The authors are very grateful to the referee for his/her critical comments.

References

1. Aghajani, A., Abbas, M., Roshan, J.R.: Common fixed point of generalized weak contractive mappings in partially ordered b-metric spaces. Math. Slovaca **64**(4), 941–960 (2014)
2. An, T.V., Dung, N.V., Hang, V.T.L.: A new approach to fixed point theorems on G-metric spaces. Topol. Appl. **160**(12), 1486–1493 (2013)
3. Bailey, D.F.: Some Theorems on contractive mappings. J. London Math. Soc. **41**, 101–106 (1996)
4. Chang, S.S., Zhong, Q.C.: On Rhoades' open questions. Proc. Am. Math. Soc. **109**(1), 269–274 (1990)
5. Lj. Ciric B.: A generalization of Banach's contraction principle. Proc. Am. Math. Soc. 45(2), 267–273 (1974)
6. Dung, N.V., Hieu, N.T., Radojevic, S.: Fixed point theorems for g-monotone maps on partially ordered S-metric spaces. Filomat **28**(9), 1885–1898 (2014)
7. Fisher, B.: Quasi-contractions on metric spaces. Proc. Am. Math. Soc. **75**(2), 321–325 (1979)
8. Gupta, A.: Cyclic contraction on S-metric space. Int. J. Anal. Appl. **3**(2), 119–130 (2013)
9. Gupta, V., Deep, R.: Some coupled fixed point theorems in partially ordered S-metric spaces. Miskolc Math. Notes **16**(1), 181–194 (2015)
10. Hieu, N.T., Ly, N.T., Dung, N.V.: A Generalization of Ciric Quasi-Contractions for Maps on S-Metric Spaces. Thai J. Math. **13**(2), 369–380 (2015)
11. Özgür, N.Y., Taş, N.: Some generalizations of fixed point theo-

rems on S-metric spaces. Essays in Mathematics and Its Applications in Honor of Vladimir Arnold, New York, Springer (2016)

12. Özgür N.Y., Taş N.A.: Some Fixed Point Theorems on S-Metric Spaces, submitted for publication
13. Rhoades, B.E.: A Comparison of various definitions of contractive mappings. Trans. Am. Math. Soc. **226**, 257–290 (1977)
14. Sedghi, S., Shobe, N., Aliouche, A.: A generalization of fixed point theorems in S-metric spaces. Mat. Vesnik **64**(3), 258–266 (2012)
15. Sedghi, S., Dung, N.V.: Fixed point theorems on S-metric spaces. Mat. Vesnik **66**(1), 113–124 (2014)

8

Upper bound of fractional differential operator related to univalent functions

Adem Kılıçman[1] · Rabha W. Ibrahim[2] · Zainab E. Abdulnaby[1,3]

Abstract In this article, we defined the generalized fractional differential Tremblay operator in the open unit disk that by usage the definition of the generalized Srivastava–Owa operator. In particular, we established a new operator denoted by $\Theta_z^{\beta,\tau,\gamma}$ based on the normalized generalized fractional differential operator and represented by convolution product. Moreover, we studied the coefficient criteria of univalence, starlikeness and convexity for the last operator mentioned.

Keywords Univalent function · Subclasses of univalent function · Generalized fractional differential operator · Convolution

Mathematics Subject Classification 30C45 · 30C55

Introduction

Let $\mathcal{A}(m)$ denoted the class of functions $\psi(z)$ of the form:

$$\psi(z) = z + \sum_{\kappa=m+1}^{\infty} a_\kappa z^\kappa \tag{1}$$

which are analytic and univalent functions in the open unit disk

$$\mathbb{U} := \{z \in \mathbb{C} : |z| < 1\}$$

and we set $\mathcal{A}(1) \equiv \mathcal{A}$ when $m = 1$. Let $\mathcal{S}(m)$ denoted the subclass of $\mathcal{A}(m)$ representing of all the univalent functions (or schlict or one-to-one) in $\mathbb{U}$ for $\psi(0) = 0$ and $\psi'(0) = 1$. The functions $\psi(z) \in \mathcal{A}(m)$ are said to be the starlike functions of order $\lambda\,(0 \le \lambda < 1)$ in $\mathbb{U}$, if it satisfies the form

$$\Re\left\{\frac{z\psi'(z)}{\psi(z)}\right\} > \lambda \quad (z \in \mathbb{U}), \tag{2}$$

we note that $\mathcal{S}_\lambda^*(m) \subseteq \mathcal{S}_0^*(m) \equiv \mathcal{S}^*(m) \subseteq \mathcal{S}(m)$. Otherwise, The functions $\psi(z) \in \mathcal{S}(m)$ are said to be convex functions of order$\lambda\,(0 \le \lambda < 1)$, if it satisfies the form

$$\Re\left\{\frac{z\psi''(z)}{\psi'(z)} + 1\right\} > \lambda \quad (z \in \mathbb{U}) \tag{3}$$

which symbolized by $\mathcal{K}_\lambda(m) \subseteq \mathcal{K}_0(m) \equiv \mathcal{K}(m)$ and $\mathcal{K}_\lambda(m) \subseteq \mathcal{S}_\lambda^*(m)$. The classes $\mathcal{S}_\lambda^*(m)$ and $\mathcal{K}_\lambda(m)$ have been discussed by many researchers (see [1, 2]). For $m = 1$, the classes $\mathcal{K}_\lambda(1)$ and $\mathcal{S}_\lambda^*(1)$ of order $\lambda\,(0 \le \lambda < 1)$ were studied before by Robertson [3], and by setting $\lambda = 0$, they are represented as equivalent form:

$$\mathcal{K}_\lambda(1) \equiv \mathcal{K}_\lambda \subseteq \mathcal{K}_0 \equiv \mathcal{K} \quad \text{and} \quad \mathcal{S}_\lambda^*(1) \equiv \mathcal{S}_\lambda^* \subseteq \mathcal{S}_0^* \equiv \mathcal{S}^*.$$

Theorem 1 (Bieberbach's Conjecture [4, 5]) *The functions $\psi(z)$ which is defined in (1), is the univalent function in class $\mathcal{S}(1)$, if $|a_\kappa| \le \kappa$ for all $\kappa \ge 2$ and its convex functions in the class $\mathcal{K}(1)$ if $|a_\kappa| \le 1$.*

Next, the concept of convolution (or Hadamard product) for two analytic and univalent functions $\psi(z)$ given by (1) and $h(z) = z + \sum_{\kappa=m+1}^{\infty} b_\kappa z^\kappa$, $m = \{1, 2, 3, \dots\}$ defined by

$$\psi * h(z) = z + \sum_{\kappa=m+1}^{\infty} a_\kappa b_\kappa z^\kappa. \tag{4}$$

✉ Adem Kılıçman
akilic@upm.edu.my; akilicman@yahoo.com

1 Department of Mathematics, Universiti Putra Malaysia (UPM), 43400 Serdang, Selangor, Malaysia

2 Faculty of Computer Science and Information Technology, University of Malaya, 50603 Kuala Lumpur, Malaysia

3 Department of Mathematics College of Science, Al-Mustansiriyah University, Baghdad, Iraq

Let us here recall some the well known geometric properties for the convolution (or Hadamard product) due to Ruscheweyh (see [6]).

Lemma 1 *From* [6, 7], *we have*

(1) *For the functions* $\psi(z)$ *and* $h(z) \in \mathcal{A}(m)$, *and c a constant, then we have*

$$c(\psi * h)(z) = c\psi * h(z) = \psi * c\,h(z).$$

(2) *The derivative convolution of two functions belong to the class* $\mathcal{A}(m)$ *is defined as:*

$$\begin{aligned} z(h*\psi)'(z) &= h * z\psi'(z) \\ &= z + \sum_{\kappa=m+1}^{\infty} \kappa a_\kappa b_\kappa z^\kappa. \end{aligned}$$

(3) *Let the functions* $\psi(z) \in \mathcal{S}^*(m)$ *and* $h(z) \in \mathcal{K}(m)$, *then* $(\psi * h)(z) \in \mathcal{S}^*(m)$.

(4) *For each functions* $\psi(z)$ *and* $h(z) \in \mathcal{K}(m)$, *then* $(\psi * h)(z) \in \mathcal{K}(m)$.

In [8, 9], Srivastava and Owa defined the fractional integral and differential operators in the complex z-plane $\mathbb{C}$ as the formula:

Definition 1 The fractional integral of order σ is defined, for a function *f*(*z*) by:

$$I_z^\sigma f(z) := \frac{1}{\Gamma(\sigma)} \int_0^z f(\zeta)(z-\zeta)^{\sigma-1} \mathrm{d}\zeta, \qquad (5)$$

where $0 \le \sigma < 1$, and the function *f*(*z*) is analytic in simply-connected region of the complex z-plane $\mathbb{C}$ containing the origin and the multiplicity of $(z-\zeta)^{\sigma-1}$ is removed by requiring $\log(z-\zeta)$ to be real when $(z-\zeta) > 0$.

Definition 2 The fractional derivative of order σ is defined, for a function *f*(*z*), by

$$D_z^\sigma f(z) := \frac{1}{\Gamma(1-\sigma)} \frac{\mathrm{d}}{\mathrm{d}z} \int_0^z f(\zeta)(z-\zeta)^{-\sigma} \mathrm{d}\zeta, \qquad (6)$$

where $0 \le \sigma < 1$, and the function *f*(*z*) is analytic in simply-connected region of the complex z-plane containing the origin and the multiplicity of $(z-\zeta)^{-\sigma}$ is removed, same in the Definition (1) above.

Tremblay defined one of the successful fractional operators in [10]. Recently, some geometric properties and applications for Termblay's operator $\mathfrak{T}_z^{\beta,\gamma}$ in complex plane and in particular on the open unit disk $\mathbb{U}$, studied and discussed by [11–13].

Definition 3 For $0 < \beta \le 1, 0 < \tau \le 1$ and $1 > \beta - \tau \ge 0$. The Tremblay operator $\mathfrak{T}_z^{\beta,\tau} f(z)$ of function $f(z) \in \mathcal{A}(1)$, for all $z \in \mathbb{U}$ is defined as:

$$\mathfrak{T}_z^{\beta,\tau} f(z) := \frac{\Gamma(\tau)}{\Gamma(\beta)} z^{1-\tau} D^{\beta-\tau} z^{\beta-1} f(z), \quad (z \in \mathbb{U}). \qquad (7)$$

Example 1 We find the fractional derivative Termblay operator $\mathfrak{T}_z^{\beta,\tau} f(z)$ in Definition 3, where the function $f(z) = z^\mu$, and $\mu \in \mathbb{R}$.

$$\mathfrak{T}_z^{\beta,\tau}\{z^\mu\} = \frac{\Gamma(\tau)\Gamma(\mu+\beta)}{\Gamma(\beta)\Gamma(\mu+\tau)}\{z^\mu\},$$

if $\mu = 1$, we have

$$\mathfrak{T}_z^{\beta,\tau}\{z\} = \frac{\Gamma(\tau)\Gamma(1+\beta)}{\Gamma(\beta)\Gamma(1+\tau)}\{z\}$$

and, if $\beta, \tau = 1$, then

$$\mathfrak{T}_z^{1,1}\{z^\mu\} = \{z^\mu\}. \qquad (8)$$

Ibrahim defined a generalization of the fractional differential and integral Srivastava–Owa operators in the open unit disk $\mathbb{U}$ as follows [14]:

Definition 4 If $0 \le \alpha < 1$, $\eta \ge 0$, then defined the generalized fractional integral Srivastava–Owa operator of order α such as

$$\mathcal{I}_z^{\alpha,\eta} f(z) := \frac{(\eta+1)^{1-\alpha}}{\Gamma(\alpha)} \int_0^z (z^{\eta+1} - \zeta^{\eta+1})^{\alpha-1} \zeta^\eta f(\zeta) \mathrm{d}\zeta, \qquad (9)$$

where *f*(*z*) function is analytic in simply-connoted region of the complex z-plane $\mathbb{C}$ containing the origin, and the multiplicity of $(z^{\eta+1} - \zeta^{\eta+1})^{-\alpha}$ is removed by requiring $\log(z^{\eta+1} - \zeta^{\eta+1})$ to be real when $(z^{\eta+1} - \zeta^{\eta+1}) > 0$.

Definition 5 If $0 \le \alpha < 1$, $\eta \ge 0$, then defined the generalized fractional derivative Srivastava–Owa operator of order α such as

$$\mathcal{D}_z^{\alpha,\eta} f(z) := \frac{(\eta+1)^\alpha}{\Gamma(1-\alpha)} \frac{\mathrm{d}}{\mathrm{d}z} \int_0^z (z^{\eta+1} - \zeta^{\eta+1})^{-\alpha} \zeta^\eta f(\zeta) \mathrm{d}\zeta, \qquad (10)$$

where $0 \le \alpha < 1$, $\eta > 0$ and *f*(*z*) function is analytic in simply-connoted region of the complex z-plane C containing the origin, and the multiplicity of $(z^{\eta+1} - \zeta^{\eta+1})^{-\alpha}$ is removed , as in Definition 4 above.

Lemma 2 *Let* $f(z) \in \mathcal{A}$; *for all* $z \in \mathbb{U}$, $\rho \in \mathbb{R}$, $0 \le \alpha < 1$ *and* $\eta \ge 0$, then

$$\mathcal{D}_z^{\alpha,\eta}\{z^\rho\} = \frac{(\eta+1)^{\alpha-1}\Gamma\left(\frac{\rho}{\eta+1}+1\right)}{\Gamma\left(\frac{\rho}{\eta+1}+1-\alpha\right)} z^{(1-\alpha)(\eta+1)+\rho-1}$$

and

$$\mathcal{I}_z^{\alpha,\eta}\{z^\rho\} = \frac{(\eta+1)^{-\alpha}\Gamma\left(\frac{\rho+\eta+1}{\eta+1}\right)}{\Gamma\left(\frac{\rho+\eta+1}{\eta+1}\right)} z^{\alpha(\eta+1)+\rho}.$$

Next, we included the Fox–Wright function, which is one of the special functions that generalize hypergeometric functions (see [10]), let denoted this function by ${}_p\Lambda_q$ and defined as:

$$\begin{aligned} {}_p\Lambda_q\left(\begin{array}{c}(\rho_1,A_1),\ldots,(\rho_p,A_p);\\ \\ (\lambda_1,B_1),\ldots,(\lambda_q,B_q);\end{array} z\right) &:= \sum_{\kappa=0}^{\infty}\frac{\Gamma(\rho_1,\kappa A_1)\ldots\Gamma(\rho_p,\kappa A_p)}{\Gamma(\lambda_1,\kappa B_1)\ldots\Gamma(\lambda_q,\kappa B_q)}\frac{z^\kappa}{(1)_\kappa} \\ &= \sum_{\kappa=0}^{\infty}\frac{\prod_{i=1}^{p}\Gamma(\rho_i,\kappa A_i)}{\prod_{j=1}^{q}\Gamma(\lambda_j,\kappa B_j)}\frac{z^\kappa}{(1)_\kappa}. \end{aligned}$$

In particular, $A_1=\cdots=A_p=B_1=\cdots=B_q=1$, then they turn into (see [15, 16])

$$\begin{aligned} {}_p\Lambda_q\left(\begin{array}{c}(\rho_1,1),\ldots,(\rho_p,1);\\ \\ (\lambda_1,1),\ldots,(\lambda_q,1);\end{array} z\right) &= \left[\frac{\prod_{i=1}^{p}\Gamma(\rho_i)}{\prod_{j=1}^{q}\Gamma(\lambda_j)}\right]{}_pF_q(\rho_1,\ldots\rho_p;\lambda_1,\ldots,\lambda_q), \end{aligned}$$

where ρ_i,λ_j are parameters in complex plan $\mathbb{C}$. $A_i>0$, $B_j>0$ for all $j=1,\ldots,q$ and $i=1,\ldots,p$, such that $0\leq 1+\sum_{j=1}^{q}B_j-\sum_{i=1}^{p}A_i$ for fitting values $|z|<1$. For all $z\in\mathbb{C}$ and $\kappa\in\{2,3,4,\ldots\}$, the Pochhammer symbol $(z)_\kappa$ defined as:

$$(z)_0=1 \quad\text{and}\quad (z)_\kappa=z(z+1)\ldots(z+\kappa-1)\quad(\kappa\in N). \tag{11}$$

where $(z)_\kappa=\frac{\Gamma(z+\kappa)}{\Gamma(z)}$ and the formula $\Gamma(z)$ is the well known gamma function. In fact, this function have many remarkable properties in complex plan, we here review some of them. For $z\in\mathbb{C}$, then

$$\Gamma(z+1):=z\Gamma(z). \tag{12}$$

and

$$z\Gamma(z-1):=\Gamma(z)\quad(z>0). \tag{13}$$

Moreover, we consider the Bloch space $\mathbb{B}(\mathbb{U})$ of all functions analytic and univalent functions f in $\mathcal{A}$ which is defined as [17]:

$$\|f\|_{\mathbb{B}}=\sup_{z\in\mathbb{U}}\left(1-|z|^2\right)|f'(z)|<\infty,\quad z(\in\mathbb{U}). \tag{14}$$

In the present paper, the generalized Tremblay operator with univalent function $\mathcal{S}(1)$, which is considered as the generalized fractional derivative operator in Definition 5, was defined. After ward, we utilized the normalized generalized Tremblay operator in a class of analytic functions $\mathcal{A}(m)$, with subclasses $\mathcal{S}(m),\mathcal{S}_\lambda^*(m)$ and $\mathcal{K}_\lambda(m)$ in the open unit disk. Furthermore, we performed some applications to prove the bound coefficient for the last operator.

Results

In this section, we defined the generalized fractional differential of the Tremblay operator in Definition 6 according to definition of the generalized fractional derivative of the Srivastava–Owa operator in complex plane $\mathbb{C}$, for the special case, $m=1$ in classes $\mathcal{A}$ and $\mathcal{S}$. Examples of power function in complex z-plane and some boundedness properties in Bloch space for the operator mentioned were presented as well.

Definition 6 Let $0\leq\beta\leq1$, $0\leq\tau\leq1$ and $\gamma\geq0$. The generalized fractional differential Tremblay operator of two parameters, is defined as

$$\mathfrak{T}_z^{\beta,\tau,\gamma}f(z):=\frac{(\gamma+1)^{\beta-\tau}\Gamma(\tau)}{\Gamma(\beta)\Gamma(1-\beta-\tau)}\left(z^{1-\tau}\frac{\mathrm{d}}{\mathrm{d}z}\right)\int_0^z\frac{\zeta^{\gamma+\beta-1}f(\zeta)}{(z^{\gamma+1}-\zeta^{\gamma+1})^{\beta-\tau}}\mathrm{d}\zeta, \tag{15}$$

where the function $f(z)$ is analytic and univalent in simple-connected region of the complex z-plane $\mathbb{C}$ containing the origin, and the multiplicity of $(z^{\gamma+1}-\zeta^{\gamma+1})^{-\beta+\tau}$ is removed by requiring $\log(z^{\gamma+1}-\zeta^{\gamma+1})$ to be non-negative when $(z^{\gamma+1}-\zeta^{\gamma+1})>0$.

Next, we provided a survey of the interest operator $\mathfrak{T}_z^{\beta,\tau,\gamma}$ to satisfy a boundedness property in the open unit disk and gave an example by using Definition 6. Note that proving the boundedness operator on Bloch space requires using expression (1), when $m=1$.

Example 2 Let $f(z):=z^\kappa$, $z\in\mathbb{U}$ and $\kappa\in N$. If $0<\beta\leq1$, $0<\tau\leq1$, $\gamma\geq0$, and $0\leq\beta-\tau<1$, then the generalized of Termblay operator with power function satisfy

$$\mathfrak{T}_z^{\beta,\tau,\gamma}\{z^\kappa\}:=\frac{(\gamma+1)^{\beta-\tau}\Gamma\left(\frac{\kappa+\beta-1}{\gamma+1}+1\right)\Gamma(\tau)}{\Gamma\left(\frac{\kappa+\beta-1}{\gamma+1}+1-\beta+\tau\right)\Gamma(\beta)}z^{(1-\beta+\tau)\gamma+\kappa},$$

note here, if $\kappa=1$, we obtain

$$\mathfrak{T}_z^{\beta,\tau,\gamma}\{z\} := \frac{(\gamma+1)^{\beta-\tau}\Gamma\left(\frac{\beta}{\gamma+1}+1\right)\Gamma(\tau)}{\Gamma\left(\frac{\beta}{\gamma+1}+1-\beta+\tau\right)\Gamma(\beta)} z^{(1-\beta+\tau)\gamma+1},$$

now go back to Example 1, we see that, if $\gamma = 0$, then

$$\mathfrak{T}_z^{\beta,\tau,0}\{z^\kappa\} := \frac{\Gamma(\tau)}{\Gamma(\beta)}\frac{\Gamma(\kappa+\beta)}{\Gamma(\kappa+\tau)}\{z^\kappa\}$$

and if $\beta = 1, \tau = 1, \gamma = 0$, we get

$$\mathfrak{T}_z^{1,1,0}\{z^\kappa\} := \{z^\kappa\}.$$

In next theorem, we considered the form of definition of the power series to prove the operator $\mathfrak{T}_z^{\beta,\tau,\gamma}$ is bounded with the univalent function $\mathcal{S}$ on Bloch space $\mathbb{B}(\mathbb{U})$ in the open unit disk.

Theorem 2 *Let the function $f \in \mathcal{S}(1) \equiv \mathcal{S}$ belongs to $\mathbb{U}$. Then the operator $\mathfrak{T}_z^{\beta,\tau,\gamma} : \mathcal{S} \to \mathcal{S}$ is bounded on the Bloch $\mathbb{B}(\mathbb{U})$, if*

$$\| \mathfrak{T}_z^{\beta,\tau,\gamma} f \|_{\mathbb{B}} \le M \, \| f \|_{\mathbb{B}}$$

where

$$M := \frac{r^{(1+\tau-\beta)\gamma}(\gamma+1)^{\beta-\tau}\Gamma(\tau)}{\Gamma(\beta)} {}_2\Lambda_1(r)$$

Proof By supposing *f*(*z*) in class of $\mathcal{S}$, we employ Lemma 1 and Example 2, we obtain

$$\begin{aligned}
&\|\mathfrak{T}_z^{\beta,\tau,\gamma} f(z)\|_{\mathbb{B}} \\
&= (1-|z|^2)\left|\left(\mathfrak{T}_z^{\beta,\tau,\gamma} f(z)\right)'\right| \\
&= (1-|z|^2)\left|\left(\frac{\Gamma(\tau)}{\Gamma(\beta)} z^{1-\tau} D_z^{\beta-\tau,\gamma} z^{\beta-1} f(z)\right)'\right| \\
&= (1-|z|^2)\left|\left(\frac{\Gamma(\tau)}{\Gamma(\beta)} z^{1-\tau} D_z^{\beta-\tau,\gamma} z^{\beta-1}\left\{\sum_{\kappa=0}^{\infty}(1)_\kappa a_\kappa \frac{z^\kappa}{\kappa!}\right\}\right)'\right| \\
&= (1-|z|^2)\left|\left(\frac{\Gamma(\tau)}{\Gamma(\beta)}\sum_{\kappa=0}^{\infty}\frac{(\gamma+1)^{\beta-\tau}\Gamma(\kappa+1)\Gamma\left(\frac{\kappa+\beta-1}{\gamma+1}+1\right)}{\Gamma\left(\frac{\kappa+\beta-1}{\gamma+1}+1+\tau-\beta\right)}\frac{a_\kappa}{(1)_\kappa} z^{(1+\tau-\beta)\gamma+\kappa}\right)'\right|
\end{aligned}$$

since $|z| < r$, for all $z \in \mathbb{U}$, then

$$\begin{aligned}
&\| \mathfrak{T}_z^{\beta,\tau,\gamma} f(z) \|_{\mathbb{B}} \\
&\le (1-|r|^2)\left|\left(\frac{r^{(1+\tau-\beta)\gamma}(\gamma+1)^{\beta-\tau}\Gamma(\tau)}{\Gamma(\beta)}\sum_{\kappa=0}^{\infty}\frac{\Gamma(\kappa+1)\Gamma\left(\frac{\kappa+\beta-1}{\gamma+1}+1\right)}{\Gamma\left(\frac{\kappa+\beta-1}{\gamma+1}+1+\tau-\beta\right)} a_\kappa \frac{r^\kappa}{(1)_\kappa}\right)'\right| \\
&= (1-|r|^2)\left|\left(\frac{r^{(1+\tau-\beta)\gamma+1}(\gamma+1)^{\beta-\tau}\Gamma(\tau)}{\Gamma(\beta)} {}_2\Lambda_1(r) * f(r)\right)'\right| \\
&= (1-|r|^2)\left|\frac{r^{(1+\tau-\beta)\gamma}(\gamma+1)^{\beta-\tau}\Gamma(\tau)}{\Gamma(\beta)} {}_2\Lambda_1(r) * f'(r)\right|, \\
&= M\|f\|_{\mathbb{B}}.
\end{aligned}$$

where $M := \frac{r^{(1+\tau-\beta)\gamma}(\gamma+1)^{\beta-\tau}\Gamma(\tau)}{\Gamma(\beta)} {}_2\Lambda_1(r)$ and

$${}_2\Lambda_1(r) := {}_2\Lambda_1\left(\begin{matrix}(1,1),\left(1+\frac{\beta}{\gamma+1}-\frac{1}{\gamma+1},\frac{1}{\gamma+1}\right); \\ \\ \left(1-\beta+\tau+\frac{\beta}{\gamma+1}-\frac{1}{\gamma+1},\frac{1}{\gamma+1}\right);\end{matrix} \; r\right).$$

Normalized operator

In this section we defined a new operator in Theorem 3, which is normalized for the generalized Tremblay operator $\mathfrak{T}_z^{\beta,\tau,\gamma} f(z)$ with an analytic function in the class $\mathcal{A}(m)$.

Theorem 3 *Let the following conditions to be realized:*

$$0 \le \beta - \tau < 1, \quad \gamma \ge 0. \tag{16}$$

Then the normalized of generalized Tremblay operator in Definition 6 is denoted by $\Theta_z^{\beta,\tau,\gamma} f(z)$ and defined as:

$$\Theta_z^{\beta,\tau,\gamma} f(z) = z + \sum_{\kappa=m+1}^{\infty} \vartheta_{\beta,\tau,\gamma}(\kappa) a_\kappa z^\kappa \quad m \in \{1,2,3,\ldots\}. \tag{17}$$

For all $f(z) \in \mathcal{A}(m)$ and $|z| < 1$, where

$$\vartheta_{\beta,\tau,\gamma}(\kappa) := \frac{\Gamma\left(\frac{\beta}{\gamma+1}+1-\beta+\tau\right)\Gamma\left(\frac{\kappa+\beta-1}{\gamma+1}+1\right)}{\Gamma\left(\frac{\beta}{\gamma+1}+1\right)\Gamma\left(\frac{\kappa+\beta-1}{\gamma+1}+1-\beta+\tau\right)}. \tag{18}$$

Proof From Definition 6, and by considering the function

$$h(z) = \frac{z^{(\beta-\tau-1)\gamma}\Gamma\left(\frac{\beta}{\gamma+1}+1-\beta+\tau\right)\Gamma(\beta)}{(\gamma+1)^{\beta-\tau}\Gamma\left(\frac{\beta}{\gamma+1}+1\right)\Gamma(\tau)},$$

we have

$$\begin{aligned}
&h(z)\,\mathfrak{T}_z^{\beta,\tau,\gamma} f(z) \\
&= \frac{z^{(\beta-\tau-1)\gamma}\Gamma\left(\frac{\beta}{\gamma+1}+1-\beta+\tau\right)\Gamma(\beta)}{(\gamma+1)^{\beta-\tau}\Gamma\left(\frac{\beta}{\gamma+1}+1\right)\Gamma(\tau)} \\
&\quad \times \left(\frac{\Gamma(\tau)}{\Gamma(\beta)} z^{1-\tau} D_z^{\beta-\tau,\gamma}\left\{z^\beta + \sum_{\kappa=m+1}^{\infty} a_\kappa z^{\kappa+\beta-1}\right\}\right)
\end{aligned} \tag{19}$$

then

$$\begin{aligned}
&= \frac{z^{(\beta-\tau-1)\gamma}\Gamma\left(\frac{\beta}{\gamma+1}+1-\beta+\tau\right)\Gamma(\beta)}{(\gamma+1)^{\beta-\tau}\Gamma\left(\frac{\beta}{\gamma+1}+1\right)\Gamma(\tau)} \\
&\left\{\frac{(\gamma+1)^{\beta-\tau-1}\Gamma(\tau)\Gamma\left(\frac{\beta}{\gamma+1}+1\right)}{\Gamma(\beta)\Gamma\left(\frac{\beta}{\gamma+1}+1-\beta+\tau\right)} z^{(1-\beta+\tau)\gamma+1}\right. \\
&\quad \left. + \sum_{\kappa=m+1}^{\infty}\frac{(\gamma+1)^{\beta-\tau-1}\Gamma(\tau)\Gamma\left(\frac{\kappa+\beta-1}{\gamma+1}+1\right)}{\Gamma(\beta)\Gamma\left(\frac{\kappa+\beta-1}{\gamma+1}+1-\beta+\tau\right)} a_\kappa z^{(1-\beta+\tau)\gamma+\kappa}\right\}
\end{aligned}$$

which equals to

$$= z + \sum_{\kappa=m+1}^{\infty} \frac{\Gamma\left(\frac{\beta}{\gamma+1}+1-\beta+\tau\right)\Gamma\left(\frac{\kappa+\beta-1}{\gamma+1}+1\right)}{\Gamma\left(\frac{\beta}{\gamma+1}+1\right)\Gamma\left(\frac{\kappa+\beta-1}{\gamma+1}+1-\beta+\tau\right)} a_\kappa z^\kappa$$

$$= z + \sum_{\kappa=m+1}^{\infty} \vartheta_{\beta,\tau,\gamma}(\kappa)\, a_\kappa z^\kappa. \tag{20}$$

Thus, the normalized for the generalized Tremblay operator is represented as the power series and preserves the class $\mathcal{A}(m)$ with their subclasses, where $m = 1, 2, \ldots$ in the open unit disk $\mathbb{U}$, as

$$\Theta_z^{\beta,\tau,\gamma} f(z) := h(z)\mathfrak{T}_z^{\beta,\tau,\gamma} f(z).$$

□

Lemma 3 Let the operator $\Theta_z^{\beta,\tau,\gamma} f(z)$ defined in the class $\mathcal{S}(m)$, $m \in N\backslash\{0\}$, for all $z \in \mathbb{U}$. Then

$$r = \left(\lim_{\kappa\to\infty} |a_\kappa|^{1/\kappa} |\vartheta_{\beta,\tau,\gamma}(\kappa)|^{1/\kappa}\right) \le 1.$$

Proof By employing the Cauchy–Hadamard formal, we find the radius of convergence of the series function in $\Theta_z^{\beta,\tau,\gamma} f(z)$. Supposing the function $f(z) \in \mathcal{S}(m)$ then the coefficient $|a_\kappa| \le \kappa$ for $\kappa \in N = \{2, 3, 4, \ldots\}$ through Theorem 1, we see that

$$\lim_{\kappa\to\infty} |a_\kappa|^{1/\kappa} \le \lim_{\kappa\to\infty} |\kappa|^{1/\kappa} \le 1,$$

and

$$\lim_{\kappa\to\infty} |\vartheta_{\beta,\tau,\gamma}(\kappa)|^{1/\kappa}$$

$$= \lim_{\kappa\to\infty} \left(\frac{\Gamma\left(\frac{\beta}{\gamma+1}+1-\beta+\tau\right)}{\Gamma\left(\frac{\beta}{\gamma+1}+1\right)}\right)^{1/\kappa} \left(\frac{\Gamma\left(\frac{\kappa+\beta-1}{\gamma+1}+1\right)}{\Gamma\left(\frac{\kappa+\beta-1}{\gamma+1}+1-\beta+\tau\right)}\right)^{1/\kappa}$$

By using the property of gamma function, we have

$$\frac{\Gamma\left(\frac{\kappa}{\gamma+1}+\frac{\beta-1}{\gamma+1}+1\right)}{\Gamma\left(\frac{\kappa}{\gamma+1}+\frac{\beta-1}{\gamma+1}+1-\beta+\tau\right)} \sim \kappa^{\beta-\tau}, \quad \kappa \to \infty$$

$$= \lim_{\kappa\to\infty} \left(\left(\frac{\kappa}{\gamma+1}\right)^{1/\kappa}\right)^{\beta-\tau}$$

$$= 1$$

thus follows $r \le 1$. □

Criteria for Hadamard product

In this section, the operator in (17) is represented as the convolution product of two univalent functions in class of $\mathcal{S}(m) \in \mathbb{U}$, in particular, when $m = 1$.

Theorem 4 *Let $f \in \mathcal{S}(1) \equiv \mathcal{S}$ be an univalent function in $\mathbb{U}$. Then we appear the operator $\Theta_z^{\beta,\tau,\gamma}$ as the convolution of two functions in $\mathcal{S}$,*

$$\Theta_z^{\beta,\tau,\gamma} f(z) = g(z) * f(z)$$

where $g(z) := \frac{\Gamma(\frac{\beta}{\gamma+1}+1-\beta+\tau)}{\Gamma(\frac{\beta}{\gamma+1}+1)} {}_2\Lambda_1(z)$.

Proof By equality (20), we have

$$\Theta_z^{\beta,\tau,\gamma} f(z) = z + \sum_{\kappa=2}^{\infty} \frac{\Gamma\left(\frac{\beta}{\gamma+1}+1-\beta+\tau\right)\Gamma\left(\frac{\kappa+\beta-1}{\gamma+1}+1\right)}{\Gamma\left(\frac{\beta}{\gamma+1}+1\right)\Gamma\left(\frac{\kappa+\beta-1}{\gamma+1}+1-\beta+\tau\right)} a_\kappa z^\kappa$$

$$= \sum_{\kappa=0}^{\infty} \frac{\Gamma(\kappa+1)\Gamma\left(\frac{\beta}{\gamma+1}+1-\beta+\tau\right)\Gamma\left(\frac{\kappa+\beta-1}{\gamma+1}+1\right)}{\Gamma\left(\frac{\beta}{\gamma+1}+1\right)\Gamma\left(\frac{\kappa+\beta-1}{\gamma+1}+1-\beta+\tau\right)} a_\kappa \frac{z^\kappa}{\kappa!}$$

$$= \frac{\Gamma\left(\frac{\beta}{\gamma+1}+1-\beta+\tau\right)}{\Gamma\left(\frac{\beta}{\gamma+1}+1\right)} {}_2\Lambda_1\left(\begin{matrix}(1,1),\left(1+\frac{\beta}{\gamma+1}-\frac{1}{\gamma+1},\frac{1}{\gamma+1}\right); \\ \left(1-\beta+\tau+\frac{\beta}{\gamma+1}-\frac{1}{\gamma+1},\frac{1}{\gamma+1}\right);\end{matrix}\; z\right) * f(z) \tag{21}$$

hence

$$\Theta_z^{\beta,\tau,\gamma} f(z) := \frac{\Gamma\left(\frac{\beta}{\gamma+1}+1-\beta+\tau\right)}{\Gamma\left(\frac{\beta}{\gamma+1}+1\right)} {}_2\Lambda_1(z) * f(z)$$

by letting

$$g(z) = \frac{\Gamma\left(\frac{\beta}{\gamma+1}+1-\beta+\tau\right)}{\Gamma\left(\frac{\beta}{\gamma+1}+1\right)} {}_2\Lambda_1(z). \tag{22}$$

Then the proof is completed. □

Based on the results, the following observations were obtained. Let $f(z) \in \mathcal{A}(1) \subseteq \mathbb{U}$.

1. If $\beta, \tau = 1, \gamma = 0$, then $\Theta_z^{1,1,0} f(z) := f(z)$ defined in (1), for $m = 1$.
2. If $\gamma = 0$, then $\Theta_z^{\beta,\tau,0} f(z) := \mathfrak{T}_z^{\beta,\tau} f(z)$ defined in (7), for $m = 1$.
3. It is clear that the operator $\Theta_z^{\beta,\tau,\gamma} f(z)$ is generalized of Carlson–Shaffer operator, when $\gamma = 0$ and $\frac{\tau}{\beta} = 1$ in (21), while the linear operator of Carlson and Shaffer defined as [18]:

   $$\mathcal{L}(a, c) f(z) = \varphi(a, c; z) * f(z), f \in \mathcal{A}$$

   where $\varphi(a, c; z) = \sum_{\kappa=0}^{\infty} \frac{(a)_\kappa}{(c)_\kappa} z^\kappa$, $z \in \mathbb{U}, a \in R$, $c \in \{1, 2, 3, \ldots\}$.

Note here the proof of the following Theorems comes immediately from Eq. (20) and Lemma 1.

Theorem 5 *Let* $0<\beta\leq 1,\ 0<\tau\leq 1$ *and the condition* (16). *If the function f(z) given by* (1) *in the class* $\mathcal{S}^*(m)$ *and the function* $g(z)$ *defined by* (22) *in* $\mathcal{K}(m)$. *Then*

$$f(z) * g(z) \in \mathcal{S}^*(m).$$

Theorem 6 *Let* $0<\beta\leq 1,\ 0<\tau\leq 1$ *and the condition* (16). *If the functions f(z) given by* (1) *and g(z) defined by* (22) *in* $\mathcal{K}(m)$. *Then*

$$f(z) * g(z) \in \mathcal{K}(m).$$

Univalency of the operator $\Theta_z^{\beta,\tau,\gamma}$

We discussed the initialization of a univalent criteria and convexity by employing the normalized Tremblay operator in the open unit disk, in particular when $m=1$.

Theorem 7 *Let* $f\in\mathcal{S}(1)$. *If the following conditions satisfied*

(i) *for* $0<\beta\leq 1,\ 0<\tau\leq 1$ *such that* $0\leq\beta-\tau<1$.
(ii) $0<\rho_i,\ i=1,\ldots,p$ *and* $0<\lambda_j,\ j=1,\ldots,q$; $p\leq q+1$,

then the operator $\Theta_z^{\beta,\tau,\gamma}f(z)\in\mathcal{S}$ *in open unite disk* $\mathbb{U}$.

$$
{}_2\Lambda_1\left(\begin{matrix}(3,1),\left(1+\dfrac{\beta}{\gamma+1}+\dfrac{1}{\gamma+1},\dfrac{1}{\gamma+1}\right); \\ \\ \left(1-\beta+\tau+\dfrac{\beta}{\gamma+1}+\dfrac{1}{\gamma+1},\dfrac{1}{\gamma+1}\right);\end{matrix}\ 1\right)
+{}_2\Lambda_1\left(\begin{matrix}(2,1),\left(1+\dfrac{\beta}{\gamma+1},\dfrac{1}{\gamma+1}\right); \\ \\ \left(1-\beta+\tau+\dfrac{\beta}{\gamma+1},\dfrac{1}{\beta+1}\right);\end{matrix}\ 1\right)
<2\left(\frac{\Gamma\left(\frac{\beta}{\gamma+1}+1\right)}{\Gamma\left(\frac{\beta}{\gamma+1}+1-\beta+\tau\right)}\right).
$$

Proof By supposing the function $f\in\mathcal{S}$ with equality (17), we have

$$\Theta_z^{\beta,\tau,\gamma}f(z)=z+\sum_{\kappa=2}^{\infty}w_\kappa z^\kappa,$$

where $w_\kappa:=\vartheta_{\beta,\tau,\gamma}(\kappa)a_\kappa$ and the function $\vartheta_{\beta,\tau,\gamma}(\kappa)$ defined in (18) satisfied the following condition in class $\mathcal{S}$ as follows:

$$\ell_1;=\sum_{\kappa=2}^{\infty}\kappa\,|w_\kappa|=\sum_{\kappa=2}^{\infty}\kappa\,\vartheta_{\beta,\tau,\gamma}(\kappa)\,|a_\kappa|<1,$$

By using Theorem 1, we give the estimate for the coefficients of an univalent function belong to $\mathcal{S}$ in $\mathbb{U}$ also, by employ this estimate, we can get another estimate for ℓ_1 in $\mathcal{S}$ as follows,

$$
\begin{aligned}
\ell_1&=\sum_{\kappa=2}^{\infty}\kappa\vartheta_{\beta,\tau,\gamma}|a_\kappa|\\
&\leq\sum_{\kappa=2}^{\infty}\kappa^2\vartheta_{\beta,\tau,\gamma}(\kappa)=\sum_{\kappa=2}^{\infty}\frac{(\kappa)^2}{\kappa!}(\vartheta_{\beta,\tau,\gamma}(\kappa)\kappa!)=\sum_{\kappa=2}^{\infty}\frac{(\kappa)^2}{\kappa!}\ell(\kappa)<1
\end{aligned}
\tag{23}
$$

where

$$\ell(\kappa)=\frac{\Gamma\left(\frac{\beta}{\gamma+1}+1-\beta+\tau\right)}{\Gamma\left(\frac{\beta}{\gamma+1}+1\right)}\frac{\Gamma\left(\frac{\kappa+\beta-1}{\gamma+1}+1\right)(1)_\kappa}{\Gamma\left(\frac{\kappa+\beta-1}{\gamma+1}+1-\beta+\tau\right)}\tag{24}$$

The series in (23) is transformed into a sum of twice the terms by employing the following relation:

$$\frac{\kappa^2}{(1)_\kappa}=\frac{\kappa}{(1)_{\kappa-1}}=\frac{1}{(1)_{\kappa-1}}+\frac{1}{(1)_{\kappa-2}}\tag{25}$$

Depending on $(1)_\kappa=\kappa!$ and $(1)_{\kappa-1}=(\kappa-1)!$, the estimate (23) becomes the next form:

$$
\begin{aligned}
\ell_1&\leq\sum_{\kappa=2}^{\infty}\frac{\kappa^2}{(1)_\kappa}=\sum_{\kappa=2}^{\infty}\left(\frac{1}{(1)_{\kappa-1}}+\frac{1}{(1)_{\kappa-2}}\right)\ell(\kappa)=\sum_{\kappa=2}^{\infty}\frac{\ell(\kappa)}{(1)_{\kappa-1}}+\frac{\ell(\kappa)}{(1)_{\kappa-2}}\\
&=\sum_{\kappa=2}^{\infty}\frac{\Gamma\left(\frac{\beta}{\gamma+1}+1-\beta+\tau\right)}{\Gamma\left(\frac{\beta}{\gamma+1}+1\right)}\frac{\Gamma\left(\frac{\kappa+\beta-1}{\gamma+1}+1\right)}{\Gamma\left(\frac{\kappa+\beta-1}{\gamma+1}+1-\beta+\tau\right)}\frac{(1)_\kappa}{(1)_{\kappa-1}}\\
&+\sum_{\kappa=2}^{\infty}\frac{\Gamma\left(\frac{\beta}{\gamma+1}+1-\beta+\tau\right)}{\Gamma\left(\frac{\beta}{\gamma+1}+1\right)}\frac{\Gamma\left(\frac{\kappa+\beta-1}{\gamma+1}+1\right)}{\Gamma\left(\frac{\kappa+\beta-1}{\gamma+1}+1-\beta+\tau\right)}\frac{(1)_\kappa}{(1)_{\kappa-2}}\\
&=\frac{\Gamma\left(\frac{\beta}{\gamma+1}+1-\beta+\tau\right)}{\Gamma\left(\frac{\beta}{\gamma+1}+1\right)}\left(\sum_{\kappa=1}^{\infty}\frac{\Gamma\left(\frac{\kappa+\beta}{\gamma+1}+1\right)}{\Gamma\left(\frac{\kappa+\beta}{\gamma+1}+1-\beta+\tau\right)}\frac{(1)_{\kappa+1}}{(1)_\kappa}\right.\\
&\left.+\sum_{\kappa=0}^{\infty}\frac{\Gamma\left(\frac{\kappa+\beta+1}{\gamma+1}+1\right)}{\Gamma\left(\frac{\kappa+\beta+1}{\gamma+1}+1-\beta+\tau\right)}\frac{(1)_{\kappa+2}}{(1)_\kappa}\right)
\end{aligned}
$$

by considering some properties of the gamma function in (13) and (12), we have

$$
\begin{aligned}
&=\frac{\Gamma\left(\frac{\beta}{\gamma+1}+1-\beta+\tau\right)}{\Gamma\left(\frac{\beta}{\gamma+1}+1\right)}\left(\sum_{\kappa=1}^{\infty}\frac{\Gamma(\kappa+2)\Gamma\left(\frac{\kappa+\beta}{\gamma}+1\right)}{\Gamma\left(\frac{\kappa+\beta}{\gamma+1}+1-\beta+\tau\right)}\frac{1}{(1)_\kappa}\right.\\
&\left.+\sum_{\kappa=0}^{\infty}\frac{\Gamma(\kappa+3)\Gamma\left(\frac{\kappa+\beta+1}{\gamma+1}+1\right)}{\Gamma\left(\frac{\kappa+\beta+1}{\gamma+1}+1-\beta+\tau\right)}\frac{1}{(1)_\kappa}\right)
\end{aligned}
$$

and by using the Fox–Wright function, we can transform the estimate ℓ_1 at $z\to 1$,

$$= {}_2\Lambda_1\left(\begin{matrix}(3,1),\left(1+\frac{\beta}{\gamma+1}+\frac{1}{\gamma+1},\frac{1}{\gamma+1}\right);\\ \left(1-\beta+\tau+\frac{\beta}{\gamma+1}+\frac{1}{\gamma+1},\frac{1}{\gamma+1}\right);\end{matrix}\; 1\right)$$

$$+ {}_2\Lambda_1\left(\begin{matrix}(2,1),\left(1+\frac{\beta}{\gamma+1},\frac{1}{\gamma+1}\right);\\ \left(1-\beta+\upsilon+\frac{\beta}{\gamma+1},\frac{1}{\gamma+1}\right);\end{matrix}\; 1\right)$$

$$-\frac{\Gamma\left(\frac{\beta}{\gamma+1}+1\right)}{\Gamma\left(\frac{\beta}{\gamma+1}+1-\beta+\tau\right)} < \frac{\Gamma\left(\frac{\beta}{\gamma+1}+1\right)}{\Gamma\left(\frac{\beta}{\gamma+1}+1-\beta+\tau\right)}.$$

We conclude from the above theorem that the operator $\Theta_z^{\beta,\tau,\gamma}f(z)$ maps preserve the property (univalent function) in class $f\in\mathcal{S}$ from a linear space to another. Further, the operator $\Theta_z^{\beta,\tau,\gamma}f(z)$ is univalent for $f\in\mathcal{S}$ for all $z\in\mathbb{U}$ in the open unit disk and $\Theta_z^{\beta,\tau,\gamma}:\mathcal{S}\to\mathcal{S}$.

Theorem 8 *Let the condition i as in the Theorem 7 is satisfied, then*

$${}_2\Lambda_1\left(\begin{matrix}(2,1),\left(1+\frac{\beta}{\gamma+1},\frac{1}{\gamma+1}\right);\\ \left(1-\beta+\tau+\frac{\beta}{\gamma+1},\frac{1}{\gamma+1}\right);\end{matrix}\; 1\right)$$

$$< 2\left(\frac{\Gamma\left(\frac{\beta}{\gamma+1}+1\right)}{\Gamma\left(\frac{\beta}{\gamma+1}+1-\beta+\tau\right)}\right).$$

then the operator maps a convex function f(z) into a univalent function that is $\Theta_z^{\beta,\tau,\gamma}:\mathcal{K}\to\mathcal{S}$.

Proof Presume that $f(z)\in\mathcal{K},\ z\in\mathcal{U}$ and the operator (17), such that

$$\Theta_z^{\beta,\tau,\gamma}f(z)=z+\sum_{\kappa=2}^{\infty}w_\kappa z^\kappa$$

where

$$w_\kappa := \vartheta_{\beta,\tau,\gamma}(\kappa)a_\kappa$$

and the function $\vartheta_{\beta,\tau,\gamma}$ is defined in inequality (18), satisfied the following condition in class $\mathcal{S}$ as follows:

$$\ell_2; = \sum_{\kappa=2}^{\infty}\kappa|w_\kappa| = \sum_{\kappa=2}^{\infty}\kappa\vartheta_{\beta,\tau,\gamma}(\kappa)|a_\kappa|<1.$$

We know That the coefficient of a convex function belong to $\mathcal{S}$ is $|a_\kappa|<1$. So we can get another estimate for ℓ_2 as follows,

$$\ell_2=\sum_{\kappa=2}^{\infty}\kappa\vartheta_{\beta,\tau,\gamma}|a_\kappa|\le\sum_{\kappa=2}^{\infty}\kappa^2\vartheta_{\beta,\tau,\gamma}(\kappa)$$
$$=\sum_{\kappa=2}^{\infty}\frac{(\kappa)^2}{\kappa!}(\vartheta_{\beta,\tau,\gamma}(\kappa)\kappa!)=\sum_{\kappa=2}^{\infty}\frac{(\kappa)^2}{\kappa!}\ell(\kappa)<1 \tag{26}$$

where

$$\ell(\kappa)=\frac{\Gamma\left(\frac{\beta}{\gamma+1}+1-\beta+\tau\right)}{\Gamma\left(\frac{\beta}{\gamma+1}+1\right)}\frac{\Gamma\left(\frac{\kappa+\beta-1}{\gamma+1}+1\right)(1)_\kappa}{\Gamma\left(\frac{\kappa+\beta-1}{\gamma+1}+1-\beta+\tau\right)}$$

where a_κ is Pochhammer symbol defined in (11),with the following relation

$$\frac{\kappa}{(1)_\kappa}=\frac{1}{(1)_{\kappa-1}}$$

and $(1)_\kappa=\kappa!$, then the estimate (26) become as the next form

$$\ell_2\le\sum_{\kappa=2}^{\infty}\frac{\kappa}{(1)_\kappa}\ell(\kappa)=\sum_{\kappa=2}^{\infty}\frac{1}{(1)_{\kappa-1}}\ell(\kappa)$$
$$=\sum_{\kappa=2}^{\infty}\frac{\ell(\kappa)}{(1)_{\kappa-1}}=\sum_{\kappa=2}^{\infty}\frac{\Gamma\left(\frac{\beta}{\gamma+1}+1-\beta+\tau\right)}{\Gamma\left(\frac{\beta}{\beta+1}+1\right)}$$
$$\frac{\Gamma\left(\frac{\kappa+\beta-1}{\gamma+1}+1\right)}{\Gamma\left(\frac{\kappa+\beta-1}{\gamma+1}+1-\beta+\tau\right)}\frac{(1)_\kappa}{(1)_{\kappa-1}}$$
$$=\frac{\Gamma\left(\frac{\beta}{\gamma+1}+1-\beta+\tau\right)}{\Gamma\left(\frac{\beta}{\gamma+1}+1\right)}\sum_{\kappa=1}^{\infty}\frac{\Gamma\left(\frac{\kappa+\beta}{\gamma+1}+1\right)}{\Gamma\left(\frac{\kappa+\beta}{\gamma+1}+1-\beta+\tau\right)}\frac{(1)_{\kappa+1}}{(1)_\kappa}$$

employ the properties of gamma function, we have

$$=\frac{\Gamma\left(\frac{\beta}{\gamma+1}+1-\beta+\tau\right)}{\Gamma\left(\frac{\beta}{\gamma+1}+1\right)}\sum_{\kappa=1}^{\infty}\frac{\Gamma(\kappa+2)\Gamma\left(\frac{\kappa+\beta}{\gamma+1}+1\right)}{\Gamma\left(\frac{\kappa+\beta}{\gamma+1}+1-\beta+\tau\right)}\frac{1}{(1)_\kappa}$$

then with the Fox–Wright function, we transform the estimate ℓ_1 at $z=1$,

$$=\frac{\Gamma\left(\frac{\beta}{\gamma+1}+1-\beta+\tau\right)}{\Gamma\left(\frac{\beta}{\gamma+1}+1\right)}$$
$${}_2\Lambda_1\left(\begin{matrix}(2,1),\left(1+\frac{\beta}{\gamma+1},\frac{1}{\gamma+1}\right);\\ \left(1-\beta+\tau+\frac{\beta}{\eta+1},\frac{1}{\eta+1}\right);\end{matrix}\; 1\right)-1<1$$

hence

$\Theta_z^{\beta,\tau,\gamma} : \mathcal{K} \to \mathcal{S}$.

We conclude from the above theorem that the maps preserve the property (univalent function) in class $\mathcal{S}$ from a linear space to another. □

Coefficients bound

Now, we study the coefficient bounds for the operator $\Theta_z^{\beta,\tau,\gamma} f(z)$, which is defined in (20), where the function $f(z)$ is in the class $\mathcal{A}(m)$, for all $z \in \mathbb{U}$. We also discuss the bounded coefficient in two subclasses $\mathcal{S}_\lambda^*(m)$ and $\mathcal{K}_\lambda(m)$, $(m = 1, 2, \ldots)$ of order λ in the open unit disk $\mathbb{U}$. In the first step, we are looking to prove that the operator $\Theta_z^{\beta,\tau,\gamma} f(z)$ in $\mathcal{S}_\lambda^*(m)$, and that by finding a coefficient bound.

Theorem 9 *Let $f(z) \in \mathcal{A}(m)$ given by (1) satisfy the condition* (16). If

$$\sum_{\kappa=m+1}^{\infty} (\kappa - \lambda)\,|a_\kappa| \le \frac{1-\lambda}{\vartheta_{\beta,\tau,\gamma}(m+1)} \quad (0 \le \lambda < 1) \tag{27}$$

where

$$\vartheta_{\beta,\tau,\gamma}(m+1) := \frac{\Gamma\left(\frac{\beta}{\gamma+1} + 1 - \beta + \tau\right)\Gamma\left(\frac{m+\beta}{\gamma+1} + 1\right)}{\Gamma\left(\frac{\beta}{\gamma+1} + 1\right)\Gamma\left(\frac{m+\beta}{\gamma+1} + 1 - \beta + \tau\right)},$$

Then the operator $\Theta_z^{\beta,\tau,\gamma} f(z) \in \mathcal{S}_\lambda^*(m)$ and satisfy the sharp result.

Proof By assuming that the function $f \in \mathcal{S}_\lambda^*(m)$, we obtain

$$\begin{aligned}
\left|\frac{z(\Theta_z^{\beta,\tau,\gamma} f(z))'}{\Theta_z^{\beta,\tau,\gamma} f(z)} - 1\right| &= \left|\frac{z(\Theta_z^{\beta,\tau,\gamma} f(z))' - \Theta_z^{\beta,\tau,\gamma} f(z)}{\Theta_z^{\beta,\tau,\gamma} f(z)}\right| \\
&= \left|\frac{\sum_{\kappa=m+1}^{\infty}(\kappa-1)\,\vartheta_{\beta,\tau,\gamma}(\kappa) a_\kappa z^\kappa}{z + \sum_{\kappa=m+1}^{\infty}\vartheta_{\beta,\tau,\gamma}(\kappa) a_\kappa z^\kappa}\right| \\
&\le \frac{\sum_{\kappa=m+1}^{\infty}(\kappa-1)\,|\vartheta_{\beta,\tau,\gamma}(\kappa)|\,|a_\kappa|\,|z|^{\kappa-1}}{1 - \sum_{\kappa=m+1}^{\infty}|\vartheta_{\beta,\tau,\gamma}(\kappa)|\,|a_\kappa|\,|z|^{\kappa-1}}, \quad |z| < 1 \\
&\le \frac{\sum_{\kappa=m+1}^{\infty}(\kappa-1)\,|\vartheta_{\beta,\tau,\gamma}(\kappa)|\,|a_\kappa|}{1 - \sum_{\kappa=m+1}^{\infty}|\vartheta_{\beta,\tau,\gamma}(\kappa)||a_\kappa|}
\end{aligned} \tag{28}$$

we see in inequality (28) is bounded by $(1-\lambda)$, if it satisfy

$$\begin{aligned}
&\sum_{\kappa=m+1}^{\infty}(\kappa-1)\,|\vartheta_{\beta,\tau,\gamma}(\kappa)|\,|a_\kappa| \\
&\quad \le (1-\lambda)\left(1 - \sum_{\kappa=m+1}^{\infty}|\vartheta_{\beta,\tau,\gamma}(\kappa)|\,|a_\kappa|\right),
\end{aligned}$$

By use the inequality $0 \le \vartheta_{\beta,\tau,\gamma}(\kappa) \le \vartheta_{\beta,\tau,\gamma}(m+1)$; for each $m+1 \le \kappa$ and for all $m = 1, 2, \ldots$, we have

$$\begin{aligned}
&\sum_{\kappa=m+1}^{\infty}(\kappa-1)\,|\vartheta_{\beta,\tau,\gamma}(\kappa)|\,|a_\kappa| \\
&\quad \le \vartheta_{\beta,\tau,\gamma}(m+1)\sum_{\kappa=m+1}^{\infty}(\kappa-1)|a_\kappa| \\
&\quad \le (1-\lambda)\left(1 - \sum_{\kappa=m+1}^{\infty}|\vartheta_{\beta,\tau,\gamma}(\kappa)||a_\kappa|\right),
\end{aligned}$$

which is on a par with

$$\sum_{\kappa=m+1}^{\infty}(\kappa-\lambda)|a_\kappa| \le \frac{1-\lambda}{\vartheta_{\beta,\tau,\gamma}(m+1)}, \tag{29}$$

hence

$$\mathcal{R}\left(\frac{z(\Theta_z^{\beta,\tau,\gamma} f(z))'}{\Theta_z^{\beta,\tau,\gamma} f(z)}\right) > \lambda.$$

From Theorem (9), we have a special case compared with the well-known results, which are reviewed in the next corollary.

Corollary 1 *Let $\Theta_z^{\beta,\tau,\gamma} f(z) \in \mathcal{S}_\lambda^*(m)$, for all $z \in \mathbb{U}$, with*

$$\sum_{\kappa=m+1}^{\infty}(\kappa-\lambda)\,|a_\kappa| \le \frac{1-\lambda}{\vartheta_{\beta,\tau,\gamma}(m+1)},$$

(i) *If $\lambda = 0, \beta = 1, \tau = 1, m = 1$ and $\gamma = 0$, we get*

$$\sum_{\kappa=2}^{\infty}\kappa|a_\kappa| \le 1$$

then $f(z) \in \mathcal{S}_0^*(1) \equiv \mathcal{S}^*$ (see [19]).

(ii) *If $\beta = 1, \tau = 1, m = 1$ and $\gamma = 0$, we get*

$$\sum_{\kappa=2}^{\infty}(\kappa-\lambda)|a_\kappa| \le (1-\lambda)$$

then $f(z) \in \mathcal{S}_\lambda^*(1) \equiv \mathcal{S}_\lambda^*$ (see [20]).

(iii) *If $\lambda = 0$, we get*

$$\sum_{\kappa=m+1}^{\infty}\kappa\,|a_\kappa| \le \frac{1}{\vartheta_{\beta,\tau,\gamma}(m+1)}. \tag{30}$$

then $\Theta_z^{\beta,\tau,\gamma} f(z) \in \mathcal{S}_0^*(m)$, (see [21]).

All these results are sharp.

Corollary 2 *Let the operator $\Theta_z^{\beta,\tau,\gamma} f(z) \in \mathcal{S}_\lambda^*(m)$. Then*

$$|a_{m+1}| \le \frac{(1-\lambda)\Gamma\left(\frac{\beta}{\gamma+1} + 1\right)\Gamma\left(\frac{m+\beta}{\gamma+1} + 1 - \beta + \tau\right)}{(m-\lambda+1)\Gamma\left(\frac{\beta}{\gamma+1} + 1 - \beta + \tau\right)\Gamma\left(\frac{m+\beta}{\gamma+1} + 1\right)}, \tag{31}$$

for $m = \{1, 2, 3, \ldots\}$.

Example 3 The function belongs to the class $\mathcal{S}^*_\lambda(m)$, is defined as

$$g_1(z) = z + \frac{(1-\lambda)\Gamma\left(\frac{\beta}{\gamma+1}+1\right)\Gamma\left(\frac{m+\beta}{\gamma+1}+1-\beta+\tau\right)}{(m-\lambda+1)\Gamma\left(\frac{\beta}{\gamma+1}+1-\beta+\tau\right)\Gamma\left(\frac{m+\beta}{\gamma+1}+1\right)} z^{m+1}.$$

We prove a bound coefficient in Theorem (10) by using similar methods in the starlike class.

Theorem 10 *Let the function $f(z) \in \mathcal{A}(m)$ and satisfied the condition* (16). *If*

$$\sum_{\kappa=m+1}^{\infty} \kappa(\kappa-\lambda)\,|a_\kappa| \leq \frac{1-\lambda}{\vartheta_{\beta,\tau,\gamma}(m+1)} \qquad m = 1, 2, \ldots. \tag{32}$$

Then $f \in \mathcal{K}_\lambda(m)$, $\lambda\,(0 \leq \lambda < 1)$, this result is sharp.

Corollary 3 *Let the operator $\Theta_z^{\beta,\tau,\gamma} f(z) \in \mathcal{K}_\lambda(m)$. Then*

$$|a_{m+1}| \leq \frac{(1-\lambda)\Gamma(\frac{\beta}{\gamma+1}+1)\Gamma(\frac{m+\beta}{\gamma+1}+1-\beta+\tau)}{(m+1)(m-\lambda+1)\Gamma(\frac{\beta}{\gamma+1}+1-\beta+\tau)\Gamma(\frac{m+\beta}{\gamma+1}+1)},$$

for $\kappa \in \{2, 3, \ldots\}$.

Example 4 The equality (32) is realized by the function

$$g_2(z) = z + \frac{(1-\lambda)\Gamma\left(\frac{\beta}{\gamma+1}+1\right)\Gamma\left(\frac{m+\beta}{\gamma+1}+1-\beta+\tau\right)}{(m+1)(m-\lambda+1)\Gamma\left(\frac{\beta}{\gamma+1}+1-\beta+\tau\right)\Gamma\left(\frac{m+\beta}{\gamma+1}+1\right)} z^{m+1}.$$

Conclusion

All results of the present work are valid in open unit disk U with respect to the fractional calculus in a complex domain. We defined a normalized fractional differential operator in the concept of the generalized Tremblay operator. Moreover, we assumed sufficient conditions for this operator to become starlike and convex functions. Finally, univalency and convolution properties are discussed.

Author's contributions All the authors jointly worked on deriving the results and approved the final manuscript.

Acknowledgments The authors would like to thank the Editor for useful suggestions.

References

1. Srivastava, H.M., Owa, S., Chatterjea, S.K.: A note on certain classes of starlike functions. Rendiconti del Seminario Matematico della Universita di Padova **77**, 115–124 (1987)
2. Kiryakova, V., Saigo, M., Srivastava, H.M.: Some criteria for univalence of analytic functions involving generalized fractional calculus operators. Fract. Calc. Appl. Anal. **1**(1), 79–104 (1998)
3. Robertson, M.I.: On the theory of univalent functions. Ann. Math. **37**(2), 374–408 (1936)
4. Graham, I., Kohr, G.: Geometric function theory in one and higher dimensions. CRC Press, Boca Raton (2003)
5. Duren, P.L.: Univalent functions, vol. 259. Springer, New York (1983)
6. Ruscheweyh, S.: New criteria for univalent functions. Proc. Am. Math. Soc. **49**(1), 109–115 (1975)
7. Ruscheweyh. S., Montreal. U.D.: Convolutions in geometric function theory (Seminaire de mathematiques superieures). Gaetan Morin Editeur Ltee, February 1982
8. Owa, S.: On the distortion theorems. I. Kyungpook Math. J. **18**(1), 53–59 (1978)
9. Owa, S., Srivastava, H.M.: Univalent and starlike generalized hypergeometric functions. Can. J. Math. **39**, 1057–1077 (1987)
10. Tremblay, R.: Some Operational Formulas Involving the Operators *xD*, *x*Δ and Fractional Derivatives. SIAM J. Math. Anal. **10**(5), 933–943 (1979)
11. Abdulnaby, Z.E., Ibrahim, R.W., Kılıçman, A.: Some properties for integro-differential operator defined by a fractional formal. Springerplus **5**(1), 1–9 (2016)
12. Kılıçman, A., Ibrahim, R.W., Abdulnaby, Z.E.: On a generalized fractional integral operator in a complex domain. Appl. Math. **10**(3), 1053–1059 (2016)
13. Ibrahim, R.W., Jahangiri, J.M.: Boundary fractional differential equation in a complex domain. Bound. Value Prob. **66**(1), 1–11 (2014)
14. Ibrahim, R.W.: On generalized Srivastava-Owa fractional operators in the unit disk. Adv. Differ. Equ. **2011**(1), 1–10 (2011)
15. Eslahchi, M.R., Masjed-Jamei, M.: Some applications of a hypergeometric identity. Math. Sci. **9**(4), 215–223 (2015)
16. Dziok, J., Srivastava, H.M.: Classes of analytic functions associated with the generalized hypergeometric function. Appl. Math. Comput. **103**, 1–13 (1999)
17. El-Sayed Ahmed, A., Bakhit, M.A.: Characterizations involving Schwarzian derivative in some analytic function spaces. Math. Sci. **7**(1), 1–8 (2013)
18. Carlson, B.C., Shaffer, D.B.: Starlike and prestarlike hypergeometric functions. SIAM J. Math. Anal. **15**(4), 737–745 (1984)
19. Goodman, A.W.: Univalent functions and nonanalytic curves. Proc. Am. Math. Soc. JSTOR. **8**(3), 598–601 (1957)
20. Silverman, H.: Univalent functions with negative coefficients. Proc. Am. Math. Soc. **51**(1), 109–116 (1957)
21. Kiryakova, V: The operators of generalized fractional calculus and their action in classes of univalentfunctions. In: Proc. Intern. Symp. on Geometric Function Theory and Applications 2010, Sofia, pp. 29–40. 27–31 August 2010

Quintic B-spline method for time-fractional superdiffusion fourth-order differential equation

Saima Arshed[1]

Abstract The main objective of this paper is to obtain the approximate solution of superdiffusion fourth-order partial differential equations. Quintic B-spline collocation method is employed for fractional differential equations (FPDEs). The developed scheme for finding the solution of the considered problem is based on finite difference method and collocation method. Caputo fractional derivative is used for time fractional derivative of order α, $1<\alpha<2$. The given problem is discretized in both time and space directions. Central difference formula is used for temporal discretization. Collocation method is used for spatial discretization. The developed scheme is proved to be stable and convergent with respect to time. Approximate solutions are examined to check the precision and effectiveness of the presented method.

Keywords Time-fractional PDE · Superdiffusion · Quintic B-spline · Collocation method · Convergence analysis · Stability analysis

Mathematics Subject Classification 35R11 · 74H15 · 35-XX

Introduction

The idea of differentiation and integration to non integer order has been studied in fractional calculus. The subject fractional calculus is the generalized form of classical calculus. It is as old as classical calculus but getting more attention these days due to its applications in various fields, such as science and engineering. Oldham and Spanier [1], Podlubny [3], and Miller and Ross [2] provide the history and a comprehensive treatment of this subject.

In recent decades, it has been observed by many mathematicians, scientists, and researchers that non-integer operators (differential and integral) play an important role in describing the properties of physical phenomena. Fractional derivatives and integrals efficiently describe the history and inherited properties of different processes and equipments. It has further been observed that fractional models are more efficient and accurate than already developed classical models. The mathematical modeling of real world problems, such as earthquake, traffic flow, fluid flow, signal processing, and viscoelastic problems, results in FPDEs.

A fourth-order linear PDE is defined as

$$\frac{\partial^2 v}{\partial t^2} + \nu \frac{\partial^4 v}{\partial z^4} = h(z,t).$$

The fourth-order problems play a significant role for the development of engineering and modern science. For example, window glasses, floor systems, bridge slabs, etc. are modeled by fourth-order PDEs. It also represents beam problem where v is the transversal displacement of the beam, ν is the ratio of flexural rigidity of the beam to its mass per unit length, t is time, and x is space variable, $h(x, t)$ is the dynamic driving force acting per unit mass.

The considered equation, is the superdiffusion fourth-order partial differential equation, which has the following form

$$\frac{\partial^\alpha v}{\partial t^\alpha} + \nu \frac{\partial^4 v}{\partial z^4} = q(z,t), \quad z \in [a,b] = \Omega, \quad t_0 < t \le t_f, 1<\alpha<2, \tag{1.1}$$

where t_0, t_f represents initial and final time, respectively.

✉ Saima Arshed
saima.math@pu.edu.pk

[1] Department of Mathematics, University of the Punjab, Lahore 54590, Pakistan

The time-fractional superdiffusion fourth-order problems play an important role in modern science and engineering. For example, airplane wings and transverse vibrations of sustained tensile beam can be modeled as plates with different boundary supports which are successfully governed by superdiffusion fourth-order differential equations.

The given problem is solved with initial conditions

$$v(z,t_0)=f_0(z), \quad v_t(z,t_0)=f_1(z), \quad a\leq z\leq b.$$

Following three boundary conditions are used for solving Eq. (1.1)

I.

$$\begin{cases} v(a,t)=v(b,t)=0, \\ v_z(a,t)=v_z(b,t)=0, \quad t_0\leq t\leq t_f.. \end{cases}$$

II.

$$\begin{cases} v(a,t)=v(b,t)=0, \\ v_{zz}(a,t)=v_{zz}(b,t)=0, \quad t_0\leq t\leq t_f.. \end{cases}$$

III

$$\begin{cases} v(a,t)=v(b,t)=0, \\ v_{zz}(a,t)-Qv_z(a,t)=v_{zz}(b,t)-Qv_z(b,t)=0, \quad t_0\leq t\leq t_f,. \end{cases}$$

where Q is an arbitrary constant.

Caputo fractional derivative is defined as

$$\frac{\partial^\alpha v(z,t)}{\partial t^\alpha}=\begin{cases} \dfrac{1}{\Gamma(2-\alpha)}\displaystyle\int_0^t \frac{\partial^2 v(z,s)}{\partial s^2}\frac{ds}{(t-s)^{\alpha-1}}, & 1<\alpha<2, \\ \dfrac{\partial^2 v(z,t)}{\partial t^2}, & \alpha=2. \end{cases}.$$

Collocation method is widely used to provide the solution of partial differential equations. Depending on the situation, some times, it is useful to find the solution of FPDEs at various locations in the given problem domain, then the spline solutions guarantee to give the information of spline interpolation between mesh points.

In many situations, it is harder to calculate the analytical solutions of fractional PDEs using analytical methods. Due to this reason, the mathematicians are motivated to develop numerical methods for the approximate solution of FPDEs. The approximate methods for solving FPDEs have become popular among the researchers in the last 10 years. A variety of approximate techniques were used for finding the solution of FPDEs, such as the FDM, FEM, FVM, and the collocation method *etc*.

Many authors applied different numerical methods to find the solution of fractional PDEs.

Siddiqi and Arshed [7] developed a numerical scheme for the solution of time-fractional fourth-order partial differential equation with fractional derivative of order α $(0<\alpha<1)$. Lin and Xu [6] developed a numerical scheme for time-fractional diffusion equation with fractional derivative of order α, $(0<\alpha<1)$. Yang et al. [11] developed a novel numerical scheme for solving fourth-order partial integro-differential equation with a weakly singular kernel. Yang et al. [8] developed a quasi-wavelet-based numerical method for solving fourth-order partial integro-differential equations with a weakly singular kernel. Khan et al. [9], proposed a numerical method for the solution of time-fractional fourth-order differential equations with variable coefficients using Adomian decomposition method (ADM) and He's variational iteration method (HVIM). Zhang and Han [10] developed a quasi-wavelet-based numerical method for time-dependent fractional partial differential equation. Zhang et al. [12] proposed quintic B-spline collocation method for the numerical solution of fourth-order partial integro-differential equations with a weakly singular kernel. Khan et al. [4] used sextic spline solution for solving fourth-order parabolic partial differential equation. Dehghan [5] used finite difference techniques for the numerical solution of partial integro-differential equation arising from viscoelasticity. In [15], Bhrawy and Abdelkawy used fully spectral collocation approximation for solving multi-dimensional fractional Schrodinger equations. Bhrawy et al. [16] proposed a new formula for fractional integrals of Chebyshev polynomials and application for solving multi-term fractional differential equations. Bhrawy [17] developed a new spectral algorithm for a time-space fractional partial differential equations with subdiffusion and superdiffusion. In [18], Bhrawy et al. gave a review of operational matrices and spectral techniques for fractional calculus. Bhrawy et al. [19] proposed Chebyshev–Laguerre Gauss–Radau collocation scheme for solving time fractional sub-diffusion equation on a semi-infinite domain. Bhrawy [20] developed a space-time collocation scheme for modified anomalous subdiffusion and nonlinear superdiffusion equations. In [21], Doha et al. developed an efficient Legendre Spectral Tau matrix formulation for solving fractional sub-diffusion and reaction sub-diffusion equations. Bhrawy [22] proposed a highly accurate collocation algorithm for $1+1$ and $2+1$ fractional percolation equations. In [23], Bhrawy and Zaky used fractional-order Jacobi tau method for a class of time-fractional PDEs with variable Coefficients.

This paper is designed to determine the approximate solution of superdiffusion fourth order PDEs. The approximate solution is based on central difference formula and B-spline collocation method.

The paper is divided into six sections. The quintic B-spline basis function is given in Sect. 2. The finite difference approximation for time discretization of the given problem is discussed in Sect. 3. The stability and convergence of temporal discretization are established. Space discretization based on B-spline collocation technique is discussed in Sect. 4. Approximate solutions are obtained in Sect. 5 which support the theoretical results. The conclusion is given is Sect. 6.

Quintic B-spline

The problem domain $[a, b]$ has been subdivided into N elements having uniform step size h with knots $z_j, j = 0, 1, 2, \ldots, N$ such that $\{a = z_0 < z_1 < z_2 < \ldots < z_N = b\}$, $h = z_j - z_{j-1}, j = 1, 2, \ldots, N$.

To define basis functions for quintic B-spline, it is first needed to introduce ten additional knots, $z_{-5}, z_{-4}, z_{-3}, z_{-2}, z_{-1}, z_{N+1}, z_{N+2}, z_{N+3}, z_{N+4}, z_{N+5}$ such that $z_{-5} < z_{-4} < z_{-3} < z_{-2} < z_{-1} < z_0$ and $z_N < z_{N+1} < z_{N+2} < z_{N+3} < z_{N+4} < z_{N+5}$.

The quintic B-spline basis functions $Q_j(z)$, $j = -2, \ldots, N+2$ are defined as

$$Q_j(z) = \begin{cases} \sum\limits_{s=j-3}^{s=j+3} \dfrac{(z_s - z)_+^5}{\prod'(z_s)}, & z \in [z_{j-3}, z_{j-3}], \\ 0, & \text{otherwise} \end{cases}$$

The approximate solution $V^{k+1}(z)$ at $k+1$ time level to the exact solution $v^{k+1}(z)$ is obtained as a linear combination of quintic B-spline basis functions $Q_j(z)$ as

$$V^{k+1}(z) = \sum_{j=-2}^{N+2} c_j(t) Q_j(z), \tag{2.1}$$

where c_j are unknowns. Using Eq. (2.1) and basis functions, the approximate solution V and its derivatives at the nodes z_j, can be calculated as

$$\begin{aligned} V(z_j) &= c_{j+2} + 26c_{j+1} + 66c_j + 26c_{j-1} + c_{j-2}, \\ V^{(1)}(z_j) &= \frac{5}{h}\left(c_{j+2} + 10c_{j+1} - 10c_{j-1} - c_{j-2}\right), \\ V^{(2)}(z_j) &= \frac{20}{h^2}\left(c_{j+2} + 2c_{j+1} - 6c_j + 2c_{j-1} + c_{j-2}\right), \\ V^{(3)}(z_j) &= \frac{60}{h^3}\left(c_{j+2} - 2c_{j+1} + 2c_{j-1} - c_{j-2}\right), \\ V^{(4)}(z_j) &= \frac{120}{h^4}\left(c_{j+2} - 4c_{j+1} + 6c_j - 4c_{j-1} + c_{j-2}\right). \end{aligned}$$

Time discretization

Caputo fractional derivative is used for temporal discretization. Let $t_k = k\Delta t$, $k = 0, 1, 2, \ldots, K$, where $\Delta t = \frac{t_f}{K}$ is the time step size and t_f be the final time. Caputo fractional derivative is discretized using central difference approximation as

$$\begin{aligned} \frac{\partial^\alpha v(z, t_{k+1})}{\partial t^\alpha} &= \frac{1}{\Gamma(2-\alpha)} \int_0^{t_{k+1}} \frac{\partial^2 v(z,s)}{\partial s^2} \frac{ds}{(t_{k+1} - s)^{\alpha-1}} \\ &= \frac{1}{\Gamma(2-\alpha)} \sum_{l=0}^{n} \int_{t_l}^{t_{l+1}} \frac{\partial^2 v(z,s)}{\partial s^2} \frac{ds}{(t_{k+1} - s)^{\alpha-1}}, \\ &= \frac{1}{\Gamma(2-\alpha)} \sum_{l=0}^{k} \frac{v(z, t_{l+1}) - 2v(z, t_l) + v(z, t_{l-1})}{\Delta t^2} \\ &\quad \int_{t_l}^{t_{l+1}} \frac{ds}{(t_{k+1} - s)^{\alpha-1}} + r_{\Delta t}^{k+1}, \\ &= \frac{1}{\Gamma(2-\alpha)} \sum_{l=0}^{k} \frac{v(z, t_{l+1}) - 2v(z, t_l) + v(z, t_{l-1})}{\Delta t^2} \\ &\quad \int_{t_{k-l}}^{t_{k+1-l}} \frac{d\tau}{\tau^{\alpha-1}} + r_{\Delta t}^{k+1}, \\ &= \frac{1}{\Gamma(2-\alpha)} \sum_{l=0}^{k} \frac{v(z, t_{k+1-l}) - 2v(z, t_{k-l}) + v(z, t_{k-l-1})}{\Delta t^2} \\ &\quad \int_{t_l}^{t_{l+1}} \frac{d\tau}{\tau^{\alpha-1}} + r_{\Delta t}^{k+1}, \\ &= \frac{1}{\Gamma(3-\alpha)} \sum_{l=0}^{k} \frac{v(z, t_{k+1-l}) - 2v(z, t_{k-l}) + v(z, t_{k-l-1})}{\Delta t^\alpha} \\ &\quad \left((l+1)^{2-\alpha} - l^{2-\alpha}\right) + r_{\Delta t}^{k+1}, \\ &= \frac{1}{\Gamma(3-\alpha)} \sum_{l=0}^{k} b_l \frac{v(z, t_{k+1-l}) - 2v(z, t_{k-l}) + v(z, t_{k-l-1})}{\Delta t^\alpha} \\ &\quad + r_{\Delta t}^{k+1}, \end{aligned} \tag{3.1}$$

where $b_l = (l+1)^{2-\alpha} - l^{2-\alpha}$ and $b_0 = 1$ and $\tau = (t_{k+1} - s)$.

and

$$-b_k + (2b_k - b_{k-1}) + \sum_{l=1}^{k-1}(-b_{l-1} + 2b_l - b_{l+1}) + (2b_0 - b_1) = 1.$$

The time discrete operator is represented by D^α and defined as

$$D^\alpha v(z, t_{k+1}) := \frac{1}{\Gamma(3-\alpha)} \sum_{l=0}^{k} b_l \frac{v(z, t_{k+1-l}) - 2v(z, t_{k-l}) + v(z, t_{k-l-1})}{\Delta t^\alpha}.$$

Then Eq. (3.1) can be written as

$$\frac{\partial^\alpha v(z, t_{k+1})}{\partial t^\alpha} = D^\alpha v(z, t_{k+1}) + r_{\Delta t}^{k+1}. \tag{3.2}$$

The second order approximation of Caputo derivative as discussed in [13] is given as

$$\left|r_{\Delta t}^{k+1}\right| \le E\Delta t^2, \tag{3.3}$$

where E is a constant.

After approximating $\frac{1}{\Gamma(2-\alpha)} \int_0^{t_{k+1}} \frac{\partial^2 v(z,s)}{\partial s^2} \frac{ds}{(t_{k+1}-s)^{\alpha-1}}$ by $D^\alpha v(z, t_{k+1})$, the finite difference scheme according to (1.1) can be written as

$$D^{\alpha}v(z,t_{k+1}) + v\frac{\partial^4 v(z,t_{k+1})}{\partial z^4} = q(z,t_{k+1}),$$

$$\frac{1}{\Gamma(3-\alpha)}\sum_{l=0}^{n} b_l \frac{v(z,t_{k+1-l}) - 2v(z,t_{k-l}) + v(z,t_{k-l-1})}{\Delta t^{\alpha}} + v\frac{\partial^4 v(z,t_{k+1})}{\partial z^4} = q(z,t_{k+1}).$$

The above equation can be rewritten as

$$\begin{aligned} v^{k+1}(z) + \alpha_0 v\frac{\partial^4 v^{k+1}}{\partial z^4} = & - b_k v^{-1}(z) + (2b_k - b_{k-1})v^0(z) + \\ & + \sum_{l=1}^{k-1}(-b_{l-1} + 2b_l - b_{l+1})v^l(z) \\ & + (2b_0 - b_1)v^k(z) \\ & + \alpha_0 q^{k+1}(z), \qquad k = 1,2,\ldots,K-1 \end{aligned} \tag{3.4}$$

where $v^{k+1}(x) = v(x,t_{k+1})$ and $\alpha_0 = \Gamma(3-\alpha)\Delta t^{\alpha}$,

Using the following initial conditions and the boundary conditions (I, II, III)

$$v(z,t_0) = f_0(z), \quad z \in [a,b], \tag{3.5}$$

$$\frac{\partial v(z,t_0)}{\partial t} = f_1(z), \quad z \in [a,b], \tag{3.6}$$

The developed method is a three time level method. To implement the developed method, it is first required to calculate (v^{-1}) and (v^0).

$$v^{-1}(z) = v(z,t_0) - \Delta t v_t(z,t_0).$$

$$v^{-1}(z) = f_0(z) - \Delta t f_1(z).$$

In particular, for $k = 0$, the scheme simply leads to

$$v^1(z) + \alpha_0 v\frac{\partial^4 v^1}{\partial z^4} = -b_0 v^{-1}(z) + 2b_0 v^0(z) + \alpha_0 q^1(z), \tag{3.7}$$

where $v^0(z) = v(z,t_0) = f_0(z)$.

Equations (3.4) and (3.7), along with initial conditions (3.5) and (3.6) and boundary conditions (I, II, III) form a whole set of the time-discrete problem of Eq. (1.1).

r^{k+1} is the error term can be defined as

$$r^{k+1} := \left(\frac{\partial^{\alpha} v(z,t_{k+1})}{\partial t^{\alpha}} - D^{\alpha}v(z,t_{k+1})\right). \tag{3.8}$$

Using Eq. (3.3),

$$|r^{k+1}| = |r^{k+1}_{\Delta t}| \le E\Delta t^2, \tag{3.9}$$

Some functional spaces endowed with standard norms and inner products, to be used hereafter, are defined as under

$$H^2(\Omega) = \left\{ w \in L^2(\Omega), \frac{dw}{dx}, \frac{d^2w}{dx^2} \in L^2(\Omega) \right\},$$

$$H_0^2(\Omega) = \left\{ w \in H^2(\Omega), w|_{\partial\Omega} = 0, \frac{dw}{dx}|_{\partial\Omega} = 0 \right\},$$

$$H^m(\Omega) = \left\{ w \in L^2(\Omega), \frac{d^k w}{dx^k} \text{ for all positive integer } k \le m \right\},$$

where $L^2(\Omega)$ is the space of measurable functions whose square is Lebesgue integrable in Ω. The inner products of $L^2(\Omega)$ and $H^2(\Omega)$ are defined, respectively, by

$$(v,w) = \int_{\Omega} wv dx, \quad (v,w)_2 = (v,w) + \left(\frac{dv}{dx}, \frac{dw}{dx}\right) + \left(\frac{d^2v}{dx^2}, \frac{d^2w}{dx^2}\right),$$

and the corresponding norms by

$$\|w\|_0 = (w,w)^{\frac{1}{2}}, \qquad \|w\|_2 = (w,w)_2^{\frac{1}{2}}.$$

The norm $\|.\|$ of the space $H^m(\Omega)$ is defined as

$$\|w\|_m = \left(\sum_{k=0}^{m}\left\|\frac{d^k w}{dx^k}\right\|_0^2\right)^{\frac{1}{2}}.$$

Instead of using the above standard H^2-norm, it is preferred to define $\|.\|_2$

$$\|w\|_2 = \left(\|w\|_0^2 + \alpha_0 v\left\|\frac{d^2 w}{dx^2}\right\|_0^2\right)^{1/2}, \tag{3.10}$$

where $\alpha_0 = \Gamma(2-\alpha)\Delta t^{\alpha}$.

For the convergence and stability analysis, the following weak formulation of Eqs. (3.4) and (3.7) is needed, i.e., finding $v^{k+1} \in H_0^2(\Omega)$, such that for all $w \in H_0^2(\Omega)$,

$$\begin{aligned} (v^{k+1},w) + \alpha_0 v\left(\frac{\partial^4 v^{k+1}}{\partial x^4}, w\right) = & - b_k(v^{-1},w) + (2b_k - b_{k-1})(v^0,w) \\ & + \sum_{l=1}^{k-1}(-b_{l-1} + 2b_l - b_{l+1})(v^j,w) \\ & + (2b_0 - b_1)(v^k,w) \\ & + \alpha_0(q^{k+1},w), \end{aligned} \tag{3.11}$$

and

$$\begin{aligned} (v^1,w) + \alpha_0 v\left(\frac{\partial^4 v^1}{\partial x^4}, w\right) = & - b_0(v^{-1},w) + 2b_0(v^0,w) \\ & + \alpha_0(q^1,w). \end{aligned} \tag{3.12}$$

Discrete Gronwall inequality

If the sequences $\{a_l\}$ and $\{z_l\}$, $jl = 1,2,\ldots, n$, satisfy inequality

$$z_l \le \left(\sum_{j=1}^{l-1} a_j z_j + b\right), l = 1, 2, \ldots, n$$

where $a_l \ge 0$, $b \ge 0$, then the inequality

$$z_l \le b.\exp\left(\sum_{j=1}^{l-1} a_j\right), l = 1, 2, \ldots, n. \tag{3.13}$$

Stability and convergence analysis

Theorem 1 *The proposed time-discrete scheme* (3.6) is stable, if for all $\Delta t > 0$, *the following relation holds*

$$\|v^{k+1}\|_2 \le c(\|f_0\|_0 + \Delta t\|f_1\|_0 + \alpha_0 q^{k+1}), k = 0, 1, 2, \ldots, K-1, \tag{3.14}$$

where $\|.\|_2$ *is defined in* (3.10).

Proof For $k = 0$, let $w = v^1$ in Eq. (3.11),

Equation (3.11) takes the following form

$$(v^1, v^1) + \alpha_0 v\left(\frac{\partial^4 v^1}{\partial x^4}, v^1\right) = -b_0(v^{-1}, v^1) + 2b_0(v^0, v^1) + \alpha_0(q^1, v^1).$$

Integrating the above equation, we have

$$(v^1, v^1) + \alpha_0 v\left(\frac{\partial^2 v^1}{\partial x^2}, \frac{\partial^2 v^1}{\partial x^2}\right) = -b_0(v^{-1}, v^1) + 2b_0(v^0, v^1) + \alpha_0(q^1, v^1) \tag{3.15}$$

Due to boundary conditions on w all the boundary contributions disappeared.

Using the inequality $\|w\|_0 \le \|w\|_2$ and Schwarz inequality Eq. (3.15) takes the following form

$$\begin{aligned}\|v^1\|_2^2 &\le \|v^{-1}\|_0\|v^1\|_0 + 2\|v^0\|_0\|v^1\|_0 + \alpha_0\|h^1\|_0\|v^1\|_0 \\ &\le \|v^{-1}\|_0\|v^1\|_2 + 2\|v^0\|_0\|v^1\|_2 + \alpha_0\|q^1\|_0\|v^1\|_2\end{aligned}$$

$$\|v^1\|_2 \le \|v^{-1}\|_0 + 2\|v^0\|_0 + \alpha_0\|q^1\|_0$$

$$\|v^1\|_2 \le (\|f_0\|_0 + \Delta t\|f_1\|_0) + 2\|f_0\|_0 + \alpha_0\|q^1\|_0$$

$$\|v^1\|_2 \le 3\|f_0\|_0 + \Delta t\|f_1\|_0 + \alpha_0\|q^1\|_0$$

hence

$$\|v^1\|_2 \le c(\|f_0\|_0 + \Delta t\|f_1\|_0 + \alpha_0\|q^1\|_0).$$

Suppose that the result hold for $w = v^l$ *i.e.*

$$\|v^l\|_2 \le c(\|f_0\|_0 + \Delta t\|f_1\|_0 + \alpha_0\|q^l\|_0) \qquad l = 2, 3, \ldots, k. \tag{3.16}$$

Taking $w = v^{k+1}$ in Eq. (3.11), it can be written as

$$\begin{aligned}&(v^{k+1}, v^{k+1}) + \alpha_0 v\left(\frac{\partial^4 v^{k+1}}{\partial x^4}, v^{k+1}\right) \\ &= -b_k(v^{-1}, v^{k+1}) + (2b_k - b_{k-1})(v^0, v^{k+1}) \\ &\quad + (2b_0 - b_1)(v^k, v^{k+1}) \\ &\quad + \sum_{l=1}^{k-1}(-b_{l-1} + 2b_l - b_{l+1})(v^l, v^{k+1}) \\ &\quad + \alpha_0(q^{k+1}, v^{k+1}).\end{aligned}$$

Integrating the above equation, we get

$$\begin{aligned}&(v^{k+1}, v^{k+1}) + \alpha_0 v\left(\frac{\partial^2 v^{k+1}}{\partial x^2}, \frac{\partial^2 v^{k+1}}{\partial x^2}\right) \\ &= -b_k(v^{-1}, v^{k+1}) + (2b_k - b_{k-1})(v^0, v^{k+1}) \\ &\quad + (2b_k - b_{k-1})(v^0, v^{k+1}) \\ &\quad + \sum_{l=1}^{k-1}(-b_{l-1} + 2b_l - b_{l+1})(u^l, u^{k+1}) \\ &\quad + \alpha_0(q^{k+1}, v^{k+1}).\end{aligned}$$

Using the Schwarz inequality and the inequality $\|w\|_0 \le \|w\|_2$ the above equation can be written as

$$\begin{aligned}\|v^{k+1}\|_2^2 \le\ & b_k\|v^{-1}\|_0\|v^{k+1}\|_0 + (2b_k - b_{k-1})\|v^0\|_0\|v^{k+1}\|_0 \\ &+ \sum_{l=1}^{k-1}(-b_{l-1} + 2b_l - b_{l+1})\|v^l\|_0\|v^{k+1}\|_0 \\ &+ (2b_0 - b_1)\|v^k\|_0\|v^{k+1}\|_0 \\ &+ \alpha_0\|q^{k+1}\|_0\|v^{k+1}\|_0,\end{aligned}$$

$$\begin{aligned}\|v^{k+1}\|_2^2 \le\ & b_k\|v^{-1}\|_0\|v^{k+1}\|_2 + (2b_k - b_{k-1})\|v^0\|_0\|v^{k+1}\|_2 \\ &+ \sum_{l=1}^{k-1}(-b_{l-1} + 2b_l - b_{l+1})\|v^l\|_2\|v^{k+1}\|_2 \\ &+ (2b_0 - b_1)\|v^k\|_2\|v^{k+1}\|_2 \\ &+ \alpha_0\|q^{k+1}\|_0\|v^{k+1}\|_2,\end{aligned}$$

$$\begin{aligned}\|v^{k+1}\|_2 \le\ & b_k\|v^{-1}\|_0 + (2b_k - b_{k-1})\|v^0\|_0 + \sum_{l=1}^{k-1} \\ &(-b_{l-1} + 2b_l - b_{l+1})\|v^l\|_2 \\ &+ (2b_0 - b_1)\|v^k\|_2 + \alpha_0\|q^{k+1}\|_0,\end{aligned}$$

Using (3.13) we have

$$\begin{aligned}\|v^{k+1}\|_2 \le\ & \left((2b_k - b_{k-1})\|v^0\|_0 + b_k\|v^{-1}\|_0 + \alpha_0\|q^{k+1}\|_0\right) \\ &\exp\left((2b_0 - b_1) + \sum_{l=1}^{k-1}(-b_{l-1} + 2b_l - b_{l+1})\right),\end{aligned}$$

$$\|v^{k+1}\|_2 \le \left(\|v^0\|_0 + \|v^{-1}\|_0 + \alpha_0\|q^{k+1}\|_0\right)\exp(1 + b_{k-1} - b_k),$$

$$\|v^{k+1}\|_2 \le \left(\|f_0\|_0+\|f_0\|_0+\Delta t\|f_1\|_0+\alpha_0\|q^{k+1}\|_0\right) \exp(1+b_{k-1}-b_k),$$

$$\|v^{k+1}\|_2 \le c\left(\|f_0\|_0+\Delta t\|f_1\|_0+\alpha_0\|q^{k+1}\|_0\right).$$

□

Theorem 2 *The numerical solution obtained by the proposed method converges the exact solution, if the following relation holds*

$$\|v(t_k)-v^k\|_2 \le E\Delta t^2, \qquad k=1,2,\ldots,K. \tag{3.17}$$

where E is a constant and $E=(1+b_k)\exp(1+b_{k-1}-b_k)$.

Proof Let $s^k=v(z,t_k)-v^k(z)$, for $k=1$, by combining Eqs. (1.1), (3.11), the error equation becomes

$$(s^1,v)+\alpha_0 v\left(\frac{\partial^2 s^1}{\partial z^2},\frac{\partial^2 v}{\partial z^2}\right) = -(s^{-1},v)+2b_0(s^0,v) + (r^1,v), \qquad \forall v\in H_0^2(\Omega).$$

using Schwarz inequality and the inequality $\|w\|_0 \le \|w\|_2$

Let $v=s^1$, noting $s^0=0$, the above equation takes the following form

$$\|s^1\|_2^2 \le \|s^{-1}\|_0\|s^1\|_0+\|r^1\|_0\|s^1\|_0.$$

$$\|s^1\|_2^2 \le \|s^{-1}\|_0\|s^1\|_2+\|r^1\|_0\|s^1\|_2.$$

$$\|s^1\|_2 \le \|s^{-1}\|_0+\|r^1\|_0.$$

since

$$\|s^{-1}\| \le \Delta t^2$$

Using Eq.(3.9), it leads to

$$\|v(t_1)-v^1\|_2 \le 2\Delta t^2. \qquad \|v(t_1)-v^1\|_2 \le E\Delta t^2. \tag{3.18}$$

Hence (3.19) is proved for the case $k=1$.

For inductive part, suppose (3.19) holds for $l=1,2,3,\ldots,k$, i.e.

$$\|v(t_j)-v^j\|_2 \le E\Delta t^2. \tag{3.19}$$

For $k=s+1$ the Eqs. (1.1), (3.11) are used and the error equation can be written, for all $v\in H_0^2(\Omega)$, as

$$\begin{aligned}(s^{k+1},v)+\alpha_0\mu\left(\frac{\partial^2 s^{k+1}}{\partial z^2},\frac{\partial^2 v}{\partial z^2}\right) &= -b_k(e^{-1},v)+(2b_k-b_{k-1})(s^0,v)\\ &+\sum_{l=1}^{k-1}(-b_{l-1}+2b_l-b_{l+1})(s^l,v)\\ &+(2b_0-b_1)(s^k,v)+\alpha_0(r^{k+1},v).\end{aligned}$$

using $\|w\|_0 \le \|w\|_2$ and Schwarz inequality the above equation, for $v=s^{k+1}$, can be written as

$$\begin{aligned}\|s^{k+1}\|_2^2 &\le b_k\|s^{-1}\|_0\|s^{k+1}\|_0+(2b_k-b_{k-1})\|s^0\|_0\|s^{k+1}\|_0\\ &+\sum_{l=1}^{k-1}(-b_{l-1}+2b_l-b_{l+1})\|s^l\|_0\|s^{k+1}\|_0\\ &+(2b_0-b_1)\|s^k\|_0\|s^{k+1}\|_0\\ &+\|r^{k+1}\|_0\|s^{k+1}\|_0,\end{aligned}$$

$$\|s^{k+1}\|_2 \le b_k\|s^{-1}\|_0+\sum_{l=1}^{k-1}(-b_{l-1}+2b_l-b_{l+1})\|s^l\|_2 + (2b_0-b_1)\|s^k\|_2+\|r^{k+1}\|_0.$$

Using (3.15)

$$\|s^{k+1}\|_2 \le \left(b_k\|s^{-1}\|_0+\|r^{k+1}\|_0\right) \exp\left[\sum_{l=1}^{k-1}(-b_{l-1}+2b_l-b_{l+1})+2b_0-b_1\right],$$

$$\|s^{k+1}\|_2 \le \left(b_k\|s^{-1}\|_0+\|r^{k+1}\|_0\right)\exp(1+b_{k-1}-b_k),$$

$$\|s^{k+1}\|_2 \le (b_k\Delta t^2+\Delta t^2)\exp(1+b_{k-1}-b_k),$$

$$\|s^{k+1}\|_2 \le E\Delta t^2.$$

□

Space discretization

Collocation method with quintic B-spline basis functions is employed for space discretization. Putting Eq. (2.1) into Eq. (3.4), we obtain the following equation

$$\begin{aligned}&\left(c_{j-2}^{k+1}+26c_{j-1}^{k+1}+66c_j^{k+1}+26c_{j+1}^{k+1}+c_{j+2}^{k+1}\right)\\ &+\alpha_0 v\frac{120}{h^4}\left(c_{j-2}^{k+1}-4c_{j-1}^{k+1}+6c_j^{k+1}-4c_{j+1}^{k+1}+c_{j+2}^{k+1}\right)\\ &=-b_k(c_{j-2}^{-1}+26c_{j-1}^{-1}+66c_j^{-1}+26c_{j+1}^{-1}+c_{j+2}^{-1})\\ &+(2b_k-b_{k-1})(c_{j-2}^0+26c_{j-1}^0+66c_j^0+26c_{j+1}^0+c_{j+2}^0)\\ &+\sum_{l=1}^{k-1}(-b_{l-1}+2b_l-b_{l+1})(c_{j-2}^l+26c_{j-1}^l+66c_j^l+26c_{j+1}^l+c_{j+2}^l)\\ &+(2b_0-b_1)(c_{j-2}^k+26c_{j-1}^k+66c_j^k+26c_{j+1}^k+c_{j+2}^k)+\alpha_0 q^{k+1}.\end{aligned}$$

The above equation can be rewritten as

$$\begin{aligned}&\left(1+\alpha_0 v\frac{120}{h^4}\right)c_{j-2}^{k+1}+\left(26-4\alpha_0 v\frac{120}{h^4}\right)c_{j-1}^{k+1}\\ &+\left(66+6\alpha_0 v\frac{120}{h^4}\right)c_j^{k+1}+\left(26-4\alpha_0 v\frac{120}{h^4}\right)c_{j+1}^{k+1}\\ &+\left(1+\alpha_0 v\frac{120}{h^4}\right)c_{j+2}^{k+1}\\ &=-b_k\left(c_{j-2}^{-1}+26c_{j-1}^{-1}+66c_j^{-1}+26c_{j+1}^{-1}+c_{j+2}^{-1}\right)\\ &+(2b_k-b_{k-1})\left(c_{j-2}^0+26c_{j-1}^0+66c_j^0+26c_{j+1}^0+c_{j+2}^0\right)\\ &+\sum_{l=1}^{k-1}(-b_{l-1}+2b_l-b_{l+1})\left(c_{j-2}^l+26c_{j-1}^l+66c_j^l+26c_{j+1}^l+c_{j+2}^l\right)\\ &+(2b_0-b_1)\left(c_{j-2}^k+26c_{j-1}^k+66c_j^k+26c_{j+1}^k+c_{j+2}^k\right)+\alpha_0 q^{k+1}.\end{aligned}$$

The above equation can be rewritten as a system of linear equations

$$\left(1+\alpha_0 v\frac{120}{h^4}\right)c_{j-2}^{k+1}+\left(26-4\alpha_0 v\frac{120}{h^4}\right)c_{j-1}^{k+1}$$
$$+\left(66+6\alpha_0 v\frac{120}{h^4}\right)c_{j}^{k+1}+\left(26-4\alpha_0 v\frac{120}{h^4}\right)c_{j+1}^{k+1}$$
$$+\left(1+\alpha_0 v\frac{120}{h^4}\right)c_{j+2}^{k+1}=F_j,$$

where

$$F_j=-b_k\left(c_{j-2}^{-1}+26c_{j-1}^{-1}+66c_j^{-1}+26c_{j+1}^{-1}+c_{j+2}^{-1}\right)$$
$$+(2b_k-b_{k-1})\left(c_{j-2}^{0}+26c_{j-1}^{0}+66c_j^{0}+26c_{j+1}^{0}+c_{j+2}^{0}\right)$$
$$+\sum_{l=1}^{k-1}(-b_{l-1}+2b_l-b_{l+1})\left(c_{j-2}^{l}+26c_{j-1}^{l}+66c_j^{l}+26c_{j+1}^{l}+c_{j+2}^{l}\right)$$
$$+(2b_0-b_1)\left(c_{j-2}^{k}+26c_{j-1}^{k}+66c_j^{k}+26c_{j+1}^{k}+c_{j+2}^{k}\right)+\alpha_0 q^{k+1}.$$

Table 1 The errors $\|s_K\|_\infty$ and $\|s_K\|_2$ for different values of time Δt with $N=100$

Δt	$\|s_K\|_\infty$	Rate	$\|s_K\|_2$	Rate
0.001	1.8221×10^{-3}		1.1556×10^{-4}	
0.0005	5.6177×10^{-4}	1.6975	2.9965×10^{-5}	1.9472
0.00025	1.5380×10^{-4}	1.8689	7.920×10^{-6}	1.9197
0.000125	3.9312×10^{-5}	1.96801	1.991×10^{-6}	1.9920

Table 2 The errors $\|s_K\|_\infty$ and $\|s_K\|_2$ for different values of Δt with $N=80$

Δt	$\|s_K\|_\infty$	Rate	$\|s_K\|_2$	Rate
0.001	6.6172×10^{-4}		5.5124×10^{-5}	
0.0005	1.7543×10^{-4}	1.9153	1.4871×10^{-5}	1.8901
0.00025	4.8575×10^{-5}	1.8526	3.8277×10^{-6}	1.9579
0.000125	1.2143×10^{-5}	1.9999	9.5947×10^{-7}	1.99617

This system has $N+1$ equations in $N+5$ unknowns $c_{-2}^{k+1}, c_{-1}^{k+1}, \ldots, c_{N+1}^{k+1}, c_{N+2}^{k+1}$. Unique solution of the above system is obtained by eliminating the unknowns $c_{-2}^{k+1}, c_{-1}^{k+1}, c_{N+1}^{k+1}$, and c_{N+2}^{k+1} using boundary conditions.

Numerical results

In this section, two examples are considered to check the accuracy and efficiency of the proposed method. The main objective of these examples is to verify the rate of convergence of the approximate results for α. From the following Tables 1 and 2, it is clear that the approximate results support the theoretical estimates.

Example 1 Consider the following superdiffusion PDE of fourth order

$$\frac{\partial^{1.75}v}{\partial t^{1.75}}+0.05\frac{\partial^4 v}{\partial z^4}=q(z,t),\quad z\in[0,4\pi],\quad t_0=0<t\le t_f=T,$$

with initial condition

$$v(z,0)=2\sin^2 z,\quad z\in[0,4\pi],$$

and boundary condition I

$$v(0,t)=v(4\pi,t)=0,$$
$$v_z(0,t)=v_z(4\pi,t)=0,\quad 0\le t\le T$$

The exact solution of the problem is

$$v(z,t)=2(t+1)\sin^2 z.$$

The error norms L_∞ and L_2 are calculated for $N=100$ and different temporal step sizes. The temporal rate of convergence at $t_f=1.0$ is presented in Table 1. The temporal rate of convergence obtained by the proposed method support the theoretical results.

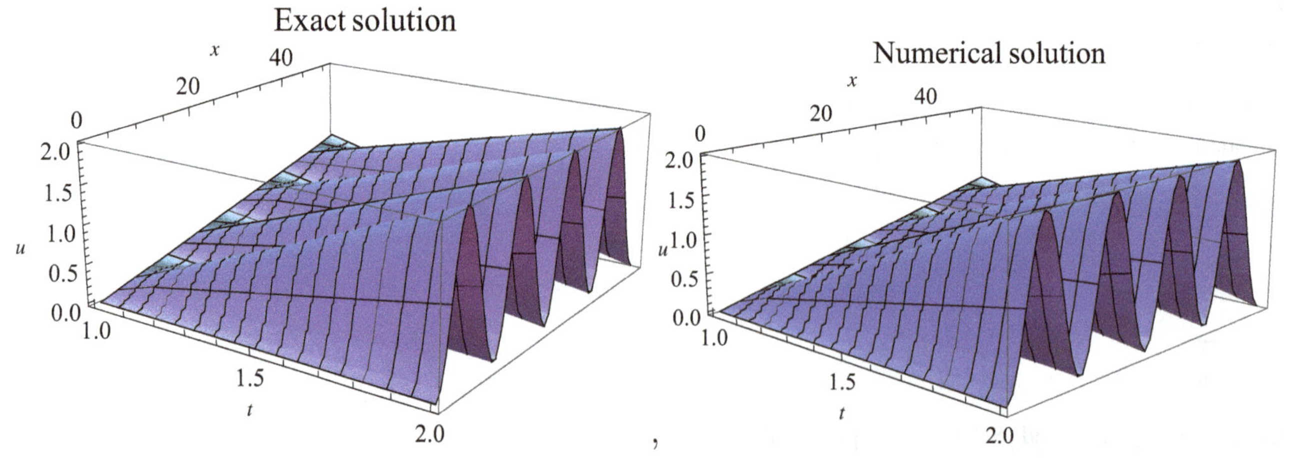

Fig. 1 The results at $N=50$, $K=1000$ and $\Delta t=0.00001$ for Example 1

To indicate the effects of the proposed method for larger ime level K, the exact solution and the approximate solution are plotted using $N = 50$, $K = 1000$ and $\Delta t =$).00001 as shown in Fig. 1. It is clear from the Fig. 1 that

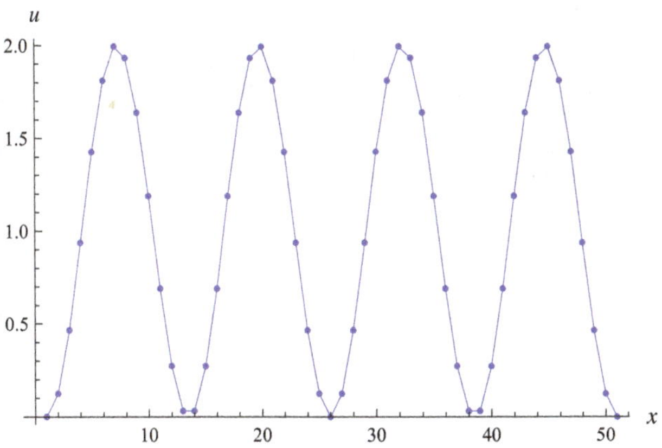

Fig. 2 The exact and numerical solutions at $K = 1000$. *Dotted line* numerical solution, *solid line* exact solution

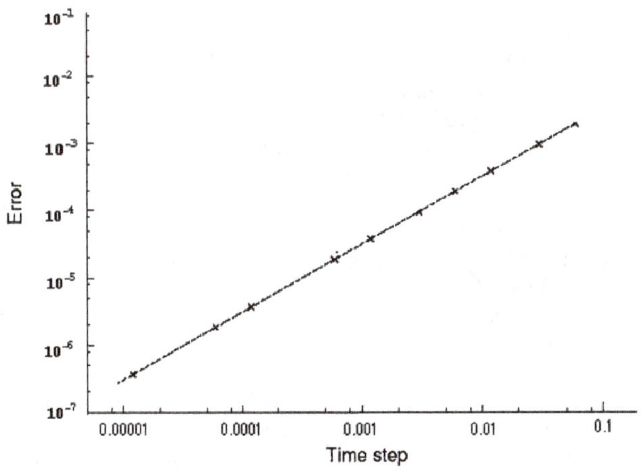

Fig. 3 Errors as a function of the time Δt for $\alpha = 1.50$

the numerical solution is highly consistent with the exact solution, which indicates that the proposed method is very effective. In Fig. 2, the exact solution is represented by solid line and the numerical solution is represented by dotted line at $K = 1000$ time level.

The temporal rate of convergence of L_2 norm error as a function of the time step Δt for $\alpha = 1.50$ is shown in Fig. 3.

Example 2 Consider the following superdiffusion PDE of fourth order

$$\frac{\partial^{1.50} v}{\partial t^{1.50}} + 0.05 \frac{\partial^4 v}{\partial z^4} = q(z,t), \quad z \in [0,1], \quad 0 < t \leq T,$$

with the initial condition

$$v(z,0) = \left(\pi^5 \sin \pi z + \frac{1}{\pi^5} \cos \pi z - \frac{1}{\pi^5} \cos 3\pi z \right), \quad z \in [0,1],$$

and boundary condition III

$$v(0,t) = v(1,t) = 0,$$

$$v_{zz}(0,t) - \frac{8}{\pi^9} v_z(0,t) = v_{zz}(1,t) - \frac{8}{\pi^9} g_z(1,t) = 0, \quad 0 \leq t \leq T.$$

The exact solution of the problem is

$$v(z,t) = (t+1) \left(\pi^5 \sin \pi z + \frac{1}{\pi^5} \cos \pi z - \frac{1}{\pi^5} \cos 3\pi z \right).$$

The error norms L_∞ and L_2 are calculated for $N = 80$ and different temporal step sizes. The temporal rate of convergence at $t_f = 1.0$ is presented in Table 2. The temporal rate of convergence obtained by the proposed method support theoretical results.

To indicate the effects of the proposed method for larger time level, the exact solution and the approximate solution are plotted using $N = 80$, $K = 1000$, and $\Delta t = 0.00001$ as shown in Fig. 4. It is clear from the Fig. 4 that the

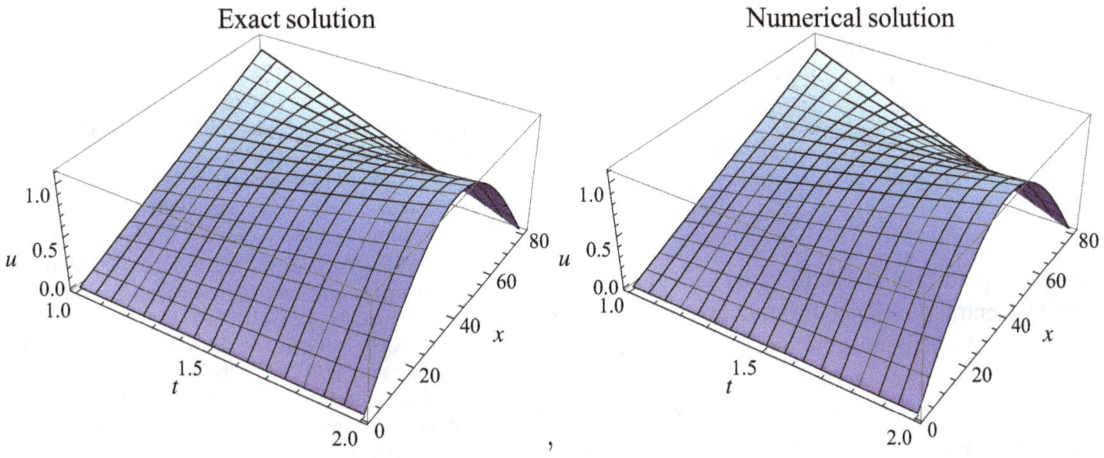

Fig. 4 The results at $N = 80$, $K = 1000$ and $\Delta t = 0.00001$ for Example 2

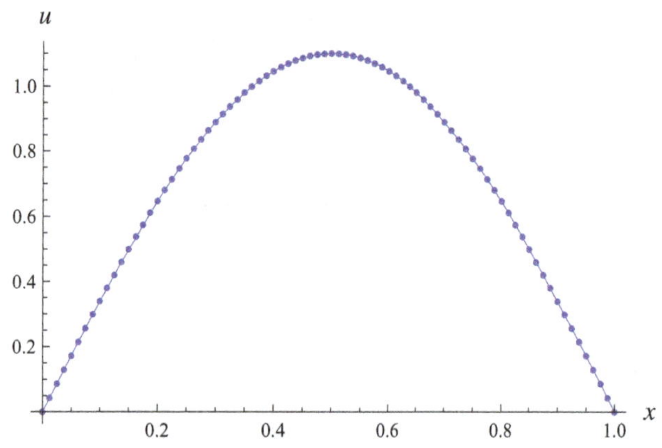

Fig. 5 The exact and numerical solutions at $K = 1000$. *Dotted line* numerical solution, *solid line* exact solution

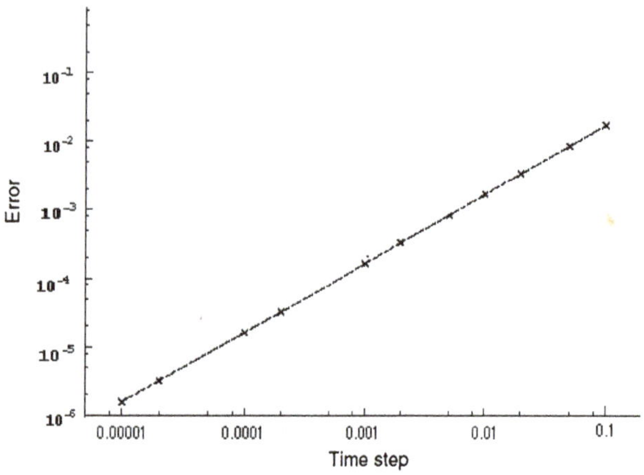

Fig. 6 Errors as a function of the time step Δt for $\alpha = 1.75$

numerical solution is highly consistent with the exact solution, which indicates that the proposed method is very effective. In Fig. 5, the exact solution is represented by solid line and the numerical solution is represented by dotted line at $K = 1000$ time level.

The temporal rate of convergence of L_2 norm error as a function of the time step Δt for $\alpha = 1.75$ is shown in Fig. 6.

Conclusion

The solution of superdiffusion fourth order equation has been developed. Central difference formula and quintic B-spline are used for discretizing time and space variables, respectively. The proposed method is stable and the approximate results approach the analytical results with order $O(\Delta t^2)$. The approximate results obtained by the suggested method support the theoretical estimates.

References

1. Oldham, K.B., Spanier, J.: The fractional calculus. Academic Press, New York (1974)
2. Miller, K.S., Ross, B.: An introduction to the fractional calculus and fractional differential equations. Wiley, New York (1993)
3. Podlubny, I.: Fractional differential equations. Academic Press, New York (1999)
4. Khan, A., Khan, I., Aziz, T.: Sextic spline solution for solving fourth-order parabolic partial differential equation. Int. J. Comput. Math. **82**, 871–879 (2005)
5. Dehghan, M.: Solution of a partial integro-differential equation arising from viscoelasticity. Int. J. Comput. Math. **83**, 123–129 (2006)
6. Lin, Y., Xu, C.: Finite difference/spectral approximations for time-fractional diffusion equation. J. Comput. Phys. **225**, 1533–1552 (2007)
7. Siddiqi, S.S., Arshed, S.: Numerical solution of time-fractional fourth-order partial differential equations. Int. J. Comput. Math. **92**(7), 1496–1518 (2015). doi:10.1080/00207160.2014.948430
8. Yang, X.H., Xu, D., Zhang, H.X.: Quasi-wavelet based numerical method for fourth-order partial integro-differential equations with a weakly singular kernel. Int. J. Comput. Math. **88**(15), 3236–3254 (2011)
9. Khan, N.A., Khan, N.U., Ayaz, M., Mahmood, A., Fatima, N.: Numerical study of time-fractional fourth-order differential equations with variable coefficients. J. King Saud Univ. (Science) **23**, 91–98 (2011)
10. Zhang, H.X., Han, X.: Quasi-wavelet method for time-dependent fractional partial differential equation. Int. J. Comput. Math. (2013). doi:10.1080/00207160.2013.786050
11. Yang, X.H., Xu, D., Zhang, H.X.: Crank–Nicolson/quasi-wavelets method for solving fourth order partial integro-differential equation with a weakly singular kernel. J. Comput. Phys. **234**, 317–329 (2013)
12. Zhang, H.X., Han, X., Yang, X.H.: Quintic B-spline collocation method for fourth order partial integro-differential equations with a weakly singular kernel. Appl. Math. Comput. **219**, 6565–6575 (2013)
13. Sousa, E.: How to approximate the fractional derivative of order $1 < \alpha \leq 2$. In: Proceedings of FDA10. The 4th IFAC workshop fractional differentiation and its applications
14. Atangana, A., Secer, A.: A note on fractional order derivatives and table of fractional derivatives of some special functions. Abstract and Applied Analysis. **2013**, 8 pages (2013). doi:10.1155/2013/279681 **(Article ID 279681)**
15. Bhrawy, A.H., Abdelkawy, M.A.: A fully spectral collocation approximation for multi-dimensional fractional Schrodinger equations. J. Comput. Phys. **294**, 462–483 (2015)
16. Bhrawy, A.H., Tharwat, M.M., Yildirim, A.: A new formula for fractional integrals of Chebyshev polynomials: application for solving multi-term fractional differential equations. Appl. Math. Model. **37**, 4245–4252 (2013)
17. Bhrawy, A.H.: A new spectral algorithm for a time-space fractional partial differential equations with subdiffusion and superdiffusion. Proc. R. Acad. A **17**, 39–46 (2016)
18. Bhrawy, A.H., Taha, T.M., Machado, J.A.T.: A review of operational matrices and spectral techniques for fractional calculus. Nonlinear Dyn. **81**, 1023–1052 (2015)
19. Bhrawy, A.H., Abdelkawy, M.A., Alzahrani, A.A., Baleanu, D., Alzahrani, E.O.: A Chebyshev–Laguerre Gauss–Radau collocation scheme for solving time fractional sub-diffusion equation on a semi-infinite domain. Proc. R. Acad. Ser. A **16**, 490–498 (2015)

20. Bhrawy, A.H.: A space-time collocation scheme for modified anomalous subdiffusion and nonlinear superdiffusion equations. Eur. Phys. J. Plus **131**, 12 (2016)
21. Doha, E.H., Bhrawy, A.H., Ezz-Eldien, S.S.: An efficient legendre spectral tau matrix formulation for solving fractional sub-diffusion and reaction sub-diffusion equations. J. Comput. Nonlinear Dyn. **10**(2), 021019 (2015)
22. Bhrawy, A.H.: A highly accurate collocation algorithm for 1+1 and 2+1 fractional percolation equations. J. Vib. Control (2015). doi:10.1177/1077546315597815
23. Bhrawy, A., Zaky, M.: A fractional-order Jacobi tau method for a class of time-fractional PDEs with variable coefficients. Math. Methods Appl. Sci. (2015). doi:10.1002/mma.3600

10

Some results on best proximity points of cyclic alpha-psi contractions in Menger probabilistic metric spaces

M. De la Sen[1] · Antonio F. Roldán[2]

Abstract This paper investigates properties of convergence of distances of p-cyclic α-ψ-type contractions on the union of the p subsets of a space X defining probabilistic metric spaces and Menger spaces. The paper also investigates the characterization of both Cauchy and G-Cauchy sequences which are convergent, in particular, to best proximity points. On the other hand, the existence and uniqueness of fixed points and best proximity points of p-cyclic α-ψ-type contractions are also investigated. The fixed points of the p-composite self-mappings, which are obtained from the p-cyclic self-mapping restricted to each of the p subsets in the cyclic disposal, are also investigated while a generalization and some illustrative examples are also given.

Keywords p-cyclic α-ψ contractions · Best proximity points · Probabilistic metric spaces · Menger spaces · Triangular norms

✉ M. De la Sen
manuel.delasen@ehu.es

Antonio F. Roldán
afroldan@ujaen.es

1 Faculty of Science and Technology, Institute of Research and Development of Processes IIDP, University of the Basque Country, PO Box 644 de Bilbao. Barrio Sarriena, 48940 Leioa, Bizkaia, Spain

2 Department of Mathematics, University of Jaén, Paraje de las Lagunillas s/n, 23071 Jaén, Spain

Introduction

Fixed point theory in the framework of probabilistic metric spaces [1–4] is receiving important research attention. See, for instance, [2–4, 7–13]. In addition, Menger probabilistic metric spaces are a special case of the wide class of probabilistic metric spaces which are endowed with a triangular norm [2, 3, 7, 9, 11, 15, 16, 30]. In probabilistic metric spaces, the deterministic notion of distance is considered to be probabilistic in the sense that, given any two points x and y of a metric space, a measure of the distance between them is a probabilistic metric $F_{x,y}(t)$, rather than the deterministic distance $d(x, y)$, which is interpreted as the probability of the distance between x and y being less than t ($t > 0$) [3].

Fixed point theorems in complete Menger spaces for probabilistic concepts of B and C-contractions can be found in [2] together with a new notion of contraction, referred to as (Ψ, C)-contraction. Such a contraction was proved to be useful for multivalued mappings while it generalizes the previous concept of C-contraction. On the other hand, 2-cyclic φ-contractions on intersecting subsets of complete Menger spaces were discussed in [7] for contractions based on control φ-functions. See also [8]. It was found that fixed points are unique. In addition, φ-contractions in complete probabilistic Menger spaces have been also studied in [11] through the use of altering distances. See also [14, 26]. On the other hand, probabilistic Banach spaces versus Fixed Point Theory were discussed in [10]. The concept of probabilistic complete metric space was adapted to the formalism of Banach spaces defined with norms being defined by triangular functions and under a suitable ordering in the considered space. In parallel, mixed monotone operators in such Banach spaces were discussed while the existence of coupled minimal and maximal fixed points for these operators was analyzed and discussed in detail. Further extensions to contractive mappings in complete fuzzy metric spaces using generalized distance distribution functions have been studied in [8, 9] and references therein. The concept of altering distances was exploited in a very general context to derive fixed point results in [14], and extended later on in [15] to Menger probabilistic metric spaces. On the other hand, general fixed point theorems have been very recently obtained in [16] for two new classes of contractive mappings in Menger probabilistic metric spaces. The results have been established for $\alpha - \psi$ contractive mappings and for a generalized β-type one. It has also to be pointed out that the parallel background literature related to results on best proximity points and fixed points in cyclic mappings in metric and Banach spaces as well as topics related to common fixed points is exhaustive including studies of fixed point theory and applications in the fuzzy framework. See, for instance, [5, 6, 13, 17–27, 31–37] as well as references therein.

This paper investigates properties of convergence of distances of p-cyclic contractions on the union of the p subsets of the abstract set X defining the probabilistic metric spaces and the Menger spaces as well as the characterization of Cauchy and G-Cauchy sequences which converge to best proximity points of p-cyclic α-ψ-type contractions. The existence and uniqueness of fixed points and best proximity points of p-cyclic α-ψ-type contractions. The fixed points of the p-composite self-mappings, which are obtained from the cyclic self-mapping restricted to each of the p subsets in the cyclic disposal, are also investigated while illustrative examples and a further generalization are also given.

Denote $\mathbf{R}_+ = \{z \in \mathbf{R} : z > 0\}$, $\mathbf{R}_{0+} = \mathbf{R}_+ \cup \{0\}$, $\mathbf{Z}_+ = \{z \in \mathbf{Z} : z > 0\}$, $\mathbf{Z}_{0+} = \mathbf{Z}_+ \cup \{0\}$, $\bar{n} = \{1, 2, \ldots, n\}$, and denote also by $\mathbf{L}$, the set of distance distribution functions $H : \mathbf{R} \to [0, 1]$, [1], which are non-decreasing and left continuous such that $H(0) = 0$ and $\sup_{t \in \mathbf{R}} H(t) = 1$. Let X be a nonempty set and let the probabilistic metric (or distance) $F : X \times X \to L$ a symmetric mapping from $X \times X$, where X is an abstract set, to the set of distance distribution functions L of the form $H : \mathbf{R} \to [0, 1]$ which are functions of elements $F_{x,y}$ for every $(x, y) \in X \times X$. Then, the ordered pair (X, F) is a probabilistic metric space (PM) [2, 3, 29] if

1. $\forall x, y \in X\big(\big(F_{x,y}(t) = 1 ; \forall t \in \mathbf{R}_+\big) \Leftrightarrow (x = y)\big)$
2. $F_{x,y}(t) = F_{y,x}(t)$; $\forall x, y \in X$, $\forall t \in \mathbf{R}$
3. $\forall x, y, z \in X$; $\forall t_1, t_2 \in \mathbf{R}_+\big(\big(F_{x,y}(t_1) = F_{y,z}(t_2) = 1\big) \Rightarrow \big(F_{x,z}(t_1 + t_2) = 1\big) \big)$ (1.1)

A particular distance distribution function $F_{x,y} \in L$ is a probabilistic metric (or distance) which takes values $F_{x,y}(t)$ identified with a probability distance density function $H : \mathbf{R} \to [0, 1]$ in the set of all the distance distribution functions $\mathbf{L}$.

A Menger PM-space is a triplet (X, F, Δ), where (X, F) is a PM-space which satisfies:

$$F_{x,y}(t_1 + t_2) \geq \Delta\left(F_{x,z}(t_1),\ F_{z,y}(t_2)\right); \quad \forall x, y, z \in X, \quad \forall t_1, t_2 \in \mathbf{R}_{0+} \tag{1.2}$$

under $\Delta : [0, 1] \times [0, 1] \to [0, 1]$ is a t-norm (or triangular norm) belonging to the set $\mathbf{T}$ of t-norms which satisfies the properties:

1. $\Delta(a, 1) = a$
2. $\Delta(a, b) = \Delta(b, a)$
3. $\Delta(c, d) \geq \Delta(a, b)$ if $c \geq a$, $d \geq b$
4. $\Delta(\Delta(a, b),\ c) = \Delta(a, \Delta(b, c))$ (1.3)

A property which follows from the above ones is $\Delta(a, 0) = 0$ for $a \in [0, 1]$. Typical continuous t-norms are the minimum t-norm defined by $\Delta_M(a, b) = \min(a, b)$, the product t-norm defined by $\Delta_P(a, b) = a.b$ and the Lukasiewicz (or nilpotent-minimum) t-norm defined by $\Delta_L(a, b) = \max(a + b - 1,\ 0)$ which are related by the inequalities $\Delta_L \leq \Delta_P \leq \Delta_M$.

The (probabilistic) diameter of a subset A of X is a function from $\mathbf{R}_{0+}$ to $[0, 1]$ defined by $D_A(z) = \sup_{t<z} \inf_{x,\, y \in A} F_{x,y}(t)$ and A is probabilistically bounded if $D_A^p = \sup_{z \in \mathbf{R}_+} D_A(z) = 1$ (D_A^p can be defined equivalently as $\lim_{z \to \infty} D_A(z)$), probabilistically semibounded if $0 < D_A^p < 1$ and probabilistically unbounded if $D_A^p = 0$ [1, 2]. The diameter of a subset $A \subset X$ in the PM-space (X, F), induced by a metric space (X, d), refers to maximum real interval measure, where the argument of the probabilistic metric is unity, that is,

$$\operatorname{diam} A = \begin{cases} \inf\left\{ t \in \mathbf{R}_+ : \left(\sup_{x,y \in A} \left(F_{x,y}(t) : d(x,y) = \sup_{x,y \in A} d(x,y) \right) \right) = 1 \right\} & \text{if } \left\{ t \in \mathbf{R}_+ : \left(\sup_{x,y \in A} \left(F_{x,y}(t) : d(x,y) = \sup_{x,y \in A} d(x,y) \right) \right) = 1 \right\} \neq \emptyset \\ \infty & \text{otherwise} \end{cases} \tag{1.4}$$

Example 1.1 Let X be an abstract nonempty set, (X, F) be a PM-space and (X, d) be a companion metric space and let A be a nonempty subset of X with $F_{x,y}(t) = \frac{\alpha(x,y)t}{\beta(x,y)t+d(x,y)}$ for $t \leq t_1$, where $t_1 = \frac{\sup_{x,y\in A} d(x,y)}{\bar{\alpha}-\bar{\beta}}$, and $F_{x,y}(t) = 1$ for $t > t_1$ with some given positive real functions subject to $\beta(x, y) = \bar{\beta}$, $\alpha(x, y) = \bar{\alpha}\left(\geq \bar{\beta}\right) \in \mathbf{R}_+$ and $\alpha(x, y) = \beta(x, y) = 1$ if $d(x, y) < \sup_{x,y\in A} d(x, y)$. In this case, $\operatorname{diam}(A) = t_1 \leq \infty$ being, in particular, infinity if $\bar{\alpha} = \bar{\beta}$ (i.e., the probability one is reached as a limit as $t \to \infty$) or if $\sup_{x,y\in A} d(x, y)$ is arbitrarily large (i.e., if A is unbounded as a subset of the metric space (X, d)). If $\sup_{x,y\in A} d(x, y) < \infty$ and $\bar{\alpha} > \bar{\beta}$ then $\operatorname{diam}(A) < \infty$.

The (probabilistic) distance in-between the subsets A and B of X defines the argument interval length of zero probability distance in-between points of two subsets A and B of X and it is defined as:

$$D = d(A,B) = \inf\left(z \in \mathbf{R}_{0+} : \sup_{x\in A,\ y\in B} F_{x,y}(z) = 0\right) \tag{1.5}$$

Definition 1.1 [7, 8, 16] Let (X, F, Δ) be a Menger PM-space. Then:

1. A sequence $\{x_n\}$ in X is said to be convergent to x in X if, for every $\varepsilon,\ \lambda \in \mathbf{R}_+$, there exists $n_0 = n_0(\varepsilon, \lambda) \in \mathbf{Z}_{0+}$ such that $F_{x_n,x}(\varepsilon) > 1 - \lambda$, whenever $n \geq n_0$.
2. A sequence $\{x_n\}$ in X is said to be a Cauchy sequence if, for every $\varepsilon,\ \lambda \in \mathbf{R}_+$, there exists $n_0 = n_0(\varepsilon, \lambda) \in \mathbf{Z}_{0+}$ such that $F_{x_n,x_m}(\varepsilon) > 1 - \lambda$, whenever $n, m \geq n_0$.
3. (X, F, Δ) is complete if every Cauchy sequence in X is convergent to a point in X.
4. A sequence $\{x_n\}$ is said to be G-Cauchy if, for every $\varepsilon \in \mathbf{R}_+$, $\lim_{n\to\infty} F_{x_n,x_{n+m}}(\varepsilon) = 1, \forall m \in \mathbf{Z}_{0+}$.
5. (X, F, Δ) is G-complete if every G-Cauchy sequence in X is convergent in X. □

Assertion 1.1 Let (X, F, Δ) be a Menger PM-space with $\Delta = \Delta_M, \Delta_P$ or Δ_L. The following properties hold:

1. $\{x_n\} \subset X$ convergent $\Rightarrow$ $\{x_n\}$ is Cauchy $\Rightarrow$ $\{x_n\}$ is G-Cauchy.
2. (X, F, Δ) G-complete $\Rightarrow (X, F, \Delta)$ is complete

Proof **Proof of (i) Step 1** We first prove that $\{x_n\}$ convergent $\Rightarrow \{x_n\}$ is Cauchy. Since $\{x_n\}$ is convergent then for every $\varepsilon,\ \lambda \in (0, 1) \in \mathbf{R}_+$, there exists $n_0 = n_0(\varepsilon, \lambda) \in \mathbf{Z}_{0+}$ such that $F_{x_m,x}(\varepsilon/2) > 1 - \lambda/2$, $\forall n$, $m(\geq n_0) \in \mathbf{Z}_{0+}$. Then, since $F : \mathbf{R} \to [0, 1]$ is non-decreasing, one gets:

$$\begin{aligned}
F_{x_n,x_m}(\varepsilon) &\geq \Delta\left(F_{x_n,x}(\varepsilon/2),\ F_{x_m,x}(\varepsilon/2)\right)\\
&\geq \min\Big(\Delta_M\left(F_{x_n,x}(\varepsilon/2),\ F_{x_m,x}(\varepsilon/2)\right),\\
&\quad \Delta_P\left(F_{x_n,x}(\varepsilon/2),\ F_{x_m,x}(\varepsilon/2)\right),\\
&\quad \Delta_L\left(F_{x_n,x}(\varepsilon/2),\ F_{x_m,x}(\varepsilon/2)\right)\Big)\\
&= \min\Big(\min\left(F_{x_n,x}(\varepsilon/2),\ F_{x_m,x}(\varepsilon/2)\right),\\
&\quad F_{x_n,x}(\varepsilon/2).F_{x_m,x}(\varepsilon/2),\\
&\quad \max\left(F_{x_n,x}(\varepsilon/2) + F_{x_m,x}(\varepsilon/2) - 1\right), 0\Big)\\
&= F_{x_n,x}(\varepsilon/2) + F_{x_m,x}(\varepsilon/2) - 1\\
&= \Delta_L\left(F_{x_n,x}(\varepsilon/2),\ F_{x_m,x}(\varepsilon/2)\right)\\
&> 2(1 - \lambda/2) - 1 = 1 - \lambda; \quad \forall n, m(\geq n_0) \in \mathbf{Z}_{0+}
\end{aligned} \tag{1.6}$$

and then $\{x_n\}$ is a Cauchy. Since the above inequalities hold for any $n, m(\geq n_0) \in \mathbf{Z}_{0+}$, it turns out that $\liminf_{n\to\infty} F_{x_n,x_{n+m}}(\varepsilon) > 1 - \lambda$, $\forall m \in \mathbf{Z}_+$.

Proof of (i) Step 2 We next prove by contradiction that $\{x_n\}$ is Cauchy $\Rightarrow$ $\{x_n\}$ is G-Cauchy. Assume that $\{x_n\}$ is Cauchy while it is not G-Cauchy. Then, $\liminf_{n\to\infty} F_{x_n,x_{n+m}}(\varepsilon_1/2) > 1 - \lambda$ and $\limsup_{n\to\infty} F_{x_n,x_{n+m}}(\varepsilon_1) < 1 - 2\lambda$, for some $m \in \mathbf{Z}_{0+}$, some $\varepsilon_1 \in \mathbf{R}_+$ and some given $\lambda = \lambda(\varepsilon_1) \in (0,\ 1/2)$. Since $F_{x,y}(t)$ is non-decreasing in t for all $x, y \in X$, so that $F_{x,y}(\varepsilon_1) \geq F_{x,y}(\varepsilon_1/2)$, then:

$$1 - 2\lambda > \liminf_{n\to\infty} F_{x_n,x_{n+m}}(\varepsilon_1) \geq \liminf_{n\to\infty} F_{x_n,x_{n+m}}(\varepsilon_1/2) > 1 - \lambda, \tag{1.7}$$

$\forall m \in \mathbf{Z}_+$. But then $\lambda < \lambda/2$. Then, $\{x_n\}$ is G-Cauchy.

Proof of (ii): Let (X, F, Δ) be G-complete and let $\{x_n\} \subset X$ be any given Cauchy sequence. Then $\{x_n\}$ is G-Cauchy, from property (1), and convergent to some $x \in X$ since (X, F, Δ) is G-complete. Since $\{x_n\}$ is an arbitrary Cauchy sequence convergent in X, it turns out that (X, F, Δ) is complete. □

The $(\varepsilon,\ \lambda)$-topology in a Menger in a PM-space (X, F, Δ) is a Hausdorff topology introduced by the family of neighborhoods N_x of a point $x \in x$ given by $N_x = \{N_x(\varepsilon, \lambda)\ :\ \varepsilon \in \mathbf{R}_+,\ \lambda \in (0, 1)\}$ where $N_x(\varepsilon, \lambda) = \{z \in X\ : F_{x,z}(\varepsilon) > 1 - \lambda\}$. In this topology, a function is continuous at $x_0 \in X$ if and only if $\{f(x_n)\} \to f(x_0)$ for every convergent sequence $\{x_n\} \to x_0$. See [1, 16] for more details. □

We next denote by $\varphi(z^+)$ and $\varphi(z^-)$, respectively, the right and left limits of $\varphi(t)$ as $t \to z$.

Definition 1.2 A function $\varphi : \mathbf{R} \to \mathbf{R}_{0+}$ is said to be a Φ_{xy}-function if, for given real constants $x, y \in \mathbf{R}_{0+}$, with $y \geq x$, it satisfies the following conditions:

1. $\varphi(t)$ is strictly increasing for $t \in [x, \infty)$
2. $\varphi(x^+) = y$
3. $\varphi(t)$ is everywhere left-continuous
4. $\varphi(t) = 0$ for $t \in (-\infty, x]$ □

The set of functions Φ_{xx} is simply denoted by Φ_x. If $\varphi : \mathbf{R} \to \mathbf{R}_{0+}$ is in Φ_x then $\varphi(x) = \varphi(x^-) = 0 \leq \varphi(x^+) = x$ and then if $\varphi \in \Phi_0$, it is continuous at $t = 0$. Note also that the particular set of functions Φ_0 coincides with the set of functions Φ of [15, 16] which have continuity at cero. Definition 1.2 will be used in the following to establish the class of contractions under investigation using functions in the sets Φ_D and $\Phi_{D\hat{D}}$, where D is the distance in-between adjacent subsets of the cyclic disposal in X.

Definition 1.3 [16] A function $\psi : \mathbf{R}_{0+} \to \mathbf{R}_{0+}$ is said to be a Ψ-function if continuous with $\psi(0) = 0$ and $\psi^n(a_n) \to 0$ when $a_n \to 0$ as $n \to \infty$.

Main results on best proximity points for p-cyclic α-ψ-type contractions

The definition of a $p(\geq 2)$-cyclic α-ψ-type contraction follows:

Definition 2.1 Let (X, F) be a PM-space and let A_i be nonempty subsets of $X, \forall i \in \bar{p}$ such that $D = d(A_i, A_{i+1})$ is the common distance in-between adjacent subsets, $\forall i \in \bar{p}$. Then $T : \bigcup_{i \in \bar{p}} A_i \to \bigcup_{i \in \bar{p}} A_i$ is a $p(\geq 2)$-cyclic α-ψ-type contraction if $T(A_i) \subseteq A_{i+1}$, $\forall i \in \bar{p}$ and there exist two functions $\alpha : \left(\bigcup_{i \in \bar{p}} A_i\right) \times \left(\bigcup_{i \in \bar{p}} A_i\right) \times \mathbf{R}_+ \to \mathbf{R}_{0+}$ and $\psi \in \Psi$ satisfying the following inequality:

$$\alpha(x, y, t)\left(\frac{1}{F_{Tx,Ty}(\varphi(Kt + (1 - K)D))} - 1\right) \leq \psi\left(\frac{1}{F_{x,y}(\varphi(t))} - 1\right), \qquad (2.1)$$

$\forall(x, y) \in A_i \times A_{i+1}$, $\forall i \in \bar{p}$, $\forall t(> D) \in \mathbf{R}_+$, where $K \in (0, 1)$, for any given $\varphi \in \Phi_{D\hat{D}}$ and any given real $\hat{D} \geq D$.

Note that if $T : \bigcup_{i \in \bar{p}} A_i \to \bigcup_{i \in \bar{p}} A_i$ is $p(\geq 2)$-cyclic and if $x \in A_i$ for some $i \in \bar{p}$ then for any $j \in \bar{p}$ and $n \in \mathbf{Z}_{0+}$, $T^{np+i+j}x \in A_{np+i+j} = A_k$ for some $k \in \bar{p}$ since if $n, m \in \mathbf{Z}_+$ and $n \equiv m(\mathrm{mod}\, p)$ then $A_m = A_n$. In particular, $A_{i+np} = A_i$, $\forall i \in \bar{p}$, $\forall n \in \mathbf{Z}_{0+}$. It can be pointed out that p-cyclic contractions include the case of cyclic self-mappings T on X such that $X = \bigcup_{i \in \bar{p}} A_i$. In this case, $\{A_i\}_{i \in \bar{p}}$ is said to be a cyclic representation of (X, T). On the other hand, note that is an α-ψ-type contraction if (2.1) holds with $D = 0$ for $t \in \mathbf{R}_+$ [16]. The distances in-between adjacent subsets are assumed to be identical just to facilitate the exposition by simplifying the contractive condition to the form (2.1) so as to make less involved their associate calculations. Note that the distances in-between adjacent subsets in non-expansive cyclic self-mappings are identical in uniformly convex Banach spaces [27].

An equivalent constraint to (2.1) is now discussed:

Proposition 2.1 *The constraint* (2.1) *is identical to:*

$$F_{Tx,Ty}(\varphi(t)) \geq \frac{\alpha(x, y, K^{-1}(t - D) + D)}{\alpha(x, y, K^{-1}(t - D) + D) + \psi\left(F_{x,y}^{-1}(\varphi(K^{-1}(t - D) + D)) - 1\right)}, \qquad (2.2)$$

$\forall(x, y) \in A_i \times A_{i+1}$, $\forall i \in \bar{p}$, $\forall t(> D) \in \mathbf{R}_+$ *which is also identical, if* $\alpha(x, y, t) \neq 0$, $\forall(x, y) \in A_i \times A_{i+1}$, $\forall i \in \bar{p}$, $\forall t(> D) \in \mathbf{R}_+$, to

$$F_{Tx,Ty}^{-1}(\varphi(t)) \leq 1 + \alpha^{-1}\left(x, y, K^{-1}(t - D) + D\right) \psi\left(F_{x,y}^{-1}\left(\varphi(K^{-1}(t - D) + D)\right) - 1\right) \qquad (2.3)$$

Proof Note that, given $K \in (0, 1)$ and $D \geq 0$, the function $\beta_{K,D} : [D, \infty) \to [D, \infty)$ defined by $\beta_{K,D}(t) = Kt + (1 - K)D$ for $t \in (D, \infty)$ is a strictly increasing, bijective and bicontinuous function of (then continuous) inverse $\beta_{K,D}^{-1}(t) = K^{-1}(t - D) + D$ for $t \in (D, \infty)$, which can be extended by continuity to $t = D$ by defining $\beta(D) = D$.Then, one gets from (2.1) that

$$\frac{F_{Tx,Ty}(\varphi(t))}{1 - F_{Tx,Ty}(\varphi(t))} \geq \frac{\alpha(x, y, K^{-1}(t - D) + D)}{\psi\left(F_{x,y}^{-1}(\varphi(K^{-1}(t - D) + D)) - 1\right)}, \qquad (2.4)$$

$\forall(x, y) \in A_i \times A_{i+1}$, $\forall i \in \bar{p}$, $\forall t(> D) \in \mathbf{R}_+$ which is identical to (2.2). Equation (2.3) follows directly from (2.1) if $\alpha(x, y, K^{-1}(t - D) + D) \neq 0$. □

Note from (2.2) that, if $\alpha(x, y, t) = 1$ and $\varphi(t) = \psi(t) = t$, $\forall x, y \in X$, $\forall t \in \mathbf{R}$, then the p-cyclic α-ψ-type contraction $T : \bigcup_{i \in \bar{p}} A_i \to \bigcup_{i \in \bar{p}} A_i$ becomes a p-cyclic B-contraction since one gets $F_{Tx,Ty}(t) \geq F_{x,y}(K^{-1}(t - D) + D)$, $\forall x, y \in X$, $\forall t \in \mathbf{R}_+$ from (2.2) if $\alpha(x, y, t) = 1$ and $\varphi(t) = \psi(t) = t$. Thus, a p-cyclic B-contraction is a particular type of p-cyclic α-ψ-type contraction. See [2] for the case $D = 0$. Some basic properties of a p-cyclic α-ψ-type contraction are now given:

Proposition 2.2 *Let* (X, F) *be a PM-space. Let* $T : \bigcup_{i \in \bar{p}} A_i \to \bigcup_{i \in \bar{p}} A_i$ *be a p-cyclic* α-ψ*-type contraction with* $A_i \subset X$ *being bounded,* $\forall i \in \bar{p}$ *with* $D = d(A_i, A_{i+1})$, $\forall i \in \bar{p}$ *and* $\bar{D} = \max_{i \in \bar{p}} \mathrm{diam}(A_i)$ *being the distance in-between adjacent subsets,* $\forall i \in \bar{p}$. *Then, the following properties hold provided that* $\varphi \in \Phi_{D\hat{D}}$ *for any given* $\hat{D} \geq D$:

1. $F_{T^n x, T^n y}(\varphi(t)) = 1, \quad \forall(x,y) \in A_i \times A_{i+1}, \quad \forall i \in \bar{p}, \forall t\,(>D+2\bar{D}) \in \mathbf{R}_+, \forall n \in \mathbf{Z}_{0+}$.
2. $\lim_{t\to\infty} F_{T^n x, T^n y}(\varphi(t)) = 1, \ \forall(x,y) \in A_i \times A_{i+1}, \ \forall i \in \bar{p}, \forall n \in \mathbf{Z}_{0+}$.
3. $F_{x,y}(\varphi(t)) = F_{x,y}(\tau) = 0, \ \forall(x,y) \in A_i \times A_{i+1}, \ \forall i \in \bar{p}, t \in (-\infty, D), \tau \in (-\infty, \hat{D})$.
4. $F_{x,y}(\varphi(D^+)) = F_{x,y}(\hat{D}) = 0; \ \forall(x,y) \in A_i \times A_{i+1}, \ \forall i \in \bar{p}$ if $\hat{D} = D, F_{x,y}(\varphi(D^+)^+) = F_{x,y}(\hat{D}^+) = 1; \ \forall(x,y) \in A_i \times A_{i+1}, \ \forall i \in \bar{p}$ if $\hat{D} = D, F_{x,y}(\varphi(D^+)) = F_{x,y}(\hat{D}) = 1;$ $\forall(x,y) \in A_i \times A_{i+1}, \ \forall i \in \bar{p}$ if $\hat{D} > D$

where $\hat{D} = \varphi(D^+) = \lim_{t\to D^+} \varphi(t)$ *and* $F_{x,y}(\hat{D}^+) = \lim_{\tau\to D^+} \lim_{t\to\tau^+} F_{x,y}(\varphi(t))$.

Proof Since A_i and A_{i+1} are bounded then the maximum distance in-between any two points of adjacent subsets is not larger than $D + 2\bar{D}$. Then, $\lim_{s\to\infty} F_{x,y}(s) = \max_{i\in\bar{p}} \max_{z\in A_i,\, \omega\in A_{i+1}} F_{z,\omega}(\varphi(t)) = 1, \forall(x,y) \in A_i \times A_{i+1}, \forall i \in \bar{p}$, if $t \in (D+2\bar{D}, \infty)$, since the distance distribution function $F : \mathbf{R} \to [0,1]$ is non-decreasing and left-continuous, and $T(A_i) \subseteq A_{i+1}$ with $D = d(A_i, A_{i+1}), \forall i \in \bar{p}$. Then,

$$F^{-1}_{x,y}\left(\varphi\left(K^{-1}(t-D)+D\right)\right) = F_{x,y}\left(\varphi\left(K^{-1}(t-D)+D\right)\right) = 1 \quad \text{for } t \in (D+2\bar{D}, \infty) \text{ so that}$$

$$\psi\left(F^{-1}_{x,y}\left(\varphi\left(K^{-1}(t-D)+D\right)\right) - 1\right) = \psi(0) = 0, \quad t \in (D+2\bar{D}, \infty)$$

since $K \in (0, 1)$, $K^{-1}(t-D)+D > 2\bar{D}+D$ and $\varphi(K^{-1}(t-D)+D) > \varphi(2\bar{D}+D) \ \forall t\,(>D+2\bar{D}) \in \mathbf{R}_+$, since $\varphi : \mathbf{R} \to \mathbf{R}_{0+}$ is strictly increasing and left-continuous in (D, ∞), $F : \mathbf{R} \to [0,1]$ is non-decreasing and left continuous and $\psi(0) = 0$. Thus, from Proposition 2.1 [Eq. (2.3)], $F_{Tx,Ty}(\varphi(t)) = F^{-1}_{Tx,Ty}(\varphi(t)) = 1$ for $t \in (D+2\bar{D}, \infty)$. Again, since $K \in (0, 1)$, $K^{-1}(t-D) + D > 2\bar{D}+D$ and $\varphi(K^{-1}(t-D)+D) > \varphi(2\bar{D}+D)$, since $\varphi : \mathbf{R} \to \mathbf{R}_{0+}$ is strictly increasing and left-continuous in (D, ∞), if $\varphi \in \Phi_D$(Definitions 2.1 and 1.2), $F : \mathbf{R} \to [0,1]$ is non-decreasing and left continuous and $\psi(0) = 0$, then one gets for $t \in (D+2\bar{D}, \infty)$, that:

$$F_{Tx,Ty}\left(\varphi\left(K^{-1}(t-D)+D\right)\right) = F^{-1}_{Tx,Ty}\left(\varphi\left(K^{-1}(t-D)+D\right)\right) = 1;$$

$$\psi\left(F^{-1}_{Tx,Ty}\left(\varphi\left(K^{-1}(t-D)+D\right)\right) - 1\right) = \psi(0) = 0$$

so that, again one gets from Proposition 2.1 [Eq. (2.3)], $F_{T^2x,T^2y}(\varphi(t)) = F^{-1}_{T^2x,T^2y}(\varphi(t)) = 1$ for $t \in (D+2\bar{D}, \infty)$. Now, proceed by complete induction by assuming that $F_{T^n x,T^n y}(\varphi(t)) = F^{-1}_{T^n x,T^n y}(\varphi(t)) = 1$ for $t \in (D+2\bar{D}, \infty)$ for any given $n \in \mathbf{Z}_{0+}$. Since $K \in (0, 1)$, $K^{-1}(t-D) + D > 2\bar{D} + D$, $\varphi(K^{-1}(t-D)+D) > \varphi(2\bar{D}+D)$ for $t \in (D+2\bar{D}, \infty)$, $F : \mathbf{R} \to [0,1]$ is non-decreasing and left continuous and $\psi(0) = 0$, one has for $t \in (D+2\bar{D}, \infty)$, $T^n x \subset A_{i+n} \equiv A_j$ and $T^n y \subset A_{i+n+1} \equiv A_{j+1}$ for a unique integer $j = j(i) \in \overline{p-1} \cup \{0\}$ fulfilling $n = qp + j - i \geq i$ for some $q = q(i) \in \mathbf{Z}_+$ and the given $i \in \bar{p}$ that:

$$F_{T^n x,T^n y}\left(\varphi\left(K^{-1}(t-D)+D\right)\right) = F^{-1}_{T^n x,T^n y}\left(\varphi\left(K^{-1}(t-D)+D\right)\right) = 1;$$
$$\psi\left(F^{-1}_{T^n x,T^n y}\left(\varphi\left(K^{-1}(t-D)+D\right)\right) - 1\right) = \psi(0) = 0.$$

Then, from Proposition 2.1 [Eq. (2.3)], $F_{T^{n+1}x,T^{n+1}y}(\varphi(t)) = F^{-1}_{T^{n+1}x,T^{n+1}y}(\varphi(t)) = 1$ for $t \in (D+2\bar{D}, \infty)$. Hence, the proofs of Properties (1) and (2) follow by complete induction.

Properties (3)–(5) follow directly from the definitions of the sets $\Phi_{D\hat{D}}$ for $\tau = \varphi(t)$, being equivalent to $t = t(\tau) = \arg\left(z \in \mathbf{R} : \varphi(z^-) = \tau \text{ for } \tau \in (\varphi(D), \hat{D})\right)$ which is point-wise unique for any $\tau \in (\varphi(D), \hat{D})$, zero for $t \in \mathbf{R}_{0-}$ if $\Phi_{D\hat{D}}$ and, left-continuous and strictly increasing for $t(\in \mathbf{R}) > D$. □

It turns out that Proposition 2.2 holds, in particular, for $\varphi \in \Phi_D$. The α-admissibility of α-ψ-type contractions is defined to state layer on the main result:

Definitions 2.2 Let (X, F) a PM-space. Then:

1. an α-ψ-type contraction $T : X \to X$ is α-admissible for a given function $\alpha : X \times X \times \mathbf{R}_+ \to \mathbf{R}_{0+}$ [16] if

$$\forall x, y \in X, \forall t \in \mathbf{R}_+[(\alpha(x,y,t) \geq 1) \Rightarrow (\alpha(Tx,Ty,t) \geq 1)],$$

2. a p-cyclic α-ψ-type contraction $T : \bigcup_{i\in\bar{p}} A_i \to \bigcup_{i\in\bar{p}} A_i$ is α-admissible for a given function $\alpha : \left(\bigcup_{i\in\bar{p}} A_i\right) \times \left(\bigcup_{i\in\bar{p}} A_i\right) \times \mathbf{R}_+ \to \mathbf{R}_{0+}$ if

$$\forall(x,y) \in A_i \times A_{i+1}, \forall i \in \bar{p}, \forall t \in \mathbf{R}_+[(\alpha(x,y,t) \geq 1) \Rightarrow (\alpha(Tx,Ty,t) \geq 1)].$$

□

Definitions 2.3 $T : \bigcup_{i\in\bar{p}} A_i \to \bigcup_{i\in\bar{p}} A_i$ be a $\alpha - \psi$-type p-cyclic mapping, $\varphi \in \Phi_{D\hat{D}}$ and $\psi \in \Psi$. Then:

1. the pair $(x, Tx) \in (clA_i, clA_{i+1})$, where $cl(.)$ stands for the closure of the (.)-set, for any given $i \in \bar{p}$ is a pair of quasi-best proximity points if $F_{x,Tx}(\hat{D} + \varepsilon) > 1 - \lambda$ for any given $\varepsilon \in \mathbf{R}_+$ and $F_{x,Tx}(D) = F_{x,Tx}(D^-) = 0$.

Each of them is a quasi-best proximity point in the corresponding subset A_i.

2. A quasi-best proximity point is a best proximity point if $F_{x,Tx}(D+\varepsilon) > 1-\lambda$ for any given $\varepsilon \in \mathbf{R}_+$ and $\lambda \in (0, 1)$. □

Remarks 2.1 Since $\varphi \in \Phi_{D\hat{D}}$ then $\hat{D} = \varphi(D^+) \geq D$. If $\varphi \in \Phi_D$, then $\hat{D} = \varphi(D^+) = D$ and, if the pair (x, Tx) is a pair of quasi-best proximity points, then $F_{x,Tx}(D) = F_{x,Tx}(\varphi(D^+)) = 0$ and $F_{x,Tx}(D^+) = 1$ [see also the two first properties of Proposition 2.2 (iv)].

If $\varphi \in \Phi_{D\hat{D}}$ with $\hat{D} = \varphi(D^+) > D$ then $F_{x,Tx}(\hat{D}+\varepsilon) > 1-\lambda$ for arbitrarily small positive real constants ε and λ so that $F_{x,Tx}(\hat{D}^+) = 1$ and $F_{x,Tx}(D) = F_{x,Tx}(D^-) = 0$ but it is not guaranteed that $F_{x,Tx}(D^+) = 1$, then, it is not guaranteed in this case that (x, Tx) is a pair of best proximity points if φ is not continuous at D. □

The most important of the main results of this paper follows below:

Theorem 2.1 *Let (X, F, Δ) be a G-complete PM-space and let $T : \bigcup_{i\in\bar{p}} A_i \to \bigcup_{i\in\bar{p}} A_i$ be a p-cyclic α-ψ-type contraction with $\varphi \in \Phi_{D\hat{D}}$ for some $\hat{D}(\in \mathbf{R}) \geq D$ satisfying the following conditions*:

1. $D = d(A_i, A_{i+1}) > 0$ and $\mathrm{diam}(A_i) > \hat{D} - D$, $\forall i \in \bar{p}$,
2. *T is α-admissible,*
3. *there exists $x_0 \in \bigcup_{i\in\bar{p}} A_i$ such that $\alpha(x_0, Tx_0, t) \geq 1$, $\forall t \in \mathbf{R}_+$,*
4. if *$\{x_n\}(\subset X) \to x$ is a Picard iteration generated as $x_{n+1} = Tx_n$, $\forall n \in \mathbf{Z}_{0+}$ with $x_0 \in \bigcup_{i\in\bar{p}} A_i$, such that $\alpha(x_n, Tx_n, t) \geq 1$, $\forall n \in \mathbf{Z}_{0+}$, $\forall t \in \mathbf{R}_+$ then $\alpha(x_n, x, t) \geq 1$, $\forall n \in \mathbf{Z}_{0+}$, $\forall t \in \mathbf{R}_+$.*

Then, the following properties hold:

1. $F_{T^{np}x_0,\, T^{(n+m)p+j}x_0}((pm+j)(\varphi(D)+\varepsilon)) > 1-\lambda$, *$\forall t(>D) \in \mathbf{R}_+$, $\forall j \in \bar{p}$ for any given real constants $\varepsilon \in \mathbf{R}_+$ and $\lambda \in (0,1)$, some $n_0 = n_0(\varepsilon, \lambda) \in \mathbf{Z}_+$ and $\forall n(\geq n_0) \in \mathbf{Z}_{0+}$, $\forall m \in \mathbf{Z}_+$, and*

$$F_{T^n x_0, T^{n+1}x_0}(\varphi(t)) \geq \frac{1}{1+\psi^n\left(F^{-1}_{x_0,Tx_0}(\varphi(K^{-n}(t-D)+D)) - 1\right)},$$
$$\forall t(>D) \in \mathbf{R}_+, \quad \forall n \in \mathbf{Z}_+.$$

2. *If $\varphi \in \Phi_0$ and $\bigcap_{i\in\bar{p}} A_i \neq \emptyset$ then*

$$\lim_{n\to\infty} F_{T^n x_0,\, T^{n+1}x_0}(t) = \lim_{n\to\infty} F_{T^{np+j}x_0,\, T^{(n+m)p+j+1}x_0}(t) = 1,$$
$$\forall t \in \mathbf{R}_+.$$

Furthermore, $\{T^n x_0\} \subset \bigcup_{i\in\bar{p}} A_i$ and $\{T^{np+j}x_0\} \subset A_{i+j}$ are both Cauchy and G-Cauchy convergent sequences to a limit point $x \in \bigcap_{i\in\bar{p}} cl A_i$. If the subsets A_i are closed for $i \in \bar{p}$ then *$x = Tx = T^p x$, that is, it is a fixed point of the self-mappings $T : \bigcup_{i\in\bar{p}} A_i \to \bigcup_{i\in\bar{p}} A_i$ and $T^p (\equiv T \circ T^{p-1}) : \bigcup_{i\in\bar{p}} A_i \big| A_k \to A_k$, $\forall k \in \bar{p}$, $\forall x_0 \in A_i$, $\forall i \in \bar{p}$, $\forall j \in \overline{p-1} \cup \{0\}$.*

3. *If A_i is closed, $\forall i \in \bar{p}$ and $\bigcap_{i\in\bar{p}} A_i = \emptyset$* then *if $x_0 \in A_i$ for any $i \in \bar{p}$ then there is a limit cycle $(\bar{x}_i, T\bar{x}_i, \ldots, T^{p-1}\bar{x}_i)$, to which the sequence $\{T^n x_0\}$ converges, with $T^p\bar{x}_i = \bar{x}_i$, $\bar{x}_i \in A_i$, $\bar{x}_{i+j} = T^j\bar{x}_i \in A_{i+j}$, $\forall i \in \bar{p}, \forall j \in \overline{p-1} \cup \{0\}$ being a fixed point of the composite self-mappings (of domain and image restricted to each of the subsets) $T^p : \bigcup_{i\in\bar{p}} A_i \big| A_k \to A_k, \forall k \in \bar{p}$, and it is also a quasi-best proximity point (in particular, a best proximity point if $\varphi \in \Phi_D \equiv \Phi_{D\hat{D}}$) of $T : \bigcup_{i\in\bar{p}} A_i \to \bigcup_{i\in\bar{p}} A_i$. The subsequence $\{T^{np+j}x_0\}$ in A_{i+j} converges to $x_{i+j} = T^j x_i, \forall j \in \bar{p}$* if *$x_0 \in A_i$ for some $i \in \bar{p}$. Furthermore:*

$$\lim_{n\to\infty} F_{T^{np}x_0,\, T^{(n+m)p}x_0}(\varepsilon) = F_{x_0,Tx_0}(\infty) = 1,$$
$$F_{T^{np}x_0,\, T^{(n+m)p}x_0}(\varepsilon) > 1-\lambda$$
$$F_{T^{npj}x_0,\, T^{npj+1}x_0}(\varphi(D)+\varepsilon) > 1-\lambda,$$
$$\lim_{n\to\infty} F_{T^{npj}x_0,\, T^{npj+1}x_0}(\varphi(D)+\varepsilon) = F_{\bar{x}_{i+j},\bar{x}_{i+j+1}} = 1$$

for any given $\varepsilon \in \mathbf{R}_+$, $\lambda \in (0, 1)$ and for some $n_0 = n_0(\varepsilon, \lambda)$, $\forall n(\geq n_0) \in \mathbf{Z}_{0+}$.

Proof Let $x_0 \in \bigcup_{i\in\bar{p}} A_i$ such that the condition (3) holds. Since $T : \bigcup_{i\in\bar{p}} A_i \to \bigcup_{i\in\bar{p}} A_i$ is a p-cyclic α-ψ-type contraction, $x_0 \in \bigcup_{i\in\bar{p}} A_i$, $\alpha(x_0, Tx_0, t) \geq 1$, $x_{n+1} = Tx_n = T^{n+1}x_0 \neq x_n$, $\forall n \in \mathbf{Z}_{0+}$, $\forall t \in \mathbf{R}_+$ and T is α-admissible, assume that $\alpha(x_j, Tx_j, t) \geq 1$, $\forall j \in \bar{n} \cup \{0\} \in \mathbf{Z}_{0+}$, $\forall t \in \mathbf{R}_+$. Since $\alpha(x_n, Tx_n, t) \geq 1 \Rightarrow \alpha(Tx_n, Tx_{n+1}, t) \geq 1$, $\forall t \in \mathbf{R}_+$, then $\alpha(x_j, Tx_j, t) \geq 1$, $\forall j \in \overline{n+1} \cup \{0\} \in \mathbf{Z}_{0+}$, $\forall t \in \mathbf{R}_+$. Then, $\alpha(x_j, Tx_j, t) \geq 1$, $\forall j \in \bar{n} \cup \{0\} \in \mathbf{Z}_{0+}$, $\forall t \in \mathbf{R}_+ \Rightarrow \alpha(x_j, Tx_j, t) \geq 1$, $\forall j \in \overline{n+1} \cup \{0\} \in \mathbf{Z}_{0+}$, $\forall t \in \mathbf{R}_+$. It has been proved by complete induction that $\alpha(x_n, Tx_n, t) \geq 1$, $\forall n \in \mathbf{Z}_{0+}$, $\forall t \in \mathbf{R}_+$ provided that $\alpha(x_0, Tx_0, t) \geq 1$, $\forall t \in \mathbf{R}_+$.Then $\alpha(x_n, Tx_n, t) \in [1, +\infty)$ and $\alpha^{-1}(x_n, Tx_n, t) \in (0, 1]$, $\forall n \in \mathbf{Z}_{0+}$, $\forall t \in \mathbf{R}_+$ if $\alpha(x_0, Tx_0, t) \geq 1, \forall t \in \mathbf{R}_+$. On the other hand, since $\varphi \in \Phi_{D\hat{D}}$ and since $T : \bigcup_{i\in\bar{p}} A_i \to \bigcup_{i\in\bar{p}} A_i$ is a p-cyclic α-ψ-type contraction, φ is strictly increasing in (D, ∞) with $\varphi(D) = \hat{D} \geq D > 0$(see Definitions 2.1 and 1.2) and then there exists $t(>D) \in \mathbf{R}_+$ such that $F_{x_0,Tx_0}(t) > 0$, $\forall t(>D) \in \mathbf{R}_+$. Since the distance distribution function is non-decreasing and left-continuous and $K^{-1} > 1$, then $F_{x_0,Tx_0}(\varphi(K^{-1}(t-D)+D)) \geq F_{x_0,Tx_0}(t) > 0$ and $0 \leq F^{-1}_{x_0,Tx_0}(\varphi(K^{-1}(t-D)+D)) - 1 < +\infty$, $\forall t(>D) \in \mathbf{R}_+$. Now, note that $\alpha^{-1}(x_n, Tx_n, t) \in (0, 1]$ since $\alpha(x_n, Tx_n, t) \in [1, \infty)$; $\forall n \in \mathbf{Z}_{0+}, \forall t \in \mathbf{R}_+$. Thus, if $\alpha^{-1}(T^n x_0, T^{n+1}x_0, K^{-1}(t-D)+D) \in (0, 1]$; $\forall n \in \mathbf{Z}_{0+}$, $\forall t(>D) \in \mathbf{R}_+$ and, furthermore,

Case a $\lim_{n\to\infty} \inf \alpha^{-1}(T^n x_0, T^{n+1}x_0, K^{-1}(t-D)+D) > 0$, $\forall t(>D) \in \mathbf{R}_+$

then one gets from (2.3) that

$$F^{-1}_{Tx_0,T^2x_0}(\varphi(t)) \leq 1 + \alpha^{-1}\left(x_0, Tx_0, K^{-1}(t-D)+D\right) \psi\left(F^{-1}_{x_0,Tx_0}(\varphi(K^{-1}(t-D)+D)) - 1\right) \leq 1 + \psi\left(F^{-1}_{x_0,Tx_0}(\varphi(K^{-1}(t-D)+D)) - 1\right), \quad \forall t(>D) \in \mathbf{R}_+, \tag{2.5}$$

equivalently,

$$F^{-1}_{Tx_0,T^2x_0}(\varphi(K^{-1}(t-D)+D)) \leq 1 + \psi\left(F^{-1}_{x_0,Tx_0}(\varphi(K^{-2}(t-D)+D)) - 1\right), \quad \forall t(>D) \in \mathbf{R}_+ \tag{2.6}$$

and replacing in (2.5) $x_0 \to Tx_0$, $Tx_0 \to T^2x_0$ with the use of (2.6) leads to:

$$F^{-1}_{T^2x_0,T^3x_0}(\varphi(t)) \leq 1 + \psi\left(F^{-1}_{Tx_0,T^2x_0}(\varphi(K^{-1}(t-D)+D)) - 1\right) \leq 1 + \psi^2\left(F^{-1}_{x_0,Tx_0}(\varphi(K^{-2}(t-D)+D)) - 1\right), \quad \forall t(>D) \in \mathbf{R}_+ \tag{2.7}$$

and proceeding recursively in the same way:

$$F^{-1}_{T^nx_0,T^{n+1}x_0}(\varphi(t)) \leq 1 + \psi\left(F^{-1}_{T^{n-1}x_0,T^nx_0}(\varphi(K^{-1}(t-D)+D)) - 1\right) \leq 1 + \psi^2\left(F^{-1}_{T^{n-2}x_0,T^{n-1}x_0}(\varphi(K^{-2}(t-D)+D)) - 1\right) \leq 1 + \psi^n\left(F^{-1}_{x_0,Tx_0}(\varphi(K^{-n}(t-D)+D)) - 1\right), \quad \forall t(>D) \in \mathbf{R}_+, \quad \forall n \in \mathbf{Z}_+, \tag{2.8a}$$

equivalently,

$$F_{T^nx_0,T^{n+1}x_0}(\varphi(t)) \geq \frac{1}{1 + \psi^n\left(F^{-1}_{x_0,Tx_0}(\varphi(K^{-n}(t-D)+D)) - 1\right)}, \quad \forall t(>D) \in \mathbf{R}_+, \quad \forall n \in \mathbf{Z}_+ \tag{2.8b}$$

Since $K^{-1} > 1$ and $\varphi \in \Phi_{D\hat{D}}$ then $\lim_{n\to+\infty}(K^{-n}(t-D)+D) = +\infty$, $\lim_{n\to\infty}\varphi(K^{-n}(t-D)+D) = +\infty$, $\forall t(>D) \in \mathbf{R}_+$ and $\lim_{n\to\infty} F^{-1}_{x_0,Tx_0}(\varphi(K^{-n}(t-D)+D)) = 1$, $\forall t(>D) \in \mathbf{R}_+$. Thus,

$$\lim_{n\to\infty} \psi^n\left(F^{-1}_{x_0,Tx_0}(\varphi(K^{-n}(t-D)+D)) - 1\right) = 0, \quad \forall t(>D) \in \mathbf{R}_+ \tag{2.9}$$

since $\lim_{n\to\infty} \psi^n\left(F^{-1}_{x_0,Tx_0}(\varphi(K^{-n}(t-D)+D)) - 1\right) = \lim_{n\to\infty} \psi^n(t_n) = 0$ if $\{t_n\} \to 0$. Then, $\lim_{n\to\infty} F_{T^nx_0,T^{n+1}x_0}(\varphi(t)) = 1, \forall t(>D) \in \mathbf{R}_+$ from (2.8b).

Case b $\lim_{n\to\infty} \alpha^{-1}(T^nx_0, T^{n+1}x_0, K^{-1}(t-D)+D) = 0$, $\forall t(>D) \in \mathbf{R}_+$

We first prove that $\lim_{n\to\infty} \sup \psi^n\left(F^{-1}_{x_0,Tx_0}(\varphi(K^{-n}(t-D)+D)) - 1\right) < \infty$, $\forall t(>D) \in \mathbf{R}_+$. The above condition is identical to $\lim_{n\to\infty} \sup \psi^n\left(F^{-1}_{x_0,Tx_0}(\varphi(t_n)) - 1\right) < \infty$ with $t_n = t_n(t) = K^{-n}(t-D)+D$ and, since $K<1$, $K^{-n} \to \infty$ as $n \to \infty$ and $\varphi(t)$ is strictly increasing then $\{t_n\} \to \infty$ and $\varphi(t_n) \to \infty$ as $n \to \infty$. Assume that this is not the case so that $\lim_{n\to\infty} \sup \psi^n\left(F^{-1}_{x_0,Tx_0}(\varphi(t_n)) - 1\right) = \infty$. Then, since the function $\psi : \mathbf{R}_{0+} \to \mathbf{R}_{0+}$ is everywhere continuous in its definition domain it can only diverge at infinity and then $\lim_{n\to\infty} F^{-1}_{x_0,Tx_0}(\varphi(t_n)) = \infty$, and equivalently, $\lim_{n\to\infty} F_{x_0,Tx_0}(\varphi(t_n)) = \lim_{t_n\to\infty} F_{x_0,Tx_0}(\varphi(t_n)) = 0$. But this would lead to the contradiction that $F_{x_0,Tx_0}(t)$ is not non-decreasing. As a result, $\lim_{n\to\infty} \sup \psi^n\left(F^{-1}_{x_0,Tx_0}(\varphi(t_n)) - 1\right) < \infty$, and then

$$\lim_{n\to\infty}\left(\alpha^{-1}\left(T^{n_1}x_0, T^{n_1+1}x_0, K^{-1}(t-D)+D\right) \psi^n\left(F^{-1}_{x_0,Tx_0}(\varphi(K^{-n}(t-D)+D)) - 1\right)\right) = 0$$

leading to $\lim_{n\to\infty} F_{T^nx_0,T^{n+1}} \; x_0^{-1}(\varphi(t)) = \lim_{n\to\infty} F_{T^nx_0,T^{n+1}x_0}(\varphi(t)) = 1$, $\forall t(>D) \in \mathbf{R}_+$ from (2.8a).

Since $\hat{D} \geq \varphi(D^+)$, one concludes for Cases a and b that $\lim_{n\to\infty} F_{T^{npj}x_0, T^{npj+1}x_0}(\varphi(D)+t) = 1$, $\forall t \in \mathbf{R}_+$ with $T^{npj}x_0 \in A_{i+j}$ and $T^{npj+1}x_0 \in A_{i+j+1}, \forall j \in \bar{p}$ if $x_0 \in A_i$ for any given $i \in \bar{p}$ since $\operatorname{diam}(A_i) > \hat{D} - D$, $\forall i \in \bar{p}$. Then, for any given real $\varepsilon \in \mathbf{R}_+$ and $\lambda \in (0,1)$, there is $n_0 = n_0(\varepsilon, \lambda) \in \mathbf{Z}_{0+}$ such that

$$F_{T^{npj}x_0, T^{npj+1}x_0}(\varphi(D)+\varepsilon) > 1 - \lambda, \quad \forall n \geq n_0, \quad \forall j \in \bar{p},$$

since the distance distribution function is non-decreasing and left-continuous. On the other hand,

$$F_{T^{np}x_0, T^{(n+m)p+j}x_0}((pm+j)(\varphi(D)+\varepsilon)) \geq \Delta_M\left(F_{T^{np}x_0, T^{np+1}x_0}(\varphi(D)+\varepsilon), F_{T^{np+1}x_0, T^{np+2}x_0}(\varphi(D)+\varepsilon), \ldots, F_{T^{(n+m)p+j-1}x_0, T^{(n+m)p+j}x_0}(\varphi(D)+\varepsilon)\right) > 1 - \lambda; \tag{2.10}$$

$\forall t(>D)\in\mathbf{R}_+$, $\forall j\in\bar{p}$ and, for any given $\varepsilon\in\mathbf{R}_+$ and $\lambda\in(0,1)$, $F_{T^{np}x_0,T^{(n+m)p+j}\ x_0}((pm+j)(\varphi(D)+\varepsilon))>1-\lambda$ for $n(\geq n_0)\in\mathbf{Z}_{0+}$, $\forall m\in\mathbf{Z}_+$. Property (1) has been proved.

Property (2) relies on the case when $\varphi\in\Phi_0$ and $\bigcap_{i\in\bar{p}}A_i\neq\emptyset$ since $D=0$. If $\varphi\in\Phi_0$ then $D=\hat{D}=\varphi(0)=0$ so that

$$F_{T^nx_0,\,T^{n+1}x_0}(\varepsilon)>1-\lambda,\quad F_{T^{np+j}x_0,\,T^{(n+1)p+j}x_0}((pm+j)\,\varepsilon)>1-\lambda,\quad \forall n(\geq n_0)\in\mathbf{Z}_{0+},\quad \forall m\in\mathbf{Z}_+, \tag{2.11}$$

and $\lim_{n\to\infty}F_{T^nx_0,\,T^{n+1}x_0}(t)=\lim_{n\to\infty}F_{T^{np+j}x_0,\,T^{(n+m)p+j+1}x_0}(t)=1$, $\forall t\in\mathbf{R}_+$. Since (X,F,Δ) is G-complete, then $\{T^{np+j}x_0\}\to\bar{x}_{i+j}(\in cl(A_{i+j}))$, $\forall j\in\overline{p-1}\cup\{0\}$, $\forall x_0\in A_i$. Assume that $\bar{x}_j\neq\bar{x}_{j+k}$ for some $j,k\in\bar{p}$. Since (X,F,Δ_M) is a Menger PM-space then from (2.10) with $D=\hat{D}=\varphi(0)=0$:

$$F_{T^{np+j+\ell}x_0,\,T^{np+j+\ell+1}x_0}(\varepsilon/k)>1-\lambda,\quad \forall n(\geq n_0)\in\mathbf{Z}_{0+};\quad \lim_{n\to\infty}F_{T^{np+j+\ell}x_0,\,T^{np+j+\ell+1}x_0}(\varepsilon/k)=1, \tag{2.12}$$

for $\ell\in\overline{k-1}\cup\{0\}$, any $\varepsilon\in\mathbf{R}_+$ and any real $\lambda\in(0,1)$ what implies that $\lim_{n\to\infty}F_{T^{np+j}x_0,\,T^{np+j+k}x_0}(t)=1$, $\forall t\in\mathbf{R}_+$ and, from the first property of (1.1), $\bar{x}_j=\bar{x}_{j+k}$, a contradiction, and then $\bar{x}_i=\bar{x}$, $\forall i\in\bar{p}$. So, the p sequences $\{T^{np+i}x_0\}$, $\forall i\in\bar{p}$ have a unique limit point in $\bigcap_{i\in\bar{p}}A_i$ provided that such a set is nonempty and from Assertion 1.1 they are Cauchy and G-Cauchy sequences and then convergent since (X,F,Δ) is G-complete. In addition, for any $\varepsilon\in\mathbf{R}_+$ and any real $\lambda\in(0,1)$ there is $n_{01}=n_{01}(\varepsilon,\lambda)\in\mathbf{Z}_{0+}$ such that from (2.2) for $D=0$ since $\{T^nx_0\}$ is Cauchy, then G-Cauchy, and convergent to x, then $\lim_{n\to\infty}\psi\left(F^{-1}_{T^nx_0,x}\ (\varphi(K^{-1}\varepsilon/3))-1\right)=\psi(0)=1$ and $\lim_{n\to\infty}F_{T^{n+1}x_0,\,Tx}(\varepsilon/3)=1$ so that

$$F_{T^{n+1}x_0,\,Tx}(\varepsilon/3)\geq\frac{\alpha(T^nx_0,\,x,\,K^{-1}\varepsilon/3)}{\alpha(T^nx_0,\,x,\,K^{-1}\varepsilon/3)+\psi\left(F^{-1}_{T^nx_0,x}(\varphi(K^{-1}\varepsilon/3))-1\right)};\quad\forall n(\in\mathbf{Z}_{0+})$$

$$F_{T^{n+1}x_0,\,Tx}(\varepsilon/3)>1-\lambda;\quad\forall n(\geq n_{01})\in\mathbf{Z}_{0+}$$

so that for some $n_0=n_0(\varepsilon,\lambda)(\geq n_{01})\in\mathbf{Z}_{0+}$ and for any arbitrary $\varepsilon\in\mathbf{R}_+$ and $\lambda\in(0,\,1)$ using the third and fourth properties of the triangular norms, one gets:

$$F_{x,Tx}(\varepsilon)\geq\min\left(F_{x,T^nx_0}(\varepsilon/3),\ F_{T^{n+1}x_0,T^nx_0}(\varepsilon/3),\ F_{T^{n+1}x_0,Tx}(\varepsilon/3)\ \right)>1-\lambda,\quad\forall n(\geq n_0)\in\mathbf{Z}_{0+}, \tag{2.13}$$

Thus, $F_{x,Tx}(0^+)=1$ so that $x=Tx$ from the property 1 of (1.1). By replacing $x\to Tx$ and $Tx\to T^2x$ in (2.13), we prove $x=T^2x$. Proceeding in the same way, it is proved that $T^ix=x$, $\forall i\in\bar{p}$. So, x is a limit point of $\{T^{np+i}x_0\}$, $\forall i\in\bar{p}$ and $\{T^nx_0\}$ which is also a fixed point of $T:\bigcup_{i\in\bar{p}}A_i\to\bigcup_{i\in\bar{p}}A_i$ and $T^p:\bigcup_{i\in\bar{p}}A_i\Big|A_j\to\bigcup_{i\in\bar{p}}A_i\Big|A_j,\forall j\in\bar{p}$ if the subsets A_i are closed $\forall i\in\bar{p}$. Hence, Property (2) has been proved.

On the other hand, it follows from (1.2) for Menger PM-spaces and the properties of (1.3) for triangular norms for the general case that $\hat{D}=\varphi(D^+)\geq D>0$ since, furthermore, $\Delta_M(x,x)\geq x$ for each $x\in[0,\,1]$, that

$$\begin{aligned}
&F_{T^{np}x_0,\,T^{(n+m)p}x_0}(t)\geq\Delta_M\left(F_{T^{np}x_0,\,T^{np+1}x_0}((1-K)t),\ F_{T^{np+1}x_0,\,T^{(n+m)p}x_0}(Kt)\right)\\
&\geq\Delta_M\left(F_{x_0,Tx_0}\left(\frac{1-K^{mp}}{1-K}K^{-np}((1-K)t-D)+D\right),\right.\\
&\quad\Delta_M\left(F_{T^{np+1}x_0,\,T^{np+2}x_0}(K(1-K)t),\,F_{T^{np+2}x_0,\,T^{(n+m)p}x_0}(K^2t)\right)\Big)\\
&\geq\Delta_M\left(F_{x_0,Tx_0}\left(\frac{1-K^{mp}}{1-K}K^{-np}((1-K)t-D)+D\right),\right.\\
&\quad\Delta_M\left(F_{x_0,Tx_0}\left(\frac{1-K^{mp}}{1-K}K^{-np-1}(K(1-K)t-D)+D\right),\right.\\
&\quad F_{T^{np+2}x_0,\,T^{(n+m)p}x_0}(K^2t)\Big)\Big)\\
&=\Delta_M\left(F_{x_0,Tx_0}\left(\frac{1-K^{mp}}{1-K}K^{-np}((1-K)t-D)+D\right),\right.\\
&\quad\Delta_M\left(F_{x_0,Tx_0}\left(\frac{1-K^{mp}}{1-K}K^{-np}((1-K)t-D)+D\right),\right.\\
&\quad F_{T^{np+2}x_0,\,T^{(n+m)p}x_0}(K^2t)\Big)\Big)\\
&\geq\Delta_M\left(F_{x_0,Tx_0}\left(\frac{1-K^{mp}}{1-K}K^{-np}((1-K)t-D)+D\right),\right.\\
&\quad\Delta_M\left(F_{x_0,Tx_0}\left(\frac{1-K^{mp}}{1-K}K^{-np}((1-K)t-D)+D\right),\right.\\
&\quad F_{T^{(n+m)p-1}x_0,\,T^{(n+m)p}x_0}(K^{mp-1}t)\Big)\Big)\\
&\geq\Delta_M\left(F_{x_0,Tx_0}\left(\frac{1-K^{mp}}{1-K}K^{-np}((1-K)t-D)+D\right),\right.\\
&\quad\Delta_M\left(F_{x_0,Tx_0}\left(\frac{1-K^{mp}}{1-K}K^{-np}((1-K)t-D)+D\right),\right.\\
&\quad F_{x_0,Tx_0}\left(K^{-(n+m)p+1}(K^{mp-1}t-D)+D\right)\Big)\Big)\\
&=\Delta_M\left(F_{x_0,Tx_0}\left(\frac{1-K^{mp}}{1-K}K^{-np}((1-K)t-D)+D\right),\right.\\
&\quad\Delta_M\left(F_{x_0,Tx_0}\left(\frac{1-K^{mp}}{1-K}K^{-np}((1-K)t-D)+D\right),\right.\\
&\quad F_{x_0,Tx_0}\left(\frac{1-K^{mp}}{1-K}K^{-np}((1-K)t-D)+D\right)\Big)\Big)\\
&\geq\Delta_M\left(F_{x_0,Tx_0}\left(\frac{1-K^{mp}}{1-K}K^{-np}((1-K)t-D)+D\right),\right.\\
&\quad F_{x_0,Tx_0}\left(\frac{1-K^{mp}}{1-K}K^{-np}((1-K)t-D)+D\right)\Big)\\
&\geq F_{x_0,Tx_0}\left(\frac{1-K^{mp}}{1-K}K^{-np}((1-K)t-D)+D\right);\quad\forall t\in\mathbf{R}_+,\\
&\forall m\in\mathbf{Z}_{0+},\quad\forall n\in\mathbf{Z}_+
\end{aligned} \tag{2.14}$$

Then, one gets for any given t, $\lambda(<1)\in\mathbf{R}_+$, some $n_0=n_0(t,\lambda)$ and $\forall n\geq n_0$.

$$\lim_{n\to\infty} F_{T^{np}x_0,\,T^{(n+m)p}x_0}(t) = F_{x_0,Tx_0}(\infty) = 1;$$
$$F_{T^{np}x_0,\,T^{(n+m)p}x_0}(t) > 1-\lambda, \tag{2.15}$$

$\forall x_0 \in \bigcup_{i\in\bar{p}} A_i$, $\forall m \in \mathbf{Z}_+$ since $F_{x_0,Tx_0}(t)$ is non-decreasing with supremum over $t \in \mathbf{R}_+$ equalizing unity. Thus, $\{T^{np+j}x_0\} \to \bar{x}_{i+j}(\in cl(A_{i+j}))$, $\forall j \in \overline{p-1} \cup \{0\}$, $\forall x_0 \in A_i$, $\forall i \in \bar{p}$. Since $D>0$ then $\bar{x}_i \neq \bar{x}_j$, $\forall i,j(\neq i) \in \bar{p}$. In addition, either

$$F_{\bar{x}_{i+j},T^p\bar{x}_{i+j}}(\varepsilon) \geq \min\Big(F_{\bar{x}_{i+j},T^{np+j}x_0}(\varepsilon/4),\ F_{T^{np+j}x_0,T^{(n+1)p+j}x_0}(\varepsilon/4),$$
$$F_{T^{(n+1)p+j}x_0,\bar{x}_{i+j}}(\varepsilon/4),\ F_{\bar{x}_{i+j},T^p\bar{x}_{i+j}}(\varepsilon/4)\Big) \tag{2.16}$$

$$\geq \min\Big(F_{\bar{x}_{i+j},T^{np+j}x_0}(\varepsilon/4),\ F_{T^{np+j}x_0,T^{(n+1)p+j}x_0}(\varepsilon/4),$$
$$F_{T^{(n+1)p+j}x_0,\bar{x}_{i+j}}(\varepsilon/4)\Big) > 1-\lambda, \tag{2.17.a}$$

$\forall n(\geq n_0) \in \mathbf{Z}_{0+}$, or $F_{\bar{x}_{i+j},T^p\bar{x}_{i+j}}(\varepsilon/4) \leq 1-\lambda$, then either $F_{\bar{x}_{i+j},T^p\bar{x}_{i+j}}(\varepsilon) \leq 1-\lambda$, $\forall n(\geq n_0) \in \mathbf{Z}_{0+}$ and

$$1-\lambda \geq F_{\bar{x}_{i+j},T^{np+j}x_0}(\varepsilon) \geq \min\big(F_{\bar{x}_{i+j},T^p\bar{x}_{i+j}}(\varepsilon/4),$$
$$F_{T^{np+j}x_0,T^{(n+1)p+j}x_0}(\varepsilon/4), F_{T^{(n+1)p+j}x_0,\bar{x}_{i+j}}(\varepsilon/4),\ F_{\bar{x}_{i+j},T^p\bar{x}_{i+j}}(\varepsilon/4)\big)$$
$$\geq \min\Big(F_{\bar{x}_{i+j},T^{np+j}x_0}(\varepsilon/4),\ F_{T^{np+j}x_0,T^{(n+1)p+j}x_0}(\varepsilon/4),$$
$$F_{T^{(n+1)p+j}x_0,\bar{x}_{i+j}}(\varepsilon/4)\Big) > 1-\lambda \tag{2.17.b}$$

for some $n(\geq n_0) \in \mathbf{Z}_{0+}$, which is a contradiction, or $F_{\bar{x}_{i+j},T^p\bar{x}_{i+j}}(\varepsilon) > 1-\lambda$. As a result (2.16) leads, in any case, to (2.17.a) and $F_{\bar{x}_{i+j},T^p\bar{x}_{i+j}}(\varepsilon) > 1-\lambda$, $\forall n(\geq n_0) \in \mathbf{Z}_{0+}$. Thus, $\bar{x}_j$ is a fixed point of the composite self-mapping $T^p : \bigcup_{i\in\bar{p}} A_i \big| A_j \to A_j$, $\forall j \in \bar{p}$. To prove that the fixed points of the composite self-mapping are quasi-best proximity points of the p-cyclic α-ψ-type contraction $T : \bigcup_{i\in\bar{p}} A_i \to \bigcup_{i\in\bar{p}} A_i$, we proceed by contradiction. Assume that this is not the case, so that there is a pair $(\bar{x}_i,\ T\bar{x}_i)$ for some $i \in \bar{p}$ such that there exist $\varepsilon \in \mathbf{R}_+, \lambda \in (0,\ 1)$ and a sequence $\{n_k\} \subset \mathbf{Z}_{0+}$ for some $n_0 = n_0(\varepsilon,\ \lambda) \in \mathbf{Z}_{0+}$ such that, for any $j \in \bar{p}$, one gets that $\lim_{k\to\infty} F_{T^{n_kp}\bar{x}_j,\ T^{(n_k+1)p}\bar{x}_j}(0^+) = 1 > \limsup_{k\to\infty} F_{T^{n_kp}\bar{x}_i,\ T^{n_kp+1}\bar{x}_i}(\hat{D}^+)$. Since (X, F, Δ_M) is a Menger space and, since any triangular norm is associative and commutative and since $\Delta_M(x,x) \geq x$, one gets the contradiction:

$$F_{\bar{x}_i,\ T^p\bar{x}_i}(\varepsilon) \geq \min_{j\in\bar{p}}\big(F_{\bar{x}_i,\ T^j\bar{x}_i}\big) = q(1-\lambda) \geq 1-\lambda \tag{2.18}$$

for some real $q = q(\varepsilon,\ \lambda) \in (0,1)$ to $\bar{x}_k$ being a fixed point of $T^p : \bigcup_{i\in\bar{p}} A_i \big| A_k \to A_k$ for $k \in \bar{p}$. Property (3) has been proved. □

Note that if the image of $\alpha(x,\ y,\ t)$ is extended to be in $cl\mathbf{R}_{0+}$ (i.e., $\alpha(x,\ y,\ t)$ can take also values at $\{+\infty\}$), then the proof of Property (1) of Theorem 2.1 is valid with a slight extension by considering also the case that $\alpha^{-1}(T^{n_1}x_0, T^{n_1+1}x_0,\ K^{-1}(t-D)+D) = 0$ for some finite $n_1 \in \mathbf{Z}_{0+}$ and $\forall t(>D) \in \mathbf{R}_+$. Since $\psi \in \Psi$, it is continuous, and then $\psi^{n_1}\Big(F^{-1}_{x_0,Tx_0}(\varphi(K^{-n_1}(t-D)+D))-1\Big) < \infty$ for any finite $n_1 \in \mathbf{Z}_+$ so that

$$\alpha^{-1}(T^{n_1-1}x_0, T^{n_1}x_0,\ K^{-1}(t-D)+D)\psi^{n_1}$$
$$\Big(F^{-1}_{x_0,Tx_0}(\varphi(K^{-n_1}(t-D)+D))-1\Big) = 0, \quad \forall t(>D) \in \mathbf{R}_+.$$

Then, we could use a similar recursive procedure as that used for the case up till the (n_1-1)-iteration, since Proposition 2.1 remains valid, see Eqs. (2.2) and (2.3), so that:

$$F^{-1}_{T^{n_1}x_0,T^{n_1+1}x_0}(\varphi(t)) \leq 1 + \alpha^{-1}(T^{n_1-1}x_0, T^{n_1}x_0,\ K^{-1}(t-D)$$
$$+D)\psi^{n_1}\Big(F^{-1}_{x_0,Tx_0}(\varphi(K^{-n_1}(t-D)+D))-1\Big) = 1.$$

Note that Theorem 2.1 generalizes some results on fixed points given in [7, 8, 15, 16, 28] for either non-cyclic self-mappings or cyclic self-mappings on union of sets which intersect to quasi-best proximity points and best proximity points in the case that such sets do not intersect. On the other hand, a direct consequence of Theorem 2.1 is the following corollary for the case that $\varphi \in \Phi_D$. The results are based on the fact that $\varphi(D^-) = D$ and $\varphi(t) = 0$ if $t \in [0, D]$ and $\varphi \in \Phi_D$ while it generalizes results on fixed points for the cases of either non-cyclic self-mappings or cyclic self-mappings with nonempty intersections of the involved subsets obtained in [7, 8, 15, 16, 28]:

Corollary 2.1 *Let (X, F, Δ) be a G-complete Menger PM-space and $T : \bigcup_{i\in\bar{p}} A_i \to \bigcup_{i\in\bar{p}} A_i$ be a p-cyclic α-ψ-type contraction satisfying the following conditions*:

1. $D = d(A_i,\ A_{i+1}) > 0, \quad \forall i \in \bar{p}$,
2. *T is α-admissible,*
3. *there exists $x_0 \in \bigcup_{i\in\bar{p}} A_i$ such that $\alpha(x_0,\ Tx_0,\ t) \geq 1$, $\forall t \in \mathbf{R}_+$,*
4. *if $\{x_n\}(\subset X) \to x$ is a Picard iteration generated as $x_{n+1} = Tx_n$, $\forall n \in \mathbf{Z}_{0+}$ with $x_0 \in \bigcup_{i\in\bar{p}} A_i$, such that $\alpha(x_n,\ Tx_n,\ t) \geq 1$, $\forall n \in \mathbf{Z}_{0+}, \forall t \in \mathbf{R}_+$ then $\alpha(x_n,\ x,\ t) \geq 1$, $\forall n \in \mathbf{Z}_{0+}$, $\forall t \in \mathbf{R}_+$.*

Then, the following properties hold:

1. *If $\varphi \in \Phi_0$ and $\bigcap_{i\in\bar{p}} A_i \neq \emptyset$ then $\{T^nx_0\} \subset \bigcup_{i\in\bar{p}} A_i$ and $\{T^{np+j}x_0\} \subset A_{i+j}$ are both Cauchy and G-Cauchy convergent sequences to a limit point $x \in \bigcap_{i\in\bar{p}} cl\,A_i$, with $cl\,A_i$ being the closure of A_i. If A_i are closed for*

$i \in \bar{p}$ then $x = Tx = T^p x$, that is, it is a fixed point of the self-mappings $T : \bigcup_{i\in\bar{p}} A_i \to \bigcup_{i\in\bar{p}} A_i$ and $T^p (\equiv T \circ T^{p-1}) : \bigcup_{i\in\bar{p}} A_i \big| A_k \to A_k$, $\forall k \in \bar{p}$, $\forall x_0 \in A_i$, $\forall i \in \bar{p}$, $\forall j \in \overline{p-1} \cup \{0\}$.

2. *If $\varphi \in \Phi_D$, A_i is closed $\forall i \in \bar{p}$ and $\bigcap_{i\in\bar{p}} A_i = \emptyset$ then, if $x_0 \in A_i$ for any $i \in \bar{p}$, there is a limit cycle $(\bar{x}_i, T\bar{x}_i, \ldots, T^{p-1}\bar{x}_i)$, to which the sequence $\{T^n x_0\}$ converges, with $T^p\bar{x}_i = \bar{x}_i$, $\bar{x}_i \in A_i$, $\bar{x}_{i+j} = T^j\bar{x}_i \in A_{i+j}$, $\forall i \in \bar{p}, \forall j \in \overline{p-1} \cup \{0\}$ being a fixed point of the composite self-mapping $T^p : \bigcup_{i\in\bar{p}} A_i \big| A_k \to A_k, \forall k \in \bar{p}$, and being also a best proximity point of $T : \bigcup_{i\in\bar{p}} A_i \to \bigcup_{i\in\bar{p}} A_i$. The subsequence $\{T^{np+j} x_0\}$ in A_{i+j} converges to $x_{i+j} = T^j x_i$, $\forall j \in \bar{p}$ if $x_0 \in A_i$ for some $i \in \bar{p}$. Furthermore,*

$$\begin{aligned}&\lim_{n\to\infty} F_{T^{np}x_0,\, T^{(n+m)p}x_0}(\varepsilon) = F_{x_0, Tx_0}(\infty) = 1,\\ &F_{T^{np}x_0,\, T^{(n+m)p}x_0}(\varepsilon) > 1 - \lambda\end{aligned} \tag{2.19}$$

$$\begin{aligned}&F_{T^{npj}x_0,\, T^{npj+1}x_0}(\varphi(D) + \varepsilon) > 1 - \lambda,\\ &\lim_{n\to\infty} F_{T^{npj}x_0,\, T^{npj+1}x_0}(\varphi(D) + \varepsilon) = F_{\bar{x}_{i+j}, \bar{x}_{i+j+1}} = 1\end{aligned} \tag{2.20}$$

for any given $\varepsilon \in \mathbf{R}_+$, $\lambda \in (0, 1) \cap \mathbf{R}$ and some $n_0 = n_0(\varepsilon, \lambda)$, $\forall n(\geq n_0) \in \mathbf{Z}_{0+}$. □

The following result is an extended version of a parallel result given in [16] for uniqueness of fixed points of (non-cyclic) self-mappings of a p-cyclic α-ψ-type contraction on X. The result is concerned with (a) the uniqueness of best proximity points of p-cyclic α-ψ-type contractions, being corresponding fixed points of the composite self-mappings restricted to each subset of the cyclic disposal, and (b) their confluence to a unique fixed point of the p-cyclic α-ψ-type contraction if the subsets intersect.

Theorem 2.2 *Assume all the hypotheses of Corollary 2.1 and the additional one which follow:*
(5) For all $x, y \in \bigcup_{i\in\bar{p}} clA_i$, there is $z \in \bigcup_{i\in\bar{p}} clA_i$ such that $\min(\alpha(x,z,t),\ \alpha(y,z,t)) \geq 1$, $\forall t\ (> D) \in \mathbf{R}_+$. *Then, there is $\bar{x}_i \in clA_i$, such that $\bar{x}_i \in A_i$ if A_i is closed, $\forall i \in \bar{p}$, which is the unique best proximity point of $T : \bigcup_{i\in\bar{p}} A_i \to \bigcup_{i\in\bar{p}} A_i$ and the unique fixed point of $T^p : \bigcup_{i\in\bar{p}} A_i \big| A_k \to A_k$, $\forall k \in \bar{p}$. If $\varphi \in \Phi_0$ with $D = 0$ and $\bigcap_{i\in\bar{p}} A_i$ is nonempty and closed then there is a unique fixed point of $T : \bigcup_{i\in\bar{p}} A_i \to \bigcup_{i\in\bar{p}} A_i$ and of the composite self-mappings $T^p : \bigcup_{i\in\bar{p}} A_i \big| A_j \to A_j, \forall j \in \bar{p}$.*

Proof Let $u \in clA_i$, $v \in clA_j$ be best proximity points for any given $i, j \in \bar{k}$ such that $u = T^{np}u = T^p u$ and $v = T^{np}v = T^p v$, $\forall n \in \mathbf{Z}_+$ so that there is $z \in \bigcup_{i\in\bar{p}} clA_i$ such that $\min(\alpha(u,z,t),\ \alpha(v,z,t)) \geq 1$, $\forall t(> D) \in \mathbf{R}_+$.

Thus, there exist integers $m = m(i,u,z) \in \overline{p-1} \cup \{0\}$ and $\ell = \ell(j,v,z) \in \overline{p-1} \cup \{0\}$, such that $T^{pn+m}z \in A_{i+1}$ and $T^{pn+\ell}z \in A_{j+1}$, and since T is α-admissible and $\min(\alpha(x,z,t),\ \alpha(y,z,t)) \geq 1$, $\forall t(> D) \in \mathbf{R}_+$, one gets:

$$\begin{aligned}&\frac{1}{F_{u, T^{pn+m}z}(\varphi(t))} - 1 \leq \sup_{n\in\mathbf{Z}_{0+}} \left[\alpha^{-1}(u, T^{np+m}z, t)\right]\\ &\psi^{np-1}\left(\frac{1}{F_{u, T^m z}(\varphi(K^{-(np+m)}(t-D)+D))} - 1\right),\\ &\forall t(> D) \in \mathbf{R}_+\end{aligned} \tag{2.21}$$

$$\begin{aligned}&\frac{1}{F_{v, T^{pn+\ell}z}(\varphi(t))} - 1 \leq \alpha^{-1}(v,z,t)\, \psi^{np-1}\\ &\left(\frac{1}{F_{v, T^\ell z}(\varphi(K^{-(np+\ell)}(t-D)+D))} - 1\right), \quad \forall t(> D) \in \mathbf{R}_+.\end{aligned} \tag{2.22}$$

Note that $\psi^{np-1}(a_{np-1}(t,m,u,z)) \to 0$ as $a_{np-1}(t,m,u,z) = \frac{1}{F_{u,T^m z}(\varphi(K^{-(np+m)}(t-D)+D))} - 1 \to 0$ as $n \to \infty$ for all $t(> D) \in \mathbf{R}_+$ which holds since $F_{u,T^m z}(\varphi(K^{-np}(t-D)+D)) \to 1$ as $n \to \infty$ for all $t(> D) \in \mathbf{R}_+$ since $(K^{-np}(t-D)+D) \to \infty$ for all $t(> D) \in \mathbf{R}_+$ as $n \to \infty$ and since $\varphi(t) > D$ from (2.21), from the properties $\varphi(t) = 0$ for $t<D$, $\varphi(D) \geq D$, and $\varphi(t) > D$ for $t > D$ since $\varphi : \mathbf{R} \to \mathbf{R}_{0+}$ is strictly increasing in $[D, \infty)$. In the same way, note that $\psi^{np-1}(b_{np-1}(t,m,v,z)) \to 0$ as $b_{np-1}(t,m,v,z) = \frac{1}{F_{u,T^m z}(\varphi(K^{-(np+m)}(t-D)+D))} - 1 \to 0$ as $n \to \infty$ for all $t(> D) \in \mathbf{R}_+$. As a result:

$$F_{u, T^{pn+m}z}(t) = F_{v, T^{pn+\ell}z}(t) = 1, \quad \forall t(> D) \in \mathbf{R}_+ \tag{2.23}$$

$$F_{u, T^{pn+m}z}(t) = F_{v, T^{pn+\ell}z}(t) = 0, \quad \forall t(\leq D) \in \mathbf{R}_+ \tag{2.24}$$

with $u \in A_i$, $v \in A_j$, $T^{pn+m}z \in A_{i+1}$, $T^{pn+\ell}z \in A_{j+1}$ for any $i, j \in \bar{p}$. Thus, if $i = j$ then $u, v \in clA_i$, $\ell = m$ and $T^{pn+m}z \in clA_{i+1}$. Assume that $u \neq v$ are best proximity points in clA_i, then $Tu, Tv \in A_{i+1}$ are corresponding adjacent best proximity points and $\{T^{pn+m}z\} \to Tu$ and $\{T^{pn+m}z\} \to Tv$. Thus, $Tu = Tv$ and $u = T^p u = T^p v = v$ which contradicts $u \neq v$. Since $i \in \bar{p}$ is arbitrary then the set of adjacent best proximity points is unique.

In the particular case that $D = 0$, one gets that $u = v$ is a unique fixed point of $T \to \bigcup_{i\in\bar{p}} A_i \to \bigcup_{i\in\bar{p}} A_i$, allocated in the nonempty set $\bigcap_{i\in\bar{p}} clA_i$, and also a fixed point of each restricted composite self-mapping $T^p : \bigcup_{i\in\bar{p}} A_i \big| A_j \to A_j, \forall j \in \bar{p}$ since Eqs. (2.23) and (2.24) result to be for $i = j$:

$$F_{u,\,T^{pn+m}z}(t) = F_{v,\,T^{pn+m}z}(t) = 1, \quad \forall t \in \mathbf{R}_+ \tag{2.25}$$

□

Example 2.1 Consider real intervals $A = -B = \{z \in \mathbf{R} : z \geq D/2\} \subset \mathbf{R}$ so that $(\mathbf{R}, d)$ is a complete metric space under some metric $d : X \times X \to \mathbf{R}_{0+}$ and $D = d(A, B)$. The sequence $\{x_n\} \subset A \cup B$ is generated by 2-cyclic self-mapping T on $A \cup B$ for some real constant $K \in [0,\ 1)$ and any given initial $x_0 \in \mathbf{R} \equiv A \cup B$, by

$$x_{n+1} = -K x_n - (1-K)(D/2)\mathrm{sgn}(x_n), \quad n \in \mathbf{Z}_+ \tag{2.26}$$

with the extended definition $\mathrm{sgn}(x) = 0$ if $x = 0$ for the case that $D = 0$. The above sequence has the two following subsequences, in A and B if $x_0 \in A$, respectively, in B and A if $x_0 \in B$:

$$x_{2n} = T^{2n}x_0 = K^{2n}x_0 + \left(1-K^{2n}\right)(D/2)\mathrm{sgn}(x_0), \quad n \in \mathbf{Z}_+, \tag{2.27a}$$

$$x_{2n+1} = Tx_{2n} = T^{2n+1}x_0 = -K^{2n+1}x_0 - \left(1-K^{2n}\right)(D/2)\mathrm{sgn}(x_0), \quad n \in \mathbf{Z}_+. \tag{2.27b}$$

Note that $T : A \cup B \to A \cup B$ is a 2-cyclic contraction, with the metric being the Euclidean norm, since

$$d(x_{n+2},\ x_{n+1}) = |x_{n+2} - x_{n+1}| = |K(x_n - x_{n+1}) + (1-K)D\mathrm{sgn}(x_n)| \leq Kd(x_{n+2},\ x_{n+1}) + (1-K)D, \quad n \in \mathbf{Z}_{0+}, \tag{2.28}$$

and it turns out that $\{x_{2n}\} \to D/2$, $\{x_{2n+1}\} \to -D/2$ if $x_0 \in A$ and $\{x_{2n}\} \to -D/2$, $\{x_{2n+1}\} \to D/2$ if $x_0 \in B$ and $x = \pm D/2$ are the unique best proximity points in A and B, respectively, and unique fixed points of $T^2 : A \to A$ and, respectively, of $T^2 : B \to B$, which are confluent at $x = 0$ if $D = 0$ and then a unique fixed point of $T : A \cup B \to A \cup B$ and $T^2 : A \cup B \to A \cup B$. Let $\mathbf{D}_D^+$ be the set of all generalized distance distribution functions of elements defined by $H_D(t) = 0$ if $t \leq D$ and $H(t) = 1$ if $t > D$. Now, define $\varphi(t) = \psi(t) = t$, and:

- a probability mapping $F : \mathbf{R} \times \mathbf{R} \to \mathbf{D}_D^+$, $F_{x,y}(t) = \frac{gt^p + D}{gt^p + d(x,y)}$ if $t(\in \mathbf{R}_+) > D$ if $x \in A$ and $y \in B$ or $x \in B$ and $y \in A$; $F_{x,y}(t) = \frac{gt^p}{gt^p + d(x,y)}$ if $t(\in \mathbf{R}_+)$ if $x, y \in A$ or $x, y \in B$; $F_{x,y}(t) = 0$ if $t(\in \mathbf{R}) \leq D$ for all $(x, y) \in A \times B \cup B \times A$, for any given $g, p \in \mathbf{R}_+$
- a weighting function $\alpha : \mathbf{R} \times \mathbf{R} \times \mathbf{R}_+ \to \mathbf{R}_{0+}$ by $\alpha(x, y, t) = \frac{\delta t^q + d(x,y)}{t + \lambda D}$ for some given real constants $\lambda \in (0,\ 1]$ and $\delta \geq 1$.

Note that $(\mathbf{R}, F, \Delta_M)$ is a G-complete generalized Menger PM-space since the distance distribution function is a generalized one, [12]. It follows from Proposition 2.1 that

$$\begin{aligned}
F_{Tx_{n+1},Tx_n}(t) &= F_{T^2x_n,Tx_n}(t) = \frac{gt^p + D}{gt^p + d(T^2x_n, Tx_n)} \\
&\geq \frac{gt^p + D}{gt^p + D + K\left(d(Tx_n, x_n) - D\right)} = \frac{1}{1 + (gt^p + D)^{-1}K\left(d(Tx_n, x_n) - D\right)} \\
&\geq \frac{1}{1 + \alpha^{-1}\left(x_n, Tx_n,\ K^{-1}(t-D) + D\right)(gt^p + D)^{-1}K\left(F^{-1}_{x_n,\,Tx_n}(K^{-1}(t-D) + D) - 1\right)\left(g(K^{-1}(t-D) + D)^p + D\right)} \\
&\geq \frac{1}{1 + \left(F^{-1}_{x_n,\,Tx_n}(K^{-1}(t-D) + D) - 1\right)} \geq F_{x_n,Tx_n}\left(K^{-1}(t-D) + D\right), \quad n \in \mathbf{Z}_{0+}, \quad \forall t(>D) \in \mathbf{R}_+
\end{aligned} \tag{2.29}$$

since $\alpha^{-1}\left(x_n, Tx_n,\ K^{-1}(t-D) + D\right) \leq 1$, $n \in \mathbf{Z}_{0+}$, $\forall t(>D) \in \mathbf{R}_+$. Then, note that T is α-admissible and since

$$\begin{aligned}
d(x_n,\ Tx_n) - D &= \left(F^{-1}_{x_n,\,Tx_n}\left(K^{-1}(t-D) + D - 1\right)\right)\left(g\left(K^{-1}(t-D) + D\right)^p + D\right) \\
&= \left(F^{-1}_{x_n,\,Tx_n}K^{-1}(t-D) + D - 1\right)\left(g\left(K^{-1}(t-D) + D\right)^p + D\right), \quad n \in \mathbf{Z}_{0+}, \quad \forall t(>D) \in \mathbf{R}_+
\end{aligned} \tag{2.30}$$

then, from Theorem 2.1, there exists the following limit

$$\begin{aligned}
1 &= \lim_{n\to\infty} F_{Tx_{n+1},Tx_n}(t) \geq \lim_{n\to\infty} F_{x_0,Tx_0}\left(K^{-n}(t-D) + D\right) \\
&= F_{x_0,Tx_0}\left(\lim_{n\to\infty} K^{-n}(t-D) + D\right) = F_{x_0,Tx_0}(+\infty) = 1, \\
&\forall t(>D) \in \mathbf{R}_+
\end{aligned} \tag{2.31}$$

so that $x = \pm D/2$ are also best proximity points in the probabilistic sense. In addition, $F_{T^{2n}x_0,\, T^{2(n+m)}x_0}(t) > 1 - \lambda$ and $F_{T^{2n}x_0,\, T^{2(n+m)+1}x_0}(t + D) > 1 - \lambda$, $\forall \lambda \in (0,1) \cap \mathbf{R}$ and $\forall t \in \mathbf{R}_+$ from Theorem 2.1[(2.10) and (2.15)] and $\{T^{2n}x_0\}$ and $\{T^{2n+1}x_0\}$ are Cauchy, G-Cauchy and convergent to $\pm x$, respectively. Note that for $D = 0$ and $A \cap B = \{0\}$ then $x = 0$ is the unique fixed point from Theorem 2.2 since $(\mathbf{R}, F, \Delta_M)$ is a G-complete Menger PM-space.

Main results on best proximity points for generalized p-cyclic α-ψ-type contractions

We generalize the concept of $p(\geq 2)$-cyclic α-ψ-type contractions as follows:

Definition 3.1 Let (X, F) be a PM-space and let A_i be nonempty subsets of X,$\forall i \in \bar{p}$ such that $D = d(A_i,\, A_{i+1})$ is the distance in-between adjacent subsets, $\forall i \in \bar{p}$. Then, $T : \bigcup_{i \in \bar{p}} A_i \to \bigcup_{i \in \bar{p}} A_i$ is a generalized $p(\geq 2)$-cyclic α-ψ-type contraction if $T(A_i) \subseteq A_{i+1}$, $\forall i \in \bar{p}$ and there exist two functions $\alpha : \left(\bigcup_{i \in \bar{p}} A_i\right) \times \left(\bigcup_{i \in \bar{p}} A_i\right) \times \mathbf{R}_+ \to \mathbf{R}_{0+}$ and $\psi \in \Psi$ satisfying the following inequality:

$$\alpha(x,y,t)\left(\frac{1}{F_{Tx,Ty}(\varphi(Kt + (1-K)D))} - 1\right) \leq \psi_0(x,y,Tx,\,Ty,\varphi(t)) \tag{3.1}$$

where

$$\begin{aligned}\psi_0(x,y,Tx,\,Ty,\varphi(t)) = \max\Bigg(&\psi\left(\frac{1}{F_{x,y}(\varphi(t))} - 1\right),\\ &\psi\left(\frac{1}{F_{x,\,Tx}(\varphi(t))} - 1\right),\ \psi\left(\frac{1}{F_{y,\,Ty}(\varphi(t))} - 1\right),\\ &\psi\left(\frac{1}{F_{x,\,Ty}(2\varphi(t))} - 1\right),\ \psi\left(\frac{1}{F_{y,\,Tx}(2\varphi(t))} - 1\right)\Bigg),\end{aligned} \tag{3.2}$$

$\forall (x,y) \in A_i \times A_{i+1}$, $\forall i \in \bar{p}$, $\forall t(>D) \in \mathbf{R}_+$, where $K \in (0,1)$ and $\varphi \in \Phi_{D\hat{D}}$. □

An extension of Proposition 2.1 which can be proved under similar arguments follows:

Proposition 3.1 *The constraint* (3.1), *subject to* (3.2), *is identical to:*

$$F_{Tx,Ty}(\varphi\,(t)) \geq \frac{\alpha\,(x,y,\,K^{-1}(t-D)\,+\,D)}{\alpha\,(x,y,\,K^{-1}(t-D)\,+\,D) + \psi_0(x,\,y,Tx,Ty,\,\varphi(K^{-1}(t-D)\,+\,D) - 1)}, \tag{3.3}$$

$\forall (x,y) \in A_i \times A_{i+1}$, $\forall i \in \bar{p}$, $\forall t(>D) \in \mathbf{R}_+$ *which is also identical, if* $\alpha(x,y,t) \neq 0$, $\forall (x,y) \in A_i \times A_{i+1}$, $\forall i \in \bar{p}$, $\forall t(>D) \in \mathbf{R}_+$, to

$$\begin{aligned}F_{Tx,Ty}^{-1}(\varphi(t)) \leq\ & 1 + \alpha^{-1}\left(x,y,\,K^{-1}(t-D)\,+\,D\right)\\ &\psi_0\left(x,\,y,Tx,Ty,\,\varphi\left(K^{-1}(t-D)\,+\,D\right) - 1\right)\end{aligned} \tag{3.4}$$

Proposition 3.2 *Let (X,F) be a PM-space and let $T : \bigcup_{i \in \bar{p}} A_i \to \bigcup_{i \in \bar{p}} A_i$ be a generalized p-cyclic α-ψ-type contraction with $A_i \subset X$ being bounded with $D = d(A_i,\, A_{i+1})$ and $\bar{D} = \max_{i \in \bar{p}} \operatorname{diam}(A_i)$,$\forall i \in \bar{p}$ being the distance in-between adjacent subsets, $\forall i \in \bar{p}$. If $\varphi \in \Phi_{D\hat{D}}$, then Proposition 2.2 holds.*

Proof It is direct since if (X,F) be a PM-space, $T : \bigcup_{i \in \bar{p}} A_i \to \bigcup_{i \in \bar{p}} A_i$ is a generalized p-cyclic α-ψ-type contraction with $A_i \subset X$ being bounded, $D = d(A_i,\, A_{i+1})$ and $\bar{D} = \max_{i \in \bar{p}} \operatorname{diam}(A_i)$ being the distance in-between adjacent subsets, $\forall i \in \bar{p}$ and $\varphi \in \Phi_{D\hat{D}}$, then:

$$\begin{aligned}&F_{x,y}^{-1}\left(\varphi\left(K^{-1}(t-D)\,+\,D\right)\right) = F_{x,y}\left(\varphi\left(K^{-1}(t-D)\,+\,D\right)\right) = 1\\ &F_{x,Tx}^{-1}\left(\varphi\left(K^{-1}(t-D)\,+\,D\right)\right) = F_{x,Tx}\left(\varphi\left(K^{-1}(t-D)\,+\,D\right)\right) = 1\\ &F_{y,Ty}^{-1}\left(\varphi\left(K^{-1}(t-D)\,+\,D\right)\right) = F_{y,Ty}\left(\varphi\left(K^{-1}(t-D)\,+\,D\right)\right) = 1\\ &F_{x,Ty}^{-1}\left(2\varphi\left(K^{-1}(t-D)\,+\,D\right)\right) = F_{x,Tx}\left(2\varphi\left(K^{-1}(t-D)\,+\,D\right)\right) = 1\\ &F_{y,Tx}^{-1}\left(2\varphi\left(K^{-1}(t-D)\,+\,D\right)\right) = F_{y,Ty}\left(2\varphi\left(K^{-1}(t-D)\,+\,D\right)\right) = 1\end{aligned} \tag{3.5}$$

for $t \in (D + 2\bar{D},\, \infty)$ so that it follows in a similar way as in the proof of Proposition 2.2 that

$$\begin{aligned}&\psi_0\left(x,y,Tx,\,Ty,\,\varphi\left(K^{-1}(t-D)\,+\,D\right)\right) = \psi_0(x,y,Tx,\,Ty,0)\\ &= 0\end{aligned} \tag{3.6}$$

for $t \in (D + 2\bar{D},\, \infty)$ since $K \in (0,\, 1)$, the distance distribution function $F : \mathbf{R} \to [0,1]$ is non-decreasing and left-continuous, $T(A_i) \subseteq T(A_{i+1})$, $\forall i \in \bar{p}$, $K^{-1}(t-D) + D > 2\bar{D} + D$, and

$$2\varphi\left(K^{-1}(t-D) + D\right) > \varphi\left(K^{-1}(t-D) + D\right) > \varphi(2\bar{D} + D) \tag{3.7}$$

since $\varphi(t)$ is strictly increasing. □

The α-admissibility property has the same sense as in the case of α-ψ-type contractions and cyclic contractions, that is:

Definitions 3.2 If (X, F) a PM-space then an α-ψ-type generalized contraction $T : X \to X$ (respectively, a generalized p-cyclic α-ψ-type contraction $T : \bigcup_{i\in\bar{p}} A_i \to \bigcup_{i\in\bar{p}} A_i$) is α-admissible if $\alpha : X \times X \times \mathbf{R}_+ \to cl\mathbf{R}_{0+}$ satisfies Definition 2.2 (1) (respectively, Definition 2.2 (2)). □

Parallel results to those in Theorem 2.1 and Corollary 2.1 are stated in the following compacted result:

Theorem 3.1 *Let (X, F, Δ) be a G-complete PM-space and let $T : \bigcup_{i\in\bar{p}} A_i \to \bigcup_{i\in\bar{p}} A_i$ be a generalized p-cyclic α-ψ-type contraction satisfying the following conditions:*

1. $D = d(A_i, A_{i+1}) > 0$ *and* $\mathrm{diam}(A_i) > \hat{D} - D$, $\forall i \in \bar{p}$,
2. T *is α-admissible,*
3. *there exists* $x_0 \in \bigcup_{i\in\bar{p}} A_i$ *such that* $\alpha(x_0, Tx_0, t) \geq 1$, $\forall t \in \mathbf{R}_+$,
4. *if* $\{x_n\}(\subset X) \to x$ *is a convergent sequence generated by the Picard iteration* $x_{n+1} = Tx_n$, $\forall n \in \mathbf{Z}_{0+}$ *for a given initial condition* $x_0 \in \bigcup_{i\in\bar{p}} A_i$, *such that* $\alpha(x_n, Tx_n, t) \geq 1, \forall n \in \mathbf{Z}_{0+}, \forall t \in \mathbf{R}_+$ *then* $\alpha(x_n, x, t) \geq 1$, $\forall n \in \mathbf{Z}_{0+}$, $\forall t \in \mathbf{R}_+$.

Then, the following properties hold:

1. *If* $\varphi \in \Phi_{D\hat{D}}$ *for some* $\hat{D}(\geq D) \in \mathbf{R}_{0+}$
2. *If* $\varphi \in \Phi_0$ *and* $\bigcap_{i\in\bar{p}} A_i \neq \emptyset$ *then* $\{T^n x_0\} \subset \bigcup_{i\in\bar{p}} A_i$ *and* $\{T^{np+j} x_0\} \subset A_{i+j}$ *are both Cauchy and G-Cauchy convergent sequences to a limit point* $x \in \bigcap_{i\in\bar{p}} cl A_i$. *If A_i are closed for $i \in \bar{p}$ then $x = Tx = T^p x$, that is, it is a fixed point of the self-mappings* $T : \bigcup_{i\in\bar{p}} A_i \to \bigcup_{i\in\bar{p}} A_i$ *and* $T^p(\equiv T \circ T^{p-1}) : \bigcup_{i\in\bar{p}} A_i \big| A_k \to A_k$, $\forall k \in \bar{p}$, $\forall x_0 \in A_i$, $\forall i \in \bar{p}$, $\forall j \in \overline{p-1} \cup \{0\}$.
3. *If A_i is closed,* $\forall i \in \bar{p}$ *and* $\bigcap_{i\in\bar{p}} A_i = \emptyset$ *then if* $x_0 \in A_i$ *for any* $i \in \bar{p}$ *then there is a limit cycle* $(\bar{x}_i, T\bar{x}_i, \ldots, T^{p-1}\bar{x}_i)$, *to which the sequence* $\{T^n x_0\}$ *converges, with* $T^p \bar{x}_i = \bar{x}_i$, $\bar{x}_i \in A_i$, $\bar{x}_{i+j} = T^j \bar{x}_i \in A_{i+j}$, $\forall i \in \bar{p}, \forall j \in \overline{p-1} \cup \{0\}$ *being a fixed point of the composite self-mapping* $T^p : \bigcup_{i\in\bar{p}} A_i \big| A_k \to A_k$, $\forall k \in \bar{p}$, *and also a quasi-best proximity point of* $T : \bigcup_{i\in\bar{p}} A_i \to \bigcup_{i\in\bar{p}} A_i$. *The subsequence* $\{T^{np+j} x_0\}$ *in* A_{i+j} *converges to* $x_{i+j} = T^j x_i$, $\forall j \in \bar{p}$ *if* $x_0 \in A_i$ *for any given* $i \in \bar{p}$. *Furthermore,*

$$\lim_{n\to\infty} F_{T^{np}x_0, T^{(n+m)p}x_0}(\varepsilon) = F_{x_0, Tx_0}(\infty) = 1, \quad F_{T^{np}x_0, T^{(n+m)p}x_0}(\varepsilon) > 1 - \lambda$$
$$F_{T^{npj}x_0, T^{npj+1}x_0}(\varphi(D) + \varepsilon) > 1 - \lambda,$$
$$\lim_{n\to\infty} F_{T^{npj}x_0, T^{npj+1}x_0}(\varphi(D) + \varepsilon) = F_{\bar{x}_{i+j}, \bar{x}_{i+j+1}} = 1,$$

for any given $\varepsilon \in \mathbf{R}_+$, $\lambda \in (0, 1)$ *and some* $n_0 = n_0(\varepsilon, \lambda)$, $\forall n(\geq n_0) \in \mathbf{Z}_{0+}$.

The following further properties hold if condition (1) is relaxed to $D = d(A_i, A_{i+1}) > 0$, $\forall i \in \bar{p}$ *and Conditions 2–4 still hold:*

4. *If* $\varphi \in \Phi_D$, $\bigcap_{i\in\bar{p}} A_i \neq \emptyset$ *then* $\{T^n x_0\} \subset \bigcup_{i\in\bar{p}} A_i$ *and* $\{T^{np+j} x_0\} \subset A_{i+j}$ *are both Cauchy and G-Cauchy convergent sequences to a limit point* $x \in \bigcap_{i\in\bar{p}} cl A_i$, *with $cl A_i$ being the closure of A_i. If A_i are closed for $i \in \bar{p}$ then $x = Tx = T^p x$, that is, it is a fixed point of the self-mappings* $T : \bigcup_{i\in\bar{p}} A_i \to \bigcup_{i\in\bar{p}} A_i$ *and* $T^p(\equiv T \circ T^{p-1}) : \bigcup_{i\in\bar{p}} A_i \big| A_k \to A_k$, $\forall k \in \bar{p}$, $\forall x_0 \in A_i$, $\forall i \in \bar{p}$, $\forall j \in \overline{p-1} \cup \{0\}$.
5. *If* $\varphi \in \Phi_D$, *A_i is closed* $\forall i \in \bar{p}$ *and* $\bigcap_{i\in\bar{p}} A_i = \emptyset$ *then, if* $x_0 \in A_i$ *for any* $i \in \bar{p}$, *there is a limit cycle* $(\bar{x}_i, T\bar{x}_i, \ldots, T^{p-1}\bar{x}_i)$, *to which the sequence* $\{T^n x_0\}$ *converges, with* $T^p \bar{x}_i = \bar{x}_i$, $\bar{x}_i \in A_i$, $\bar{x}_{i+j} = T^j \bar{x}_i \in A_{i+j}$, $\forall i \in \bar{p}, \forall j \in \overline{p-1} \cup \{0\}$ *being a fixed point of the composite self-mapping* $T^p : \bigcup_{i\in\bar{p}} A_i \big| A_k \to A_k$, $\forall k \in \bar{p}$, *and being also a best proximity point of* $T : \bigcup_{i\in\bar{p}} A_i \to \bigcup_{i\in\bar{p}} A_i$. *The subsequence* $\{T^{np+j} x_0\}$ in A_{i+j} *converges to* $x_{i+j} = T^j x_i$, $\forall j \in \bar{p}$ *if* $x_0 \in A_i$ *for some* $i \in \bar{p}$. *Furthermore:*

$$\lim_{n\to\infty} F_{T^{np}x_0, T^{(n+m)p}x_0}(\varepsilon) = F_{x_0, Tx_0}(\infty) = 1, \quad F_{T^{np}x_0, T^{(n+m)p}x_0}(\varepsilon) > 1 - \lambda \tag{3.8}$$

$$F_{T^{npj}x_0, T^{npj+1}x_0}(\varphi(D) + \varepsilon) > 1 - \lambda, \quad \lim_{n\to\infty} F_{T^{npj}x_0, T^{npj+1}x_0}(\varphi(D) + \varepsilon) = F_{\bar{x}_{i+j}, \bar{x}_{i+j+1}} = 1 \tag{3.9}$$

for any given $\varepsilon \in \mathbf{R}_+$, $\lambda \in (0, 1)$ *and some* $n_0 = n_0(\varepsilon, \lambda)$, $\forall n(\geq n_0) \in \mathbf{Z}_{0+}$. □

Proof Since $T : \bigcup_{i\in\bar{p}} A_i \to \bigcup_{i\in\bar{p}} A_i$ is a p-cyclic α-ψ-type generalized contraction, $x_0 \in \bigcup_{i\in\bar{p}} A_i$, $\alpha(x_0, Tx_0, t) \geq 1$, it is proved by complete induction as in Theorem 2.1 that $\alpha(x_n, Tx_n, t) \geq 1$, $\forall n \in \mathbf{Z}_{0+}$, $\forall t \in \mathbf{R}_+$ since $\varphi \in \Phi_{D\hat{D}}$, φ is strictly increasing in (D, ∞) with $\varphi(D) = \hat{D} \geq D > 0$ and then there exists such that $\min\left(F_{x_0,Tx_0}(\varphi(t)), F_{Tx_0,T^2x_0}(\varphi(t)), F_{x_0,T^2x_0}(\varphi(t))\right) > 0$, $\forall t(> D) \in \mathbf{R}_+$. Since the distance distribution function is non-decreasing and left-continuous and $K^{-1} > 1$,

$$\min\left(F_{x_0,Tx_0}(\varphi(t)), F_{Tx_0,T^2x_0}(\varphi(t)), F_{x_0,T^2x_0}(\varphi(t))\right) \geq \min\left(F_{x_0,Tx_0}(\varphi(t)), F_{Tx_0,T^2x_0}(\varphi(t)), F_{x_0,T^2x_0}(\varphi(t))\right) > 0 \tag{3.10}$$

with $t' = K^{-1}(t-D) + D,\ \forall t(>D) \in \mathbf{R}_+$ and taking inverses in (3.10):

$$0 \le \psi_0\left(x_0, Tx_0, Tx_0, T^2x_0, \varphi(t)\right) < +\infty, \quad \forall t(>D) \in \mathbf{R}_+ \tag{3.11}$$

The cases a–c of the proof of Theorem 2.1 are re-addressed via the changes:

$$\psi^n\left(F^{-1}_{x_0,Tx_0}\left(\varphi\left(K^{-n+1}(t-D)+D\right)\right)-1\right) \to \psi_0^n\left(x_0, Tx_0, Tx_0, T^2x_0, \varphi\left(K^{-n+1}(t-D)+D\right)\right) \tag{3.12}$$

$$\begin{aligned}&\psi_0\left(T^{n-1}x_0, T^nx_0, T^nx_0, T^{n+1}x_0, \varphi(t)\right)\\ &= \max\left(\psi\left(\frac{1}{F_{T^{n-1}x_0,T^nx_0}(\varphi(t))}-1\right), \psi\left(\frac{1}{F_{T^{n-1}x_0,T^nx_0}(\varphi(t))}-1\right),\right.\\ &\times\psi\left(\frac{1}{F_{T^nx_0,T^{n+1}x_0}(\varphi(t))}-1\right), \psi\left(\frac{1}{F_{T^{n-1}x_0,T^{n+1}x_0}(2\varphi(t))}-1\right),\\ &\left.\psi\left(\frac{1}{F_{T^nx_0,T^nx_0}(2\varphi(t))}-1\right)\right)\\ &= \max\left(\psi\left(\frac{1}{F_{T^{n-1}x_0,T^nx_0}(\varphi(t))}-1\right), \psi\left(\frac{1}{F_{T^nx_0,T^{n+1}x_0}(\varphi(t))}-1\right),\right.\\ &\left.\psi\left(\frac{1}{F_{T^{n-1}x_0,T^{n+1}x_0}(2\varphi(t))}-1\right)\right),\end{aligned} \tag{3.13}$$

$\forall n \in \mathbf{Z}_+, \forall t(>D) \in \mathbf{R}_+$, from (3.2), and also

$$\begin{aligned}\psi_0\left(x_0, Tx_0, T^2x_0, \varphi(t)\right) &= \max\left(\psi\left(\frac{1}{F_{x_0,Tx_0}(\varphi(t))}-1\right),\right.\\ &\left.\psi\left(\frac{1}{F_{Tx_0,T^2x_0}(\varphi(t))}-1\right), \psi\left(\frac{1}{F_{x_0,T^2x_0}(2\varphi(t))}-1\right)\right)\end{aligned} \tag{3.14}$$

since

$$\psi\left(\frac{1}{F_{T^nx_0,T^nx_0}(2\varphi(t))}-1\right) = \psi\left(\frac{1}{F_{Tx_0,Tx_0}(2\varphi(t))}-1\right) = \psi(0) = 0, \quad \forall t \in \mathbf{R}_{0+} \tag{3.15}$$

since $F_{x,x}(t) = 1,\ \forall t \in \mathbf{R}_+$, since (X,F) is a probabilistic metric space, and $\varphi : \mathbf{R} \to \mathbf{R}_{0+}$ is nonzero, $\forall t \in \mathbf{R}_+$ for all $x \in X$ can be removed from the evaluation of the maximum in (3.13). Equations (2.8a)–(2.8b) are changed to:

$$\begin{aligned}F^{-1}_{T^nx_0,T^{n+1}x_0}(\varphi(t)) &\le 1 + \psi_0\left(T^{n-1}x_0,\ T^nx_0,\ T^{n+1}x_0,\right.\\ &\left.\left(\varphi\left(K^{-1}(t-D)+D\right)\right)-1\right)\\ &\le 1 + \psi_0^n\left(x_0,\ Tx_0,\ T^2x_0, \varphi\left(K^{-n+1}(t-D)+D\right)\right),\\ &\forall t(>D) \in \mathbf{R}_+, \quad \forall n \in \mathbf{Z}_+\end{aligned} \tag{3.16a}$$

using (3.2) with $x = T^{n-1}x_0,\ y = Tx = T^nx_0,\ \forall n \in \mathbf{Z}_+$, equivalently,

$$\begin{aligned}&F_{T^nx_0,T^{n+1}x_0}(\varphi(t))\\ &\ge \frac{1}{1+\psi_0^n(x_0,\ Tx_0,\ T^2x_0, \varphi(K^{-n+1}(t-D)+D))},\\ &\forall t(>D) \in \mathbf{R}_+, \quad \forall n \in \mathbf{Z}_+\end{aligned} \tag{3.16b}$$

Since $K^{-1} > 1$, $\lim_{t\to\infty}\varphi(t) = \infty$ then $\lim_{n\to+\infty}(K^{-n+1}(t-D)+D) = +\infty$ and $\lim_{n\to\infty}\varphi(K^{-n+1}(t-D)+D) = +\infty$, $\forall t(>D) \in \mathbf{R}_+$, $\lim_{n\to\infty}F^{-1}_{x_0,Tx_0}(\varphi(K^{-n+1}(t-D)+D)) = 1$, $\forall t(>D) \in \mathbf{R}_+$. Thus, from (3.15),

$$\lim_{n\to\infty}\psi_0^n\left(x_0,\ Tx_0,\ T^2x_0, \varphi\left(K^{-n+1}(t-D)+D\right)\right) = 0, \quad \forall t(>D) \in \mathbf{R}_+ \tag{3.17}$$

since

$$\begin{aligned}&F_{x_0,Tx_0}\left(\varphi\left(K^{-n+1}(t-D)+D\right)\right) \to 1,\\ &F_{x_0,T^2x_0}\left(\varphi\left(K^{-n+1}(t-D)+D\right)\right) \to 1,\\ &F_{T^2x_0,Tx_0}\left(\varphi\left(K^{-n+1}(t-D)+D\right)\right) \to 1\end{aligned} \tag{3.18a}$$

$$\begin{aligned}&\psi^n\left(F^{-1}_{x_0,Tx_0}\left(\varphi\left(K^{-n+1}(t-D)+D\right)\right)-1\right) \to 0,\\ &\psi^n\left(F^{-1}_{x_0,T^2x_0}\left(\varphi\left(K^{-n+1}(t-D)+D\right)\right)-1\right) \to 0,\\ &\psi^n\left(F^{-1}_{T^2x_0,Tx_0}\left(\varphi\left(K^{-n+1}(t-D)+D\right)\right)-1\right) \to 0\end{aligned} \tag{3.18b}$$

as $n \to \infty$, $\forall t(>D) \in \mathbf{R}_+$, since $\psi^n(t_n) \to 0$ if $\{t_n\} \to 0$. Since $\hat{D} = \varphi(D^+)$, one has from Eqs. (3.16a) and (3.16b) that $\lim_{n\to\infty}F_{T^{npj}x_0,T^{npj+1}x_0}(\varphi(D)+t) = 1$, $\forall t \in \mathbf{R}_+$ with $T^{npj}x_0 \in A_{i+j}$ and $T^{npj+1}x_0 \in A_{i+j+1}$, $\forall j \in \bar{p}$ if $x_0 \in A_i$ for any given $i \in \bar{p}$ since $\operatorname{diam}(A_i) > \hat{D} - D$, $\forall i \in \bar{p}$. Then, for any given real constants $\varepsilon \in \mathbf{R}_+$ and $\lambda \in (0,1)$, there is some $n_0 = n_0(\varepsilon, \lambda) \in \mathbf{Z}_{0+}$ such that $F_{T^{npj}x_0,T^{npj+1}x_0}(\varphi(D)+\varepsilon) > 1-\lambda$, $\forall n \ge n_0$, $\forall j \in \bar{p}$ since the distance distribution function is non-decreasing and left-continuous. Thus, (2.10), for $\forall t(>D) \in \mathbf{R}_+$, (2.11), for $D = \hat{D} = \varphi(0) = 0$, and (2.12), for $\ell \in \overline{k-1} \cup \{0\}$, and (2.13) obtained in the proof of Theorem 2.1 also hold, for any $\varepsilon \in \mathbf{R}_+$ and $\lambda \in (0,1)$, $\forall j \in \bar{p}$, $\forall n(\ge n_0) \in \mathbf{Z}_{0+}$, $\forall m \in \mathbf{Z}_+$ and some $n_0 = n_0(\varepsilon, \lambda) \in \mathbf{Z}_{0+}$. Then, the p sequences $\{T^{np+i}x_0\}$, $\forall i \in \bar{p}$ have a unique limit point in $\bigcap_{i\in\bar{p}} A_i$ provided that such a set is nonempty and closed (otherwise, it is allocated in the intersection of the corresponding closures) and, from Assertion 1.1, they are both Cauchy and G-Cauchy sequences. The limit point is also proved to be a fixed point of $T:\bigcup_{i\in\bar{p}}A_i\to\bigcup_{i\in\bar{p}}A_i$ and of $T^p:\bigcup_{i\in\bar{p}}A_i\big|A_j\to A_j,\forall j\in\bar{p}$. Hence, Property (2) has been proved. Property (3) is proved from the still valid formulas (2.14)–(2.18). Properties (4) and (5) follow from their still applicable counterparts of Corollary 2.1 for the cases when $\varphi\in\Phi_{D\hat{D}}$, $\bigcap_{i\in\bar{p}}A_i\neq\emptyset$ and, respectively, $\varphi\in\Phi_D$,A_i is closed,$\forall i\in\bar{p}$ and $\bigcap_{i\in\bar{p}}A_i=\emptyset$. □

Theorem 3.2 *Let (X,F,Δ) a G-complete PM-space and let $T:\bigcup_{i\in\bar{p}}A_i\to\bigcup_{i\in\bar{p}}A_i$ be a p-cyclic α-ψ-type generalized contraction satisfying the following conditions:*

1. $D=d(A_i,\ A_{i+1})>0,\quad\forall i\in\bar{p}$,
2. T *is* α*-admissible,*
3. *there exists* $x_0\in\bigcup_{i\in\bar{p}}A_i$ *such that* $\alpha(x_0,\ Tx_0,\ t)\geq 1$, $\forall t\in\mathbf{R}_+$,
4. *if* $\{x_n\}(\subset X)\to x$ *is a sequence generated as* $x_{n+1}=Tx_n$, *such that* $\alpha(x_n,\ Tx_n,\ t)\geq 1,\forall n\in\mathbf{Z}_{0+},\forall t\in\mathbf{R}_+$ *then* $\alpha(x_n,\ x,\ t)\geq 1,\ \forall n\in\mathbf{Z}_{0+},\forall t\in\mathbf{R}_+$.
5. *For all* $x,y\in\bigcup_{i\in\bar{p}}clA_i$, *there is* $z\in\bigcup_{i\in\bar{p}}clA_i$ *such that* $\min\left(\alpha\left(x,T^kz,t\right),\ \alpha\left(y,T^kz,t\right)\right)\geq 1,\ \forall t(>D)\in\mathbf{R}_+$ *for some* $k=k(i,j)\in\overline{p-1}\cup\{0\}$ *if* $x\in clA_i$, $y\in clA_j$ *for any given* $i,j\in\bar{p}$.

Then, there is $\bar{x}_i\in clA_i$, such that $\bar{x}_i\in A_i$ if A_i is closed, $\forall i\in\bar{p}$, which is the unique best proximity point of $T:\bigcup_{i\in\bar{p}}A_i\to\bigcup_{i\in\bar{p}}A_i$ and the unique fixed point of $T^p:\bigcup_{i\in\bar{p}}A_i\big|A_k\to A_k,\forall k\in\bar{p}$. If, in addition, $\varphi\in\Phi_0$ with $D=0$ and $\bigcap_{i\in\bar{p}}A_i$ is nonempty and closed then there is a unique fixed point of $T:\bigcup_{i\in\bar{p}}A_i\to\bigcup_{i\in\bar{p}}A_i$ and the composite self-mappings $T^p:\bigcup_{i\in\bar{p}}A_i\big|A_k\to A_k,\ \forall k\in\bar{p}$.

Proof Let $u\in clA_i$, $v\in clA_j$ be best proximity points for any given $i,j\in\bar{k}$ such that $u=T^{np}u=T^pu$ and $v=T^{np}v=T^pv$, $\forall n\in\mathbf{Z}_+$ so that there is $z\in\bigcup_{i\in\bar{p}}clA_i$ such that $\min(\alpha(u,z,t),\ \alpha(v,z,t))\geq 1,\ \ \forall t(>D)\in\mathbf{R}_+$. Thus, note that $T^{pn+k}z\notin clA_i\cup clA_j$ and $T^{k+1}z\notin A_l$ if $z\in A_l$ for any $\ell\in\bar{p}$ if $D>0$.

Note that $\psi^{np-1}(a_{np-1}(k))\to 0$ as $a_{np-1}(t,k,\ u,z)=\frac{1}{F_{u,T^mz}(\varphi(K^{-(np+k)}(t-D)+D))}-1\to 0$ as $n\to\infty$ for all $t(>D)\in\mathbf{R}_+$ which holds since $F_{u,T^mz}(\varphi(K^{-np}(t-D)+D))\to 1$ as $n\to\infty$ for all $\forall t(>D)\in\mathbf{R}_+$ since $(K^{-np}(t-D)+D)\to\infty$ for all $t(>D)\in\mathbf{R}_+$ as $n\to\infty$ and since $\varphi(t)>D$ from (2.21), from the properties $\varphi(t)=0$ for $t<D$, $\varphi(D)\geq D$, and $\varphi(t)>D$ for $\forall t(>D)\in\mathbf{R}_+$ since $\varphi:\mathbf{R}_{0+}\to\mathbf{R}_{0+}$ is strictly increasing in $[D,\infty)$. As a result, $F_{u,T^{pn+m}z}(t)=F_{v,T^{pn+\ell}z}(t)=1,\ \forall t(>D)\in\mathbf{R}_+$. Let $u,v\in clA_i$ be best proximity points for any given $i\in\bar{k}$ such that $u=T^pu$ and $v=T^pv$. Then, one gets from (3.19) to (3.20) that, if $\alpha(u,v,t)\geq 1$ for any $t\in\mathbf{R}_+$ then since and since T is α-admissible and $\min\left(\alpha\left(x,T^kz,t\right),\ \alpha\left(y,T^kz,t\right)\right)\geq 1$,

$$\frac{1}{F_{T^{np}u,T^{np}v}(\varphi(t))}-1=\frac{1}{F_{T^pu,T^pv}(\varphi(t))}-1=\frac{1}{F_{u,v}(\varphi(t))}-1$$
$$\leq\sup_{n\in Z_{0+}}\left[\alpha^{-1}\left(u,T^{np+k}z,t\right)\right]\psi_0^{np-1}\left(u,T^kz,Tu,\ T^{k+1}v,\ \varphi(K^{-np}(t-D)+D)\right),\quad\forall t(>D)\in\mathbf{R}_+. \tag{3.19}$$

Note also that

$$\limsup_{n\to\infty}\psi_0^{np-1}\left(u,z,Tu,T^kz,\varphi(K^{-np}(t-D)+D)\right)$$
$$=\max\left[\limsup_{n\to\infty}\psi\left(\frac{1}{F_{u,z}(\varphi(t_n'))}-1\right)\right.,\ \limsup_{n\to\infty}\psi\left(\frac{1}{F_{u,Tu}(\varphi(t_n'))}-1\right),\ \limsup_{n\to\infty}\psi\left(\frac{1}{F_{z,Tz}(\varphi(t_n'))}-1\right),$$
$$\left.\limsup_{n\to\infty}\psi\left(\frac{1}{F_{u,Tz}(2\varphi(t_n'))}-1\right),\ \limsup_{n\to\infty}\psi\left(\frac{1}{F_{Tz,Tu}(2\varphi(t_n'))}-1\right)\right]$$
$$\leq\max\left[\limsup_{n\to\infty}\psi\left(\frac{1}{F_{u,z}(\varphi(t_n'))}-1\right)\right.,\ \limsup_{n\to\infty}\psi\left(\frac{1}{F_{u,Tu}(\varphi(t_n'))}-1\right),\ \limsup_{n\to\infty}\psi\left(\frac{1}{F_{z,Tz}(\varphi(t_n'))}-1\right),$$
$$\left.\limsup_{n\to\infty}\psi\left(\frac{1}{F_{u,Tz}(2\varphi(t_n'))}-1\right),\ \limsup_{n\to\infty}\psi\left(\frac{1}{F_{Tz,Tu}(2\varphi(t_n'))}-1\right)\right]$$
$$\leq\max\left[\limsup_{n\to\infty}\psi\left(\frac{1}{F_{u,z}(\varphi(t_n'))}-1\right)\limsup_{n\to\infty}\psi\left(\frac{1}{F_{u,Tu}(\varphi(t_n'))}-1\right),\right.$$
$$\left.\left.\limsup_{n\to\infty}\psi\left(\frac{1}{F_{z,Tz}(\varphi(t_n'))}-1\right)\right]\right]=\psi(0)=0,\quad\forall t(>D)\in\mathbf{R}_+ \tag{3.20}$$

where $t_n'=t_n'(t)=K^{-np}(t-D)+D$ since

$$F_{u,Tz}(2\varphi(t_n'))\geq\Delta_M\left(F_{u,z}(\varphi(t_n')),\ \ F_{Tz,z}(\varphi(t_n'))\right)$$
$$\geq\min\left(F_{u,z}(\varphi(t_n')),\ F_{Tz,z}(\varphi(t_n'))\right) \tag{3.21}$$

$$F_{Tu,Tz}(2\varphi(t_n'))\geq\Delta_M\left(F_{u,Tu}(\varphi(t_n')),\ F_{Tz,u}(\varphi(t_n'))\right)$$
$$\geq\min\left(F_{u,Tu}(\varphi(t_n')),\ F_{Tz,u}(\varphi(t_n'))\right) \tag{3.22}$$

and, in the same way,

$$\limsup_{n\to\infty}\psi_0^{np-1}\left(v,z,Tv,\ T^kz,\ \varphi(K^{-np}(t-D)+D)\right)=0,$$
$$\forall t(>D)\in\mathbf{R}_+ \tag{3.23}$$

so that $t_n'\to+\infty$ as $n\to\infty$, $\forall t(>D)\in\mathbf{R}_+$.Then, it follows from (3.19) and (3.20) that

$$F_{u,T^{pn+k}z}(t)=F_{v,T^{pn+k}z}(t)=1,\quad\forall t(>D)\in\mathbf{R}_+ \tag{3.24}$$

$$F_{u,T^{pn+k}z}(t)=F_{v,T^{pn+k}z}(t)=0,\quad\forall t(\leq D)\in\mathbf{R}_+ \tag{3.25}$$

since $\psi^{np-1}(a_{np-1}(t,k,u,z)) \to 0$ as $a_{np-1}(t,k,u,z) = \frac{1}{F_{u,T^{k}z}(\varphi(K^{-(np+k)}(t-D)+D))} - 1 \to 0$ as $n \to \infty$ for all $t(>D) \in R_+$ which holds since $F_{u,T^{k}z}(\varphi(K^{-np}(t-D)+D)) \to 1$ as $n \to \infty$ for all $\forall t(>D) \in \mathbf{R}_+$ since $(K^{-np}(t-D)+D) \to \infty$ for all $t(>D) \in \mathbf{R}_+$ as $n \to \infty$ and since $\varphi(t) > D$ from (2.21), from the properties $\varphi(t) = 0$ for $t<D$, $\varphi(D) \geq D$, and $\varphi(t) > D$ for $t > D$ since $\varphi : \mathbf{R} \to \mathbf{R}_{0+}$ is strictly increasing in $[D,\infty)$. The same conclusion arises by replacing u by v. Thus, $F_{u,v}(\varphi(t)) = F_{T^{p}u,T^{p}v}(\varphi(t)) = 1, \forall t \in \mathbf{R}_+$ and then $u = v$ is a unique fixed point of $T^p : \bigcup_{j\in\bar{p}} A_j \Big| A_i \to A_i$ for any arbitrary given $i \in \bar{p}$, $u \in clA_i$ and $Tu \in clA_{i+1}$ are unique adjacent best proximity points. If the intersection of the closures of the subsets A_i is nonempty then the unique adjacent best proximity points coincide in a unique fixed point which follows in the same way as its counterpart in the proof of Theorem 2.2. □

Example 3.1 Example 2.1 is revisited under the constraint (3.1) subject to (3.2) with the same distance distribution function and identical $\alpha(x,y,t)$ with $\varphi(t) = \psi(t) = t$ and making $x \to x_0$, $y \to Tx_0$, $Ty \to Tx_0^2$ for any arbitrary $x_0 \in \mathbf{R}$, we get $\{x_{2n} = T^{2n}x_0\} \to x$, $\{x_{2n+1} = T^{2n+1}x_0\} \to -x$ with $x = D$ if $x_0 \in A$ and $x = -D$ if $x_0 \in B$, where

$$\begin{aligned}\psi_0(x_0, Tx_0, Tx_0, T^2x_0, t) = \max\Bigg(&\psi\left(\frac{1}{F_{x_0,Tx_0}(t)} - 1\right),\\ &\psi\left(\frac{1}{F_{Tx_0,T^2x_0}(t)} - 1\right), \psi\left(\frac{1}{F_{x_0,T^2x_0}(2t)} - 1\right),\\ &\psi\left(\frac{1}{F_{Tx_0,Tx_0}(2t)} - 1\right)\Bigg), \quad \forall t(>D) \in R_+\end{aligned} \tag{3.26}$$

$$\begin{aligned}&\psi_0(T^n x_0, T^{n+1}x_0, T^{n+1}x_0, T^{n+2}x_0, t)\\ &= \max\Bigg(\psi\left(\frac{1}{F_{T^n x_0,T^{n+1}x_0}(t)} - 1\right), \psi\left(\frac{1}{F_{T^{n+1}x_0,T^{n+2}x_0}(t)} - 1\right),\\ &\psi\left(\frac{1}{F_{T^n x_0,T^{n+2}x_0}(2t)} - 1\right), \quad \psi\left(\frac{1}{F_{T^{n+1}x_0,T^{n+1}x_0}(2t)} - 1\right)\Bigg),\\ &\forall n \in \mathbf{Z}_{0+}, \quad \forall t(>D) \in \mathbf{R}_+\end{aligned} \tag{3.27}$$

Since $\{T^n x_0\}, \{T^{n+2}x_0\} \to \pm x$ and $\{T^{n+1}x_0\} \to \mp x$, $\lim_{n\to\infty} \psi_0(T^n x_0, T^{n+1}x_0, T^{n+1}x_0, T^{n+2}x_0, t) = 0$ $\forall t(\in \mathbf{R}_+) > D$, since

$$\begin{aligned}&\lim_{n\to\infty} F_{T^n x_0,T^{n+1}x_0}(t) = \lim_{n\to\infty} F_{T^{n+1}x_0,T^{n+2}x_0}(t)\\ &= \lim_{n\to\infty} F_{T^{n+1}x_0,T^{n+2}x_0}(t') = \lim_{n\to\infty} F_{T^{n+1}x_0,T^{n+1}x_0}(t') = 1,\\ &\forall t(>D) \in \mathbf{R}_+, \quad \forall t' \in \mathbf{R}_+\end{aligned} \tag{3.28}$$

Authors' contributions Both authors contributed equally to all the parts of this manuscript.

Acknowledgements The first author is grateful to the Spanish Government for its support with Grant DPI2012-30651, and to the Basque Government for its support through Grant IT378-10.

Compliance with ethical standards

Competing interests The authors declare that they have no competing interests.

References

1. Schweizer, B., Sklar, A.: Probabilistic Metric Spaces. North-Holland, Amsterdam (1983)
2. Pap, E., Hadzic, O., Mesiar, R.: A fixed point theorem in probabilistic metric spaces and an application. J. Math. Anal. Appl. **202**, 433–439 (1996)
3. Sehgal, V.M., Bharucha-Reid, A.T.: Fixed points of contraction mappings on probabilistic metric spaces. Theory Comput. Syst. **6**(1), 97–102 (1972)
4. Schweizer, B., Sklar, A.: Statistical metric spaces. Pac. J. Math. **10**, 313–334 (1960)
5. Eldred, A.A., Veeramani, P.: Existence and convergence of best proximity points. J. Math. Anal. Appl. **323**, 1001–1006 (2006)
6. De la Sen, M.: Linking contractive self-mappings and cyclic Meir-Keeler contractions with Kannan self-mappings. Fixed Point Theory Appl. **1**, 572057 (2010). doi:10.1155/2010/572057. **(Article ID: 572057)**
7. Choudhury, B.S., Das, K., Bhandari, S.K.: Fixed point theorem for mappings with cyclic contraction in Menger spaces. Int. J. Pure Appl. Sci. Technol. **4**(1), 1–9 (2011)
8. Choudhury, B.S., Das, K., Bhandari, S.K.: Cyclic contraction result in 2- Menger space. Bull. Int. Math. Virtual Inst. **2**(1), 223–234 (2012)
9. Beg, I., Abbas, M.: "Fixed point and best approximation in Menger convex metric spaces", *Archivum Mathematicum* (BRNO). Tomus **41**, 389–397 (2005)
10. Beg, I., Latif, A., Abbas, M.: Coupled fixed points of mixed monotone operators on probabilistic Banach spaces. Arch. Math. **37**, 1–8 (2001)
11. Mihet, D.: Altering distances in probabilistic Menger spaces. Nonlinear Anal. **71**, 2734–2738 (2009)
12. Mihet, D.: A Banach contraction theorem for fuzzy metric spaces. Fuzzy Sets Syst. **144**, 431–439 (2004)
13. Sedghi, S., Choudhury, B.S., Shobe, N.: Strong common coupled fixed point result in fuzzy metric spaces. J. Phys. Sci. **17**, 1–9 (2013)
14. Wairojjana, N., Dosenovic, T., Rakic, D., Gopal, D., Kumam, P.: An altering distance function in fuzzy metric fixed point theorems. Fixed Point Theory Appl. **2015**, 69 (2015)
15. Choudhury, B.S., Das, K.P.: A new contraction principle in Menger spaces. Acta Math.Sin. (Engl. Ser.) **24**(8), 1379–1386 (2008)
16. Gopal, D., Abbas, M., Vetro, C.: Some new fixed point theorems in Menger PM-spaces with application to Volterra type integral equation. Appl. Math. Comput. **232**, 955–967 (2014)
17. De la Sen, M., Agarwal, R.P., Nistal, N.: "Non-expansive and potentially expansive properties of two modified p-cyclic self-maps in metric spaces. J. Nonlinear Convex Anal. **14**(4), 661–686 (2013)
18. De la Sen, M., Agarwal, R.P.: Fixed point-type results for a class

of extended cyclic self-mappings under three general weak contractive conditions of rational type. Fixed Point Theory Appl. **2011**, 102 (2011). doi:10.1186/1687-1812-2011-102
19. De la Sen, M.: On best proximity point theorems and fixed point theorems for p-cyclic hybrid self-mappings in Banach spaces. Abstr. Appl. Anal. (2013). doi:10.1155/2013/1831747. **(Article ID: 183174)**
20. Karpagam, S., Agrawal, S.: Best proximity point theorems for p-cyclic Meir-Keeler contractions. Fixed Point Theory Appl. (2009) (**Article ID: 197308**)
21. Suzuki, T.: Some notes on Meir-Keeler contractions and L-functions. Bull. Kyushu Inst. Technol. **53**, 12–13 (2006)
22. Di Bari, C., Suzuki, T., Vetro, C.: Best proximity points for cyclic Meir-Keeler contractions. Nonlinear Anal. Theory Methods Appl. **69**(11), 2790–3794 (2008)
23. Rezapour, S., Derafshpour, M., Shahzad, N.: On the existence of best proximity points of cyclic contractions. Adv. Dyn. Syst. Appl. **6**(1), 33–40 (2011)
24. Derafshpour, M., Rezapour, S., Shahzad, N.: Best proximity points of cyclic φ-contractions on reflexive Banach spaces. Fixed Point Theory Appl. **2010**, 33–40 (2010). **(Article ID: 946178)**
25. Al-Thagafi, M.A., Shahzad, N.: Convergence and existence results for best proximity points. Nonlinear Anal. Theory Methods Appl. **70**(10), 3665–3671 (2009)
26. De la Sen, M.: Stable iteration procedures in metric spaces which generalize a Picard-type iteration. Fixed Point Theory Appl. (2010). doi:10.1155/2010/953091. **(Article ID: 953091)**
27. Karpagam, S., Agrawal, S.: Best proximity point theorems for p-cyclic Meir-Keeler contractions. Fixed Point Theory Appl. (2009) (**Article ID 197308**)
28. Dutta, P.N., Choudhury, B.S., Das, K.: Some fixed point results in Menger spaces using a control function. Surv. Math. Appl. **4**, 41–52 (2009)
29. Chang, S.S., Cho, Y.J., Kang, S.M.: Nonlinear Operator Theory in Probabilistic Metric Spaces. Nova Science Publishers, New York (2001)
30. De la Sen, M., Karapınar, E.: Some results on best proximity points of cyclic contractions in probabilistic metric spaces. J. Funct. Spaces **2015**, 11 (2015). **(Article ID: 470574)**
31. Chen, C.M., Chang, T.H., Juang, K.S.: Common fixed point theorems for the stronger Meir-Keeler cone-type function in cone-ball metric spaces. Appl. Math. Lett. **25**(4), 692–697 (2012)
32. Chen, C.M.: Common fixed point theorems in complete generalized metric spaces. J. Appl. Math. **2012**, 14 (2012). **(Article ID: 945915)**
33. Alotaibi, A., Mursaleen, M., Sharma, S.K., Mohiuddine, S.A.: Sequence spaces of fuzzy numbers defined by a Musielak-Orlicz function. Filomat **29**(7), 1461–1468 (2015)
34. Wang, S.: On φ-contractions in partially ordered fuzzy metric spaces. Fixed Point Theory Appl. **2015**, 233 (2015)
35. Batool, A., Kamran, T., Jang, S.Y., Park, C.: Generalized φ-weak contractive fuzzy mappings and related fixed point results on complete metric space. J. Comput. Anal. Appl. **21**(4), 729–737 (2016)
36. Shen, Y., Wang, F.: A fixed point approach to the Ulam stability of fuzzy differential equations under generalized differentiability. J. Intell. Fuzzy Syst. **30**(6), 3253–3260 (2016)
37. Alotaibi, A., Mursaleen, M., Mohiuddine, S.A.: Some fixed point theorems for Meir-Keeler condensing operators with applications to integral equations. Bull. Belg. Math. Soc. Simon Stevin **22**(4), 529–541 (2015)

11

An approximate solution for a fractional model of generalized Harry Dym equation

Kamel Al-Khaled · Marwan Alquran

Abstract The nonlinear partial differential equation of Harry Dym is generalized by replacing the first-order time derivative by a fractional derivative of order α, $0 \leq \alpha \leq 2$. The aim of the present paper is to obtain an approximate solution of time fractional generalized Harry Dym equation using Adomian Decomposition Method (ADM). The fractional derivative is described in the Caputo sense. Numerical examples are given to show the application of the present technique. The results show that the solution of ADM is in good agreement with the exact solution when $\alpha = 1$, also reveal that the method is very simple and effective.

Keywords Generalized Harry Dym equation · Approximate solution · Fractional calculus · Adomian decomposition · Caputa derivative

Introduction

Nonlinear partial differential equations appear in many branches of physics, engineering and applied mathematics. In recent years, there has been a growing interest in the field of fractional calculus. Oldham and Spanier [1], Miller and Ross [2] and Podlubny [3] provide the history and a comprehensive treatment of this subject. Fractional calculus is the field of mathematical analysis, which deals with the investigation and applications of integrals and derivatives of arbitrary order, which can be real or complex. The idea appeared in a letter by Leibniz to L'Hospital in (1695). The subject of fractional calculus have gained importance during the past three decades and popularity, mainly due to its demonstrated applications in different areas of physics and engineering. Several fields of applications of fractional differentiation and fractional integration are already well established, some others have just started. Many applications of fractional calculus can be found in turbulence and fluid dynamics, stochastic dynamical systems, plasma physics and controlled thermonuclear fusion, nonlinear control theory, image processing, nonlinear biological systems; for more details see [4] and the references therein. Indeed, it provides several potentially useful tools for solving differential equations. The most important advantage of using fractional differential equations is their nonlocal property. It is known [5] that the integer order differential operator is nonlocal. This means that the next state of a system depends not only upon its current state but also upon all of its historical states. Hence, the importance of investigating fractional equations arises from the necessity to sharpen the concepts of equilibrium, stability states, and time evolution in the long time limit. There have been few attempts to solve linear problems with multiple fractional derivatives. In [6], an approximate solution based on the decomposition method is given for the generalized fractional diffusion-wave equation. Not much work has been done for nonlinear problems, and only a few numerical schemes have been proposed to solve nonlinear fractional differential equations. More recently, applications have included classes of nonlinear equations with multi-order fractional derivative, and this motivates us to develop a numerical scheme for their solutions.

K. Al-Khaled · M. Alquran
Department of Mathematics and Statistics, Jordan University of Science and Technology, P.O. Box: 3030, Irbid 22110, Jordan

K. Al-Khaled · M. Alquran (✉)
Department of Mathematics and Statistics, Sultan Qaboos University, Al-Khod, P.O. Box: 36, PC 123 Muscat, Oman
e-mail: marwan04@just.edu.jo; marwan04@squ.edu.om

Table 1 Approximate solution of equation (1.1) at $t = 0.25$ for some values of the order α

$x \vert \alpha$	0.5	0.75	1	1.5	1.75	2
0.0	2.16862	2.27296	2.35999	2.35641	2.35919	2.36071
0.1	2.10187	2.20634	2.29446	2.29093	2.29374	2.29527
0.2	2.03425	2.13872	2.22798	2.22450	2.22735	2.22890
0.3	1.96575	2.07005	2.16049	2.15708	2.15996	2.16153
0.4	1.89634	2.00026	2.09193	2.08860	2.09152	2.09311
0.5	1.82601	1.92927	2.02224	2.01900	2.02196	2.02357
0.6	1.75475	1.85702	1.95133	1.94820	1.95119	1.95282
0.7	1.68258	1.78340	1.87911	1.87610	1.87914	1.88079
0.8	1.60951	1.70831	1.80548	1.80262	1.80570	1.80737
0.9	1.53561	1.63165	1.73033	1.72764	1.73076	1.73246
1.0	1.46098	1.55329	1.65352	1.65104	1.65420	1.65593

The main objective of the present paper is to use the Adomian decomposition method [7, 8] to calculate the approximate solutions of the Dym equation that is named for Harry Dym and occurs in the study of Solitons. The Dym equation first appeared in Kruskal [9] and is attributed to an unpublished paper by Harry Dym. Harry Dym is a completely integrable nonlinear evolution equation that may be solved by means of the inverse scattering transform. It is interesting because it obeys an infinite number of conservation laws and it does not possess the Painleve property. The Dym equation has strong links to the Korteweg–de Vries equation. As mentioned in [10], the Harry Dym has been recently generalized in different ways. The authors in [11] introduced and investigated the coupled Harry Dym equations, and showed that these equations are concerned with the new iso-spectral flows. Different generalizations of the Harry Dym equation have been constructed in [12]. In [13] group analysis of time fractional Harry Dym equation was performed and some group invariant solutions are obtained. An efficient approach based on Homotopy perturbation method and Sumudu transform were proposed in [14] to solve fractional Harry Dym equation for $0 \le \alpha \le 1$. Beside these generalizations, we would like to present a new generalization of the nonlinear partial differential equation of Harry Dym by replacing the first-order time derivative by a fractional derivative of order α, $0 \le \alpha \le 2$, and takes the form

$$\frac{\partial^\alpha u(x,t)}{\partial t^\alpha} = u^3(x,t)\frac{\partial^3 u(x,t)}{\partial x^3} \tag{1.1}$$

where α is a parameter describing the order of the fractional derivative, x and t are the space and time variables, and $u(x,t)$ is the field defined in the space domain $(-\infty, \infty)$. Theoretically, α can be any positive number. Note that for $\alpha = 1$, Eq. (1.1) represents the standard Harry Dym equation, which has exact solution (see, [5]) $u(x,t) = \left(a - \frac{3\sqrt{b}}{2}(x+ct)\right)^{2/3}$, where a, b and c are suitable constants. This type of equation has been investigated by several authors. In [5], an approximate analytical solution of time fractional Harry Dym equation is obtained using Homotopy perturbation method only for $0 \le \alpha \le 1$. Explicit solutions for Harry Dym equation was obtained by Fuchssteinert et al. [12]. Soliton solutions of the $2+1$ dimensional Harry Dym equation was found by Halim in [17].

The plan of this work is the following. A brief review of the fractional calculus theory is given in Sect. 2. In Sect. 3, we use the decomposition method to construct our numerical solutions for Eq. (1.1). The general response expression contains a parameter describing the order of the fractional derivative that can be varied to obtain various responses. In Sect. 4, we present some numerical results (Graphs and Tables) to show the nature of the solution as the fractional derivative parameter is changed.

Basic definition of fractional calculus

This section is devoted to a description of the operational properties of the purpose of acquainting with sufficient fractional calculus theory, to enable us to follow the solution of the generalized Harry Dym equation. Many definitions and studies of fractional calculus have been proposed in the last two centuries. These definitions include, Riemman–Liouville, Weyl, Reize, Campos, Caputa, and Nishimoto fractional operator. Mainly, in this paper, we will re-introduce section 2 of [6]. The Riemann–Liouville definition of fractional derivative operator J_a^α is defined as follows:

Definition 2.1 Let $\alpha \in \mathbb{R}_+$. The operator J^α, defined on the usual Lebesque space $L_1[a,b]$ by

$$J_a^\alpha f(x) = \frac{1}{\Gamma(\alpha)}\int_a^x (x-t)^{\alpha-1} f(t)\mathrm{d}t$$
$$J_a^0 f(x) = f(x)$$

for $a \le x \le b$, is called the Riemann–Liouville fractional integral operator of order α.

Properties of the operator J^α can be found in [1], we mention the following: for $f \in L_1[a,b], \alpha, \beta \ge 0$ and $\gamma > -1$

1. $J_a^\alpha f(x)$ exists for almost every $x \in [a,b]$.
2. $J_a^\alpha J_a^\beta f(x) = J_a^{\alpha+\beta} f(x)$
3. $J_a^\alpha J_a^\beta f(x) = J_a^\beta J_a^\alpha f(x)$
4. $J_a^\alpha x^\gamma = \frac{\Gamma(\gamma+1)}{\Gamma(\alpha+\gamma+1)}(x-a)^{\alpha+\gamma}$.

The Riemann–Liouville derivative has certain disadvantages when trying to model real-world phenomena with fractional differential equations. Therefore, we shall introduce now a modified fractional differentiation operator D^{α} proposed by Caputo in his work on the theory of visco-elasticity [19].

Definition 2.2 The fractional derivative of $f(x)$ in the Caputo sense is defined as

$$D^{\alpha}f(x) = J^{m-\alpha}D^{m}f(x) = \frac{1}{\Gamma(m-\alpha)}\int_0^x (x-t)^{m-\alpha-1}f^{(m)}(t)\mathrm{d}t, \tag{2.1}$$

$m-1<\alpha\leq m, \quad m\in\mathbb{N}, \quad x>0$

Also, we need here two of its basic properties.

Lemma 2.1 *If* $m-1<\alpha\leq m$, *and* $f\in L_1[a,b]$, *then* $D_a^{\alpha}J_a^{\alpha}f(x)=f(x)$, *and*

$$J_a^{\alpha}D_a^{\alpha}f(x)=f(x)-\sum_{k=0}^{m-1}f^{(k)}(0^-)\frac{(x-a)^k}{k!}, \quad x>0.$$

The fractional derivative is considered in the Caputo sense. To solve differential equations, we need to specify additional conditions to produce a unique solution. For the case of Caputo fractional differential equations, these additional conditions are just the traditional conditions, which are taken to those of classical differential equations, and are, therefore, familiar to us. In contrast, for Riemann–Liouville fractional differential equations, these additional conditions constitute certain fractional derivatives of the unknown solution at the initial point $x=0$, which are functions of x. The unknown function $u=u(x,t)$ is assumed to be a causal function of time, i.e., vanishing for $t<0$. Also, the initial conditions are not physical; furthermore, it is not clear how much quantities are to be measured from experiment, say, so that they can be appropriately assigned in an analysis. For more details on the geometric and physical interpretation for fractional derivatives of both Riemann–Liouville and Caputo types see [18].

Definition 2.3 For m to be the smallest integer that exceeds α, the Caputo fractional derivatives of order $\alpha>0$ is defined as

$$D^{\alpha}u(x,t)=\frac{\partial^{\alpha}u(x,t)}{\partial t^{\alpha}} = \begin{cases} \dfrac{1}{\Gamma(m-\alpha)}\displaystyle\int_0^t (t-\tau)^{m-\alpha-1}\frac{\partial^m u(x,\tau)}{\partial \tau^m}\mathrm{d}\tau, & m-1<\alpha<m \\ \dfrac{\partial^m u(x,t)}{\partial t^m}, & \alpha=m\in N \end{cases}$$

For mathematical properties of fractional derivatives and integrals, one can consult the above-mentioned references.

Analysis of the method

To illustrate the basic idea of the ADM for fractional Harry Dym equation, in an operator form, Eq. (1.1) can be written as

$$D_t^{\alpha}=u^3u_{xxx}, \quad x\in\mathbb{R},\ t>0 \tag{3.1}$$

where the fractional differential operator D_t^{α} is defined as $D_t^{\alpha}=\frac{\partial^{\alpha}}{\partial t^{\alpha}}$, so that D_t^{α} is the operator defined in (2.1). In this study, we shall consider Eq. (1.1) subject to the initial conditions

$$u(x,0)=g_1(x),\ u_t(x,0)=g_2(x),\ x\in\mathbb{R} \tag{3.2}$$

To solve the nonlinear fractional Eq. (1.1), we apply the operator J^{α}, the inverse of the operator D_t^{α}, on both sides of Eq. (3.1) and using the initial condition (3.2) yields

$$u(x,t)=\sum_{k=0}^{m-1}\frac{\partial^k u}{\partial t^k}(x,0^+)\frac{t^k}{k!}+J^{\alpha}(\phi(u)) \tag{3.3}$$

where $\phi(u)=u^3u_{xxx}$. Following Adomian [7, 8], we expect the decomposition of the solution into a sum of components to be defined by the decomposition series

$$u(x,t)=\sum_{n=0}^{\infty}u_n(x,t) \tag{3.4}$$

where the components $u_n(x,t)$ will be determined recursively. The nonlinear function $\phi(u)$ is then written in the decomposed form

$$\phi(u)=\sum_{n=0}^{\infty}A_n(u_0,u_1,\ldots,u_n), \tag{3.5}$$

where A_n are called the Adomian polynomials, these polynomials can be calculated for all forms of nonlinearity according to specific algorithms constructed by Adomian [7, 8]. In this specific nonlinearity, we use the general form of formula for A_n polynomials as

$$A_n=\frac{1}{n!}\left[\frac{\mathrm{d}^n}{\mathrm{d}\lambda^n}\phi\left(\sum_{k=0}^{\infty}\lambda^k u_k\right)\right]_{\lambda=0}, \quad n\geq 0 \tag{3.6}$$

This formula is easy to set computer code to get as many polynomials as we need in the calculation of the numerical solution as well as explicit solutions. To be easy to follow by the reader, we can give the first few Adomian polynomials for $\phi(u)=u^3u_{xxx}$ of the nonlinearity as

$$
\begin{aligned}
A_0 &= \phi(u_0) = u_0^3(u_0)_{xxx}, \\
A_1 &= u_1\phi'(u_0) = 3u_0^2u_1(u_0)_{xxx} + (u_1)_{xxx}u_0^3, \\
A_2 &= u_2\phi'(u_0) + \frac{u_1^2}{2!}\phi''(u_0) = 3u_0u_1^2(u_0)_{xxx} + 3u_0^2u_2(u_0)_{xxx} \\
&\quad +3u_0^2u_1(u_1)_{xxx} + u_0^3(u_2)_{xxx} \\
A_3 &= u_3\phi'(u_0) + u_1u_2\phi''(u_0) + \frac{u_1^3}{3!}\phi'''(u_0)
\end{aligned} \tag{3.7}
$$

and so on, the rest of the polynomials can be constructed in a similar manner. Wazwaz, in [21] developed an alternative approach for the construction of these polynomials. Substituting (3.4) and (3.5) into both sides of Eq. (3.3) gives

$$\sum_{n=0}^{\infty} u_n(x,t) = \sum_{n=0}^{m-1} \frac{\partial^k u}{\partial t^k}(x,0^+)\frac{t^k}{k!} + J^\alpha\left(\sum_{n=0}^{\infty} A_n\right) \tag{3.8}$$

Following the decomposition method, we introduce the recursive relations as

$$u_0(x,t) = \sum_{k=0}^{m-1} \frac{\partial^k u}{\partial t^k}(x,0^+)\frac{t^k}{k!} \tag{3.9}$$

$$u_{n+1}(x,t) = J^\alpha(A_n), \quad n \geq 0 \tag{3.10}$$

and the recurrence relation given as in (3.10) determines the exact solutions, while the truncated series

$$\Phi_n(x,t) = \lim_{n\to\infty} \sum_{k=0}^{n-1} u_k(x,t), \quad n \geq 0 \tag{3.11}$$

give an approximate solution to the nonlinear fractional Harry Dym equation. Moreover, the decomposition series solutions (3.11) generally converge very rapidly in real physical problems. The convergence of the decomposition series have been investigated by several authors [15, 16].

Numerical applications and results

In this section, we present some numerical results to demonstrate the behavior of the solution as the order of the time fractional derivative is changed. We shall illustrate the numerical scheme by two cases. The first case is somewhat artificial in the sense that the exact answer, for the special case $\alpha = 1$, is known in advance, and the initial condition is directly taken from this answer. Nonetheless, such an approach is needed to evaluate the accuracy of the numerical scheme, and to examine the effect of varying the order of the time fractional derivative on the behavior of the solution. To calculate the terms of the decomposition series (3.11) for $u(x,t)$, we shall mention that a second initial condition $u_t(x,0) = g(x)$, for $1 < \alpha \leq 2$ is assumed to ensure continuous dependence of the solution on the parameter α in the transition from $\alpha = 1^-$ to $\alpha = 1^+$. Therefore, in (3.9), we need to distinguish two cases:

1. Case I: For $0 < \alpha \leq 1$, in this case, we choose $m = 1$ in (3.9), and so upon using Mathematica, the solution can be obtained as

$$
\begin{aligned}
u_0(x,t) &= \sum_{k=0}^{0} \frac{\partial^k u}{\partial t^k}(x,0^+)\frac{t^k}{k!} = \left(a - \frac{3\sqrt{b}}{2}x\right)^{2/3} \\
u_1(x,t) &= J^\alpha(A_0) = J^\alpha\left(u_0^3(u_0)_{xxx}\right) \\
&= \frac{-b^{3/2}}{(a - \frac{3\sqrt{b}}{2}x)^{1/3}} \frac{t^\alpha}{\Gamma(\alpha+1)} \\
u_2(x,t) &= J^\alpha(A_1) = J^\alpha\left(3u_0^2u_1(u_0)_{xxx} + (u_1)_{xxx}u_0^3\right) \\
&= \frac{-b^3}{2(a - \frac{3\sqrt{b}}{2}x)^{4/3}} \frac{t^{2\alpha}}{\Gamma(2\alpha+1)} \\
u_3(x,t) &= J^\alpha(A_2) = J^\alpha\Big(3u_0u_1^2(u_0)_{xxx} + 3u_0^2u_2(u_0)_{xxx} \\
&\quad + 3u_0^2u_1(u_1)_{xxx} + u_0^3(u_2)_{xxx}\Big) \\
&= \frac{-b^{9/2}}{(a - \frac{3\sqrt{b}}{2}x)^{7/3}} \left(\frac{15\Gamma(2\alpha+1)}{2(\Gamma(\alpha+1))^2} - 16\right) \frac{t^{3\alpha}}{\Gamma(3\alpha+1)} \\
&\ \vdots
\end{aligned}
$$

and so on, in this manner the other components of the decomposition series can easily be obtained. Substituting $u_0, u_1, u_2, u_3, \ldots$ into Eq. (3.3) gives the solution $u(x,t)$ in a series form solution by

$$u(x,t) = \sum_{i=0}^{\infty} u_i(x,t) \tag{4.1}$$

So, the approximate solution for the Harry Dym equation (when $\alpha = 1$ and $n = 4$) is given by

$$u(x,t) = \sum_{i=0}^{3} u_i(x,t) \tag{4.2}$$

Figure 1, shows, respectively, the approximate and exact solution to this case when $\alpha = 1$. Figure 2, shows the obtained error compared with the exact. Figure 3, shows profile solutions when $t = 1$ for different values of α, i.e. $\alpha = 0.1,\ 0.2,\ 0.4,\ 0.9,\ 1$.

It can be seen from Fig. 1 that the solution obtained by the present method is nearly identical to the exact. Also, Fig. 1 shows the exact solution $u(x,t)$ of the regular Harry Dym equation ($\alpha = 1$) given by Mokhtari [20] for constants value of $a = 4,\ b = 1,\ c = 1$.

It is known that using ADM, the series solution converges very rapidly. The rapid convergence means only few terms are required to get analytical function. From Fig. 1, it is to be noted that only four terms of the decomposition series were used for our approximations. It is evident that the solution can be improved by adding more terms of the decomposition series.

Fig. 1 The approximate and exact solutions, respectively, for Case 1 when $0<x<1$ and $0<t<1$ and $\alpha=1$

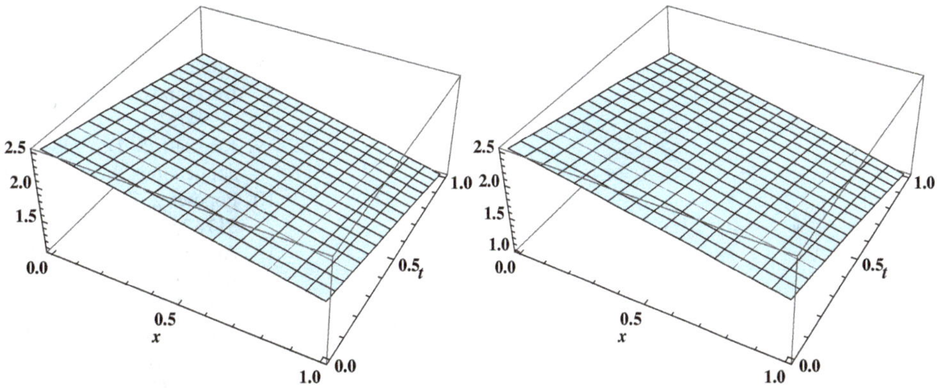

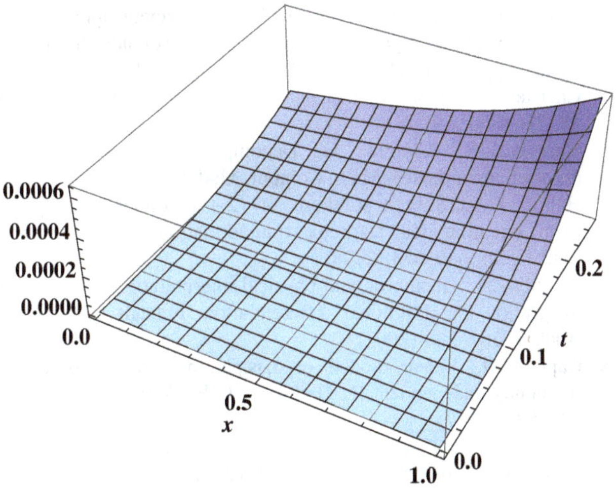

Fig. 2 The obtained absolute error for case 1, $\alpha=1$

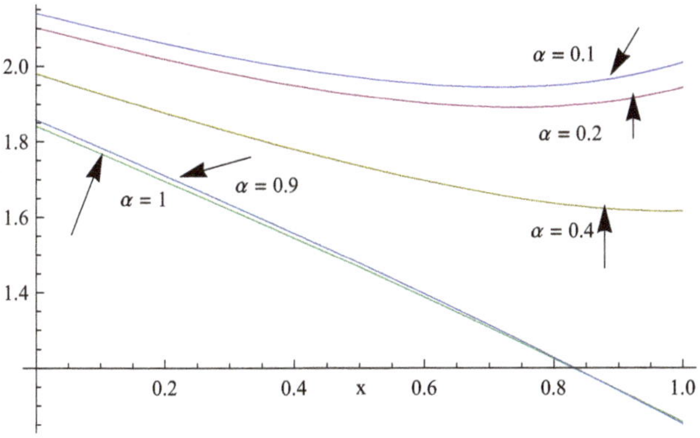

Fig. 3 Profile solutions for case 1 for different values of α

Figures 3 and 5, show that the solution propagates and diffuses with time, which means that the solution continuously depends on the fractional derivatives.

2. Case II: For $1<\alpha\leq2$, in this case, we choose $m=2$ in (3.9) and the components are determined by

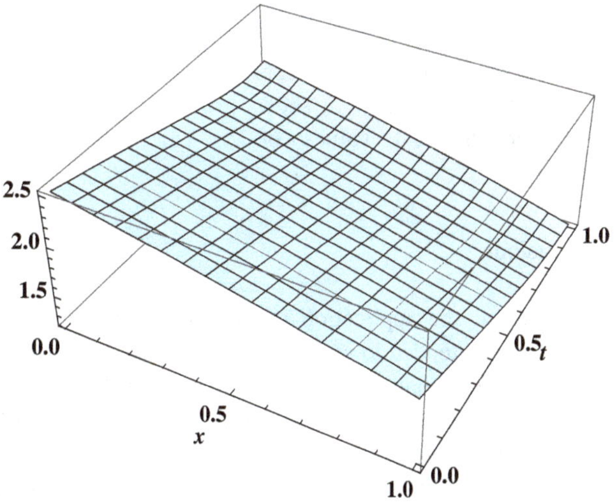

Fig. 4 The approximate solution for Case 2 when $0<x<1$ and $0<t<1$ and $\alpha=2$, $N=3$

$$u_0(x,t) = \sum_{k=0}^{1} \frac{\partial^k u}{\partial t^k}(x,0^+)\frac{t^k}{k!} = \left(a - \frac{3\sqrt{b}}{2}x\right)^{2/3}$$

$$- \sqrt{bc}\left(a - \frac{3\sqrt{b}x}{2}\right)^{-1/3}t$$

$$u_1(x,t) = J^\alpha(A_0) = J^\alpha\left(u_0^3(u_0)_{xxx}\right)$$

$$= \frac{2^{1/3}b^{3/2}t^{1+\alpha}(2a+\sqrt{b}(7ct-3x))(-2a+\sqrt{b}(2ct+3x))^3}{(2a-3\sqrt{b}x)^{13/3}\Gamma(2+\alpha)}$$

$$u_2(x,t) = J^\alpha(A_1) = J^\alpha\left(3u_0^2u_1(u_0)_{xxx} + (u_1)_{xxx}u_0^3\right)$$

$$u_3(x,t) = J^\alpha(A_2) = J^\alpha\left(3u_0u_1^2(u_0)_{xxx} + 3u_0^2u_2(u_0)_{xxx}\right.$$

$$\left. + 3u_0^2u_1(u_1)_{xxx} + u_0^3(u_2)_{xxx}\right)$$

$$\vdots$$

Accordingly, the approximate solution for the Harry Dym equation (when $\alpha=2$ and $n=4$) is given by

$$u(x,t) = \sum_{i=0}^{3} u_i(x,t) \tag{4.3}$$

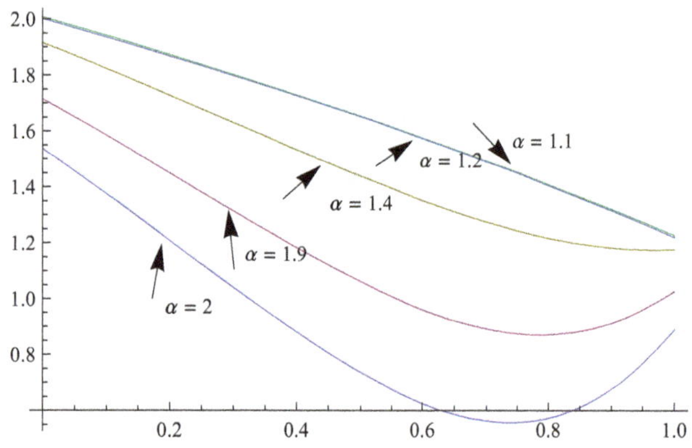

Fig. 5 Profile solutions for case 2 for different values of α

Figure 4, shows the obtained approximate solution to this case when $\alpha = 2$. Figure 5, shows profile solutions when $t = 1$ for different values of α, i.e. $\alpha = 1.1,\ 1.2,\ 1.4,\ 1.9,\ 2$.

Conclusions

In this paper, the application of Adomian decomposition method was extended to explicit and numerical solution of time fractional nonlinear PDEs in mathematical physics with initial conditions. It may be concluded that the decomposition methodology is a very powerful and efficient technique that provides more realistic solutions. It also provides series solutions which generally converge very rapidly in real physical problems. The decomposition method, which is proved to be efficient in solving differential equations, can be applied to various types of fractional differential equations [6]. Finally, the recent appearance of nonlinear fractional partial differential equations as adequate models in science and engineering makes it necessary to investigate the method of solutions for such equations.

References

1. Oldham, K.B., Spanier, J.: The Fractional Calculus. Academic Press, New York (1974)
2. Miller, K.S., Ross, B.: An introduction to the Fractional Calculus and Fractional Differential Equations. John Wiley and Sons Inc., New York (1993)
3. Podlubny, I.: Fractional Differential Equations. Academic Press, New York (1999)
4. Gepreel, K.: Adomian decomposition method to find approximate solutions for the fractional PDEs. WSEAS Trans. Math. **11**(7), 652–657 (2012)
5. Kumar, S., Tripathi, M.P., Singh, O.P.: A fractional model of Harry Dym equation and its approximate solution. Ain Shams Eng. J. **4**, 111–115 (2013)
6. Al-Khaled, K., Momani, S.: An approximate solution for a fractional diffusion wave equation using the decomposition method. Appl. Math. Comput. **165**(2), 473–483 (2005)
7. Adomian, G.: A review of the decomposition method in applied mathematics. J. Math. Anal. Appl. **135**, 501–544 (1988)
8. Adomian, G.: Solving Frontier Problems of Physics. The Decomposition Method. Kluwer Academic Publishers, Boston (1994)
9. Kruskal, M.D., Moster, J.: Dynamical systems, theory and applications. Lecturer notes physics. Springer, Berlin (1975)
10. Popowicz, Z.: The generalized Harry Dym equation. Phy. Lett. A **317**, 260–264 (2003)
11. Antonowicz, M., Fordy, A.P.: Coupled Harry Dym equations with multi-Hamiltonian structures. J. Phys. A Math. Gen. **21**(5), L269–L75 (1988)
12. Fuchssteinert, B., Schulzet, T., Carllot, S.: Explicit solutions for Harry Dym equation. J. Phys. Math. Gen. **25**, 223–230 (1992)
13. Huang, Q., Zhdanov, R.: Symmetries and exact solutions of the time fractional Harry-Dym equation with Riemann–Liouville derivative. Phys. A **409**, 110–118 (2014)
14. Kumar, D., Singh, J., Kilicman, A.: An efficient approach for fractional Harry Dym equation by using Sumudu transform. Abstr. Appl. Anal. **2013**, 1–8. Art. ID 608943 (2013)
15. Cherrualt, Y.: Convergence of Adomian's method. Kybernetes **18**, 31–38 (1989)
16. Cherrualt, Y., Adomian, G.: Decomposition methods: a new proof of convergence. Math. Comput. Model. **18**, 103–106 (1993)
17. Halim, A.A.: Soliton solutions of $(2 + 1)$ dimensional Harry Dym equation via Darboux transformation. Chaos. Soliton Fractals **36**, 646–653 (2008)
18. Podlubny, I.: Geometric and physical interpretation of fractional integration and fractional differentiation. Fractals Calc. Appl. Anal. **5**, 367–386 (2002)
19. Caputo, M.: Linear models of dissipation whose Q is almost frequency independent-II. Geophys. J. R. Astron. Soc. **13**(5), 529–539 (1967)
20. Mokhtari, R.: Exact solutions of the Harry Dym equation. Commun. Theor. Phys. **55**, 204–208 (2011)
21. Wazwaz, A.M.: A new algorithm for calculating Adomian polynomials for nonlinear operators. Appl. Math. Comput. **111**, 53–69 (2000)

12

Bernstein Multiscaling polynomials and application by solving Volterra integral equations

M. Mohamadi[1] · E. Babolian[1] · S. A. Yousefi[2]

Abstract In this paper, we present a direct computational method to solve Volterra integral equations. The proposed method is a direct method based on approximate functions with the Bernstein Multiscaling polynomials. In this method, using operational matrices, the integral equation turns into a system of equations. Our approach can solve nonlinear integral equations of the first kind and the second kind with piecewise solution. The computed operational matrices in this article are exact and new. The comparison of obtained solutions with the exact solutions shows that this method is acceptable. We also compared our approach with two direct and expansion–iterative methods based on the block-pulse functions. Our method produces a system, which is more economical, and the solutions are more accurate. Moreover, the stability of the proposed method is studied and analyzed by examining the noise effect on the data function. The appropriateness of noisy solutions with the amount of noise approves that the method is stable.

Keywords Volterra integral equation · Bernstein polynomials (BPs) · Bernstein Multiscaling polynomials (BMSPs) · Transformation matrices · Operational matrices

✉ E. Babolian
babolian@khu.ac.ir

M. Mohamadi
m.mohamadi@amoliau.ac.ir; m.moahamdi2000@gmail.com

S. A. Yousefi
s-yousefi@sbu.ac.ir

[1] Department of Mathematics, Science and Research Branch, Islamic Azad University, Tehran, Iran

[2] Department of Mathematics, Shahid Beheshti University, G. C. Tehran, Iran

Introduction

Most of integral equations of the first kind are ill-posed problems. Many authors paid attention to solve these equations and presented some methods [1–8]. These problems have application in mathematics, physics, and engineering. Recently, using polynomials have been common to solve these equations, see [9–27].

BMSPs are more general forms of BPs. One of the advantages of BMSPs compared to BPs is that they can approximate piecewise functions. In addition, using BMSP basis, we will have two degrees of freedom which increase accuracy of the method. One of these parameters is m, the degree of polynomials, and the other one k, which corresponds to the number of partitions in the interval $[a, b)$.

In this paper, we review BP properties and preliminary theorems in Subsect. 1 of Sect. "Review of Bernstein polynomials". In Subsects. "Tranformation matrices" and "Operational matrices", transformation matrices and operational matrices for BPs are computed. In Sect. "Bernstein Multiscaling polynomials", BMSPs are defined. Transformational matrices and operational matrices for BMSPs are obtained in Sects. "Transformation matrices" and "BMSPs operational matrices", respectively. In Sect. "Solution of Volterra integral equation", by applying obtained matrices and functions approximation, the integral equations are turned into a system of equations. We present some numerical examples to illustrate the accuracy and ability of this method in Sect. "Numerical examples". In next section, we compare our approach with two direct and expansion–iterative methods based on block-pulse functions. In Sect. "Stability of method", stability is shown and we end this paper with a short conclusion in Sect. "Conclusion".

Review of Bernstein polynomials

Preliminaries

Definition 1. Suppose m is a positive integer number, BPs of degree m on the interval $[a, b]$ are defined as

$$B_{i,m}(x) = \binom{m}{i} \frac{(x-a)^i (b-x)^{m-i}}{(b-a)^m}, \quad i = 0, \ldots, m.$$

In addition, $B_{i,m}(x) = 0$ if $i < 0$ or $i > m$.

For convenience, we consider $[a, b] = [0, 1]$, namely

$$B_{i,m}(x) = \binom{m}{i} x^i (1-x)^{m-i}, \quad i = 0, \ldots, m.$$

We denote Φ_m, an $m+1-$column vector, as follows: $\Phi_m(x) = [\phi_0(x), \ldots, \phi_m(x)]^T$, where $\phi_i(x) = B_{i,m}(x)$, $i = 0, \ldots, m$.

The BPs have many interesting properties [24–30]. However, here, some of them that are useful in our work are stated:

$$(\boldsymbol{p1}) \quad B_{i,m}(x) B_{j,m}(x) = \frac{\binom{m}{i}\binom{m}{j}}{\binom{2m}{i+j}} B_{i+j,2m}(x), \quad i,j = 0, \ldots, m.$$

$$(\boldsymbol{p2}) \quad B_{i,m}(x) = \sum_{j=0}^{k} \frac{\binom{m}{i}\binom{k}{j}}{\binom{m+k}{i+j}} B_{i+j,m+k}(x), \quad j,k = 0, \ldots, m.$$

$$(\boldsymbol{p3}) \quad \frac{B_{i,m}(x)}{\binom{m}{i}} = \frac{B_{i,m+1}(x)}{\binom{m+1}{i}} + \frac{B_{i+1,m+1}(x)}{\binom{m+1}{i+1}}, \quad i = 0, \ldots, m.$$

The following theorems are a fundamental tool that justifies the use of polynomials.

Theorem A [31] *Suppose $H = l^2([a, b])$ is a Hilbert space with the inner product defined by $\langle f, g\rangle = \int_a^b f(t)g(t)\mathrm{d}t$, and in addition, $Y = \mathrm{Span}\{B_{0,m}(x), B_{1,m}(x), \ldots, B_{m,m}(x)\}$ is the span space by Bernstein's polynomials of degree m. Let f be an arbitrary element in H. Since Y is a finite dimensional and closed subspace, it is a complete subset of H. Therefore, f has the unique best approximation out of Y, such that y_0*

$$\exists y_0 \in Y; \forall y \in Y : \|f - y_0\|_2 \leq \|f - y\|_2.$$

Therefore, there are the unique coefficients $\alpha_j, j = 0, 1, \ldots, m$, such that

$$f(t) \approx y_0(t) = \sum_{j=0}^{m} \alpha_j B_{j,m}(t) = \alpha^T \Phi_m,$$

where $\alpha = [\alpha_0\ \alpha_0 \ldots \alpha_m]^T$ can be obtained by

$$\alpha = \frac{\langle f(t), \Phi_m(t)\rangle}{\langle \Phi_m(t), \Phi_m(t)\rangle},$$

such that $\langle f(t), \Phi_m(t)\rangle = \int^b f(t)\Phi_m(t)\,\mathrm{d}t$.

In above theorem, we[a]denote $Q = \langle \Phi_m(t), \Phi_m(t)\rangle$ as dual matrix. Furthermore, it is easy to see

$$Q_{i,j} = \frac{\binom{m}{i-1}\binom{m}{j-1}}{(2m+1)\binom{2m}{i+j-2}}, \quad i,j = 1, \ldots, m+1.$$

Next theorem indicates that dual matrix is symmetric and invertible.

Theorem B [31] *Elements $y_1, y_2, \ldots, y_n$ of a Hilbert space H constitute a linearly independent set in H if and only if $G(y_1, y_2, \ldots, y_n) \neq 0$.*

where $G(y_1, y_2, \ldots, y_n)$ is the Gram determinant of $y_1, y_2, \ldots, y_n$ defined by

$$G(y_1, y_2, \ldots, y_n) = \begin{vmatrix} \langle y_1, y_1\rangle & \langle y_1, y_2\rangle & \cdots & \langle y_1, y_n\rangle \\ \langle y_2, y_1\rangle & \langle y_2, y_2\rangle & \cdots & \langle y_2, y_n\rangle \\ \vdots & \vdots & \ddots & \vdots \\ \langle y_n, y_1\rangle & \langle y_1, y_1\rangle & \cdots & \langle y_n, y_n\rangle \end{vmatrix}.$$

For a two-dimensional function $k(x, t) \in l^2([0, 1] \times [0, 1])$, it can be similarly expanded with respect to BPs, such as

$$k(x, t) \simeq \Phi_m^T(x) K \Phi_m(t),$$

and K is the $(m+1) \times (m+1)$ BP coefficient matrix with

$$K_{i,j} \simeq Q^{-1}\left(\int_0^1 (Q^{-1}\int_0^1 k(x,t)\phi_i(t)\,\mathrm{d}t)\phi_j(x)\,\mathrm{d}x\right),$$

$i,j = 0, \ldots, m.$

Transformation matrices

Transformation matrix is used to change the dimension of the problem. In other words, this matrix can convert Φ_m to Φ_n and vice versa.

Suppose m is less than n, T_m^n is an $(m+1) \times (n+1)$ matrix, called increasing transformation matrix, that converts Φ_m to Φ_n. In other words, $\Phi_m = T_m^n.\Phi_n$.

The increasing transformation matrix can be computed as follows:

$$[T_m^n]_{i,j} = \begin{cases} 0, \text{if } i<j \text{ or } j>i+k, \\ \dfrac{\binom{m}{i-1}\binom{k}{j-1}}{\binom{m+k}{i+j-2}}, \text{otherwise.} \end{cases}$$

It is sufficient to use $p3$, k times, where $k = n - m$.

In addition, decreasing transformation matrix is an $(n+1) \times (m+1)$ matrix, which is shown by T_n^m, and converts Φ_n to Φ_m, where n is greater than m. In other words, $\Phi_n = T_n^m \Phi_m$.

The ith row of decreasing transformation matrix can be calculated as follows:

$$[T_n^m]_{i+1} = \frac{1}{m+n+1} \begin{bmatrix} \dfrac{1}{\binom{m+n}{i}} & \dfrac{1}{\binom{m+n}{i+1}} & \cdots & \dfrac{1}{\binom{m+n}{i+m}} \end{bmatrix}^T Q^{-1}, \quad i = 0, \ldots, n.$$

Operational matrices

Operational matrix is a matrix that works on basis, such as an operator; in other words, if Λ is an operator, an operational matrix is a matrix, such as P, such that $\Lambda(\Phi) \simeq P\Phi$.

Operational matrix of integration

Lemma 3 *Suppose $\Phi_m(x) = [\phi_0(x), \ldots, \phi_m(x)]^T$,, then*

$$\int_0^x \Phi_m(x)\mathrm{d}x = M\Phi_m(x). \tag{1}$$

where M is called operational matrix of integration.

Proof With a simple calculation can be seen

$$B_{i,m}(x) = \int_0^x m\big(B_{i-1,m-1}(t) - B_{i,m-1}(t)\big)\mathrm{d}t.$$

Assume $0 \le k \le m$,

$$\sum_{i=k}^{m} B_{i,m}(x) = \sum_{i=k}^{m} \int_0^x m\big(B_{i-1,m-1}(t) - B_{i,m-1}(t)\big)\mathrm{d}t$$
$$= m\int_0^x B_{k-1,m-1}(t)\mathrm{d}t.$$

Therefore,

$$\int_0^x B_{k,m}(t)\mathrm{d}t = \frac{1}{m+1}\sum_{i=k+1}^{m+1} B_{i,m+1}(x) = M_k^T \Phi_{m+1},$$

where

$$M_k = \frac{1}{m+1}\Big[\overbrace{0\cdots0}^{k+1} \quad \overbrace{0\cdots1}^{m+1-k}\Big]^T, \quad k = 0, \ldots, m.$$

It is obvious, $im = \begin{bmatrix} M_0^T \\ M_1^T \\ \vdots \\ M_m^T \end{bmatrix}$ is an $(m+2) \times (m+1)$ matrix. Accordingly, $M = im\, T_{m+1}^m$.

Operational matrix of product

Lemma 4 *Let C be an $(m+1) \times m+1$ matrix, then*

$$\Phi_m^T(x) C \Phi_m(x) = \hat{C}^T \Phi_{2m}(x), \tag{2}$$

where

$$\hat{C}_k = \sum_{j=0}^{k} \frac{\binom{m}{k-j}\binom{m}{j}}{\binom{2m}{k}} C_{k-j,j}, \quad k = 0, 1, \ldots, 2m.$$

Proof. Let $\phi_i^*(x) = B_{i,2m}(x)$, for $i = 0, \ldots, 2m$.

$$\Phi_m^T(x) C \Phi_m(x) = \sum_{i=0}^{m}\sum_{j=0}^{m} c_{ij}\phi_i\phi_j.$$

Using $p1$ gives:

$$\Phi_m^T(x) C \Phi_m(x) = \sum_{i=0}^{m}\sum_{j=0}^{m} c_{ij} \frac{\binom{m}{i}\binom{m}{j}}{\binom{2m}{i+j}} \phi_{i+j}^*$$
$$= \left[\frac{\binom{m}{0}\binom{m}{0}}{\binom{2m}{0}} c_{00} \quad \sum_{j=0}^{1} \frac{\binom{m}{1-j}\cdot\binom{m}{j}}{\binom{2m}{1}} . c_{1-j,j} \quad \cdots \quad \sum_{j=0}^{k} \frac{\binom{m}{k-j}\cdot\binom{m}{j}}{\binom{2m}{k}} . C_{k-j,j} \cdots \frac{\binom{m}{m}\binom{m}{m}}{\binom{2m}{2m}} c_{mm} \right] \Phi_{2m} = \hat{C}^T \Phi_{2m}(x).$$

Lemma 5. *Let u be an arbitrary $(m+1)$ vector, then*

$$\Phi_m(x)\Phi_m^T(x)u = \tilde{u}\Phi_{2m}(x). \quad (3)$$

where $\tilde{u}$ is an $(m+1)\times(2m+1)$ matrix with elements

$$\tilde{u}_{i+1,j+1} = \begin{cases} 0, & \text{if } j\langle i \text{ or } j\rangle m+i, \\ \dfrac{\binom{m}{i}\cdot\binom{m}{j}}{\binom{2m}{i+j}}u_j, & \text{otherwise,} \end{cases}$$

$i=0,\ldots,m,\ j=0,\ldots,2m.$

Proof. Property $p1$ implies

$$\Phi_m(x)\Phi_m^T(x)u = \left[\sum_{j=0}^{m}\frac{\binom{m}{0}\binom{m}{j}}{\binom{2m}{j}}u_j\phi_j^* \quad \sum_{j=0}^{m}\frac{\binom{m}{1}\binom{m}{j}}{\binom{2m}{j+1}}u_j\phi_{j+1}^* \quad \cdots \quad \sum_{j=0}^{m}\frac{\binom{m}{m}\binom{m}{j}}{\binom{2m}{j+m}}u_j\phi_{j+m}^*\right]^T.$$

Now, the ith entry of the above matrix can be rewritten as follows:

$$\sum_{j=0}^{m}\frac{\binom{m}{i-1}\binom{m}{j}}{\binom{2m}{j+i-1}}u_j\phi_{j+i-1}^* = \left[\underbrace{0\ldots0}_{i-1} \quad \frac{\binom{m}{i-1}\binom{m}{0}}{\binom{2m}{i-1}}u_0 \quad \cdots \quad \frac{\binom{m}{i-1}\binom{m}{m}}{\binom{2m}{m+i-1}}u_m \quad \underbrace{0\ldots0}_{m-i+1}\right]\Phi_{2m}.$$

Bernstein Multiscaling polynomials

Definition 3. Suppose $B_{i,m}(x)$ is the ith BPs of degree m on interval $[0,1]$, Bernstein Multiscaling polynomials on $[0,1)$ define as follows:

$$\psi_{i,j}(t) = \begin{cases} B_{i,m}(kt-j), & \frac{j}{k}\le t<\frac{j+1}{k}, \\ 0, & \text{other wise,} \end{cases} \quad (4)$$

where $k\geq 1$ is the number of partitions on $[0,1]$ and $i=0,\ldots,m$, and in addition, $j=0,\ldots,k-1$.

Now, every function $f\in\mathcal{L}^2([0,1))$ has the unique best approximation with respect to span space by BMSPs as follows:

$$f(x) = \sum_{j=0}^{k-1}\sum_{i=0}^{m}c_{i,j}\psi_{i,j} = C^T.\Psi,$$

where $C^T = [c_{0,0}, c_{1,0}, \ldots, c_{m-1,0}, c_{m,0}, \ldots, c_{0,k-1}, c_{1,k-1}, \ldots, c_{m-1,k-1}, c_{m,k-1}]$.

We denote

$\Psi = [\Psi_0\cdots\Psi_{k-2}\quad\Psi_{k-1}]^T$ where $\Psi_i = [\psi_{0i}\ \psi_{1i}\cdots\psi_{mi}]^T,\quad i=0,\ldots,k-1.$

Equation (4) implies that $\psi_i^{'}$s are disjoint. In other words when

$$i\neq j:\psi_i^T\psi_j = 0. \quad (5)$$

In addition, Eq. (4) and $p3$ imply

$$\psi_{i,j(x)}\psi_{p,q(x)} = \begin{cases} 0, & j\neq q, \\ \dfrac{\binom{m}{i}\cdot\binom{m}{j}}{\binom{2m}{i+j}}\psi_{i+p,j(x)}^*, & j=q, \end{cases} \text{ where } \psi_{i+p,j}^*(x) \text{ is a elemet of } \Psi_{2m}. \quad (6)$$

Lemma 6. *Suppose $\bar{Q}$ is the dual operational matrix of BMSPs, then*

$$\overline{Q}=\frac{1}{k}\begin{bmatrix} Q & \bar{0} & \cdots & \bar{0} \\ \bar{0} & Q & \ddots & \vdots \\ \vdots & \ddots & \ddots & \bar{0} \\ \bar{0} & \cdots & \bar{0} & Q \end{bmatrix},$$

where $\bar{0}$ is an $(m+1)\times(m+1)$ zero matrix, and Q is the dual operational matrix of BPs.

Proof. With respect to (5) and (6), it is obvious.For any arbitrary function $f(x)\in\mathcal{L}^2([0,1))$, there is an expansion with respect to BMSPs, such that

$$f(x)\simeq\sum_{j=0}^{k-1}\sum_{i=0}^{m}f_{i,j}\psi_{i,j}=F^T\Psi=\Psi^TF, \tag{7}$$

where F is a $k(m+1)$ – vector

Theorem A indicates

$$F\simeq\overline{Q}^{-1}\int_0^1 f(x)\Psi(x)\mathrm{d}x.$$

Let $k(x,t)\in\mathcal{L}^2([0,1)\times[0,1))$ be a two-dimensional function. With respect to BMSPs, k has the following expansion:

$$k(x,t)\simeq\Psi^T(t)K\Psi(x), \tag{8}$$

where

$$K_{i+1,j+1}=\overline{Q}^{-1}\left(\int_0^1\left(\overline{Q}^{-1}\int_0^1 k(x,t)\phi_i(t)\mathrm{d}t\right)\phi_j(x)\mathrm{d}x\right),i,j$$
$$=0,\ldots,k(m+1).$$

Transformation matrices

Let Ψ_m and Ψ_n be two different BMSPs and $m\leq n$. There are two matrices, τ_m^n and τ_n^m, such that $\Psi_m=\tau_m^n.\Psi_n$ and $\Psi_n=\tau_n^m.\Psi_m$. These transformation matrices have dimensions $k(m+1)\times k(n+1)$ and $k(n+1)\times k(m+1)$, respectively:

$$\tau_m^n=\begin{bmatrix} T_m^n & \bar{0} & \cdots & \bar{0} \\ \bar{0} & T_m^n & \ddots & \vdots \\ \vdots & \ddots & \ddots & \bar{0} \\ \bar{0} & \cdots & \bar{0} & T_m^n \end{bmatrix} \text{ and } \tau_n^m=\begin{bmatrix} T_n^m & \bar{0} & \cdots & \bar{0} \\ \bar{0} & T_n^m & \ddots & \vdots \\ \vdots & \ddots & \ddots & \bar{0} \\ \bar{0} & \cdots & \bar{0} & T_n^m \end{bmatrix},$$

where T_m^n and T_n^m are increasing transformation matrix and decreasing transformation matrix, respectively.

BMSP operational matrices

Operational matrices for BMSPs are obtained by BPs operational matrices and results are similar.

Operational matrices of product

Lemma 7. *Assume A is a $k(m+1)\times k(m+1)$ matrix, then*

$$\Psi^TA\Psi=\hat{A}^T\Psi^*, \tag{9}$$

where $\hat{A}$ is a $k(2m+1)$ – vector.

Proof. Let C is a $(m+1)\times(m+1)$ matrix, Eq. (5) implies

$$\Psi_i^TC\Psi_j=0, \tag{10}$$

where $i\neq j$.

Furthermore, Lemma 4 gives:

$$\Psi_i^TC\Psi_i=\hat{C}^T\Psi_i^*, \tag{11}$$

where $\hat{C}$ is a $(2m+1)$ – vector.

Now, consider A as follows:

$$A=\begin{bmatrix} A_{0,0} & \cdots & A_{0,k-1} \\ \vdots & \ddots & \vdots \\ A_{k-1,0} & \cdots & A_{k-1,k-1} \end{bmatrix},$$

where $A_{i,j}(i,j=0,\ldots,k-1)$ is an $(m+1)\times(m+1)$ matrix.

Equation (10) implies:

$$\Psi^TA\Psi=\sum_{i=0}^{k-1}\Psi_i^TA_{ii}\Psi_i.$$

Equation (11) gives

$$\Psi^TA\Psi=\sum_{i=0}^{k-1}\hat{A}_{ii}^T.\Psi_i^*.$$

Therefore,

$$\Psi^TA\Psi=[\hat{A}_{0,0}\ldots\hat{A}_{k-1,k-1}]\Psi^*=\hat{A}^T\Psi^*,$$

where $\hat{A}_{i,i}$ is a $2m+1$ vector.

$$\sum_{i=0}^{k-1}\hat{A}_{ii}^T\Psi_i^*=[\hat{A}_{0,0}\ldots\hat{A}_{k-1,k-1}]\Psi^*=\hat{A}^T\Psi^*.$$

Lemma 8. *Suppose C is an arbitrary $k(m+1)$ vector, then,*

$$\Psi\Psi^T C = \tilde{C}\Psi^*, \tag{12}$$

where $\tilde{C}$ is a $k(m+1)\times k(2m+1)$ matrix.

Proof. Suppose

$$C^T = [\, C_0 \quad C_1 \quad \cdots \quad C_{k-1} \,],$$

where $C_i(i = 0, \ldots, k-1)$ is an $m+1$ – vector.

Equation (3) implies:

$$\Psi_i\Psi_i^T C_i = \tilde{C}_i\Psi_i^*, i = 0, \ldots, k-1, \tag{13}$$

where $\tilde{C}_i$ is an $(m+1)\times(2m+1)$ matrix.

Now

$$\Psi.\Psi^T.C = \begin{bmatrix} \Psi_0 \\ \vdots \\ \Psi_{k-1} \end{bmatrix} . [\Psi_0^T \ \cdots \ \Psi_{k-1}^T] . \begin{bmatrix} C_0 \\ \vdots \\ C_{k-1} \end{bmatrix}$$

$$= \begin{bmatrix} \Psi_0.\Psi_0^T.C_0 & 0 & & 0 \cdots 0 \\ 0 & \Psi_1.\Psi_1^T.C_1 & & \ddots \ \vdots \\ \vdots \ \ddots & & \ddots & 0 \\ 0 \ \cdots & & 0 & \Psi_{k-1}.\Psi_{k-1}^T.C_i \end{bmatrix}.$$

Equation (12) gives:

$$\Psi\Psi^T C = \begin{bmatrix} \tilde{C}_0.\Psi_0^* & 0 & \cdots & 0 \\ 0 & \ddots & \cdots & \vdots \\ \vdots & \ddots & \ddots & 0 \\ 0 & \cdots & 0 & \tilde{C}_{k-1}.\Psi_{k-1}^* \end{bmatrix}$$

$$= \begin{bmatrix} \tilde{C}_0 & 0 & \cdots & 0 \\ 0 & \tilde{C}_0 & \ddots & \vdots \\ \vdots & \ddots & \ddots & 0 \\ 0 & \cdots & 0 & \tilde{C}_{k-1} \end{bmatrix} \Psi^* = \tilde{C}\Psi^*.$$

Operational matrix of integration

Lemma 7. *Assume* $\mathcal{M}$ be the operational matrix of integration, then

$$\int_0^x \Psi(t)\mathrm{d}t = \mathcal{M}\Psi(x).$$

Proof Lemma 1 implies:

$$\int_0^x \Psi_0(t)\mathrm{d}t = \begin{cases} \frac{1}{k}M\Psi_0(t), & 0 \le x < \frac{1}{k}, \\ \frac{1}{k(m+1)}, & x \ge \frac{1}{k}. \end{cases}$$

For $i = 1, \ldots, k-2$, Eq. (1) gives:

$$\int_0^x \Psi_i(t)\mathrm{d}t = \begin{cases} 0, & 0 \le x < \frac{i}{k}, \\ \frac{1}{k}M\Psi_i(t), & \frac{i}{k} \le x < \frac{i+1}{k}, \\ \frac{1}{k(m+1)}, & x \ge \frac{i+1}{k}, \end{cases}.$$

and

$$\int_0^x \Psi_{k-1}(t)\mathrm{d}t = \begin{cases} 0, & 0 \le x < \frac{k-1}{k}, \\ \frac{1}{k}M\Psi_{k-1}(t), & \frac{k-1}{k} \le x < 1. \end{cases}$$

Consequently,

$$\mathcal{M} = \begin{bmatrix} \frac{1}{k}M & \frac{1}{k(m+1)}\bar{1} & \cdots & \frac{1}{k(m+1)}\bar{1} \\ 0 & \frac{1}{k}M & \ddots & \vdots \\ \vdots & \ddots & \ddots & \frac{1}{k(m+1)}\bar{1} \\ 0 & \cdots & 0 & \frac{1}{k}M \end{bmatrix},.$$

where $\bar{1}$ is an $(m+1)\times(m+1)$ matrix that all entries are one.

Solution of Volterra integral equation

In this section, we are going to convert an integral equation to a system.

Linear Volterra integral equation of the first kind

Consider the following Volterra integral equation of the first kind:

$$f(x) = \int_0^x k(x,t)u(t)\mathrm{d}t. \tag{14}$$

where f and k are known, but u is not. Moreover, $k(x,t) \in l^2([0,1) \times [0,1))$ and $f(t) \in l^2([0,1))$.

Approximating functions f, u, and k with respect to BMSPs gives:

$$f(x) = F^T\Psi(x) = \Psi^T(x)F,$$

$$u(t) = U^T\Psi(t) = \Psi^T(t)U,$$

$$k(x,t) \simeq \Psi^T(x)K\Psi(t), \quad (15)$$

where the vectors F, U, and matrix K are BMSP coefficients of $f(x), u(t)$, and $k(x,t)$, respectively.

Now, replacing (15) into (14) gives:

$$F^T.\Psi_m(x) = \int_0^x \Psi_m^T(x)K\Psi_m(t)\Psi_m^T(t)U\mathrm{d}t$$

$$= \Psi_m^T(x)K\int_0^x \Psi_m(t)\Psi_m^T(t)U\mathrm{d}t.$$

Using (12) follows:

$$F^T\Psi_m(x) = \Psi_m^T(x)K\int_0^x \tilde{U}\Psi_{2m}(t)\mathrm{d}t$$

$$= \Psi_m^T(x)K\tilde{U}\int_0^x \Psi_{2m}(t)\mathrm{d}t. \quad (16)$$

Using operational matrix of integration $\mathcal{M}$, in Eq. (16) gives

$$F^T\Psi_m(x) = \Psi_m^T(x)K\tilde{U}M\Psi_{2m}(x). \quad (17)$$

Let $U^* = K\tilde{U}M\tau_{2m}^m$, where U^* is a $k(m+1) \times k(m+1)$.

Equation (17) changes to:

$$F^T\Psi_m(x) = \Psi_m^T(x)U^*\Psi_m(x) \quad (18)$$

Using Eq. (11) in (18) gives:

$$F^T\Phi_m(x) = \Phi_m^T U^*\Phi_m = \hat{U^*}^T\Phi_{2m}(x).$$

Using decreasing transformation matrix τ_{2m}^m, gives the final system:

$$\overline{U} = F, \quad (19)$$

where $\overline{U}^T = \hat{U^*}^T\tau_{2m}^m$.

Nonlinear Volterra integral equation of the first kind

Consider the following nonlinear Volterra integral equation:

$$f(x) = \int_0^x k(x,t)g(u(t))\mathrm{d}t, \quad (20)$$

Put $w(x) = g(u(x))$ and subsequently $w(x) = W^T\Psi_m(x)$. (21)

where W is an unknown $k(m+1)$ vector. Following the same procedure, the final system is as follows:

$$\overline{W} = F,$$

Finally, $u(x) = g^{-1}(w(x))$ is the desire solution.

One advantage of this method is solving linear or nonlinear Volterra integral equation of the second kind with piecewise functions. In these equations, solution, kernel, or data function can be piecewise. It is essential k, the number of partitions, be chosen, such that discontinuity points lie on boundary point of partitions.

Linear Volterra integral equation of the second kind

Consider linear Volterra integral equations as the following form:

$$u(x) = f(x) + \int_0^x k(x,t)u(t)\mathrm{d}t. \quad (22)$$

Substituting (15) into (22) and a process similar to the previous state final linear system is

$$U - \overline{U} = F.$$

Nonlinear Volterra integral equation of the second kind

Consider the following nonlinear Volterra equation of the second kind:

$$u(x) = f(x) + \int_0^x k(x,t)g(u(t))\mathrm{d}t \quad (23)$$

Table 1 Results of Example 1

x	$m = 8$, $k = 4$	Absolute error	Exact solution
0	1.000000000	0.00000000	1.00000000
0.1	0.9048374873	6.93×10^{-8}	0.904837418
0.2	0.8187326019	1.81×10^{-6}	0.8187307531
0.3	0.740829897	1.16×10^{-5}	0.7408182207
0.4	0.670361092	4.1×10^{-5}	0.670320046
0.5	0.606644000	1.13×10^{-4}	0.6065306597
0.6	0.549040000	2.28×10^{-4}	0.5488116361
0.7	0.497190000	6.04×10^{-4}	0.4965853038
0.8	0.449100000	3.28×10^{-4}	0.4493289641
0.9	0.405500000	3.06×10^{-4}	0.4065696597

Table 2 Results of Example 2

x	$m = 4, k = 4$	Absolute error	$m = 8, k = 4$	Absolute error	Exact solution
0	0.0006126852892	5.91×10^{-4}	0.0001240979	1.24×10^{-4}	0.0000000000
0.1	0.1998624698	1.95×10^{-4}	0.1998263752	1.59×10^{-4}	0.1996668333
0.2	0.3978000958	4.61×10^{-4}	0.3971608136	1.77×10^{-4}	0.3973386616
0.3	0.5906752400	3.65×10^{-4}	0.5911505000	1.10×10^{-4}	0.5910404134
0.4	0.7785010600	3.35×10^{-4}	0.7788225000	1.41×10^{-5}	0.7788366846
0.5	0.9606750920	1.82×10^{-3}	0.9590100000	1.58×10^{-4}	0.9588510772
0.6	1.130310725	1.02×10^{-3}	1.1293000000	1.50×10^{-5}	1.129284947
0.7	1.286968954	1.46×10^{-3}	1.2890000000	5.64×10^{-4}	1.288435374
0.8	1.443023053	8.31×10^{-3}	1.4345600000	1.52×10^{-4}	1.434712182
0.9	1.56155138	5.10×10^{-3}	1.566430000	2.23×10^{-4}	1.566653819

Table 3 Results of Example 3

x	$m = 5, k = 4$	Absolute error	Exact solution
0.00	0.0000207940	2.0×10^{-5}	0.00000000
0.15	0.1500076065	7.6×10^{-6}	0.15000000
0.30	0.3000093281	9.3×10^{-6}	0.30000000
0.45	0.4499909631	9.0×10^{-6}	0.45000000
0.60	0.5999915857	8.4×10^{-6}	0.60000000
0.75	0.7499869395	1.3×10^{-5}	0.75000000
0.90	0.9000073666	7.3×10^{-6}	0.90000000

where $g(x)$ is one to one on $[0, 1]$.

Let $w(t) = g(u(t))$, substituting (23) gives

$$w(x) = g\left(f(x) + \int_0^x k(x,t)u(t)dt\right). \quad (24)$$

Now, substituting (15) and (20) into (24) implies:

$$W^T\Psi = g\left(F^T\Psi + \Psi_m^T(x)K\int_0^x \Psi(t)\Psi^T(t)Wdt\right).$$

Equations (16)–(19) give:

$W = g(F + \bar{W})$.

Numerical examples

Now, we test our method on some numerical examples; in every example, we use a table to show approximations, exact solution, and absolute errors in some points.

Example 1 Suppose $u(x) = e^{-x}$ be the exact solution of the following Volterra integral equation of the first kind:

$$xe^x = \int_0^x e^{x+t}u(t)dt.$$

Table 1 shows results of Example 1.

Example 2 Consider the following integral equation:

$$x\sin x = \int_0^x \cos(x-t)u(t)dt$$

with the exact solution $u(x) = 2\sin x$.

Table 2 shows approximated solutions, absolute errors, and exact solution in some points.

Example 3 $x = \int_0^x (x-t+1)e^{-u(t)}dt$

Table 3 shows results of Example 3.

Example 4 Consider the following linear integral equation $u(x) = f(x) + \int_0^x (t-x)u(t)dt$, with exact solution:

$$u(x) = \begin{cases} x, & 0<x<0.5, \\ x^2, & 0.5<x<1, \end{cases}$$

and the nonsmooth data function:

$$f(x) = \begin{cases} x + \frac{1}{6}x^3, & 0<x<0.5, \\ \frac{-5}{192} + \frac{1}{12}x + x^2 + \frac{1}{12}x^4, & 0.5<x<1. \end{cases}$$

Results of Example 4 are presented in Table 4.

Example 5 Assume $u(x) = \begin{cases} 1-x, & 0<x<0.5, \\ \sqrt{x}, & 0.5<x<1, \end{cases}$ be the exact solution of the following nonlinear Volterra integral equation of the second kind:

$$u(x) = f(x) + \int_0^x xtu^2(t)\mathrm{d}t,$$

where,

$$f(x) = \begin{cases} 0, & x<0, \\ 1 - x - \frac{x^3}{2} - \frac{2}{3}x^4 - \frac{x^5}{4}, & 0<x\leq 0.5, \\ \sqrt{x} - 0.01x - \frac{1}{3}x^4, & 0.5\leq x<1, \\ -0.35x, & x>1. \end{cases}$$

Table 4 Results of Example 4

x	$m = 4, k = 4$	Absolute error	Exact solution
0	0.0000000000	1.1×10^{-5}	0.0000000000
0.1	0.0999999999	3.5×10^{-6}	0.0998334166
0.2	0.2000000000	4.7×10^{-6}	0.1986693308
0.3	0.3000000000	0.9×10^{-7}	0.2955202067
0.4	0.4000000001	2.1×10^{-6}	0.3894183423
0.5	0.2500000000	3.2×10^{-9}	0.2500000000
0.6	0.3600000002	3.9×10^{-9}	0.3600000000
0.7	0.4900000460	4.6×10^{-9}	0.490000000
0.8	0.6399999998	5.4×10^{-9}	0.640000000
0.9	0.8100000002	6.3×10^{-9}	0.810000000

Table 5 Results of Example 5

x	$m = 3, k = 4$	Absolute error	Exact solution
0	1.0000000000	0.0000000	1.0000000000
0.1	0.8999990300	907×10^{-7}	0.9000000000
0.2	0.7999912000	8.8×10^{-6}	0.8000000000
0.3	0.6999800000	2.0×10^{-5}	0.7000000000
0.4	0.5999939000	6.1×10^{-5}	0.6000000000
0.5	0.7069967812	1.1×10^{-4}	0.7071067812
0.6	0.7744366692	1.6×10^{-4}	0.7745966692
0.7	0.8364200265	2.4×10^{-4}	0.8366600265
0.8	0.894097191	3.3×10^{-4}	0.8944271910
0.9	0.9482032981	4.8×10^{-4}	0.9486832981

Table 5 shows results of Example 5.

Comparison

Block-pulse functions are a special case of BMSPs. However, our method is different from the methods, as presented in [32] and [33]. Consider Example 1, in Table 6, the expansion–iterative method and direct method are compared with our method. In Table 6, we presented mean-absolute errors for the expansion–iterative method and direct method and absolute error of BMSPs for two different values of m and k. With respect to dimensions of the final system, our method is more accurate than the expansion–iterative method and direct method.

Consider Example 2, Table 7 shows a comparison between BMSPs method and the expansion–iterative method and direct method with block-pulse functions. Mean-absolute errors for methods are presented in Table 7.

Stability of method

To demonstrate the stability of the method, we review effect of noise on data function. In other words, we replace $f(x)$ by $(1 + \varepsilon p)f(x)$ into integral equation. Where p is a real random number between -1 and 1, and ε is percent of noise. Now, we want to show that our method is stable and noise is proportional to the variations of solutions.

Table 6 Results of comparison in Example 1

Method	Mid-points, $k = 32$	Mid-points, $k = 64$	Ten points, $k = 32$	Ten points, $k = 64$
Expansion–iterative method	6.6×10^{-4}	1.9×10^{-4}	5.2×10^{-3}	2.6×10^{-3}
Direct method	3.3×10^{-3}	1.6×10^{-3}	5.9×10^{-3}	2.9×10^{-3}
	$m = 5, k = 4$	$m = 3, k = 1$		
BMSPs method	1.26×10^{-4}	2.01×10^{-3}		

Table 7 Results of comparison in Example 2

Method	Mid-points, $k = 64$	Mid-points, $k = 128$	Ten points, $k = 64$	Ten points, $k = 128$
Expansion–iterative method	4.9×10^{-4}	1.4×10^{-4}	6.5×10^{-3}	3.3×10^{-3}
Direct method	5.2×10^{-3}	2.6×10^{-3}	8.2×10^{-3}	4.1×10^{-3}
	$m = 4, k = 4$	$m = 8, k = 4$		
BMSPs method	1.96×10^{-4}	1.70×10^{-5}		

Table 8 Results of Example 6

x	$m = 5, k = 4$	$m = 5, k = 4, \varepsilon = 0.01$	$m = 5, k = 4, \varepsilon = 0.02$	$m = 5, k = 4\varepsilon = 0.03$	Exact solution
0.0	0.00000	0.01558624677	0.01538079594	0.01523810893	0.00000
0.1	0.01000	0.00740436021	0.0747198483	0.007231113603	0.01000
0.2	0.04000	0.040255392	0.039719848	0.0393334275	0.04000
0.3	0.09000	0.089864898	0.088666389	0.0878033869	0.09000
0.4	0.16000	0.159423786	0.157302906	0.1557740489	0.16000
0.5	0.25000	0.255541400	0.252166300	0.2496420000	0.25000
0.6	0.36000	0.360208100	0.3553999000	0.3519660000	0.36000
0.7	0.49000	0.490870000	0.4833260000	0.4786120000	0.49000
0.8	0.64000	0.633360000	0.6249109639	0.6189400000	0.64000
0.9	0.81000	0.804420000	0.7936019242	0.7858300000	0.81000

Example 6 Consider the following Volterra integral equation of first kind:

$$\frac{7}{12}x^4 = \int_0^x (x + t)u(t)\mathrm{d}t$$

with exact solution $u(x) = x^2$.

Table 8 shows approximated solution, noisy solutions, and exact solution.

Conclusion

BMSPs that we use to solve Volterra integral equations have acceptable accuracy. Operational matrices, which we have computed, are exact. These exact matrices lead to fewer errors in our computations. In addition, BMSPs can solve piecewise Volterra integral equations of the second kind. Effect of noise on data function shows that our method is reliable and ill-posedness does not occur. This method with respect to complexity of computations and desirable accuracy is recommended. Furthermore, this method can be used to solve optimal control equations, differential equations, and systems of integral or differential equations.

References

1. Delves, L.M., Walsh, J.: Numerical solution of integral equations. Clarendon Press, Oxford (1974)
2. Delves, L.M., Mohamed, J.L.: Computational methods for integral equations. Cambridge University Press, Cambridge (1985)
3. Golberg, M.A.: Numerical solution of integral equations. Plenum Press, Berlin (1990)
4. Atkinson, K.E.: The numerical solution of integral equations of the second kind. Cambridge University Press, Cambridge (1997)
5. Collins, P.J.: Differential and integral equations. Oxford University Press, Oxford (2006)
6. Rahman, M.: Integral equations and their applications. WIT Press, Southampton (2007)
7. Wazwaz, A.: Linear and nonlinear integral equations methods and applications. Higher Education Press, Beijing (2011)
8. Chakrabarti, A., Martha, S.C.: Methods of solution of singular integral equations. Math. Sci. **6**, 15 (2012)
9. Rashidinia, J., Najafi, E., Arzhang, A.: An iterative scheme for numerical solution of Volterra integral equations using collocation method and Chebyshev polynomials. Math. Sci. **6**, 60 (2012)
10. Maleknejad, K., Hashemizadeh, E., Ezzati, R.: A new approach to the numerical solution of Volterra integral equations by using Bernstein's approximation, Commun. Nonlinear Sci. Numer. Simul. **16**, 647–655 (2011)
11. Mandal, B.N., Bhattacharya, S.: Numerical solutions of some classes of integral equations using Bernstein polynomials. Appl. Math. Comput. **190**, 1707–1716 (2007)
12. Shahsavaran, A.: Numerical approach to solve second kind Volterra integral equations of Abel type using Block-Pulse functions and Taylor expansion by collocation method. Appl. Math. Sci. **5**, 685–696 (2011)
13. Bellour, A., Rawashdeh, A.E.: Numerical solution of first kind integral equations by using Taylor polynomials. J. Inequal. Spec. Func. **1**, 23–29 (2011)
14. Wang, W.: A mechanical algorithm for solving the Volterra integral equation. Appl. Math. Comput. **172**, 1323–1341 (2006)
15. Shirin, A., Islam, M.S.: Numerical solutions of Fredholm integral equations using Bernstein polynomials. J. Sci. Res. **2**(2), 264–272 (2010)
16. Sannuti, P.: Analysis and synthesis of dynamic systems via block pulse functions. Proc. Inst. Electr. Eng. **124**, 569–571 (1977)
17. Hwang, C., Shih, Y.P.: Solution of integral equations via Laguerre polynomials. J. Optim. Theory Appl. **45**, 101–112 (1985)
18. Paraskevopoulos, P.N., Sparis, P.D., Mouroutsos, S.G.: The Fourier series operational matrix of integration. Int. J. Syst. Sci. **16**, 171–176 (1985)
19. Tsay, S.C., Lee, T.T.: Solution of the integral equations via Taylor series. Int. J. Control **44**, 701–709 (1986)
20. Paraskevopoulos, P. N.: The operational matrices of integration and differentiation for Fourier sine-cosine and exponential series. IEEE Trans. Autom. Control. **32**, 648–651 (1987)
21. Paraskevopoulos, P.N., Sklavounos, P., Georgiou, G.C.: The operation matrix of integration for Bessel functions. J. Franklin Inst. **327**, 329–341 (1990)
22. Babolian, E., Mokhtari, R., Salmani, M.: Using direct method for solving variational problems via triangular orthogonal functions. Appl. Math. Comput. **191**, 206–217 (2007)
23. Babolian, E., Marzban, H.R., Salmani, M.: Using triangular orthogonal functions for solving Fredholm integral equations of

the second kind. Appl. Math. Comput. **201**, 452–464 (2008)
24. Yousefi, S.A., Behroozifar, M.: Operational matrices of Bernstein polynomials and their applications. Int. J. Syst. Sci. **41**(6), 709–716 (2010)
25. Yousefi, S.A., Behroozifar, M., Dehghan, M.: The operational matrices of Bernstein polynomials for solving the parabolic equation subject to specification of the mass. J. Comput. Appl. Math. **235**, 5272–5283 (2011)
26. Maleknejad, K., Basirat, B., Hashemizadeh, E.: A Bernstein operational matrix approach for solving a system of high order linear Volterra-Fredholm integro-differential equations. Math. Comput. Model. **55**, 1363–1372 (2012)
27. Yousefi, S.A., Behroozifar, M., Dehghan, M.: Numerical solution of the nonlinear age-structured population models by using the operational matrices of Bernstein polynomials. Appl. Math. Model. **36**, 945–963 (2012)
28. Joy, K. I.: Bernstein polynomials. On-line Geom. Model. Notes. (2000). http://www.idav.ucdavis.edu/education/CAGDNotes/Bernstein-Polynomials/Bernstein-Polynomials.html. Accessed 28 Nov 2000
29. Phillips, G.M.: Interpolations and approximation by polynomials. Springer, New York (2003)
30. Parand, K., Kaviani, S.A.: Application of the exact operational matrices based on the Bernstein polynomials. J. Math. Comput. Sci. **6**, 36–59 (2013)
31. Kreyszig, E.: Introduction functional analysis with applications. Wiley, New Jersey (1978)
32. Babolian, E., Masouri, Z.: Direct method to solve Volterra integral equation of the first kind using operational matrix with block-pulse functions. J. Comput. Appl. Math. **220**, 51–57 (2008)
33. Masouri, Z., Babolian, E., Varmazyar, S.H.: An expansion iterative method for numerically solving Volterra integral equation of the first kind. Comput. Math Appl. **59**, 1491–1499 (2010)

13

Digital barrier options pricing: an improved Monte Carlo algorithm

Kazem Nouri[1] · Behzad Abbasi[1] · Farahnaz Omidi[1] · Leila Torkzadeh[1]

Abstract A new Monte Carlo method is presented to compute the prices of digital barrier options on stocks. The main idea of the new approach is to use an exceedance probability and uniformly distributed random numbers in order to efficiently estimate the first hitting time of barriers. It is numerically shown that the answer of this method is closer to the exact value and the first hitting time error of the modified Monte Carlo method decreases much faster than of the standard Monte Carlo methods.

Keywords Digital option · Double barrier · Monte Carlo simulation · Uniform distribution

Introduction

Derivative securities have witnessed incredible innovation over the past years. In particular, path-dependent options are successful, and most of them comprise barrier options to reduce the cost of hedging [4, 8, 22]. For these derivatives, exact valuation expressions are seldom available, thus one resorts to simulations multiple times. In this manuscript a new Monte Carlo method is proposed in order to efficiently compute the prices of digital barrier options based on an exceedance probability.

Binary options, a.k.a. digital options, are popular in the over-the-counter (OTC) markets for hedging and speculation. In addition, they are important to financial engineers as building blocks for constructing more complex derivatives products. A binary option is a type of option where the payoff is either some fixed amount of some asset or nothing at all. Therefore, binaries are considered to be one of the fastest growing simplified trading products out there, where the trader knows their exact exposure and potential gains at the time of placing a trade. The two main types of binary options are the cash-or-nothing and the asset-or-nothing options, the expiration values of the European asset-or-nothing and cash-or-nothing binary calls are shown in Fig. 1. The options are digital in nature because there are only two possible outcomes, they are also called all-or-nothing options and fixed return options (FROs), on the American Stock Exchange (ASE). Binary options are usually European-style options. In May 2008, ASE for the first time launched exchange trading European cash or nothing-digital options, which were soon followed in June 2008 by the Chicago Board Options Exchange. Binary contracts are available on a variety of underlying assets: stocks, commodities, currencies and indices. Since the binaries are popular options, much research work has been done on them. For example, Palan [20] has tested experimentally whether digital options can reduce price bubbles in a laboratory setting, and Appolloni et al. [1], proposed an efficient lattice procedure which permits to obtain European and American option prices under the Black and Scholes model for digital options with barrier features. Hyong-Chol et al. [10], have considered a special binary option called integral of i-th binary or nothing and then obtain the pricing formulae. In addition, Ballestra [3] considered the problem of pricing vanilla and digital options under the Black–Scholes model, and showed that, if the payoff functions are dealt with properly, then errors close to the machine precision are obtained in only some hundredths of a second.

✉ Kazem Nouri
knouri@semnan.ac.ir

1 Department of Mathematics, Faculty of Mathematics, Statistics and Computer Sciences, Semnan University, P.O. Box 35195-363, Semnan, Iran

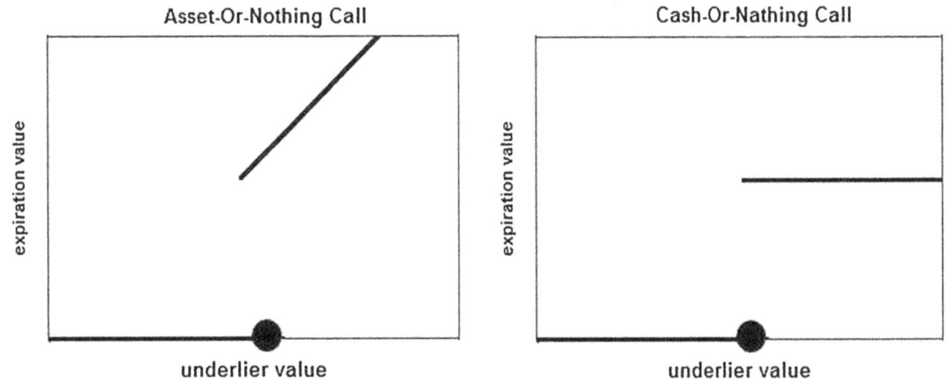

Fig. 1 The expiration values for European asset-or-nothing and cash-or-nothing binary calls

Barrier options are similar to vanilla options except that the option is knocked out or in, if the underlying asset price hits the barrier price B, before expiration date. Since 1967, barrier options have been traded in the OTC market and nowadays are the most popular class of exotic options. A step further along the option evolution path is where we combine barrier and binary options to obtain binary barrier options and binary double barrier options. Accordingly, it is quite important to develop accurate and efficient methods to evaluate barrier digital option prices in financial derivative markets.

Most research done to date have focused on option pricing with various methods, for example, Mehrdoust [17] has proposed an efficient algorithm for pricing arithmetic Asian options based on the AV and the MCV procedures, and Jerbi et al. [13], have calculated the conditional expectation using the Malliavin approach and shown that with this formula, the American option under J-process can be performed using the Monte Carlo simulation. In addition, Zhang et al. [23], have presented the total least squares quasi-Monte Carlo approach for valuing American barrier options, and Jasra and Del Moral provided a review and development of sequential Monte Carlo (SMC) methods for option pricing [12], and in Kim et al. [15], have considered Heston's stochastic volatility model and derive exact analytic expressions for the prices of fixed strike and floating-strike geometric Asian options with continuously sampled averages.

The Monte Carlo method is very popular and robust numerical method, since it is not only easily extended to multiple underlying assets but also is stochastic and amenable to coding. On the other hand, one of main drawbacks of the Monte Carlo method is slow convergence. The statistical error of the Monte Carlo method is of order $O(\frac{1}{\sqrt{M}})$ with M simulations. In particular, for continuously monitored barrier options, the hitting time error is of order $O(\frac{1}{\sqrt{N}})$ with N time steps, see [7], while the European vanilla options have no time discretization error. In this study, to efficiently reduce the hitting time error near the barrier price, inspired by [16], at each finite time step, we suggest the use of a uniformly distributed random variable and a conditional exceedance probability to correctly check whether the continuous underlying asset price hits the barrier or not. Numerical results show that the new Monte Carlo method converges much faster than the standard Monte Carlo method [18]. This idea of using exceedance probability for stopped diffusion is well known in the physics community [11, 16].

The outline of the paper is as follows: in "Digital options" section, we introduce digital options and their pricing formulas and we estimate it by using standard Monte Carlo. In "Modified Monte Carlo algorithm" section, we propose the new Monte Carlo method based on the idea of using uniformly distributed random variable and the conditional exceedance probability. In "Digital barrier options" section, we present numerical results for digital barrier options with one underlying assets and compare the accuracy and efficiency between the standard and the new Monte Carlo methods. In "Double-barrier digital options" section, we present numerical results for pricing double barrier digital options and see the efficiency of the new Monte Carlo method. Finally, we summarize our conclusions and give some direction for future work.

Digital options

The purpose of this section is to introduce two main types of digital options and express their pricing formula.

Cash-or-nothing options

The cash-or-nothing options pay an amount of cash x at expiration if the option is in-the-money. The payoff from a call is 0 if $S_T \leq K$ and x if $S_T > K$, and the payoff from a put is 0 if $S_T \geq K$ and x if $S_T < K$, where S_T and K are stock price at maturity and strike price, respectively. Valuation of cash-or-nothing call and put options can be made using the formula described by Rubinstein and Reiner [21]:

$$c = xe^{-rT}N(d), \tag{1}$$

$$p = xe^{-rT}N(-d), \tag{2}$$

with

$$d = \frac{\ln(S/K) - (\sigma^2/2)T}{\sigma\sqrt{T}},$$

where S is the price of the underlying asset, r is a risk-free interest rate, σ is a volatility, T is the exercise date and $N(\cdot)$ denotes the cumulative function for the standard normal distribution. For example, the value of a cash-or-nothing put option with 9 months to expiration, futures price 100, strike price 80, cash payout 10, risk-free interest rate 6 % per year, and the volatility 35 % per year is $p = 10e^{-0.06\times0.75}N(-0.5846) = 2.6710$. The simulation of standard Monte Carlo that conducted on Matlab for this example has the answer 2.23.

Asset-or-nothing options

At expiration, the asset-or-nothing call option pays 0 if $S_T \le K$ and S_T if $S_T > K$. Similarly, a put option pays 0 if $S_T \ge K$ and S_T if $S_T < K$. The option can be valued using the Cox and Rubinstein formula [6]:

$$c = Se^{-rT}N(d), \tag{3}$$

$$p = Se^{-rT}N(-d), \tag{4}$$

where

$$d = \frac{\ln(S/K) + (\sigma^2/2)T}{\sigma\sqrt{T}}.$$

Consider an asset-or-nothing put option with six months to expiration, $S = 70, K = 65, r = 7\%$ and $\sigma = 27\%$. Valuation of this asset-or-nothing option is $p = 70e^{-0.07\times0.5}N(-0.4836) = 21.2461$, whereas simulation of standard Monte Carlo by Matlab for this example has the answer 21.45.

Modified Monte Carlo algorithm

Let us assume that $(\Omega, \mathcal{F}, Q)$ is a probability space and the evolution of the underlying asset price follows the geometric Brownian motion with a constant expected rate of return $r > 0$, and a constant volatility $\sigma > 0$ of the asset price, i.e.,

$$dS_t = rS_t dt + \sigma S_t dW_t, \tag{5}$$

where W_t is the standard Brownian motion. Equations of the form (5) are powerful tools to description of many real-life phenomena with uncertainty, and there are some studies on the numerical solutions of them [5, 19]. From the Ito's formula, the analytic solution of (5) satisfies

$$S_t = S_0 e^{\left(r-\frac{\sigma^2}{2}\right)t+\sigma W_t}. \tag{6}$$

Using the Monte Carlo method, the expected value of the discounted terminal payoff is approximated under a risk-neutral measure Q, by a sample average of M simulations

$$V(s,t) = E^Q[\Lambda(S_\tau, \tau) \mid S_t = s] \approx \widetilde{V}(s,t) := \frac{1}{M}\sum_{j=1}^{M} \Lambda(S_{\widetilde{\tau}}, \widetilde{\tau}; \omega_j), \tag{7}$$

where $\Lambda(S_\tau, \tau)$ is a discounted payoff function and $\widetilde{\tau}$ is an approximation of the hitting time τ. The global error can be split into the first hitting time error and statistical error,

$$\begin{aligned} \varepsilon := \mid V(s,t) - \widetilde{V}(s,t) \mid &= \left(E^Q[\Lambda(S_\tau,\tau) - \Lambda(S_{\widetilde{\tau}},\widetilde{\tau}) \mid S_t = s]\right) \\ &+ \left(E^Q[\Lambda(S_{\widetilde{\tau}},\widetilde{\tau})] - \frac{1}{M}\sum_{j=1}^{M}\Lambda(S_{\widetilde{\tau}},\widetilde{\tau};\omega_j)\right) \\ &= \varepsilon_T + \varepsilon_S. \end{aligned} \tag{8}$$

From the central limit theorem, the statistical error ε_S in (8), has the following upper bound

$$\mid \varepsilon_S \mid \le c_0 \frac{b_M}{\sqrt{M}}, \tag{9}$$

where b_M is a sample standard deviation of the function values $\Lambda(S_{\widetilde{\tau}}, \widetilde{\tau})$, and c_0 is a positive constant related to confidence interval. For instance, $c_0 = 1.96$ for 95 % of confidence interval. On the other hand, the first hitting time error ε_T in (8), is approximated using an exceedance probability given the asset prices at each time step.

Let us first discretize the time interval [0, T] into N uniform subinterval $0 = t_0 < t_1 < \cdots < t_N = T$. Then compute $S_{n+1} := S_{t_{n+1}}$ at each time step for $n = 0, ..., N-1$ by

$$S_{n+1} = S_n e^{\left(r-\frac{\sigma^2}{2}\right)\Delta t_n + \sigma\Delta W_n}, \tag{10}$$

where Δt_n and ΔW_n denote the time increments $\Delta t_n = t_{n+1} - t_n$ and the Wiener increments $\Delta W_n = W_{t_{n+1}} - W_{t_n}$ for $n = 0, \ldots, N-1$. Also, for the up-and-out barrier case, the approximation of the first hitting time $\widetilde{\tau}$ can be defined by

$$\widetilde{\tau} := \inf\{t_n, n = 1, \ldots, N : S_n \ge B\},$$

with the given barrier price B. The idea is to use an exceedance probability at each time step. Let p_n denotes the probability that a diffusion process X exits of domain D at $t \in [t_n, t_{n+1}]$ by given values X_n and X_{n+1}. In one dimensional half interval case, $D = (-\infty, B)$ for a constant B, the probability p_n has a simple expression using the law of Brownian bridge, see [14]. So,

$$p_{n+1} = \mathbb{P}\left[\max_{t\in[t_n,t_{n+1}]} X_t \geq B \mid X_n = x_1, X_{n+1} = x_2\right]$$
$$= \exp\left(-2\frac{(B-x_1)(B-x_2)}{\beta(x_1)^2\Delta t_n}\right), \quad n = 0,1,\ldots,N-1, \tag{11}$$

where $\beta(x_1)$ is the diffusion part of X_n with $x_1 < B$ and $x_2 < B$. For more general domain in higher dimension, the probability can be approximated by an asymptotic expansion in Δt_n [2]. For up-and-out barrier option, at each time interval $t \in [t_n, t_{n+1}]$, we compute S_n and S_{n+1} by (10), though S_n and S_{n+1} do not hit the barrier, i.e., $S_n < B$ and $S_{n+1} < B$, the continuous path S_t, may hit the barrier at some time $\tau \in [t_n, t_{n+1}]$. To approximate this hitting event, we generate an uniformly distributed random variable u_n and compare with the exceedance probability p_n in (11). If $p_n < u_n$, then we accept that the continuous path S_t does not hit the barrier during this time interval $t \in [t_n, t_{n+1}]$, since the exceedance probability is very small, i.e., the hitting event is rare to occur. On the other hand if $p_n \geq u_n$, then the probability that the continuous path S_t hits the barrier is high therefore we regard that $S_\tau \geq B$ at $\tau \in [t_n, t_{n+1}]$. Therefore, we have the rebate R and start the next sample path, i.e., the value of the barrier option of this path is $V(S_0, 0) = Re^{-r\tau}$, where R is a prescribed cash rebate. In this case, as an approximation of the first hitting time τ, we may choose the midpoint $\tilde{\tau} = (t_n + t_{n+1})/2$.

Digital barrier options

The digital barrier options can be divided into two main categories:

1. Cash-or-nothing barrier options. These payout either a prespecified cash amount or nothing, depending on whether the asset price has hit the barrier or not.
2. Asset-or-nothing barrier options. These payout the value of the asset or nothing, depending on whether the asset price has hit the barrier or not.

Rubinstein and Reiner present the set of formulas which can be used to price twenty eight different types of so-called binary barrier options [21].

Example 1 Consider a down-and-out cash-or-nothing put option with 6 months to expiration. The asset price is $S = 105$, the strike price is $K = 102$, the barrier is $B = 100$, the cash payout is $x = 15$, the risk-free interest rate is $r = 10\,\%$ per year, and the volatility is $\sigma = 20\,\%$ per year. Using below equations, the value of this barrier digital option is 0.0361.

$$z_1 = \frac{\ln(S/K)}{\sigma\sqrt{T}} + \frac{\sigma\sqrt{T}}{2}, \quad z_2 = \frac{\ln(S/B)}{\sigma\sqrt{T}} + \frac{\sigma\sqrt{T}}{2},$$
$$y_1 = \frac{\ln(B^2/SK)}{\sigma\sqrt{T}} + \frac{\sigma\sqrt{T}}{2}, \quad y_2 = \frac{\ln(B/S)}{\sigma\sqrt{T}} + \frac{\sigma\sqrt{T}}{2},$$
$$V = xe^{-rT}(N(-z_1+\sigma\sqrt{T}) - N(-z_2+\sigma\sqrt{T}) + (S/B)N(y_1-\sigma\sqrt{T}) - (S/B)N(y_2-\sigma\sqrt{T})).$$

Simulation of the standard Monte Carlo for this example has the answer 0.42, and simulation of the new Monte Carlo, that conducted on Matlab with $M = 10,000$, has the answer 0.0088. Figure 2 shows comparison between the exact value and the new Monte Carlo values for this example and Fig. 3 displays comparison between the standard MC and the improve MC errors.

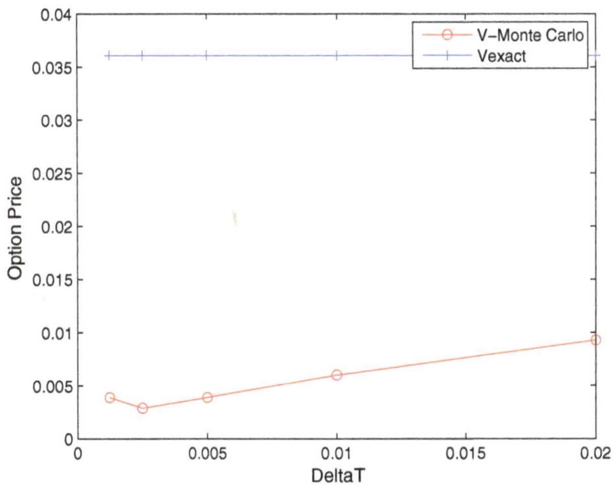

Fig. 2 The exact and new Monte Carlo values for Example 1

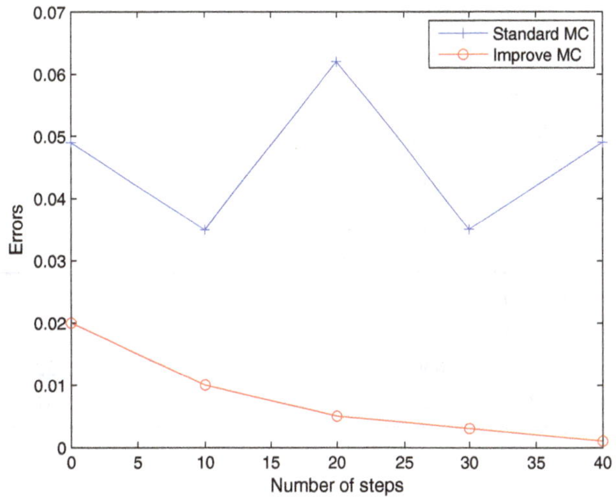

Fig. 3 Comparison of approximation errors between the standard MC and the improve MC for Example 1

Double-barrier digital options

Hui has published closed-form formulas for the valuation of one-touch double-barrier binary options [9]. A knock-in one-touch double-barrier pays off a cash amount x at maturity if the asset price touches the lower L or upper U barriers before expiration. The option pays off zero if the barriers are not hit during the lifetime of the option. Similarly, a knock-out pays out a predefined cash amount x at maturity if the lower or upper barriers are not hit during the lifetime of the option. If the underlying asset price touches any of barriers during the option's life, the option vanishes. Using the Fourier sine series, we can show that the risk natural value of double barrier cash or nothing knock-out is:

$$C = \sum_{i=1}^{\infty} \frac{2\pi i x}{Z^2} \left[\frac{\left(\frac{S}{L}\right)^{1/2} - (-1)^i \left(\frac{S}{U}\right)^{1/2}}{\left(\frac{1}{2}\right)^2 + \left(\frac{i\pi}{Z}\right)^2} \right] \times \sin\left(\frac{i\pi}{Z} \ln\left(\frac{S}{L}\right)\right) \exp\left(\frac{-1}{2}\left[\left(\frac{i\pi}{Z}\right)^2 - \beta\right]\sigma^2 T\right),$$

where

$$Z = \ln\left(\frac{U}{L}\right), \quad \beta = -\frac{1}{4} - 2\frac{r}{\sigma^2}.$$

Example 2 Table 1 gives examples of values for knock-out double-barrier binary options for different choices of barriers and volatilities and the value of them simulation with $M = 10,000$ using the new Monte Carlo in Matlab. Also, Fig. 4 shows comparison between the exact value and the new Monte Carlo values on this example with $\sigma = 0.1$, and Fig. 5 displays comparison between the standard MC and the improve MC errors.

Table 1 Comparison of numerical approximations using the improve MC for Example 2

Double-barrier binary option parameters $S = 100$, $T = 0.25$, $r = 0.05$, $x = 10$

L	U	$\sigma = 0.1$		$\sigma = 0.2$	
		Exact	New MC	Exact	New MC
80	120	9.873	9.864	8.977	8.898
85	115	9.815	9.770	7.268	7.250
90	110	8.977	8.825	3.685	3.622
95	105	3.667	3.598	0.091	0.081

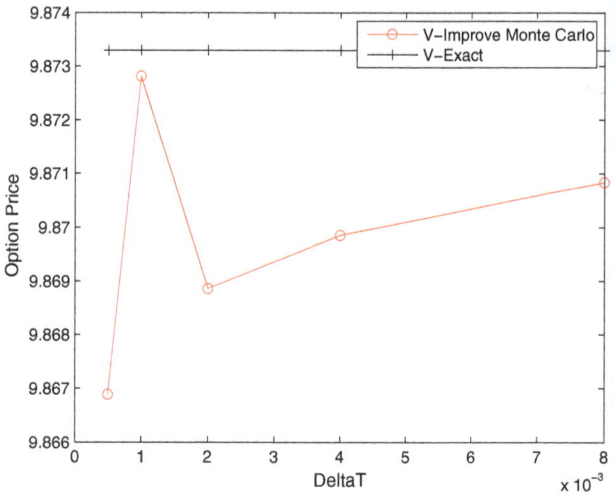

Fig. 4 The exact value and the new MC values for Example 2 with $\sigma = 0.1$

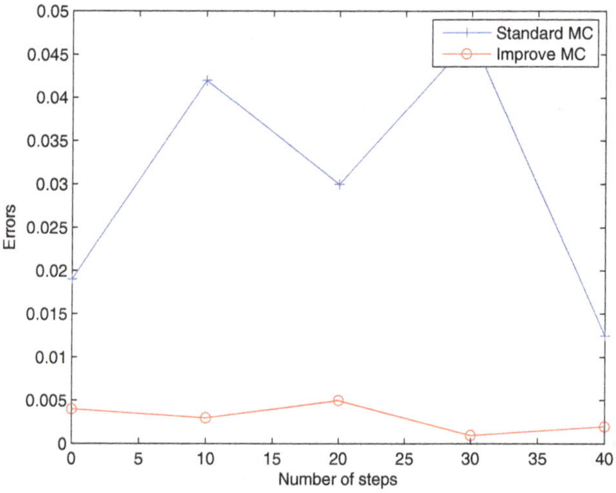

Fig. 5 Comparison of approximation errors between the standard MC and the improve MC for Example 2 with $\sigma = 0.1$

Conclusion

In this paper, we have proposed a new efficient Monte Carlo approach for estimate values of the digital barrier and double barrier options, to correctly compute the first hitting time of the barrier price by the underlying asset. The approximate error of the new method converges much faster than the standard Monte Carlo method. Future work will be devoted to extend this idea to more general diffusion problems, and theoretically study the rate of convergence of the approximate errors, and also pricing digital barrier options by other methods such as SMC and comparing results.

Acknowledgments The authors are grateful to the referees for their careful reading, insightful comments and helpful suggestions which have led to improvement of the paper.

References

1. Appolloni, E., Ligori, A.: Efficient tree methods for pricing digital barrier options (2014). arxiv.org/pdf/1401.2900
2. Baldi, P.: Exact asymptotics for the probability of exit from a domain and applications to simulation. Ann. Probab. **23**, 1644–1670 (1995)
3. Ballestra, L.V.: Repeated spatial extrapolation: an extraordinarily efficient approach for option pricing. J. Comput. Appl. Math. **256**, 83–91 (2014)
4. Bingham, N., Kiesel, R.: Risk-Neutral Valuation: Pricing and Hedging of Financial Derivatives. Springer, New York (2004)
5. Cortes, J.C., Jodar, L., Villafuerte, L.: Numerical solution of random differential equations: a mean square approach. Math. Comput. Model. **45**, 757–765 (2007)
6. Cox, J.C., Rubinstein, M.: Options Markets. Prentice Hall, New Jersey (1985)
7. Gobet, E.: Weak approximation of killed diffusion using Euler schemes. Stoch. Process. Appl. **87**, 167–197 (2000)
8. Haug, E.G.: Option Pricing Formulas. McGraw-Hill Companies, New York (2007)
9. Hui, C.H.: One-touch double barrier binary option values. Appl. Financ. Econ. **6**, 343–346 (1996)
10. Hyong-Chol, O., Dong-Hyok, K., Jong-Jun, J., Song-Hun, R.: Integrals of higher binary options and defaultable bonds with discrete default information. Electron. J. Math. Anal. Appl. **2**, 190–214 (2014)
11. Jansons, K.M., Lythe, G.D.: Efficient numerical solution of stochastic differential equations using exponential time stepping. J. Stat. Phys. **100**, 1097–1109 (2000)
12. Jasra, A., Del Moral, P.: Sequential Monte Carlo methods for option pricing. Stoch. Anal. Appl. **29**, 292–316 (2011)
13. Jerbi, Y., Kharrat, M.: Conditional expectation determination based on the J-process using Malliavin calculus applied to pricing American options. J. Stat. Comput. Simul. **84**, 2465–2473 (2014)
14. Karatzas, I., Shreve, S.E.: Brownian Motion and Stochastic Calculus. Springer, New York (1991)
15. Kim, B., Wee, I.S.: Pricing of geometric Asian options under Heston's stochastic volatility model. Quant. Fin. **14**, 1795-1809 (2014)
16. Mannella, R.: Absorbing boundaries and optimal stopping in a stochastic differential equation. Phys. Lett. A **254**, 257–262 (1999)
17. Mehrdoust, F.: A new hybrid Monte Carlo simulation for Asian options pricing. J. Stat. Comput. Simul. **85**, 507–516 (2015)
18. Moon, K.: Efficient Monte Carlo algorithm for pricing barrier options. Comm. Korean Math. Soc. **23**, 285–294 (2008)
19. Nouri, K., Ranjbar, H.: Mean square convergence of the numerical solution of random differential equations. Mediter. J. Math. **12**, 1123–1140 (2015)
20. Palan, S.: Digital options and efficiency in experimental asset markets. J. Econ. Behav. Organ. **75**, 506–522 (2010)
21. Rubinstein, M., Reiner, E.: Unscrambling the binary code. Risk Mag. **4**, 75–83 (1991)
22. Wilmott, P.: Derivatives: The Theory and Practice of Financial Engineering. Wiley, New York (1998)
23. Zhang, L., Zhang, W., Xu, W., Shi, X.: A modified least-squares simulation approach to value American barrier options. Comput. Econ. **44**, 489–506 (2014)

14

Optimal homotopy analysis method for nonlinear partial fractional differential equations

Khaled A. Gepreel[1,2] · **Taher A. Nofal**[1,3]

Abstract The main objective of this paper is to improve the optimal homotopy analysis method to find the approximate solutions for the linear and nonlinear partial fractional differential equations. The fractional derivatives are described in the Caputo sense. The optimal homotopy analysis method in applied mathematics can be used for obtaining the analytic approximate solutions for some nonlinear partial fractional differential equations such as the time and space fractional nonlinear Schrödinger partial differential equation and the time and space fractional telegraph partial differential equation. The optimal homotopy analysis method contains the h parameter which controls the convergence of the approximate solution series. Also, this method determines the optimal value of h as the best convergence of the series of solutions.

Keywords Homotopy analysis method · Fractional derivatives in Caputo sense · Nonlinear fractional Schrodinger equations · Telegraph fractional equation

✉ Khaled A. Gepreel
kagepreel@yahoo.com

Taher A. Nofal
nofal_ta@yahoo.com

1 Mathematics Department, Faculty of Science, Taif University, Taif, Kingdom of Saudi Arabia

2 Mathematics Department, Faculty of Science, Zagazig University, Zagazig, Egypt

3 Mathematics Department, Faculty of Science, El-Minia University, El-Minia, Egypt

Introduction

In recent years, there has been a great deal of interest in fractional differential equations. First there were almost no practical applications of fractional calculus, and it was considered by many as an abstract area containing only mathematical manipulations of little or no use. Nearly 30 years ago, the paradigm began to shift from pure mathematical formulations to applications in various fields, during the last decade, such as fractional systems, astrophysics [1–11]. Historical summaries of the Calculus have been applied to almost every field of science, engineering, and mathematics. Several fields of application of fractional differentiation and fractional integration are already well established; some others have just started. Many applications of fractional calculus can be found in turbulence and fluid dynamics, stochastic dynamical system, plasma physics and controlled thermonuclear fusion, nonlinear control theory, image processing; nonlinear biological developments of fractional calculus can be found in [1–3]. There has been some attempt to solve linear problems with multiple fractional derivatives (the so-called multi-term equations) [2, 12]. Not much work has been done for nonlinear problems and only a few numerical schemes have been proposed to solve nonlinear fractional differential equations. More recently, applications have included classes of nonlinear equation with multi-order fractional derivative and this motivates us to develop a numerical scheme for their solutions [13]. Numerical and analytical methods have included Adomian decomposition method (ADM) [14–17], variational iteration method (VIM) [18–20], homotopy analysis method [21–23], and the fractional complex transformation [24] to get some special exact solutions for nonlinear partial fractional differential equations. The optimal homotopy analysis approach contains convergence control parameters that are rather efficient to nonlinear differential equations [25]. As the main objective of this paper, we will use the optimal homotopy analysis method for calculating the analytic approximate solutions for some nonlinear partial fractional differential equations via the nonlinear Schrödinger partial fractional differential equation and the telegraph partial fractional differential equation. Also, we make the figures to compare between the approximate solutions and the exact for nonlinear partial fractional differential equations when $\alpha,\ \beta \to 1$.

Preliminaries and notations

We give some basic definitions and properties of the fractional calculus theory which are used further in this paper [2, 3]. For the finite derivative $[a, b]$, we define the following fractional integral and derivatives.

Definition 2.1 If $f(t) \in L_1(a,\ b)$, the set of all integrable functions, and $\alpha > 0$ then the Riemann–Liouville fractional integral of order α, denoted by $I^{\alpha}_{\alpha+}$ is defined by

$$I^{\alpha}_{\alpha+}f(t) = \frac{1}{\Gamma(\alpha)}\int_a^t (t-\tau)^{\alpha-1}f(\tau)d\tau$$

Definition 2.2 For $\alpha > 0$, the Caputo fractional derivative of order α, denoted by ${}^{C}D^{\alpha}_{a+}$, is defined by

$${}^{C}D^{\alpha}_{a+}f(t) = \frac{1}{\Gamma(n-\alpha)}\int_a^t (t-\tau)^{n-\alpha-1}D^{n}f(\tau)d\tau,$$

where n is such that $n-1<\alpha<n$ and $D = \frac{d}{d\tau}$

If α is an integer, then this derivative takes the ordinary derivative

$${}^{C}D^{\alpha}_{a+} = D^{\alpha}, \quad \alpha = 1, 2, 3, \ldots$$

Finally the Caputo fractional derivative on the whole space $\Re$ is defined by,

Definition 2.3 For $\alpha > 0$ the Caputo fractional derivative of order α on the whole space, denoted by ${}^{C}D^{\alpha}_{+}$, is defined by

$${}^{C}D^{\alpha}_{+}f(x) = \frac{1}{\Gamma(n-\alpha)}\int_{-\infty}^{x} (x-\xi)^{n-\alpha-1}D^{n}f(\xi)d\xi.$$

Basic idea of the optimal homotopy analysis method (OHAM)

To describe the basic ideas of the HAM, we consider the following differential equation:

$$N\left[D^{\alpha}_{t}u(x,t)\right] = 0, \tag{3.1}$$

where N is a nonlinear operator for this problem, D^{α}_{t} stands for the fractional derivative, x, t denotes the independent variables, and $u(x,\ t)$ is an unknown function.

By means of the HAM, one first constructs zero-order deformation equation

$$(1-q)\ell(\phi(x,t;q) - u_0(x,t)) = qhH(t)N\left[D^{\alpha}_{t}\phi(x,t,q)\right], \tag{3.2}$$

where $q \in [0,\ 1]$ is the embedding parameter, $h \neq 0$ is an auxiliary parameter, $H(t) \neq 0$ is an auxiliary function, ℓ is an auxiliary linear operator, and $u_0(x,t)$ is an initial guess. Obviously, when $q = 0$ and $q = 1$, it holds

$$\phi(x,t;0) = u_0(x,t), \quad \phi(x,t;1) = u(x,t). \tag{3.3}$$

Liao [22, 23] expanded $\phi(x,t;q)$ in Taylor series with respect to the embedding parameter q, as follows:

$$\phi(x,t;q) = u_0(x,t) + \sum_{m=1}^{\infty} u_m(x,t)q^m, \tag{3.4}$$

where

$$u_m(x,t) = \frac{1}{m!}\frac{\partial^m \phi(x,t;q)}{\partial q^m}\Big|_{q=0} \tag{3.5}$$

Assume that the auxiliary linear operator, the initial guess, the auxiliary parameter h, and the auxiliary function $H(t)$ are selected such that the series (3.4) is convergent at $q = 1$, then we have from (3.4)

$$u(x,t) = u_0(x,t) + \sum_{m=1}^{\infty} u_m(x,t). \tag{3.6}$$

Let us define the vector

$$\vec{u_n}(t) = \{u_0(x,t)u_1(x,t)u_2(x,t), \ldots, u_n(x,t)\}. \tag{3.7}$$

Differentiating (3.2) m times with respect to q, then setting $q = 0$ and dividing then by $m!$, we have the mth-order deformation equation

$$\ell(u_m(x,t) - \chi_m u_{m-1}(x,t)) = hH(t)R_m\left(\vec{u}_{m-1}\right), \tag{3.8}$$

where

$$R_m\left(\vec{u}_{m-1}\right) = \frac{1}{(m-1)!}\frac{\partial^{m-1}N[\phi(x,t;q)]}{\partial q^{m-1}}\Big|_{q=0}, \tag{3.9}$$

and

$$\chi_m = \begin{cases} 0 & m \leq 1, \\ 1 & m > 1. \end{cases} \tag{3.10}$$

Applying the Riemann–Liouville integral operator I^{α} on both sides of (3.8), we have

$$u_m(x,t) = \chi_m u_{m-1}(x,t) - \chi_m \sum_{i=0}^{n-1} u^{i}_{m-1}(0^+)\frac{t^i}{i!} + hH(t)I^{\alpha}R_m\left(\vec{u}_{m-1}\right), \tag{3.11}$$

the mth-order deformation Eq. (3.8) which is linear and thus can be easily solved, especially by means of symbolic computation softwares such as Mathematica, Maple, and MathLab. For the convergence of the above method we refer the reader to Liao's work.

Liao [25], Yabushita et al. [26] and Abbasbandy et al. [27] applied the homotopy analysis method to nonlinear ODE's and suggested the so-called optimization method to find out the optimal convergence control parameters by minimum of the square residual error integrated in the whole region having physical meaning. Their approach is based on the square residual error

$$\Delta(h) = \int_{\Omega} \left(N \sum_{i=0}^{M} u_i(\tau) \right)^2 d\tau, \tag{3.12}$$

of a nonlinear Eq. (3.1), where $\sum_{i=0}^{M} u_i(\tau)$ gives the Mth-order homotopy analysis method approximation. Obviously $\Delta(h) \to 0$, as $M \to \infty$ corresponds to a convergent series solution. For given order M of approximation, the optimal value of h is given by a nonlinear algebraic equation

$$\frac{d\Delta(h)}{dh} = 0 \tag{3.13}$$

Approximate solution for the time–space fractional nonlinear Schrodinger equation with the trapping potential equation

In this section, to demonstrate the effectiveness of our approach, we will apply the OHAM to construct approximate solutions for the time–space fractional nonlinear Schrodinger equation with the trapping potential equation

$$\begin{aligned} i_t^C D_{0+}^{\alpha} u &= -\frac{1}{2} {}_x^C D_+^{\beta} \left({}_x^C D_+^{\beta} u \right) + u \cos^2 x + |u|^2 u, \quad t > 0, \quad 0 < \alpha, \beta \le 1, \\ u(x, 0) &= f(x) = \sin x, \end{aligned} \tag{4.1}$$

where $i = \sqrt{-1}$. The fractional Schorodinger equations can be represented some dynamical system in quantum mechanics especially in exact solutions for nonlinear Schrodinger equations of motion, also in fluid and non-linear dynamical system [17]. When $\beta \to 1, \alpha \to 1$, the exact solution for the nonlinear partial Schorodinger equation takes the following form [28]:

$$u_{ex} = e^{-\frac{3}{2}it} \quad \sin x. \tag{4.2}$$

By means of the OHAM, we choose the linear operator

$$\mathcal{L}[\phi(x, t; q)] = \frac{\partial^{\alpha} \phi(x, t; q)}{\partial t^{\alpha}}. \tag{4.3}$$

with property $\mathcal{L}[c] = 0$, where c is a constant. We define a nonlinear operator as

$$\begin{aligned} N\Bigg[\phi(x, t; q) &= \frac{\partial^{\alpha} \phi(x, t; q)}{\partial t^{\alpha}} - \frac{i}{2} \frac{\partial^{2\beta} \phi(x, t; q)}{\partial x^{2\beta}} \\ &+ i\phi(x, t; q) \cos^2 x + i\phi^2(x, t; q) \overline{\phi}(x, t; q) \Bigg]. \end{aligned} \tag{4.4}$$

We construct the zeroth-order deformation equation

$$(1 - q)\mathcal{L}[\phi(x, t; q) - u_0(x, t)] = qhH(t)N[\phi(x, t; q)]. \tag{4.5}$$

For $q = 0$ and $q = 1$ it holds

$$\phi(x, t; 0) = u_0(x, t), \quad \phi(x, t; 1) = u(x, t). \tag{4.6}$$

Thus, we obtain the mth-order deformation equations

$$\mathcal{L}[u_m(x, t) - \chi_m u_{m-1}(x, t)] = hH(t)\mathcal{R}_m \tag{4.7}$$

where

$$\begin{aligned} \mathcal{R}_m &= \frac{\partial^{\alpha} u_{m-1}}{\partial t^{\alpha}} - \frac{i}{2} \frac{\partial^{2\beta} u_{m-1}}{\partial x^{2\beta}} + i u_{m-1} \cos^2 x \\ &+ i \sum_{l=0}^{m-1} \sum_{j=0}^{l} u_j u_{l-j} u_{m-1-l}, \end{aligned} \tag{4.8}$$

and

$$\chi_m = \begin{cases} 1 & m > 1, \\ 0 & m \le 1. \end{cases} \tag{4.9}$$

To obey both the rule of solution expression and the rule of the coefficient ergodicity [22, 23], having the freedom to choose the auxiliary parameter h, the auxiliary function $H(t)$, the initial approximate $u_0(x, t)$, and the auxiliary linear operator $\mathcal{L}$, we can assume that all of them are properly chosen so that the solution $\phi(x, t; q)$ exists for $0 \le q \le 1$. $H(t) = 1$ is properly chosen so that the power series (3.4) is convergent at $q = 1$. The convergence of the homotopy analysis method for solving the Schordinger equations is discussed in [29]. Now the solution of the mth-order deformation Eq. (4.7) for $m \ge 1$ becomes

$$u_m(x, t) = \chi_m u_{m-1}(x, t) + hJ^{\alpha} \mathcal{R}_m(u_{m-1}) \tag{4.10}$$

and so on, we substitute the initial condition in (4.1) into the system (4.10) with the aid of Maple; the approximate solutions of Eq. (4.1) take the following form

$$\begin{aligned} u_0 &= \sin x, \\ u_1 &= -\frac{it^{\alpha} h}{\Gamma(\alpha + 1)} \left\{ \frac{1}{2} \sin\left(x + \frac{2\pi\beta}{2} \right) - \sin x \right\}, \\ u_2 &= -(1 + h) \frac{it^{\alpha} h}{\Gamma(\alpha + 1)} \left\{ \frac{1}{2} \sin\left(x + \frac{2\pi\beta}{2} \right) - \sin x \right\}, \\ &+ \frac{h^2 (-1)^2 t^{2\alpha}}{\Gamma(2\alpha + 1)} \left\{ \frac{1}{4} \sin\left(x + \frac{4\pi\beta}{2} \right) - \sin\left(x + \frac{2\pi\beta}{2} \right) + \sin x \right\}, \end{aligned} \tag{4.11}$$

and

$$
\begin{aligned}
u_3 =&(1+h)u_2 + \frac{(1+h)h^2 i^2 t^{2\alpha}}{\Gamma(2\alpha+1)}\left\{\frac{1}{4}\sin\left(x+\frac{4\pi\beta}{2}\right)\right.\\
&\left.-\frac{1}{2}\sin\left(x+\frac{2\pi\beta}{2}\right)\right\}\\
&-\frac{h^3(i)^3 t^{3\alpha}}{\Gamma(3\alpha+1)}\left\{\frac{1}{8}\sin\left(x+\frac{6\pi\beta}{2}\right)\right.\\
&\left.-\frac{3}{4}\sin\left(x+\frac{4\pi\beta}{2}\right)+\frac{3}{2}\sin\left(x+\frac{2\pi\beta}{2}\right)-\sin x\right\}\\
&-\frac{h^2(1+h)(i)^2 t^{2\alpha}}{\Gamma(2\alpha+1)}\left\{\frac{1}{2}\sin\left(x+\frac{2\pi\beta}{2}\right)-\sin x\right\}\\
&+\frac{ih^3\sin x t^{3\alpha}\Gamma(2\alpha+1)}{\Gamma^2(\alpha+1)\Gamma(3\alpha+1)}\left\{\frac{1}{2}\sin\left(x+\frac{2\pi\beta}{2}\right)-\sin x\right\}^2\\
&+\frac{2h^3(i)^3\sin^2 x t^{3\alpha}}{\Gamma(3\alpha+1)}\left\{\frac{1}{4}\sin\left(x+\frac{4\pi\beta}{2}\right)\right.\\
&\left.-\sin\left(x+\frac{2\pi\beta}{2}\right)+\sin x\right\},
\end{aligned}
$$

and so on.

In this case the approximate solution to the time–space fractional nonlinear Schrodinger equation with the trapping potential of Eq. (5.1) takes the following form

$$
\begin{aligned}
u_{app} =&\sin x - \frac{it^\alpha h}{\Gamma(\alpha+1)}\left\{\frac{1}{2}\sin\left(x+\frac{2\pi\beta}{2}\right)-\sin x\right\}\\
&-(1+h)\frac{it^\alpha h}{\Gamma(\alpha+1)}\left\{\frac{1}{2}\sin\left(x+\frac{2\pi\beta}{2}\right)-\sin x\right\}\\
&+\frac{h^2(i)^2 t^{2\alpha}}{\Gamma(2\alpha+1)}\left\{\frac{1}{4}\sin\left(x+\frac{4\pi\beta}{2}\right)\right.\\
&\left.-\sin\left(x+\frac{2\pi\beta}{2}\right)+\sin x\right\}\\
&+(1+h)u_2+\frac{(1+h)h^2 i^2 t^{2\alpha}}{\Gamma(2\alpha+1)}\left\{\frac{1}{4}\sin\left(x+\frac{4\pi\beta}{2}\right)\right.\\
&\left.-\frac{1}{2}\sin\left(x+\frac{2\pi\beta}{2}\right)\right\}\\
&-\frac{h^3(i)^3 t^{3\alpha}}{\Gamma(3\alpha+1)}\left\{\frac{1}{8}\sin\left(x+\frac{6\pi\beta}{2}\right)-\frac{3}{4}\sin\left(x+\frac{4\pi\beta}{2}\right)\right.\\
&\left.+\frac{3}{2}\sin\left(x+\frac{2\pi\beta}{2}\right)-\sin x\right\}\\
&-\frac{h^2(1+h)(i)^2 t^{2\alpha}}{\Gamma(2\alpha+1)}\left\{\frac{1}{2}\sin\left(x+\frac{2\pi\beta}{2}\right)-\sin x\right\}\\
&+\frac{ih^3\sin x t^{3\alpha}\Gamma(2\alpha+1)}{\Gamma^2(\alpha+1)\Gamma(3\alpha+1)}\left\{\frac{1}{2}\sin\left(x+\frac{2\pi\beta}{2}\right)-\sin x\right\}^2\\
&+\frac{2h^3(i)^3\sin^2 x\, t^{3\alpha}}{\Gamma(3\alpha+1)}\left\{\frac{1}{4}\sin\left(x+\frac{4\pi\beta}{2}\right)\right.\\
&\left.-\sin\left(x+\frac{2\pi\beta}{2}\right)+\sin x\right\}\ldots,
\end{aligned}
\tag{4.12}
$$

Abbasbandy et al. [30] have proved in the general case that the h-curve is the main trait of the homotopy analysis method and plays an important role in obtaining the convergence of the series solutions. Moreover, the h curve can be used in predicting multiple solutions for Boundary value problems. Abbasbandy et al. [30] have shown why the line property is important in the graph of the h-curve, which is the plot of the series solution via convergence control parameter.

So that to investigate the influence of h on the convergence of the solution series given by the HAM, we first plot the so-called h-curves of (4.12) when $x = 0.5, t = 0.5, \alpha = 0.6, \beta = 0.5$. According to the h-curves, it is easy to discover the valid region of h. We use the first three terms in evaluating the approximate solution (4.12). Note that the solution series contains the auxiliary parameter h which provides us with a simple way to adjust and control the convergence of the solution series [27]. In general, by means of the so-called h-curve, i.e., a curve of a versus h. As pointed by Liao [22, 23, 25], the valid region of h is a horizontal line segment. Therefore, it is straightforward to choose an appropriate range for h which ensures the convergence of the solution series. We stretch the h-curve of $u(0.7, 0.3)$ in Fig. 1, which shows that the solution series is convergent when $-0.5 < h < -1.5$.

As mentioned in Sect. 2 the optimal value of h is determined by the minimum of Δ_4, corresponding to the nonlinear algebraic equation $\frac{d\Delta_4}{dh} = 0$. Our calculations showed that Δ_4 has its minimum value at -1.021628429.

In this case, we compare between the approximate solution (4.12) and the exact solution (4.2) at the optimal value of h ($h = -1.021628429$).

Remark 1 Figures 2, 3, and 4 present the behavior of the approximate solution which is similar to the exact solution for some different values. Consequently, the approximate solutions are rabidly convergence as the exact solution.

Approximate solution for the time–space fractional telegraph equation

In this section, to demonstrate the effectiveness of our approach, we will apply the OHAM to construct approximate solutions for the time–space fractional telegraph equation

$$
\begin{aligned}
{}^C_x D^\alpha_+\left({}^C_x D^\alpha_+ u\right) =& {}^C_t D^\beta_+\left({}^C_t D^\beta_+ u\right) + 4\,{}^C_t D^\beta_+ u + 4u, \quad x>0, \quad 0<\alpha,\beta\le 1,\\
u(0,t) =& e^{-2t}+1, \quad {}^C_x D^\alpha_+ u(0,t) = 2.
\end{aligned}
\tag{5.1}
$$

The fractional telegraph equations investigate several efforts to better understand the anomalous diffusion

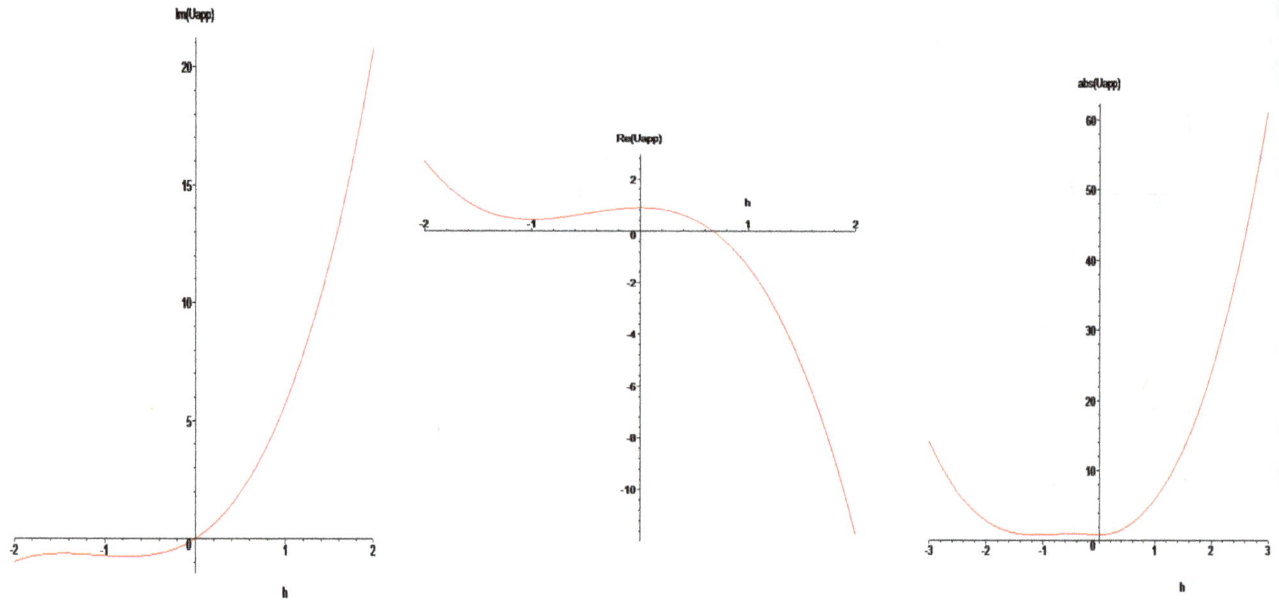

Fig. 1 The h-curves of the fourth-order approximation to $\mathrm{Im}(u(x,t))$, $\mathrm{Re}(u(x,t))$, and $\mathrm{abs}(u(x,t))$, when $x=0.5, t=0.5, \alpha=0.6, \beta=0.5$

processes observed in blood flow experiments [31]. When $\beta \to 1, \alpha \to 1$ the exact solution for the nonlinear partial Telegraph equation takes the following form [32]:

$$u_{ex} = e^{-2t} + e^{2x}. \tag{5.2}$$

By means of the HAM, we choose the linear operator

$$\mathcal{L}[\phi(x,t;q)] = \frac{\partial^{2\alpha}\phi(x,t;q)}{\partial x^{2\alpha}} \tag{5.3}$$

with property $\mathcal{L}[c]=0$, where c is a constant. We define a linear operator as

$$N[\phi(x,t;q)] = \frac{\partial^{2\alpha}\phi(x,t;q)}{\partial x^{2\alpha}} - \frac{\partial^{2\beta}\phi(x,t;q)}{\partial t^{2\beta}} - 4\frac{\partial^{\beta}\phi(x,t;q)}{\partial t^{\beta}} - 4\phi(x,t;q). \tag{5.4}$$

We construct the zeroth-order deformation equation

$$(1-q)\mathcal{L}[\phi(x,t;q) - u_0(x,t)] = qhH(t)N[\phi(x,t;q)]. \tag{5.4}$$

For $q=0$ and $q=1$ it holds

$$\phi(x,t;0) = u_0(x,t), \quad \phi(x,t;1) = u(x,t). \tag{5.5}$$

Thus, we obtain the mth-order deformation equations

$$\mathcal{L}[u_m(x,t) - \chi_m u_{m-1}(x,t)] = hH(t)\mathcal{R}_m \tag{5.6}$$

where

$$\mathcal{R}_m = \frac{\partial^{2\alpha}u_{m-1}}{\partial x^{2\alpha}} - \frac{\partial^{2\beta}u_{m-1}}{\partial t^{2\beta}} - 4\frac{\partial^{\beta}u_{m-1}}{\partial t^{\beta}} - 4u_{m-1} \tag{5.7}$$

and

$$\chi_m = \begin{cases} 1 & m > 1, \\ 0 & m \leq 1. \end{cases} \tag{5.8}$$

To obey both the rule of solution expression and the rule of the coefficient ergodicity [22, 23], the auxiliary function can be determined uniquely $H(t)=1$. Now the solution of the mth-order deformation Eq. (5.6) for $m \geq 1$ becomes

$$u_m(x,t) = \chi_m u_{m-1}(x,t) + hI^{\alpha}(I^{\alpha}\mathcal{R}_m(u_{m-1})) \tag{5.9}$$

and so on; we substitute the initial condition in (5.1) into the system (5.9) then approximate solutions of Eq. (5.1) take the following form

$$u_0(x,t) = 1 + e^{-2t} + 2\frac{x^{\alpha}}{\Gamma(\alpha+1)},$$

$$u_1(x,t) = \frac{hx^{2\alpha}}{\Gamma(2\alpha+1)}\left\{e^{-2t}\left[-(-2)^{2\beta} - 4(-2)^{\beta} - 4 - 4\right]\right\} - \frac{8hx^{3\alpha}}{\Gamma(3\alpha+1)},$$

$$u_2(x,t) = (1+h)u_1(x,t) - \frac{h^2x^{4\alpha}}{\Gamma(4\alpha+1)} \times \left\{e^{-2t}\left[-(-2)^{4\beta} - 8(-2)^{3\beta} - 24(-2)^{2\beta} - 32(-2)^{\beta} - 16\right] - 16\right\} + \frac{32x^{5\alpha}}{\Gamma(5\alpha+1)},$$

$$u_3(x,t) = (1+h)u_2(x,t) + \frac{h^3x^{6\alpha}}{\Gamma(6\alpha+1)}\left\{e^{-2t}\left[-(-2)^{6\beta} - 12(-2)^{5\beta} - 60(-2)^{4\beta} - 160(-2)^{3\beta} - 240(-2)^{2\beta} - 192(-2)^{\beta} - 64\right] - 64\right\}$$
$$- \frac{128h^3x^{7\alpha}}{\Gamma(7\alpha+1)} + -(1+h)h^2\left\{\frac{x^{4\alpha}}{\Gamma(4\alpha+1)}\left[e^{-2t}\left(-(-2)^{4\beta} - 8(-2)^{3\beta} - 24(-2)^{2\beta} - 32(-2)^{\beta} - 16\right) - 16\right] - \frac{32x^{5\alpha}}{\Gamma(5\alpha+1)}\right\}, \ldots \quad (5.10)$$

In this case, the approximate solution to the time–space fractional telegraph equation takes the following form

$$u(x,t) = 1 + e^{-2t} + 2\frac{x^{\alpha}}{\Gamma(\alpha+1)} + \frac{hx^{2\alpha}}{\Gamma(2\alpha+1)}\left\{e^{-2t}\left[-(-2)^{2\beta} - 4(-2)^{\beta} - 4\right] - 4\right\}$$
$$- \frac{8hx^{3\alpha}}{\Gamma(3\alpha+1)} + (1+h)u_1(x,t)$$
$$- \frac{h^2x^{4\alpha}}{\Gamma(4\alpha+1)}\left\{e^{-2t}\left[-(-2)^{4\beta} - 8(-2)^{3\beta} - 24(-2)^{2\beta} - 32(-2)^{\beta} - 16\right] - 16\right\},$$
$$+ \frac{32x^{5\alpha}}{\Gamma(5\alpha+1)}(1+h)u_2(x,t) + \frac{h^3x^{6\alpha}}{\Gamma(6\alpha+1)}\left\{e^{-2t}\left[-(-2)^{6\beta} - 12(-2)^{5\beta} - 60(-2)^{4\beta} - 160(-2)^{3\beta}\right.\right.$$
$$\left.\left. - 240(-2)^{2\beta} - 192(-2)^{\beta} - 64\right] - 64\right\} - \frac{128h^3x^{7\alpha}}{\Gamma(7\alpha+1)} +$$
$$- (1+h)h^2\left\{\frac{x^{4\alpha}}{\Gamma(4\alpha+1)}\left[e^{-2t}\left(-(-2)^{4\beta} - 8(-2)^{3\beta} - 24(-2)^{2\beta} - 32(-2)^{\beta} - 16\right) - 16\right] - \frac{32x^{5\alpha}}{\Gamma(5\alpha+1)}\right\} + \ldots \quad (5.11)$$

To investigate the influence of h on the convergence of the solution series given by the HAM, we first plot the so-

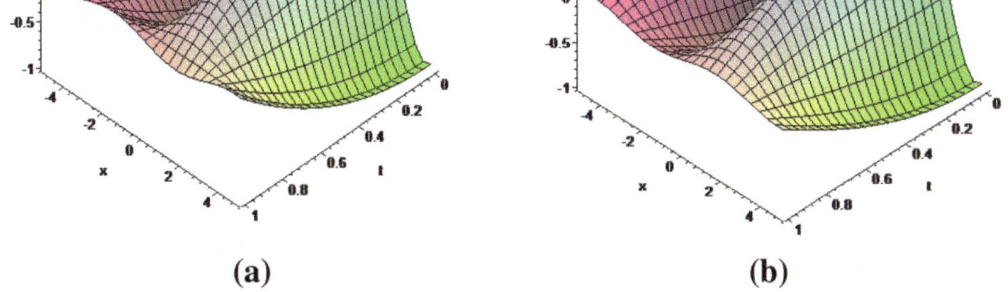

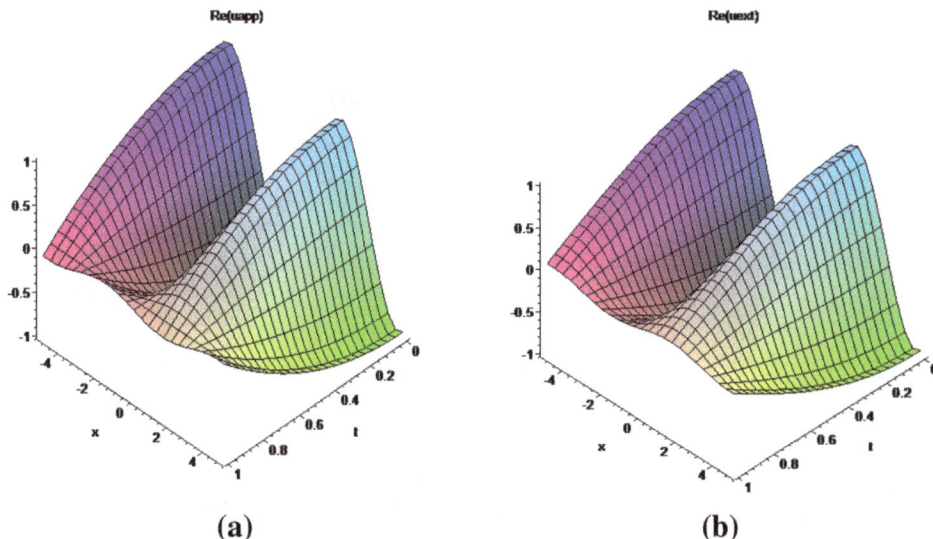

Fig. 2 The real part of the approximate solution (4.12) shown in the **a** in comparison with that of the exact solution (4.2) shown in **b** when $\alpha = 1$, $\beta = 1$, $h = -1.021628429$, $-5<x<5$, and $0<t<1$

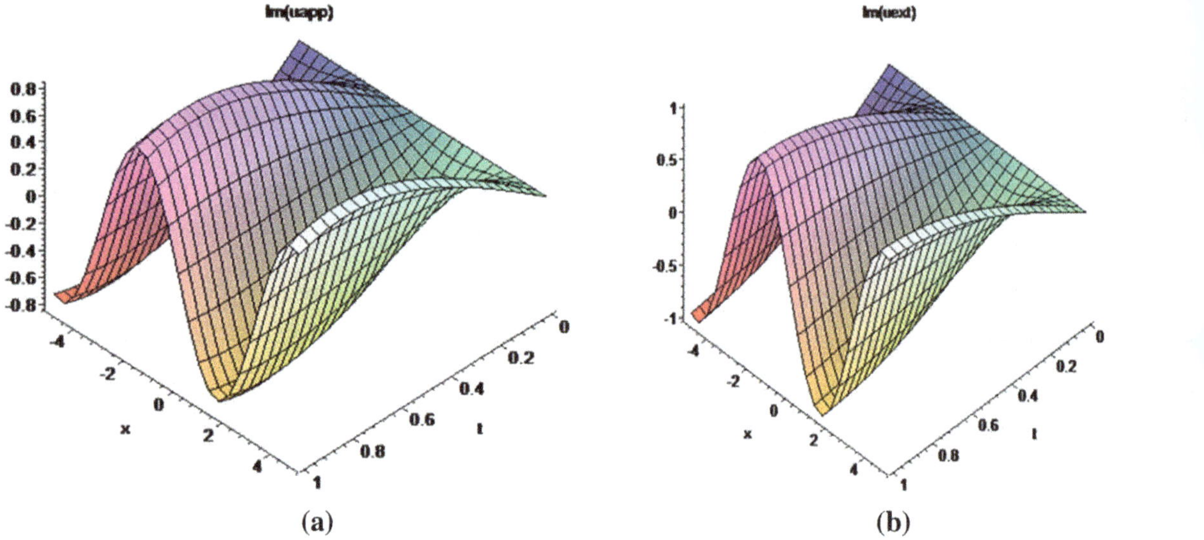

Fig. 3 The imaginary part of the approximate solution (4.12) shown in the **a** in comparison with that of the exact solution (4.2) shown in **b** when $\alpha = 1$, $\beta = 1$, $h = -1.021628429$, $-5<x<5$, and $0<t<1$

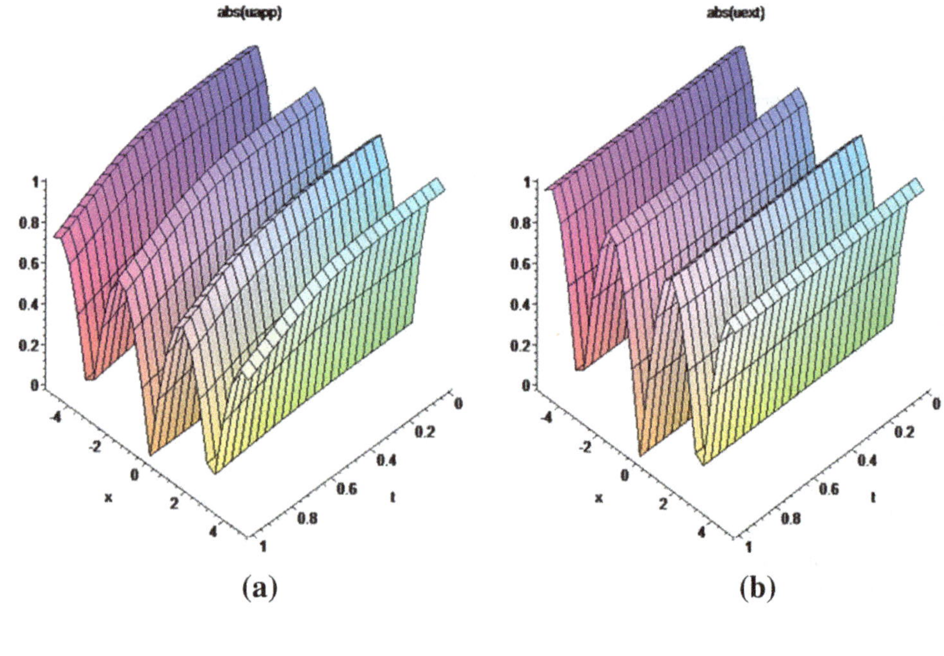

Fig. 4 The absolute value of the approximate solution (4.12) shown in the **a** in comparison with that of the exact solution (4.2) shown in **b** when $\alpha = 1$, $\beta = 1$, $h = -1.021628429$, $-5<x<5$, and $0<t<1$

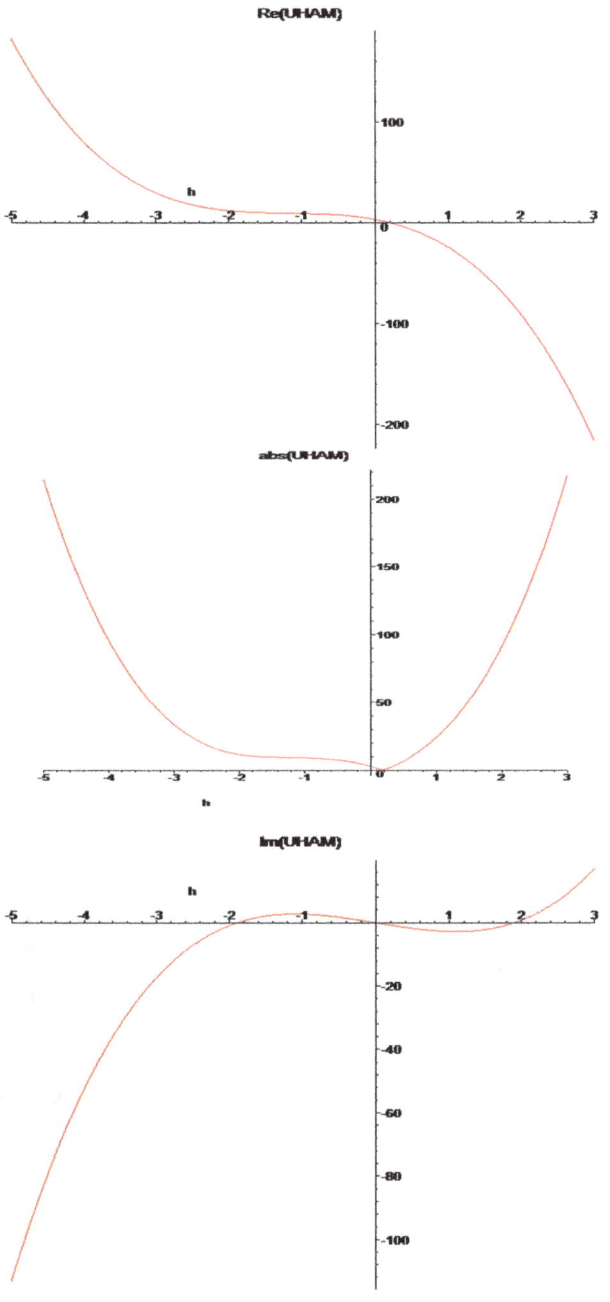

Fig. 5 The h-curves of the four-order approximation to $\mathrm{Im}(u(x,t))$, $\mathrm{Re}(u(x,t))$, and $\mathrm{abs}(u(x,t))$, when $x = 0.5$, $t = 0.5$, $\alpha = 0.5$, and $\beta = 0.3$

called h-curves of (5.11) when $x = t = \alpha = 0.5, \beta = 0.3$ According to the h-curves, it is easy to discover the valid region of h. We use the first three terms in evaluating the approximate solution (5.11). Note that the solution series contains the auxiliary parameter h which provides us with a simple way to adjust and control the convergence of the solution series [25]. In general, by means of the so-called h-curve, i.e., a curve of a versus h. As pointed by Liao [22, 23], the valid region of h is a horizontal line segment. Therefore, it is straightforward to choose an appropriate range for h which ensure the convergence of the solution series. We stretch the h-curve of $u(0.5, 0.5)$ in Fig. 5, which shows that the solution series is convergent when $-0.5 < h < -1.5$.

As mentioned in Sect. 2 the optimal value of h is determined by the minimum of Δ_4, corresponding to the nonlinear algebraic equation $\frac{d\Delta_4}{dh} = 0$. Our calculations showed that Δ_4 has its minimum value at -0.902644.

In this case, we compare between the approximate solution (5.11) and the exact solution (5.2) at the optimal value of h ($h = -0.902644$).

Remark 2 Figure 6 presents the behavior of the approximate solution using the optimal homotopy analysis method which is similar to that of the exact solution to some different values of partial fractional Telegraph equation when $\alpha = 1$ and $\beta = 1$. Consequently, the approximate solutions are rapidly convergent as the exact solution.

Conclusion

In this paper, we used the one-step homotopy analysis method to obtain the analytic approximate solutions for linear and nonlinear partial fractional differential equations. The optimal homotopy analysis method investigates the influence of h on the convergence of the approximate solution. The solution series contains the auxiliary parameter h which provides us with a simple way to adjust and control the convergence of the solution series [25]. In general, by means of the so-called h-curve, i.e., a curve of a versus h. As pointed by Liao [22, 23], the valid region of h

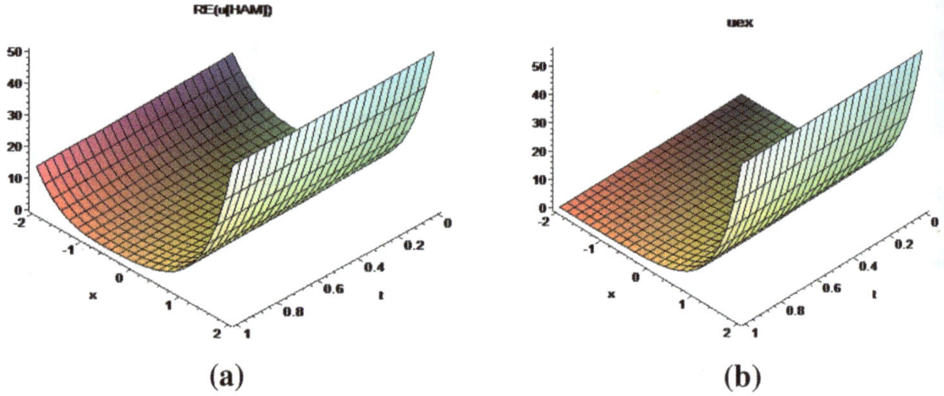

Fig. 6 The approximate solution (5.11) shown in the **a** in comparison with the exact solution (5.2) shown in **b** when $\alpha = 1$, $\beta = 1$, $h = -0.9$, $-2 < x < 2$, and $0 < t < 1$

is a horizontal line segment. Therefore, it is straightforward to choose an appropriate range for h which ensures the convergence of the solution series. This method determined the optimal value of h as the best convergence of the series of solutions.

Author contribution statement Khaled A. Gepreel and T. A. Nofal carried out the computational analysis of this method and wrote the manuscript. Khaled A. Gepreel and T. A. Nofal made helpful suggestions to the Manuscript. Khaled A. Gepreel made a figure to show the comparison between solutions and T. A. Nofal made a discussion between solutions. Khaled A. Gepreel and T. A. Nofal designed, led and coordinated the entire study.

References

1. Kilbas, A.A., Srivastava, H.M., Trujillo, J.J.: Theory and applications of fractional differential equations (north-holland mathematical studies), vol. 204. Elsevier, Amsterdam (2006)
2. Podlubny, I.: Fractional differential equation. Academic Press, San Diego (1999)
3. Samko, S.G., Kilbas, A.A., Marichev, O.I.: Fractional integrals and derivatives: theory and applications. Gordon and Breach, Langhorne (1993)
4. El-Sayed, A.M.A.: Int. J. Theor. Phys. **35**, 311 (1996)
5. Gepreel, K.A.: Appl. Math. Lett. **24**, 1428 (2011)
6. Herzallah, M.A.E., Muslih, S., Baleanu, D., Rabei, E.M.: Nonlinear Dyn. **66**, 459 (2012)
7. Magin, R.L.: Fractional Calculus in Bioengineering. Begell House Publisher, Inc., Connecticut (2006)
8. West, B.J., Bologna, M., Grigolini, P.: Physics of Fractal operators. Springer, New York (2003)
9. Jesus, I.S., Machado, J.A.T.: Nonlinear Dyn. **54**, 263 (2008)
10. Agrawal, O.P., Baleanu, D.: J. Vibr. Contr. **13**, 1269 (2007)
11. Tarasov, V.E.: Ann. Phys. **323**, 2756 (2008)
12. He, J.H.: Bull. Sci. Tecknol. **15**, 86 (1999)
13. Erturk, V.S., Momani, Sh, Odibat, Z.: Commun. Nonlinear Sci. Numer. Simulat. **13**, 1642 (2008)
14. Daftardar-Gejji, V., Bhalekar, S.: Appl. Math. Comput. **202**, 113 (2008)
15. Daftardar-Gejji, V., Jafari, H.: Appl. Math. Comput. **189**, 541 (2007)
16. Zayed, E.M.E., Nofal, T.A., Gepreel, K.A.: Commu. Appl. Nonlinear Anal. **15**, 57 (2008)
17. Herzallah, M.A.E., Gepreel, K.A.: Appl. Math. Model. **36**, 5678 (2012)
18. Sweilam, N.H., Khader, M.M., Al-Bar, R.F.: Phys. Lett. A **371**, 26 (2007)
19. He, J.H.: Int. J. Nonlinear Mech. **34**, 699 (1999)
20. Wu, G., Lee, E.W.M.: Phys. Lett. A **374**, 2506 (2010)
21. Gepreel, K.A., Mohamed, S.M.: Chin. Phys. B **22**, 010201 (2013)
22. Liao, S.J.: The proposed homotopy analysis technique for the solution of nonlinear problem (Ph.D. thesis, Shanghai Jiao Tong University) (1992)
23. Liao, S.J.: Int J Nonlinear Mech **30**, 371 (1995)
24. Gepreel, K.A., Omran, S.: Chin. Phys. B **21**, 110204 (2011)
25. Liao, S.J.: Commun. Nonlinear Sci. Numer. Simul. **15**, 2016 (2010)
26. Yabushita, K., Yamashita, M., Tsuboi, K.: J. Phys. A: Math. Gen. **40**, 8403 (2007)
27. Abbasbandy, S., Jalili, M.: Determination of optimal convergence control parameter value in homotopy analysis method. Numer. Algor. doi:10.1007/s11075-012-9680-9 (in press) (2013)
28. Kanth, A.S.V., Aruna, K.: Chaos Solitons Fract. **41**, 2277 (2009)
29. Faribozi, M., Fallahzhed, A.: J. Basic Appl. Sci. Res. **2**, 6076 (2012)
30. Abbasbandy, S., Shivanian, E., Varjarvelu, K.: Commun. Nonliear Sci. Numer. Simul. **16**, 4268 (2012)
31. Cascaral, R., Eckstein, E., Frota, C., Goldstein, J.: J. Math. Anal. Appl. **276**, 145 (2002)
32. Biazar, J., Eslami, M.: Phys. Lett. A **374**, 2904 (2010)

15

New existence results on positive solutions of four-point integral type BVPs for coupled multi-term fractional differential equations

Xiaohui Yang[1] · Yuji Liu[2]

Abstract In this article, we establish some new existence results on positive solutions of a four-point integral boundary value problem for coupled nonlinear multi-term fractional differential equations. Our analysis rely on the well known fixed point theorems. Numerical examples are given to illustrate the main theorems.

Keywords Four-point integral boundary value problem · Multi-term fractional differential system · Non-Carathéodory function · Fixed-point theorem

Mathematics Subject Classification 92D25 · 34A37 · 34K15

Introduction

Fractional differential systems have many applications in modeling of physical and chemical processes and in engineering [3, 14, 19], and have been of great interest recently. In its turn, mathematical aspects of studies on fractional differential systems were discussed by many authors, see the text books [5, 15] and papers [1, 6, 8, 13, 16, 17, 20, 21, 23–26]. A survey concerning the studies on solvability of two-point or four-point boundary value problems for fractional differential systems was given in [11].

In this paper, we discuss the existence of positive solutions of the following four-point integral type boundary value problem for the multi-term fractional differential system

$$\begin{cases} D_{0^+}^{\alpha}u(t)+p(t)f(t,v(t),D_{0^+}^{n}v(t))=0, a.e., t\in(0,1),\\ D_{0^+}^{\beta}v(t)+q(t)g(t,u(t),D_{0^+}^{m}u(t))=0, a.e., t\in(0,1),\\ \lim\limits_{t\to 0}t^{2-\alpha}u(t)-au(\xi)=\int_0^1\phi_1(t,v(t),D_{0^+}^{n}v(t))dt,\\ u(1)-bu(\eta)=\int_0^1\psi_1(t,v(t),D_{0^+}^{n}v(t))dt,\\ \lim\limits_{t\to 0}t^{2-\beta}v(t)-cv(\xi)=\int_0^1\phi_2(t,u(t),D_{0^+}^{m}u(t))dt,\\ v(1)-dv(\eta)=\int_0^1\psi_2(t,u(t),D_{0^+}^{m}u(t))dt, \end{cases} \tag{1}$$

where

(i) $1<\alpha,\beta\le 2$, $\alpha-1<m<\alpha$ and $\beta-1<n<\beta$, $D_{0^+}^{*}$ is the standard Riemann–Liouville differential derivative of order $*>0$ with the starting point 0,

(ii) $0<\xi\le\eta<1$ and $a,b,c,d\ge 0$,

(iii) $p,q:(0,1)\to\mathbb{R}$, p satisfies that there exist numbers k_1,l_1 such that $k_1>-1$, $\alpha-m+l_1>0$, $2+k_1+l_1>0$ and $|p(t)|<t^{k_1}(1-t)^{l_1}$ for $t\in(0,1)$, q satisfies that there exist numbers k_2,l_2 such that $k_2>-1$, $\beta-n+l_2>0$, $2+k_2+l_2>0$ and $|q(t)|<t^{k_2}(1-t)^{l_2}$ for $t\in(0,1)$, with $p(t)\not\equiv 0$ and $q(t)\not\equiv 0$ on (0, 1),

(iv) $f,g,\phi_i,\psi_i:(0,1)\times[0,+\infty)\times\mathbb{R}\to[0,+\infty)$, f is a strong (n,β)-Carathéory function and g is a strong

---

Supported by the Natural Science Foundation of Guangdong province (No. S2011010001900) and Natural science research project for colleges and universities of Guangdong Province (No: 2014KTSCX126).

✉ Yuji Liu
liuyuji888@sohu.com

Xiaohui Yang
xiaohuiyang@sohu.com

1 Department of Computer, Guangdong Police College, Guangzhou 510230, People's Republic of China

2 Department of Mathematics, Guangdong University of Finance and Economics, Guangzhou 510320, People's Republic of China

(m, α)-Carathéory function with $f(t,0,0) \not\equiv 0$ and $g(t,0,0) \not\equiv 0$ on (0, 1), ϕ_1, ψ_1 are (n, β)-Carathéory functions and ϕ_2, ψ_2 are (m, α)-Carathéory functions.

A pair of functions (x, y) is called a solution of BVP (1) if $x, y \in C^0(0,1]$ and x, y satisfy all equations in (1). We obtain the results on solutions of BVP(1) using Schauder's fixed point theorem in Banach spaces. The salient features of this study are as follows:

(a) the fractional differential equations in (1) are multi-term ones and their nonlinearities depend on the lower order fractional derivatives with order greater than $\alpha - 1$ and $\beta - 1$;
(b) instead of the condition $u(0) = 0, v(0) = 0$ we consider integral boundary conditions which are more suitable as $D^{\alpha}_{0+}x(t) = 0$ with $\alpha \in (1,2)$ implies $x(t) = ct^{\alpha-1}$ and obviously x is not continuous at $t = 0$ while $\lim_{t\to 0} t^{2-\alpha}x(t)$ exists;
(c) BVP(1) is a generalized form of known ones in references [4, 7, 9, 10, 21], the positive solutions of BVP(1) obtained are unbounded (discontinuous at $t = 0$) which are different from those ones (continuous on [0,1]) in [1, 8, 23, 24];
(d) this paper is a complement of [11] in which the existence of positive solutions of BVP(1) was studied under the assumptions $m \in (0, \alpha - 1]$ and $n \in (0, \beta - 1]$ while $m \in (\alpha - 1, \alpha), n \in (\beta - 1, \beta)$ are supposed in this paper.

The remainder of this paper is arranged as follows: in Sect. 2, we present preliminary results; in Sect. 3, the main results are presented; and two examples are given in Sect. 4 to illustrate the main results.

Preliminary results

For the convenience of readers, we present here the necessary definitions from fixed point theory and fractional calculus theory.

Definition 2.1 [2] Let X be a Banach space. An operator $T : X \to X$ is completely continuous if it is continuous and maps bounded sets into pre-compact sets (or relatively compact sets).

Definition 2.2 [15] The left Riemann–Liouville fractional integral (left forward) of order $\alpha > 0$ of a function $f : (0, \infty) \to \mathbb{R}$ is given by

$$I^{\alpha}_{0+}f(t) = \frac{1}{\Gamma(\alpha)} \int_0^t (t-s)^{\alpha-1} f(s)\mathrm{d}s, \quad t > 0$$

provided that the right-hand side exists.

Definition 2.3 [15] The left Riemann–Liouville fractional derivative (left farward) of order $\alpha > 0$ of a continuous function $f : (0, \infty) \to \mathbb{R}$ is given by

$$D^{\alpha}_{0+}f(t) = \frac{1}{\Gamma(n-\alpha)} \frac{\mathrm{d}^n}{\mathrm{d}t^n} \int_0^t \frac{f(s)}{(t-s)^{\alpha-n+1}} \mathrm{d}s, \quad t > 0$$

where $n - 1 < \alpha < n$, provided that the right-hand side exists.

Definition 2.4 $h : (0,1) \times \mathbb{R} \times \mathbb{R} \to \mathbb{R}$ is called a $(m, \alpha)-$Carathédory function if it satisfies

(i) $t \to h(t, t^{\alpha-2}x, t^{2+m-\alpha}y)$ is measurable on (0, 1) for all $(x, y) \in \mathbb{R}^2$,
(ii) $(x, y) \to h(t, t^{\alpha-2}x, t^{2+m-\alpha}y)$ is continuous for a.e. $t \in (0,1)$,
(iii) for each $r > 0$, there exists nonnegative number M_r such that $|u|, |v| \le r$ imply

$$\left|h\left(t, t^{\alpha-2}x, t^{2+m-\alpha}y\right)\right| \le M_r, \quad a.e.t \quad \in (0,1).$$

Definition 2.5 $h : (0,1) \times \mathbb{R} \times \mathbb{R} \to \mathbb{R}$ is called a (m, α)-Carathédory function if it satisfies

(i) $t \to h(t, t^{\alpha-2}x, t^{2+m-\alpha}y)$ is measurable on (0, 1) for all $(x, y) \in \mathbb{R}^2$,
(ii) $(x, y) \to h(t, t^{\alpha-2}x, t^{2+m-\alpha}y)$ is continuous for a.e. $t \in (0,1)$,
(iii) for each $r > 0$, there exists nonnegative function $\phi_r \in L^1(0,1)$ such that $|u|, |v| \le r$ imply

$$\left|h\left(t, t^{\alpha-2}x, t^{2+m-\alpha}y\right)\right| \le \phi_r(t), \quad a.e.t \quad \in (0,1).$$

Lemma 2.1 [15] *Let* $n - 1 \le \alpha < n$, $u \in C^0(0, \infty) \bigcap L^1(0, \infty)$. *Then*

$$I^{\alpha}_{0+}D^{\alpha}_{0+}u(t) = u(t) + C_1 t^{\alpha-1} + C_2 t^{\alpha-2} + \cdots + C_n t^{\alpha-n},$$

where $C_i \in R$, $i = 1, 2, \ldots n$

Choose

$$X = \left\{ x : (0,1] \to \mathbb{R} \begin{array}{l} x,\ D^{m}_{0+}x \in C^0(0,1] \text{ the following limits exist} \\ \lim_{t\to 0} t^{2-\alpha}x(t),\ \lim_{t\to 0} t^{2+m-\alpha}D^{m}_{0+}x(t) \end{array} \right\}$$

with the norm

$$||x|| = ||x||_X = \max\left\{ \sup_{t\in(0,1]} t^{2-\alpha}|x(t)|,\ \sup_{t\in(0,1]} t^{2+m-\alpha}\left|D^{m}_{0+}x(t)\right| \right\}$$

for $x \in X$. *It is easy to show that* X *is a real Banach space.*

Choose

$$Y = \left\{ y : (0,1] \to \mathbb{R} \begin{array}{l} y,\ D^{n}_{0+}y \in C^0(0,1] \text{ the following limits exist} \\ \lim_{t\to 0} t^{2-\beta}y(t),\ \lim_{t\to 0} t^{2+n-\beta}D^{n}_{0+}y(t) \end{array} \right\}$$

with the norm

$$\|y\| = \|y\|_Y = \max\left\{\sup_{t\in(0,1]} t^{2-\beta}|y(t)|,\ \sup_{t\in(0,1]} t^{2+n-\beta}|D_{0^+}^n y(t)|\right\}$$

for $y \in Y$. It is easy to show that Y is a real Banach space.

Thus, $(X\times Y, \|\cdot\|)$ is Banach space with the norm defined by

$$\|(x,y)\| = \max\{\|x\| = \|x\|_X,\ \|y\| = \|y\|_Y\} \quad \text{for } (x,y)\in X\times Y.$$

For a function $x:(0,1]\to\mathbb{R}$, a number m and a function $F:(0,1)\times\mathbb{R}^2\to\mathbb{R}$, denote $F_{m,x}(t) = F(t, x(t), D_{0^+}^m x(t))$.

Denote

$$\begin{aligned}
&\mu_1 = a\xi^{\alpha-1},\quad \upsilon_1 = 1 - a\xi^{\alpha-2},\quad \omega_1 = 1 - b\eta^{\alpha-1},\\
&\lambda_1 = 1 - b\eta^{\alpha-2},\quad \Delta = \mu_1\lambda_1 + \upsilon_1\omega_1,\\
&\mu_2 = c\xi^{\beta-1},\quad \upsilon_2 = 1 - c\xi^{\beta-2},\quad \omega_2 = 1 - d\eta^{\beta-1},\\
&\lambda_2 = 1 - d\eta^{\beta-2},\quad \nabla = \mu_2\lambda_2 + \upsilon_2\omega_2. \qquad (2)
\end{aligned}$$

Lemma 2.2 (Lemma 2.6 in [11]) *Suppose that $\Delta \neq 0$ and* **(B0)** *$h\in C^0(0,1)$ and there exist $k > -1$ and $l \le 0$ such that $2+l+k>0$ and $|h(t)| \le t^k(1-t)^l$ for all $t\in(0,1)$.*

Then $x\in X$ is a solution of problem

$$\begin{cases} D^\alpha x(t) + h(t) = 0, 0<t<1,\\ \lim_{t\to 0} t^{2-\alpha}x(t) - ax(\xi) = M,\ x(1) - bx(\eta) = N \end{cases} \qquad (3)$$

if and only if $x\in X$ satisfies

$$\begin{aligned}
x(t) =& \frac{\upsilon_1 t^{\alpha-1} + \mu_1 t^{\alpha-2}}{\Delta}N + \frac{\omega_1 t^{\alpha-2} - \lambda_1 t^{\alpha-1}}{\Delta}M\\
&- \int_0^t \frac{(t-s)^{\alpha-1}}{\Gamma(\alpha)}h(s)ds + \frac{\upsilon_1 t^{\alpha-1} + \mu_1 t^{\alpha-2}}{\Delta}\\
&\times \int_0^1 \frac{(1-s)^{\alpha-1}}{\Gamma(\alpha)}h(s)ds\\
&- \frac{b\upsilon_1 t^{\alpha-1} + b\mu_1 t^{\alpha-2}}{\Delta}\int_0^\eta \frac{(\eta-s)^{\alpha-1}}{\Gamma(\alpha)}h(s)ds\\
&+ \frac{a\lambda_1 t^{\alpha-1} - a\omega_1 t^{\alpha-2}}{\Delta}\int_0^\xi \frac{(\xi-s)^{\alpha-1}}{\Gamma(\alpha)}h(s)ds. \qquad (4)
\end{aligned}$$

Lemma 2.3 (Lemma 2.7 in [11]) *Suppose that $\nabla\neq 0$ and* **(B0)** *holds. Then $y\in Y$ is a solution of problem*

$$\begin{cases} D^\beta y(t) + h(t) = 0, 0<t<1,\\ \lim_{t\to 0} t^{2-\beta}y(t) - cy(\xi) = M,\ y(1) - dy(\eta) = N \end{cases} \qquad (5)$$

if and only if $y\in Y$ satisfies

$$\begin{aligned}
y(t) =& \frac{\upsilon_2 t^{\beta-1} + \mu_2 t^{\beta-2}}{\nabla}N + \frac{\omega_2 t^{\beta-2} - \lambda_2 t^{\beta-1}}{\nabla}M\\
&- \int_0^t \frac{(t-s)^{\beta-1}}{\Gamma(\beta)}h(s)ds + \frac{\upsilon_2 t^{\beta-1} + \mu_2 t^{\beta-2}}{\nabla}\\
&\times \int_0^1 \frac{(1-s)^{\beta-1}}{\Gamma(\beta)}h(s)ds\\
&- \frac{d\lambda_2 t^{\beta-1} + d\mu_2 t^{\beta-2}}{\nabla}\int_0^\eta \frac{(\eta-s)^{\beta-1}}{\Gamma(\beta)}h(s)ds\\
&+ \frac{c\lambda_2 t^{\beta-1} - c\omega_2 t^{\beta-2}}{\nabla}\int_0^\xi \frac{(\xi-s)^{\beta-1}}{\Gamma(\beta)}h(s)ds. \qquad (6)
\end{aligned}$$

Define the operator T on $X\times Y$, for $(x,y)\in X\times Y$, by $T(x,y)(t) = ((T_1 y)(t), (T_2 x)(t))$ with

$$\begin{aligned}
(T_1 y)(t) =& \frac{\upsilon_1 t^{\alpha-1} + \mu_1 t^{\alpha-2}}{\Delta}\int_0^1 \psi_{1n,y}(s)ds + \frac{\omega_1 t^{\alpha-2} - \lambda_1 t^{\alpha-1}}{\Delta}\\
&\times \int_0^1 \phi_{1n,y}(s)ds\\
&- \int_0^t \frac{(t-s)^{\alpha-1}}{\Gamma(\alpha)}p(s)f_{n,y}(s)ds + \frac{\upsilon_1 t^{\alpha-1} + \mu_1 t^{\alpha-2}}{\Delta}\\
&\times \int_0^1 \frac{(1-s)^{\alpha-1}}{\Gamma(\alpha)}p(s)f_{n,y}(s)ds\\
&- \frac{b\upsilon_1 t^{\alpha-1} + b\mu_1 t^{\alpha-2}}{\Delta}\int_0^\eta \frac{(\eta-s)^{\alpha-1}}{\Gamma(\alpha)}p(s)f_{n,x}(s)ds\\
&+ \frac{a\lambda_1 t^{\alpha-1} - a\omega_1 t^{\alpha-2}}{\Delta}\int_0^\xi \frac{(\xi-s)^{\alpha-1}}{\Gamma(\alpha)}p(s)f_{n,y}(s)ds
\end{aligned}$$

and

$$\begin{aligned}
(T_2 x)(t) =& \frac{\upsilon_2 t^{\beta-1} + \mu_2 t^{\beta-2}}{\nabla}\int_0^1 \psi_{2m,x}(s)ds + \frac{\omega_2 t^{\beta-2} - \lambda_2 t^{\beta-1}}{\nabla}\\
&\times \int_0^1 \phi_{2m,x}(s)ds\\
&- \int_0^t \frac{(t-s)^{\beta-1}}{\Gamma(\beta)}q(s)g_{m,x}(s)ds + \frac{\upsilon_2 t^{\beta-1} + \mu_2 t^{\beta-2}}{\nabla}\\
&\times \int_0^1 \frac{(1-s)^{\beta-1}}{\Gamma(\beta)}q(s)g_{m,x}(s)ds\\
&- \frac{d\upsilon_2 t^{\beta-1} + d\mu_2 t^{\beta-2}}{\nabla}\int_0^\eta \frac{(\eta-s)^{\beta-1}}{\Gamma(\beta)}q(s)g_{m,x}(s)ds\\
&+ \frac{c\lambda_2 t^{\beta-1} - c\omega_2 t^{\beta-2}}{\nabla}\int_0^\xi \frac{(\xi-s)^{\beta-1}}{\Gamma(\beta)}q(s)g_{m,x}(s)ds.
\end{aligned}$$

By Lemmas 2.2 and 2.3, we have that $(x,y)\in X\times Y$ is a solution of BVP(8) if and only if $(x,y)\in X\times Y$ is a fixed point of T.

Lemma 2.4 *Suppose that (i)–(iv) defined in Sect. 1 hold, $\Delta\neq 0$ and $\nabla\neq 0$. Then $T: X\times Y\to X\times Y$ is completely continuous.*

Proof We will prove that both T_1 and T_2 are completely continuous. The proof of the completeness of T_1 is divided into four steps and similarly we can prove that T_2 is completely continuous.

Step 1 Suppose that $\alpha-1<m<\alpha$. We prove that both $T_1 : Y \to X$ is well defined.

For $y \in Y$, there exits $r>0$ such that

$$\max\left\{\sup_{t\in(0,1]} t^{2-\beta}|y(t)|,\ \sup_{t\in(0,1]} t^{2+n-\beta}|D_{0^+}^n y(t)||\right\}<r.$$

Then (iii) and (iv) imply that there exists a number $M_r>0$ and $\phi_0, \psi_0 \in L^1(0,1)$ such that

$$\begin{aligned}|f(t,y(t),D_{0^+}^n y(t)| &= \left|f\left(t, t^{\beta-2}t^{2-\beta}y(t), t^{\beta-n-2}t^{2+n-\beta}D_{0^+}^n y(t)\right)\right| \le M_r,\\ |\phi_1(t,y(t),D_{0^+}^n y(t)| &\le \phi_0(t),\ |\psi_1(t,y(t),D_{0^+}^n y(t)| \le \psi_0(t)\end{aligned} \tag{7}$$

for all $t\in(0,1)$. Then

$$\begin{aligned}\left|\int_0^t \frac{(t-s)^{\alpha-1}}{\Gamma(\alpha)}p(s)f_{n,y}(s)\mathrm{d}s\right| &\le \int_0^t \frac{(t-s)^{\alpha-1}}{\Gamma(\alpha)}s^{k_1}(1-s)^{l_1}M_r\mathrm{d}s\\ &\le M_r\int_0^t \frac{(t-s)^{\alpha+l_1-1}}{\Gamma(\alpha)}s^{k_1}\mathrm{d}s\\ &= M_r t^{\alpha+k_1+l_1}\int_0^1 \frac{(1-w)^{\alpha+l_1-1}}{\Gamma(\alpha)}w^{k_1}\mathrm{d}w\\ &= \frac{\mathbf{B}(\alpha+l_1,k_1+1)M_r}{\Gamma(\alpha)}t^{\alpha+k_1+l_1},\end{aligned}$$

$$\begin{aligned}\left|\int_0^t \frac{(t-s)^{\alpha-m-1}}{\Gamma(\alpha-m)}p(s)f_{n,y}(s)\mathrm{d}s\right| &\le \int_0^t \frac{(t-s)^{\alpha-m-1}}{\Gamma(\alpha-m)}s^{k_1}(1-s)^{l_1}M_r\mathrm{d}s\\ &\le M_r\int_0^t \frac{(t-s)^{\alpha-m+l_1-1}}{\Gamma(\alpha-m)}s^{k_1}\mathrm{d}s\\ &= M_r t^{\alpha-m+k_1+l_1}\int_0^1 \frac{(1-w)^{\alpha-m+l_1-1}}{\Gamma(\alpha)}w^{k_1}\mathrm{d}w\\ &= \frac{\mathbf{B}(\alpha-m+l_1,k_1+1)M_r}{\Gamma(\alpha-m)}t^{\alpha-m+k_1+l_1}.\end{aligned}$$

On the other hand, note $D_{0^+}^m t^\mu = \frac{\Gamma(\mu+1)}{\Gamma(\mu+1-m)}t^{\mu-m}$, $\Gamma(0)=\infty$ with $\frac{1}{\Gamma(0)}=0$, we have

$$\begin{aligned}t^{2-\alpha}(T_1y)(t) =& \frac{\upsilon_1 t+\mu_1}{\Delta}\int_0^1 \psi_{1n,y}(s)\mathrm{d}s + \frac{\omega_1-\lambda_1 t}{\Delta}\int_0^1 \phi_{1n,y}(s)\mathrm{d}s\\ &- t^{2-\alpha}\int_0^t \frac{(t-s)^{\alpha-1}}{\Gamma(\alpha)}p(s)f_{n,y}(s)\mathrm{d}s + \frac{\upsilon_1 t+\mu_1}{\Delta}\\ &\times\int_0^1 \frac{(1-s)^{\alpha-1}}{\Gamma(\alpha)}p(s)f_{n,y}(s)\mathrm{d}s\\ &- \frac{b\upsilon_1 t+b\mu_1}{\Delta}\int_0^\eta \frac{(\eta-s)^{\alpha-1}}{\Gamma(\alpha)}p(s)f_{n,x}(s)\mathrm{d}s\\ &+ \frac{a\lambda_1 t-a\omega_1}{\Delta}\int_0^\xi \frac{(\xi-s)^{\alpha-1}}{\Gamma(\alpha)}p(s)f_{n,y}(s)\mathrm{d}s,\end{aligned}$$

$$\begin{aligned}&t^{2+m-\alpha}D_{0^+}^m(T_1y)(t)\\ =& \frac{\upsilon_1\frac{\Gamma(\alpha)}{\Gamma(\alpha-m)}t+\mu_1\frac{\Gamma(\alpha-1)}{\Gamma(\alpha-m-1)}}{\Delta}\int_0^1 \psi_{1n,y}(s)\mathrm{d}s\\ &+ \frac{\omega_1\frac{\Gamma(\alpha-1)}{\Gamma(\alpha-m-1)}-\lambda_1\frac{\Gamma(\alpha)}{\Gamma(\alpha-m)}t}{\Delta}\int_0^1 \phi_{1n,y}(s)\mathrm{d}s - t^{2+m-\alpha}\\ &\times\int_0^t \frac{(t-s)^{\alpha-m-1}}{\Gamma(\alpha-m)}p(s)f_{n,y}(s)\mathrm{d}s\\ &+ \frac{\upsilon_1\frac{\Gamma(\alpha)}{\Gamma(\alpha-m)}t+\mu_1\frac{\Gamma(\alpha-1)}{\Gamma(\alpha-m-1)}}{\Delta}\int_0^1 \frac{(1-s)^{\alpha-1}}{\Gamma(\alpha)}p(s)f_{n,y}(s)\mathrm{d}s\\ &- \frac{b\upsilon_1\frac{\Gamma(\alpha)}{\Gamma(\alpha-m)}t+b\mu_1\frac{\Gamma(\alpha-1)}{\Gamma(\alpha-m-1)}}{\Delta}\int_0^\eta \frac{(\eta-s)^{\alpha-1}}{\Gamma(\alpha)}p(s)f_{n,x}(s)\mathrm{d}s\\ &+ \frac{a\lambda_1\frac{\Gamma(\alpha)}{\Gamma(\alpha-m)}t-a\omega\omega_1\frac{\Gamma(\alpha-1)}{\Gamma(\alpha-m-1)}}{\Delta}\int_0^\xi \frac{(\xi-s)^{\alpha-1}}{\Gamma(\alpha)}p(s)f_{n,y}(s)\mathrm{d}s.\end{aligned}$$

It is easy to show that $T_1y\in X$. So $T_1 : Y\to X$ is well defined.

Step 2 Suppose that $\alpha-1<m<\alpha$. Prove that T_1 is continuous.

Let $\{y_i \in Y\}$ be a sequence such that $y_i \to y_0$ as $i\to+\infty$ in Y. Then there exists $r>0$ such that

$$\max\left\{\sup_{t\in(0,1]} t^{2-\beta}|y_i(t)|,\ \sup_{t\in(0,1]} t^{2+n-\beta}|D_{0^+}^n y_i(t)|\right\}\le r$$

holds for $i=0,1,2,\ldots$.

Then (iii) and (iv) imply that there exists a number $M_r>0$ and $\phi_0,\psi_0\in L^1(0,1)$ such that

$$\begin{aligned}|f(t,y_i(t),D_{0^+}^n y_i(t)| &= \left|f\left(t, t^{\beta-2}t^{2-\beta}y_i(t), t^{\beta-n-2}t^{2+n-\beta}D_{0^+}^n y_i(t)\right)\right| \le M_r,\\ |\phi_1(t,y_i(t),D_{0^+}^n y_i(t)| &\le \phi_0(t),\\ |\psi_1(t,y_i(t),D_{0^+}^n y_i(t)| &\le \psi_0(t)\end{aligned} \tag{8}$$

for all $t\in(0,1)$. By a direct computation, we get $(T_1y_i)(t)$ and $D_{0^+}^m(T_1y_i)(t)$. One sees that

$$t^{2-\alpha}\left|\int_0^t \frac{(t-s)^{\alpha-1}}{\Gamma(\alpha)}p(s)f_{n,y_i}(s)ds - \int_0^t \frac{(t-s)^{\alpha-1}}{\Gamma(\alpha)}p(s)f_{n,y_i}(s)ds\right|$$
$$\le 2t^{2-\alpha}\int_0^t \frac{(t-s)^{\alpha-1}}{\Gamma(\alpha)}s^{k_1}(1-s)^{l_1}M_r ds$$
$$= \frac{2M_r\mathbf{B}(\alpha+l_1,k_1+1)}{\Gamma(\alpha)}t^{2+k_1+l_1} \le \frac{2M_r\mathbf{B}(\alpha+l_1,k_1+1)}{\Gamma(\alpha)},$$

$$t^{2+m-\alpha}\left|\int_0^t \frac{(t-s)^{\alpha-m-1}}{\Gamma(\alpha-m)}p(s)f_{n,y_i}(s)ds - \int_0^t \frac{(t-s)^{\alpha-m-1}}{\Gamma(\alpha-m)}p(s)f_{n,y_i}(s)ds\right|$$
$$\le 2M_r t^{2+m-\alpha}\int_0^t \frac{(t-s)^{\alpha-m-1}}{\Gamma(\alpha-m)}s^{k_1}(1-s)^{l_1}ds$$
$$= 2M_r t^{2+m-\alpha}t^{\alpha-m+k_1+l_1}\int_0^1 \frac{(1-w)^{\alpha-m+l_1-1}}{\Gamma(\alpha-m)}w^{k_1}ds$$
$$\le \frac{2M_r\mathbf{B}(\alpha-m+l_1,k_1+1)}{\Gamma(\alpha-m)}t^{2+k_1+l_1}$$
$$\le \frac{2M_r\mathbf{B}(\alpha-m+l_1,k_1+1)}{\Gamma(\alpha-m)}.$$

We can show using the dominant convergence theorem that $T_1y_i \to T_1y_0$ as $i \to +\infty$. Then T_1 is continuous.

Now we prove that T_1 maps bounded sets in Y into relatively compact sets in X. Let $\Omega \subset Y$ be a bounded subset. Then there exists $r > 0$ such that

$$\max\left\{\sup_{t\in(0,1]} t^{2-\beta}|y(t)|,\ \sup_{t\in(0,1]} t^{2+n-\beta}|D^n_{0^+}y(t)|\right\} \le r$$

holds for all $y \in \Omega$. Then (iii) and (iv) imply that there exists a number $M_r > 0$ and $\phi_0, \psi_0 \in L^1(0,1)$ such that

$$|f(t,y(t),D^n_{0^+}y(t)| = \left|f(t,t^{\beta-2}t^{2-\beta}y(t), t^{\beta-n-2}t^{2+n-\beta}D^n_{0^+}y(t))\right| \le M_r,$$
$$|\phi_1(t,y(t),D^n_{0^+}y(t)| \le \phi_0(t),\quad |\psi_1(t,y(t),D^n_{0^+}y(t)| \le \psi_0(t) \tag{9}$$

for all $t \in (0,1)$.

Step 3 Suppose that $\alpha-1<m<\alpha$. Prove that $\{T_1y : y \in \Omega\}$ is a bounded set in X.

Similar to Step 1 and Step 2, we can show that

$$t^{2-\alpha}|(T_1y)(t)|$$
$$\le \frac{|1-a\xi^{\alpha-2}|+|a|\xi^{\alpha-1}}{|\Delta|}\int_0^1 \psi_0(s)ds + \frac{|1-b\eta^{\alpha-1}|+|1-b\eta^{\alpha-2}|}{|\Delta|}\int_0^1 \phi_0(s)ds$$
$$+\frac{M_r\mathbf{B}(\alpha+l_1,k_1+1)}{\Gamma(\alpha)} + \frac{|1-a\xi^{\alpha-2}|+|a|\xi^{\alpha-1}}{|\Delta|}\times\frac{2M_r\mathbf{B}(\alpha+l_1,k_1+1)}{\Gamma(\alpha)}$$
$$+\frac{|b(1-a\xi^{\alpha-2})|+|ab|\xi^{\alpha-1}}{|\Delta|}\eta^{-2-k_1-l_1}\frac{M_r\mathbf{B}(\alpha+l_1,k_1+1)}{\Gamma(\alpha)}$$
$$+\frac{|a(1-b\eta^{\alpha-2})|+|a(1-b\eta^{\alpha-1})|}{|\Delta|}\xi^{-2-k_1-l_1}\frac{M_r\mathbf{B}(\alpha+l_1,k_1+1)}{\Gamma(\alpha)},$$

$$t^{2+m-\alpha}D^m_{0^+}(T_1y)(t)$$
$$\le \frac{|1-a\xi^{\alpha-2}|\frac{\Gamma(\alpha)}{\Gamma(\alpha-m)}+|a|\xi^{\alpha-1}\frac{\Gamma(\alpha-1)}{|\Gamma(\alpha-m-1)|}}{|\Delta|}\times\int_0^1 \psi_0(s)ds$$
$$+\frac{|1-b\eta^{\alpha-1}|\frac{\Gamma(\alpha-1)}{|\Gamma(\alpha-m-1)|}+|1-b\eta^{\alpha-2}|\frac{\Gamma(\alpha)}{\Gamma(\alpha-m)}}{|\Delta|}\int_0^1 \phi_0(s)ds$$
$$+\frac{M_r\mathbf{B}(\alpha-m+l_1,k_1+1)}{\Gamma(\alpha-m)}$$
$$+\frac{|1-a\xi^{\alpha-2}|\frac{\Gamma(\alpha)}{\Gamma(\alpha-m)}+|a|\xi^{\alpha-1}\frac{\Gamma(\alpha-1)}{|\Gamma(\alpha-m-1)|}}{|\Delta|}\times\frac{M_r\mathbf{B}(\alpha-m+l_1,k_1+1)}{\Gamma(\alpha-m)}$$
$$+\frac{|b(1-a\xi^{\alpha-2})|\frac{\Gamma(\alpha)}{\Gamma(\alpha-m)}+|ab|\xi^{\alpha-1}\frac{\Gamma(\alpha-1)}{|\Gamma(\alpha-m-1)|}}{|\Delta|}\eta^{m-2-k_1-l_1}\times\frac{M_r\mathbf{B}(\alpha-m+l_1,k_1+1)}{\Gamma(\alpha-m)}$$
$$+\frac{|a(1-b\eta^{\alpha-2})|\frac{\Gamma(\alpha)}{\Gamma(\alpha-m)}+|a(1-b\eta^{\alpha-1})|\frac{\Gamma(\alpha-1)}{|\Gamma(\alpha-m-1)|}}{|\Delta|}\times\xi^{m-2-k_1-l_1}\frac{M_r\mathbf{B}(\alpha-m+l_1,k_1+1)}{\Gamma(\alpha-m)}.$$

So T_1 maps bounded sets into bounded sets in X.

Step 4 Suppose that $\alpha-1<m<\alpha$. Prove that $\{T_1y: y\in\Omega\}$ is a relatively compact set in X.

We prove first that both $\{t^{2-\alpha}(T_1y)(t): y\in\Omega\}$ and $\{t^{2+m-\alpha}D_{0^+}^m(T_1y)(t): y\in\Omega\}$ are equi-continuous on (0, 1]. By the definition of T_1, it suffices to show that both

$$\left\{t^{2-\alpha}\int_0^t(t-s)^{\alpha-1}p(s)f_{n,y}(s)\mathrm{d}s: y\in\Omega\right\} \text{ and } \left\{t^{2+m-\alpha}\int_0^t(t-s)^{\alpha-m-1}p(s)f_{n,y}(s)\mathrm{d}s: y\in\Omega\right\}$$

are equi-continuous on (0, 1] (we can prove that the other parts of $\{t^{2-\alpha}(T_1y)(t): y\in\Omega\}$ and $\{t^{2+m-\alpha}D_{0^+}^m(T_1y)(t): y\in\Omega\}$ are equi-continuous on (0, 1] similar to [1]). Then, we prove that both $\{t^{2-\alpha}(T_1y)(t): y\in\Omega\}$ and $\{t^{2+m-\alpha}D_{0^+}^n(T_1y)(t): y\in\Omega\}$ are equi-convergent as $t\to 0$. By the definition of T_1, it suffices to show that both

$$\left\{t^{2-\alpha}\int_0^t(t-s)^{\alpha-1}p(s)f_{n,y}(s)\mathrm{d}s: y\in\Omega\right\} \text{ and } \left\{t^{2+m-\alpha}\int_0^t(t-s)^{\alpha-m-1}p(s)f_{n,y}(s)\mathrm{d}s: y\in\Omega\right\}$$

are equi-convergent as $t\to 0$.

First, let $t_1,t_2\in[e,f]\subset(0,1]$ with $t_1<t_2$, $0<e<f\le 1$, and $y\in\Omega$. Then we have

$$\begin{aligned}
&\left|t_1^{2-\alpha}\int_0^{t_1}(t_1-s)^{\alpha-1}p(s)f_{n,y}(s)\mathrm{d}s-t_2^{2-\alpha}\int_0^{t_2}(t_2-s)^{\alpha-1}p(s)f_{n,y}(s)\mathrm{d}s\right|\\
&\le|t_1^{2-\alpha}-t_2^{2-\alpha}|\int_0^{t_2}(t_2-s)^{\alpha-1}|p(s)f_{n,y}(s)|\mathrm{d}s\\
&\quad+t_1^{2-\alpha}\int_{t_1}^{t_2}(t_2-s)^{\alpha-1}|p(s)f_{n,y}(s)|\mathrm{d}s\\
&\quad+t_1^{2-\alpha}\int_0^{t_1}|(t_1-s)^{\alpha-1}-(t_2-s)^{\alpha-1}||p(s)f_{n,y}(s)|\mathrm{d}s\\
&\le M_r|t_1^{2-\alpha}-t_2^{2-\alpha}|\int_0^{t_2}(t_2-s)^{\alpha-1}s^{k_1}(1-s)^{l_1}\mathrm{d}s+M_rt_1^{2-\alpha}\int_{t_1}^{t_2}(t_2-s)^{\alpha-1}s^{k_1}(1-s)^{l_1}\mathrm{d}s\\
&\quad+M_rt_1^{2-\alpha}\int_0^{t_1}|(t_1-s)^{\alpha-1}-(t_2-s)^{\alpha-1}|s^{k_1}(1-s)^{l_1}\mathrm{d}s\\
&\le M_r|t_1^{2-\alpha}-t_2^{2-\alpha}|t_2^{\alpha+k_1+l_1}\mathbf{B}(\alpha+l_1,k_1+1)+M_rt_1^{2-\alpha}t_2^{\alpha+l_1+k_1}\int_{\frac{t_1}{t_2}}^1(1-w)^{\alpha+l_1-1}w^{k_1}\mathrm{d}w\\
&\quad+M_rt_2^{2-\alpha}\int_0^1[(t_2-s)^{\alpha-1}-(t_1-s)^{\alpha-1}]s^{k_1}(1-s)^{l_1}\mathrm{d}s
\end{aligned}$$

$\to 0$ uniformly in Ω as $t_1\to t_2$ on $[e,f]$.

Second, let $t_1,t_2\in[e,f]\subset(0,1]$ with $0<e\le t_1<t_2\le f\le 1$ and $y\in\Omega$. Then we have

$$\begin{aligned}
&\left|t_1^{2+m-\alpha}\int_0^{t_1}(t_1-s)^{\alpha-m-1}p(s)f_{n,y}(s)\mathrm{d}s-t_2^{2+m-\alpha}\int_0^{t_2}(t_2-s)^{\alpha-m-1}p(s)f_{n,y}(s)\mathrm{d}s\right|\\
&\le|t_1^{2+m-\alpha}-t_2^{2+m-\alpha}|\int_0^{t_2}(t_2-s)^{\alpha-m-1}|p(s)f_{n,y}(s)|\mathrm{d}s\\
&\quad+t_1^{2+m-\alpha}\int_{t_1}^{t_2}(t_2-s)^{\alpha-m-1}|p(s)f_{n,y}(s)|\mathrm{d}s\\
&\quad+t_1^{2+m-\alpha}\int_0^{t_1}|(t_1-s)^{\alpha-m-1}-(t_2-s)^{\alpha-m-1}||p(s)f_{n,y}(s)|\mathrm{d}s\\
&\le M_r|t_1^{2+m-\alpha}-t_2^{2+m-\alpha}|\int_0^{t_2}(t_2-s)^{\alpha-m-1}s^{k_1}(1-s)^{l_1}\mathrm{d}s\\
&\quad+M_rt_1^{2+m-\alpha}\int_{t_1}^{t_2}(t_2-s)^{\alpha-m-1}s^{k_1}(1-s)^{l_1}\mathrm{d}s\\
&\quad+M_rt_1^{2+m-\alpha}\int_0^{t_1}[(t_1-s)^{\alpha-m-1}-(t_2-s)^{\alpha-m-1}]s^{k_1}(1-s)^{l_1}\mathrm{d}s\\
&\le M_r|t_1^{2+m-\alpha}-t_2^{2+m-\alpha}|t_2^{\alpha-m+k_1+l_1}\mathbf{B}(\alpha-m+l_1,k_1+1)\\
&\quad+M_rt_1^{2+m-\alpha}t_2^{\alpha-m+k_1+l_1}\int_{\frac{t_1}{t_2}}^1(1-w)^{\alpha-m-1+l_1}w^{k_1}\mathrm{d}w\\
&\quad+M_rt_1^{2+m-\alpha}\int_0^{t_1}(t_1-s)^{\alpha-m-1}s^{k_1}(t_1-s)^{l_1}\mathrm{d}s\\
&\quad-M_rt_1^{2+m-\alpha}\int_0^{t_1}(t_2-s)^{\alpha-m-1}s^{k_1}(t_2-s)^{l_1}\mathrm{d}s\\
&\le M_r|t_1^{2+m-\alpha}-t_2^{2+m-\alpha}|t_2^{\alpha-m+k_1+l_1}\mathbf{B}(\alpha-m+l_1,k_1+1)\\
&\quad+M_r\max\{e^{\alpha-m+k_1+l_1},f^{\alpha-m+k_1+l_1}\}\int_{\frac{t_1}{t_2}}^1(1-w)^{\alpha-m-1+l_1}w^{k_1}\mathrm{d}w\\
&\quad+M_rt_1^{2+m-\alpha}t_1^{\alpha-m+k_1+l_1}\int_0^1(1-w)^{\alpha-m+l_1-1}w^{k_1}\mathrm{d}w\\
&\quad-M_rt_1^{2+m-\alpha}t_2^{\alpha-m+k_1+l_1}\int_0^{\frac{t_1}{t_2}}(1-w)^{\alpha-m+l_1-1}w^{k_1}\mathrm{d}w\\
&\le M_r|t_1^{2+m-\alpha}-t_2^{2+m-\alpha}|t_2^{\alpha-m+k_1+l_1}\mathbf{B}(\alpha-m+l_1,k_1+1)\\
&\quad+M_r\max\{e^{\alpha-m+k_1+l_1},f^{\alpha-m+k_1+l_1}\}\int_{\frac{t_1}{t_2}}^1(1-w)^{\alpha-m-1+l_1}w^{k_1}\mathrm{d}w\\
&\quad+M_rt_1^{2+m-\alpha}\left(t_1^{\alpha-m+k_1+l_1}-t_2^{\alpha-m+k_1+l_1}\right)\int_0^1(1-w)^{\alpha-m+l_1-1}w^{k_1}\mathrm{d}w\\
&\quad+M_rt_1^{2+m-\alpha}t_2^{\alpha-m+k_1+l_1}\int_{\frac{t_1}{t_2}}^1(1-w)^{\alpha-m+l_1-1}w^{k_1}\mathrm{d}w
\end{aligned}$$

$\to 0$ uniformly in Ω as $t_1\to t_2$ on $[e,f]$.

Third, we have

$$\left|t^{2-\alpha}(T_1y)(t)-\left(\frac{a\xi^{\alpha-1}}{\Delta}\int_0^1\psi_{1n,y}(s)\mathrm{d}s+\frac{(1-b\eta^{\alpha-1})}{\Delta}\right.\right.$$
$$\times\int_0^1\phi_{1n,y}(s)\mathrm{d}s..$$
$$+\frac{a\xi^{\alpha-1}}{\Delta}\int_0^1\frac{(1-s)^{\alpha-1}}{\Gamma(\alpha)}p(s)f_{n,y}(s)\mathrm{d}s-\frac{ab\xi^{\alpha-1}}{\Delta}$$
$$\times\int_0^\eta\frac{(\eta-s)^{\alpha-1}}{\Gamma(\alpha)}p(s)f_{n,x}(s)\mathrm{d}s$$
$$\left.\left.-\frac{a(1-b\eta^{\alpha-1})}{\Delta}\int_0^\xi\frac{(\xi-s)^{\alpha-1}}{\Gamma(\alpha)}p(s)f_{n,y}(s)\mathrm{d}s\right)\right|$$
$$\le\frac{|1-a\xi^{\alpha-2}|t}{|\Delta|}\int_0^1\psi_0\mathrm{d}s+\frac{|1-b\eta^{\alpha-2}|t}{|\Delta|}\int_0^1\phi_0(s)\mathrm{d}s$$
$$+t^{2-\alpha}\int_0^t\frac{(t-s)^{\alpha-1}}{\Gamma(\alpha)}|p(s)f_{n,y}(s)|\mathrm{d}s+\frac{|1-a\xi^{\alpha-2}|t}{|\Delta|}$$
$$\times\int_0^1\frac{(1-s)^{\alpha-1}}{\Gamma(\alpha)}|p(s)f_{n,y}(s)|\mathrm{d}s$$
$$+\frac{|b(1-a\xi^{\alpha-2})|t}{|\Delta|}\int_0^\eta\frac{(\eta-s)^{\alpha-1}}{\Gamma(\alpha)}|p(s)f_{n,x}(s)|\mathrm{d}s$$
$$+\frac{|a(1-b\eta^{\alpha-2})|t}{|\Delta|}\int_0^\xi\frac{(\xi-s)^{\alpha-1}}{\Gamma(\alpha)}|p(s)f_{n,y}(s)|\mathrm{d}s$$
$$\le\frac{|1-a\xi^{\alpha-2}|t}{|\Delta|}\int_0^1\psi_0\mathrm{d}s+\frac{|1-b\eta^{\alpha-2}|t}{|\Delta|}\int_0^1\phi_0(s)\mathrm{d}s$$
$$+t^{2+k_1+l_1}\frac{\mathbf{B}(\alpha+l_1,k_1+1)}{\Gamma(\alpha)}M_r+\frac{|1-a\xi^{\alpha-2}|t}{|\Delta|}$$
$$\times\frac{\mathbf{B}(\alpha+l_1,k_1+1)}{\Gamma(\alpha)}M_r$$
$$+\frac{|b(1-a\xi^{\alpha-2})|t}{|\Delta|}\frac{\mathbf{B}(\alpha+l_1,k_1+1)}{\Gamma(\alpha)}M_r$$
$$+\frac{|a(1-b\eta^{\alpha-2})|t}{|\Delta|}\frac{\mathbf{B}(\alpha+l_1,k_1+1)}{\Gamma(\alpha)}M_r$$

$\to 0$ uniformly on Ω as $t\to 0$.

Fourth, we have

$$\left|t^{2+m-\alpha}D_{0^+}^m(T_1y)(t)-\left(\frac{a\xi^{\alpha-1}\frac{\Gamma(\alpha-1)}{\Gamma(\alpha-m-1)}}{\Delta}\int_0^1\psi_{1n,y}(s)\mathrm{d}s\right.\right.$$
$$+\frac{(1-b\eta^{\alpha-1})\frac{\Gamma(\alpha-1)}{\Gamma(\alpha-m-1)}}{\Delta}\int_0^1\phi_{1n,y}(s)\mathrm{d}s$$
$$+\frac{a\xi^{\alpha-1}\frac{\Gamma(\alpha-1)}{\Gamma(\alpha-m-1)}}{\Delta}$$
$$\times\int_0^1\frac{(1-s)^{\alpha-1}}{\Gamma(\alpha)}p(s)f_{n,y}(s)\mathrm{d}s$$
$$-\frac{ab\xi^{\alpha-1}\frac{\Gamma(\alpha-1)}{\Gamma(\alpha-m-1)}}{\Delta}\int_0^\eta\frac{(\eta-s)^{\alpha-1}}{\Gamma(\alpha)}p(s)f_{n,x}(s)\mathrm{d}s$$
$$\left.\left.-\frac{a(1-b\eta^{\alpha-1})\frac{\Gamma(\alpha-1)}{\Gamma(\alpha-m-1)}}{\Delta}\int_0^\xi\frac{(\xi-s)^{\alpha-1}}{\Gamma(\alpha)}p(s)f_{n,y}(s)\mathrm{d}s\right)\right|$$
$$\le\frac{|1-a\xi^{\alpha-2}|\frac{\Gamma(\alpha)}{\Gamma(\alpha-m)}t}{|\Delta|}\int_0^1\psi_0(s)\mathrm{d}s+\frac{|1-b\eta^{\alpha-2}|\frac{\Gamma(\alpha)}{\Gamma(\alpha-m)}t}{|\Delta|}$$
$$\times\int_0^1\phi_0(s)\mathrm{d}s$$
$$+t^{2+m-\alpha}\int_0^t\frac{(t-s)^{\alpha-m-1}}{\Gamma(\alpha-m)}s^{k_1}(1-s)^{l_1}M_r\mathrm{d}s$$
$$+\frac{|1-a\xi^{\alpha-2}|\frac{\Gamma(\alpha)}{\Gamma(\alpha-m)}t}{|\Delta|}\int_0^1\frac{(1-s)^{\alpha-1}}{\Gamma(\alpha)}s^{k_1}(1-s)^{l_1}M_r\mathrm{d}s$$
$$+\frac{|b(1-a\xi^{\alpha-2})|\frac{\Gamma(\alpha)}{\Gamma(\alpha-m)}t}{|\Delta|}\int_0^\eta\frac{(\eta-s)^{\alpha-1}}{\Gamma(\alpha)}s^{k_1}(1-s)^{l_1}M_r\mathrm{d}s$$
$$+\frac{|a(1-b\eta^{\alpha-2})|\frac{\Gamma(\alpha)}{\Gamma(\alpha-m)}t}{|\Delta|}\int_0^\xi\frac{(\xi-s)^{\alpha-1}}{\Gamma(\alpha)}s^{k_1}(1-s)^{l_1}M_r\mathrm{d}s$$
$$\le\frac{|1-a\xi^{\alpha-2}|\frac{\Gamma(\alpha)}{\Gamma(\alpha-m)}t}{|\Delta|}\int_0^1\psi_0(s)\mathrm{d}s+\frac{|1-b\eta^{\alpha-2}|\frac{\Gamma(\alpha)}{\Gamma(\alpha-m)}t}{|\Delta|}$$
$$\times\int_0^1\phi_0(s)\mathrm{d}s$$
$$+t^{2+k_1+l_1}\frac{\mathbf{B}(\alpha-m+l_1,k_1+1)}{\Gamma(\alpha-m)}M_r$$
$$+\frac{|1-a\xi^{\alpha-2}|\frac{\Gamma(\alpha)}{\Gamma(\alpha-m)}t}{|\Delta|}\frac{\mathbf{B}(\alpha+l_1,k_1+1)}{\Gamma(\alpha)}M_r$$
$$+\frac{|b(1-a\xi^{\alpha-2})|\frac{\Gamma(\alpha)}{\Gamma(\alpha-m)}t}{|\Delta|}\eta^{\alpha-2}\frac{\mathbf{B}(\alpha+l_1,k_1+1)}{\Gamma(\alpha)}M_r$$
$$+\frac{|a(1-b\eta^{\alpha-2})|\frac{\Gamma(\alpha)}{\Gamma(\alpha-m)}t}{|\Delta|}\xi^{\alpha-2}\frac{\mathbf{B}(\alpha+l_1,k_1+1)}{\Gamma(\alpha)}M_r$$

$\to 0$ uniformly on Ω as $t\to 0$.

Therefore, $T_1\Omega$ is relatively compact.

From above discussion, T_1 is completely continuous. The proof is completed. □

Define

$$G(t,s)=\frac{1}{\Gamma(\alpha)\Delta}\begin{cases}(\upsilon_1 t^{\alpha-1}+\mu_1 t^{\alpha-2})(1-s)^{\alpha-1}\\ +(\lambda_1 a t^{\alpha-1}-\omega_1 a t^{\alpha-2})(\xi-s)^{\alpha-1}\\ -(\upsilon_1 b t^{\alpha-1}+b\mu_1 t^{\alpha-2})(\eta-s)^{\alpha-1}\\ -(\mu_1\lambda_1+\omega_1\upsilon_1)(t-s)^{\alpha-1}, & 0\le s\le \min\{t,\xi\},\\[4pt] (\upsilon_1 t^{\alpha-1}+\mu_1 t^{\alpha-2})(1-s)^{\alpha-1}\\ -(\upsilon_1 b t^{\alpha-1}+b\mu_1 t^{\alpha-2})(\eta-s)^{\alpha-1}\\ -(\mu_1\lambda_1+\omega_1\upsilon_1)(t-s)^{\alpha-1}, & \xi<s\le \min\{t,\eta\},\\[4pt] (\upsilon_1 t^{\alpha-1}+\mu_1 t^{\alpha-2})(1-s)^{\alpha-1}\\ -(\upsilon_1 b t^{\alpha-1}+b\mu_1 t^{\alpha-2})(\eta-s)^{\alpha-1}, & \max\{t,\xi\}<s\le\eta,\\[4pt] (\upsilon_1 t^{\alpha-1}+\mu_1 t^{\alpha-2})(1-s)^{\alpha-1}\\ +(\lambda_1 a t^{\alpha-1}-\omega_1 a t^{\alpha-2})(\xi-s)^{\alpha-1}\\ -(\upsilon_1 b t^{\alpha-1}+b\mu_1 t^{\alpha-2})(\eta-s)^{\alpha-1}, & t<s\le\xi,\\[4pt] (\upsilon_1 t^{\alpha-1}+\mu_1 t^{\alpha-2})(1-s)^{\alpha-1}\\ -(\mu_1\lambda_1+\omega_1\upsilon_1)(t-s)^{\alpha-1}, & \eta<s\le t,\\[4pt] (\upsilon_1 t^{\alpha-1}+\mu_1 t^{\alpha-2})(1-s)^{\alpha-1}, & \max\{\eta,t\}<s\le 1,\end{cases}$$

and

$$H(t,s)=\frac{1}{\Gamma(\beta)\nabla}\begin{cases}(\upsilon_2 t^{\beta-1}+\mu_2 t^{\beta-2})(1-s)^{\beta-1}\\ +(\lambda_2 c t^{\beta-1}-\omega_2 c t^{\beta-2})(\xi-s)^{\beta-1}\\ -(\mu_2\lambda_2+\omega_2\upsilon_2)(t-s)^{\beta-1}, & 0\le s\le \min\{t,\xi\},\\[4pt] (\upsilon_2 t^{\beta-1}+\mu_2 t^{\beta-2})(1-s)^{\beta-1}\\ -(\upsilon_2 d t^{\beta-1}+d\mu_2 t^{\beta-2})(\eta-s)^{\beta-1}\\ -(\mu_2\lambda_2+\omega_2\upsilon_2)(t-s)^{\beta-1}, & \xi<s\le \min\{t,\eta\},\\[4pt] (\upsilon_2 t^{\beta-1}+\mu_2 t^{\beta-2})(1-s)^{\beta-1}\\ -(\upsilon_2 d t^{\beta-1}+d\mu_2 t^{\beta-2})(\eta-s)^{\beta-1}, & \max\{t,\xi\}<s\le\eta,\\[4pt] (\upsilon_2 t^{\beta-1}+\mu_2 t^{\beta-2})(1-s)^{\beta-1}\\ +(\lambda_2 c t^{\beta-1}-\omega_2 c t^{\beta-2})(\xi-s)^{\beta-1}\\ -(\upsilon_2 d t^{\beta-1}+d\mu_2 t^{\beta-2})(\eta-s)^{\beta-1}, & t<s\le\xi,\\[4pt] (\upsilon_2 t^{\beta-1}+\mu_2 t^{\beta-2})(1-s)^{\beta-1}\\ -(\mu_2\lambda_2+\omega_2\upsilon_2)(t-s)^{\beta-1}, & \eta<s\le t,\\[4pt] (\upsilon_2 t^{\beta-1}+\mu_2 t^{\beta-2})(1-s)^{\beta-1}, & \max\{\eta,t\}<s\le 1.\end{cases}$$

Now, we rewrite

$$\begin{aligned}(T(x,y))(t)&=((T_1y)(t),(T_2x)(t))\\ &=\Bigg(\frac{\upsilon_1 t^{\alpha-1}+\mu_1 t^{\alpha-2}}{\Delta}\int_0^1 \psi_{1n,y}(s)\mathrm{d}s\\ &\quad+\frac{\omega_1 t^{\alpha-2}-\lambda_1 t^{\alpha-1}}{\Delta}\int_0^1 \phi_{1n,y}(s)\mathrm{d}s\\ &\quad+\int_0^1 G(t,s)p(s)f_{n,y}(s)\mathrm{d}s,\\ &\quad\frac{\upsilon_2 t^{\beta-1}+\mu_2 t^{\beta-2}}{\nabla}\int_0^1 \psi_{2m,x}(s)\mathrm{d}s\\ &\quad+\frac{\omega_2 t^{\beta-2}-\lambda_2 t^{\beta-1}}{\nabla}\int_0^1 \phi_{2m,x}(s)\mathrm{d}s\\ &\quad+\int_0^1 H(t,s)g_{m,x}(s)\mathrm{d}s\Bigg).\end{aligned}$$

Lemma 2.5 (Lemma 2.9 in [11]) *Suppose that* $a,b,c,d\geq 0$, *and*

$$\begin{aligned}&\Delta>0,\ 0\le a<\frac{1}{\xi^{\alpha-2}(1-\xi)},\quad 0\le b<\frac{1}{\eta^{\alpha-1}},\\ &\nabla>0,\ 0\le c<\frac{1}{\xi^{\beta-2}(1-\xi)},\quad 0\le d<\frac{1}{\eta^{\beta-1}}.\end{aligned}\tag{10}$$

Then

$$G(t,s)\geq 0 \text{ for all } t,s\in(0,1),\quad H(t,s)\geq 0 \text{ for all } t,s\in(0,1).\tag{11}$$

Main results

In this section, we prove existence result on solutions of BVP(1). Let $\mu_i,\upsilon_i,\omega_i,\lambda_i(i=1,2)$ and Δ,∇ be defined by (10). For $\Phi\in L^1(0,1)$, denote $||\Phi||_1=\int_0^1|\Phi(s)|\mathrm{d}s$. The following assumption will be used in the main theorem.

A function $\Phi:[0,\infty)\times[0,\infty)\to[0,\infty)$ is called a bi-increasing function if both $u\to\Phi(u,v)$ and $v\to\Phi(u,v)$ are increasing. We now list the following assumption:

(B1) there exist $\overline{\phi}_i,\overline{\psi}_i\in L^1(0,1)(i=1,2)$ and bi-increasing functions $\Phi,\Psi,\Phi_i,\Psi_i(i=1,2)$ such that

$$\begin{aligned}&\left|f\left(t,\frac{u}{t^{2-\beta}},\frac{v}{t^{2+n-\beta}}\right)\right|\le\Phi(|u|,|v|),t\in(0,1),u,v\in\mathbb{R},\\ &\left|g\left(t,\frac{u}{t^{2-\alpha}},\frac{v}{t^{2+m-\alpha}}\right)\right|\le\Psi(|u|,|v|),t\in(0,1),u,v\in\mathbb{R},\\ &\left|\phi_1\left(t,\frac{u}{t^{2-\beta}},\frac{v}{t^{2+n-\beta}}\right)\right|\le\overline{\phi}_1(t)\Phi_1(|u|,|v|),t\in(0,1),u,v\in\mathbb{R},\\ &\left|\psi_1\left(t,\frac{u}{t^{2-\beta}},\frac{v}{t^{2+n-\beta}}\right)\right|\le\overline{\psi}_1(t)\Psi_1(|u|,|v|),t\in(0,1),u,v\in\mathbb{R},\\ &\left|\phi_2\left(t,\frac{u}{t^{2-\alpha}},\frac{v}{t^{2+m-\alpha}}\right)\right|\le\overline{\phi}_2(t)\Phi_2(|u|,|v|),t\in(0,1),u,v\in\mathbb{R},\\ &\left|\psi_2\left(t,\frac{u}{t^{2-\alpha}},\frac{v}{t^{2+m-\alpha}}\right)\right|\le\overline{\psi}_2(t)\Psi_2(|u|,|v|),t\in(0,1),u,v\in\mathbb{R}.\end{aligned}$$

For ease expression, denote

$$M_1=\left[\frac{\upsilon_1+\mu_1}{\Delta}||+\frac{\upsilon_1\frac{\Gamma(\alpha)}{\Gamma(\alpha-m)}+\mu_1\frac{\Gamma(\alpha-1)}{\Gamma(\alpha-m-1)}}{\Delta}\right]||\overline{\psi}_1||_1,$$

$$N_1=\left[\frac{\omega_1+\lambda_1}{\Delta}||+\frac{\omega_1\frac{\Gamma(\alpha-1)}{\Gamma(\alpha-m-1)}+\lambda_1\frac{\Gamma(\alpha)}{\Gamma(\alpha-m)}}{\Delta}\right]||\overline{\phi}_1||_1,$$

$$\begin{aligned}Q_1=&\left[1+\frac{\upsilon_1+\mu_1}{\Delta}+\frac{b\upsilon_1+b\mu_1}{\Delta}\eta^{\alpha+k_1+l_1}\right.\\ &\left.+\frac{a\lambda_1+a\omega_1}{\Delta}\xi^{\alpha+k_1+l_1}\right]\frac{\mathbf{B}(\alpha+l_1,k_1+1)}{\Gamma(\alpha)}\\ &+\frac{\mathbf{B}(\alpha-m+l_1,k_1+1)}{\Gamma(\alpha-m)}+\frac{\upsilon_1\frac{\Gamma(\alpha)}{\Gamma(\alpha-m)}+\mu_1\frac{\Gamma(\alpha-1)}{|\Gamma(\alpha-m-1)|}}{\Delta}\\ &\times\frac{\mathbf{B}(\alpha+l_1,k_1+1)}{\Gamma(\alpha)}\\ &+\frac{b\upsilon_1\frac{\Gamma(\alpha)}{\Gamma(\alpha-m)}+b\mu_1\frac{\Gamma(\alpha-1)}{|\Gamma(\alpha-m-1)|}}{\Delta}\eta^{\alpha+k_1+l_1}\frac{\mathbf{B}(\alpha+l_1,k_1+1)}{\Gamma(\alpha)}\\ &+\frac{a\lambda_1\frac{\Gamma(\alpha)}{\Gamma(\alpha-m)}+a\omega_1\frac{\Gamma(\alpha-1)}{|\Gamma(\alpha-m-1)|}}{\Delta}\xi^{\alpha+k_1+l_1}\frac{\mathbf{B}(\alpha+l_1,k_1+1)}{\Gamma(\alpha)},\end{aligned}$$

and

$$M_2 = \left[\frac{\upsilon_2+\mu_2}{\nabla}|| + \frac{\upsilon_2\frac{\Gamma(\beta)}{\Gamma(\beta-n)} + \mu_2\frac{\Gamma(\beta-1)}{\Gamma(\beta-n-1)}}{\nabla}\right]||\overline{\psi}_2||_1,$$

$$N_2 = \left[\frac{\omega_2+\lambda_2}{\nabla}|| + \frac{\omega_2\frac{\Gamma(\beta-1)}{\Gamma(\beta-n-1)} + \lambda_2\frac{\Gamma(\beta)}{\Gamma(\beta-n)}}{\nabla}\right]||\overline{\phi}_2||_1,$$

$$Q_2 = \left[1 + \frac{\upsilon_2+\mu_2}{\nabla} + \frac{c\upsilon_2+d\mu_2}{\nabla}\eta^{\beta+k_2+l_2} + \frac{c\lambda_2+c\omega_2}{\nabla}\xi^{\beta+k_2+l_2}\right]\frac{\mathbf{B}(\beta+l_2,k_2+1)}{\Gamma(\beta)} + \frac{\mathbf{B}(\beta-n+l_2,k_2+1)}{\Gamma(\beta-n)} + \frac{\upsilon_2\frac{\Gamma(\beta)}{\Gamma(\beta-n)}+\mu_2\frac{\Gamma(\beta-1)}{|\Gamma(\beta-n-1)|}}{\nabla} \times \frac{\mathbf{B}(\beta+l_2,k_2+1)}{\Gamma(\beta)} + \frac{d\upsilon_2\frac{\Gamma(\beta)}{\Gamma(\beta-n)}+d\mu_2\frac{\Gamma(\beta-1)}{|\Gamma(\beta-n-1)|}}{\nabla}\frac{\eta^{\beta+k_2+l_2}\mathbf{B}(\beta+l_2,k_2+1)}{\Gamma(\beta)} + \frac{c\lambda_2\frac{\Gamma(\beta)}{\Gamma(\beta-n)}+c\omega_2\frac{\Gamma(\beta-1)}{|\Gamma(\beta-n-1)|}}{\nabla}\frac{\xi^{\beta+k_2+l_2}\mathbf{B}(\beta+l_2,k_2+1)}{\Gamma(\beta)}.$$

Theorem 3.1 *Suppose that* (12) *holds,* (i)–(iv) *defined in Sect.* 1 *and (B1) hold. Then BVP(1) has at least one positive solution if*

$$\begin{aligned} &M_1\Psi_1(r_2,r_2)+N_1\Phi_1(r_2,r_2)+Q_1\Phi(r_2,r_2)\le r_1,\\ &M_2\Psi_2(r_1,r_1)+N_2\Phi_1(r_1,r_1)+Q_2\Psi(r_1,r_1)\le r_2 \end{aligned} \tag{12}$$

has a solution (r_1,r_2) *satisfying* $r_1>0, r_2>0$.

Proof From Lemmas 2.2 and 2.3, we know that (x, y) is a solution of BVP(1) if and only if (x, y) is a fixed point of T. From Lemma 2.4, $T : X\times Y \to X\times Y$ is completely continuous. By Lemma 2.5 and (i)–(iv), (x, y) is a positive solution if (x, y) is a solution of BVP(1).

To get a fixed point of T, we apply the Schauder's fixed point theorem. We should define a closed convex bounded subset Ω of E such that $T(\Omega)\subseteq\Omega$. It is easy to see that $\Omega=\{(x,y)\in E: ||x||\le r_1, ||y||\le r_2\}$ is a closed convex bounded subset Ω of E.

For $(x,y)\in\Omega$, we get $||x||\le r_1,\ \ ||y||\le r_2$. Furthermore, we have

$$\begin{aligned} \left|f\left(t,y(t),D^n_{0^+}y(t)\right)\right| &\le \Phi(t^{2-\beta}|y(t)|, t^{2+n-\beta}|D^n_{0^+}y(t)|)\\ &\le \Phi(r_2,r_2), t\in(0,1),\\ \left|g\left(t,x(t),D^m_{0^+}x(t)\right)\right| &\le \Psi(t^{2-\alpha}|x(t)|, t^{2+m-\alpha}|D^m_{0^+}x(t)|)\\ &\le \Psi(r_1,r_1), t\in(0,1),\\ \left|\phi_1\left(t,y(t),D^n_{0^+}y(t)\right)\right| &\le \overline{\phi}_1(t)\Phi_1(r_2,r_2), t\in(0,1),\\ \left|\psi_1\left(t,y(t),D^n_{0^+}y(t)\right)\right| &\le \overline{\psi}_1(t)\Psi_1(r_2,r_2), t\in(0,1),\\ \left|\phi_2\left(t,x(t),D^m_{0^+}x(t)\right)\right| &\le \overline{\phi}_2(t)\Phi_2(r_1,r_1), t\in(0,1),\\ \left|\psi_2\left(t,x(t),D^m_{0^+}x(t)\right)\right| &\le \overline{\psi}_2(t)\Psi_2(r_1,r_1), t\in(0,1). \end{aligned}$$

By the definition of T, we have

$$\begin{aligned} t^{2-\alpha}|(T_1y)(t)| &\le \frac{\upsilon_1+\mu_1}{\Delta}||\overline{\psi}_1||_1\Psi_1(r_2,r_2) + \frac{\omega_1+\lambda_1}{\Delta}||\overline{\phi}_1||_1\Phi_1(r_2,r_2)\\ &\quad + t^{2-\alpha}\int_0^t\frac{(t-s)^{\alpha-1}}{\Gamma(\alpha)}s^{k_1}(1-s)^{l_1}\mathrm{d}s\Phi(r_2,r_2)\\ &\quad + \frac{\upsilon_1+\mu_1}{\Delta}\int_0^1\frac{(1-s)^{\alpha-1}}{\Gamma(\alpha)}s^{k_1}(1-s)^{l_1}\mathrm{d}s\Phi(r_2,r_2)\\ &\quad + \frac{b\upsilon_1+b\mu_1}{\Delta}\int_0^\eta\frac{(\eta-s)^{\alpha-1}}{\Gamma(\alpha)}s^{k_1}(1-s)^{l_1}\mathrm{d}s\Phi(r_2,r_2)\\ &\quad + \frac{a\lambda_1+a\omega_1}{\Delta}\int_0^\xi\frac{(\xi-s)^{\alpha-1}}{\Gamma(\alpha)}s^{k_1}(1-s)^{l_1}\mathrm{d}s\Phi(r_2,r_2)\\ &\le \frac{\upsilon_1+\mu_1}{\Delta}||\overline{\psi}_1||_1\Psi_1(r_2,r_2) + \frac{\omega_1+\lambda_1}{\Delta}||\overline{\phi}_1||_1\Phi_1(r_2,r_2)\\ &\quad + \left[1+\frac{\upsilon_1+\mu_1}{\Delta}+\frac{b\upsilon_1+b\mu_1}{\Delta}\eta^{\alpha+k_1+l_1} + \frac{a\lambda_1+a\omega_1}{\Delta}\xi^{\alpha+k_1+l_1}\right]\frac{\mathbf{B}(\alpha+l_1,k_1+1)}{\Gamma(\alpha)}\Phi(r_2,r_2) \end{aligned}$$

and similarly we get

$$\begin{aligned} t^{2+m-\alpha}|D^m_{0^+}(T_1y)(t)| &\le \frac{\upsilon_1\frac{\Gamma(\alpha)}{\Gamma(\alpha-m)}+\mu_1\frac{\Gamma(\alpha-1)}{\Gamma(\alpha-m-1)}}{\Delta}||\overline{\psi}_1||_1\Psi_1(r_2,r_2) + \frac{\omega_1\frac{\Gamma(\alpha-1)}{\Gamma(\alpha-m-1)}+\lambda_1\frac{\Gamma(\alpha)}{\Gamma(\alpha-m)}}{\Delta}||\overline{\phi}_1||_1\Phi_1(r_2,r_2)\\ &\quad + t^{2+m-\alpha}\int_0^t\frac{(t-s)^{\alpha-m-1}}{\Gamma(\alpha-m)}s^{k_1}(1-s)^{l_1}\mathrm{d}s\Phi(r_2,r_2)\\ &\quad + \frac{\upsilon_1\frac{\Gamma(\alpha)}{\Gamma(\alpha-m)}+\mu_1\frac{\Gamma(\alpha-1)}{|\Gamma(\alpha-m-1)|}}{\Delta}\times\int_0^1\frac{(1-s)^{\alpha-1}}{\Gamma(\alpha)}s^{k_1}(1-s)^{l_1}\mathrm{d}s\Phi(r_2,r_2)\\ &\quad + \frac{b\upsilon_1\frac{\Gamma(\alpha)}{\Gamma(\alpha-m)}+b\mu_1\frac{\Gamma(\alpha-1)}{|\Gamma(\alpha-m-1)|}}{\Delta}\times\int_0^\eta\frac{(\eta-s)^{\alpha-1}}{\Gamma(\alpha)}s^{k_1}(1-s)^{l_1}\mathrm{d}s\Phi(r_2,r_2)\\ &\quad + \frac{a\lambda_1\frac{\Gamma(\alpha)}{\Gamma(\alpha-m)}+a\omega_1\frac{\Gamma(\alpha-1)}{|\Gamma(\alpha-m-1)|}}{\Delta} \end{aligned}$$

$$
\begin{aligned}
&\times \int_0^{\xi} \frac{(\xi - s)^{\alpha-1}}{\Gamma(\alpha)} s^{k_1}(1-s)^{l_1} ds \Phi(r_2, r_2) \\
&\le \frac{\upsilon_1 \frac{\Gamma(\alpha)}{\Gamma(\alpha-m)} + \mu_1 \frac{\Gamma(\alpha-1)}{\Gamma(\alpha-m-1)}}{\Delta} ||\overline{\psi}_1||_1 \Psi_1(r_2, r_2) \\
&+ \frac{\omega_1 \frac{\Gamma(\alpha-1)}{\Gamma(\alpha-m-1)} + \lambda_1 \frac{\Gamma(\alpha)}{\Gamma(\alpha-m)}}{\Delta} ||\overline{\phi}_1||_1 \Phi_1(r_2, r_2) \\
&+ \left[\frac{\mathbf{B}(\alpha - m + l_1, k_1 + 1)}{\Gamma(\alpha - m)} + \frac{\upsilon_1 \frac{\Gamma(\alpha)}{\Gamma(\alpha-m)} + \mu_1 \frac{\Gamma(\alpha-1)}{|\Gamma(\alpha-m-1)|}}{\Delta} \right. \\
&\times \frac{\mathbf{B}(\alpha + l_1, k_1 + 1)}{\Gamma(\alpha)} \\
&+ \frac{b\upsilon_1 \frac{\Gamma(\alpha)}{\Gamma(\alpha-m)} + b\mu_1 \frac{\Gamma(\alpha-1)}{|\Gamma(\alpha-m-1)|}}{\Delta} \frac{\eta^{\alpha+k_1+l_1} \mathbf{B}(\alpha + l_1, k_1 + 1)}{\Gamma(\alpha)} \\
&+ \frac{a\lambda_1 \frac{\Gamma(\alpha)}{\Gamma(\alpha-m)} + a\omega_1 \frac{\Gamma(\alpha-1)}{|\Gamma(\alpha-m-1)|}}{\Delta} \\
&\left. \frac{\xi^{\alpha+k_1+l_1} \mathbf{B}(\alpha + l_1, k_1 + 1)}{\Gamma(\alpha)} \Gamma(\alpha) \right] \Phi(r_2, r_2).
\end{aligned}
$$

We get

$$
\begin{aligned}
||T_1 y|| \le &\frac{\upsilon_1 + \mu_1}{\Delta} ||\overline{\psi}_1||_1 \Psi_1(r_2, r_2) + \frac{\omega_1 + \lambda_1}{\Delta} ||\overline{\phi}_1||_1 \Phi_1(r_2, r_2) \\
&+ \left[1 + \frac{\upsilon_1 + \mu_1}{\Delta} + \frac{b\upsilon_1 + b\mu_1}{\Delta} \eta^{\alpha+k_1+l_1} \right. \\
&\left. + \frac{a\lambda_1 + a\omega_1}{\Delta} \xi^{\alpha+k_1+l_1} \right] \frac{\mathbf{B}(\alpha + l_1, k_1 + 1)}{\Gamma(\alpha)} \Phi(r_2, r_2) \\
&+ \frac{\upsilon_1 \frac{\Gamma(\alpha)}{\Gamma(\alpha-m)} + \mu_1 \frac{\Gamma(\alpha-1)}{\Gamma(\alpha-m-1)}}{\Delta} ||\overline{\psi}_1||_1 \Psi_1(r_2, r_2) \\
&+ \frac{\omega_1 \frac{\Gamma(\alpha-1)}{\Gamma(\alpha-m-1)} + \lambda_1 \frac{\Gamma(\alpha)}{\Gamma(\alpha-m)}}{\Delta} ||\overline{\phi}_1||_1 \Phi_1(r_2, r_2) \\
&+ \left[\frac{\mathbf{B}(\alpha - m + l_1, k_1 + 1)}{\Gamma(\alpha - m)} \right. \\
&+ \frac{\upsilon_1 \frac{\Gamma(\alpha)}{\Gamma(\alpha-m)} + \mu_1 \frac{\Gamma(\alpha-1)}{|\Gamma(\alpha-m-1)|}}{\Delta} \frac{\mathbf{B}(\alpha + l_1, k_1 + 1)}{\Gamma(\alpha)} \\
&+ \frac{b\upsilon_1 \frac{\Gamma(\alpha)}{\Gamma(\alpha-m)} + b\mu_1 \frac{\Gamma(\alpha-1)}{|\Gamma(\alpha-m-1)|}}{\Delta} \frac{\eta^{\alpha+k_1+l_1} \mathbf{B}(\alpha + l_1, k_1 + 1)}{\Gamma(\alpha)} \\
&+ \frac{a\lambda_1 \frac{\Gamma(\alpha)}{\Gamma(\alpha-m)} + a\omega_1 \frac{\Gamma(\alpha-1)}{|\Gamma(\alpha-m-1)|}}{\Delta} \\
&\left. \times \frac{\xi^{\alpha+k_1+l_1} \mathbf{B}(\alpha + l_1, k_1 + 1)}{\Gamma(\alpha)} \Gamma(\alpha) \right] \Phi(r_2, r_2) \\
&= M_1 \Psi_1(r_2, r_2) + N_1 \Phi_1(r_2, r_2) + Q_1 \Phi(r_2, r_2).
\end{aligned}
$$

Similarly, we get

$$||T_2 x|| \le M_2 \Psi_2(r_1, r_1) + N_2 \Phi_1(r_1, r_1) + Q_2 \Psi(r_1, r_1).$$

Since (13) has positive solution $r_1 > 0, r_2 > 0$, we choose $\Omega = \{(x, y) \in E : ||x|| \le r_1, ||y|| \le r_2\}$. Then we get $T(\Omega) \subset \Omega$. Hence, the Schauder's fixed point theorem implies that T has a fixed point $(x, y) \in \Omega$. So (x, y) is a positive solution of BVP(1).

The proof of Theorem 3.1 is completed. □

Theorem 3.2 *Suppose*

(B2) *there exists* $\overline{\phi}_i, \overline{\psi}_i \in L^1(0,1)(i = 1, 2)$ *and non-negative constants* $M_\Phi, M_\Psi, M_{\Phi_i}, M_{\Psi_i}(i = 1, 2)$ *such that*

$$
\begin{aligned}
&\left| f\left(t, \frac{u}{t^{2-\beta}}, \frac{v}{t^{2+n-\beta}}\right) \right| \le M_\Phi, t \in (0, 1), u, v \in \mathbb{R}, \\
&\left| g\left(t, \frac{u}{t^{2-\alpha}}, \frac{v}{t^{2+m-\alpha}}\right) \right| \le M_\Psi, t \in (0, 1), u, v \in \mathbb{R}, \\
&\left| \phi_1\left(t, \frac{u}{t^{2-\beta}}, \frac{v}{t^{2+n-\beta}}\right) \right| \le \overline{\phi}_1(t) M_{\Phi_1}, t \in (0, 1), u, v \in \mathbb{R}, \\
&\left| \psi_1\left(t, \frac{u}{t^{2-\beta}}, \frac{v}{t^{2+n-\beta}}\right) \right| \le \overline{\psi}_1(t) M_{\Psi_1}, t \in (0, 1), u, v \in \mathbb{R}, \\
&\left| \phi_2\left(t, \frac{u}{t^{2-\alpha}}, \frac{v}{t^{2+m-\alpha}}\right) \right| \le \overline{\phi}_2(t) M_{\Phi_2}, t \in (0, 1), u, v \in \mathbb{R}, \\
&\left| \psi_2\left(t, \frac{u}{t^{2-\alpha}}, \frac{v}{t^{2+m-\alpha}}\right) \right| \le \overline{\psi}_2(t) M_{\Psi_2}, t \in (0, 1), u, v \in \mathbb{R}.
\end{aligned}
$$

Then BVP(1) *has at least one positive solution.*

Proof Let $M_i, N_i, Q_i (i = 1, 2)$ be defined in Theorem 3.1. Choose $\Phi(u, v) = M_\Phi$, $\Psi(u, v) = M_\Psi$, $\Phi_i(u, v) = M_{\Phi_i}$ and $\Psi_i(u, v) = M_{\Psi_i}(i = 1, 2)$. We see that (13) has positive solution

$$
\begin{aligned}
r_1 &= M_1 M_{\Psi_1} + N_1 M_{\Phi_1} + Q_1 M_\Phi, \\
r_2 &= M_2 M_{\Psi_2} + N_2 M_{\Phi_1} + Q_2 M_\Psi.
\end{aligned}
$$

The results follows from Theorem 3.1 directly. □

Numerical examples

In this section, we present two examples for the illustration of our main result (Theorems 3.1 and 3.2).

Example 4.1 We consider the following boundary value problem

$$
\begin{cases}
D_{0^+}^{\frac{19}{10}} u(t) + t^{-\frac{1}{10}}(1-t)^{-\frac{17}{20}} f(t, v(t), D_{0^+}^{\frac{39}{40}} v(t)) = 0, & t \in (0, 1), \\
D_{0^+}^{\frac{39}{20}} v(t) + t^{-\frac{1}{10}}(1-t)^{-\frac{13}{20}} g(t, u(t), D_{0^+}^{\frac{19}{20}} u(t)) = 0, & t \in (0, 1), \\
\lim_{t\to 0} t^{\frac{1}{5}} u(t) - \frac{1}{2} u(1/2) = 0, \\
u(1) - \frac{1}{2} u(3/4) = 0, \\
\lim_{t\to 0} t^{\frac{1}{9}} v(t) - \frac{1}{2} v(1/2) = 0. \\
v(1) - \frac{1}{2} v(3/4) = 0,
\end{cases}
\tag{13}
$$

Then

(i) BVP(13) has at least one positive solution if there exists a constant $H > 0$ such that

$$|f(t,t^{-\frac{1}{20}}u,t^{-\frac{41}{40}}v)| \le H, \quad t\in(0,1), u,v\in \mathbb{R},$$

$$|g(t,t^{-\frac{1}{10}}u,t^{-\frac{21}{20}}v)| \le H, \quad t\in(0,1), u,v\in \mathbb{R}.$$

(ii) BVP(13) has at least one positive solution if

$$|f(t,t^{-\frac{1}{20}}u,t^{-\frac{41}{40}}v)| \le c_1 + b_1|u|^{\epsilon_1} + a_1|v|^{\delta_1}, \; c_1,b_1,a_1 \ge 0, \; \epsilon_1,\delta_1 > 0,$$

$$|g(t,t^{-\frac{1}{10}}u,t^{-\frac{21}{20}}v)| \le c_2 + b_2|u|^{\sigma_1} + a_2|v|^{\gamma_1}, \; c_2,b_2,a_2 \ge 0, \; \sigma_1,\gamma_1 > 0$$

and one of the followings holds:

(a) $\max\{\epsilon_1,\delta_1\}\max\{\sigma_1,\gamma_1\}<1$;

(b) $\max\{\epsilon_1,\delta_1\}\max\{\sigma_1,\gamma_1\}=1$ with $(38.1089b_1)^{1/\sigma_1}$ $34.0678b_2<1$ or $38.1089b_1(34.0678b_2)^{1/\tau_1}<1$

(c) $\max\{\epsilon_1,\delta_1\}\max\{\sigma_1,\gamma_1\}>1$ for sufficiently small b_1,a_1,b_2,a_2.

Proof Corresponding to BVP(1), we have $\alpha=\frac{19}{10}, \beta=\frac{39}{20}$, $m=\frac{19}{20}$ and $n=\frac{39}{40}$, $\xi=\frac{1}{2}$, $\eta=\frac{3}{4}$, $a=b=c=d=\frac{1}{2}$ and $\phi_i(t,u,v)=\psi_i(t,u,v)\equiv 0 (i=1,2)$ and $p(t)=t^{-\frac{1}{10}}(1-t)^{-\frac{17}{20}}$, $q(t)=t^{-\frac{1}{10}}(1-t)^{-\frac{13}{20}}$.

It is easy to see that (i)–(iv) hold with $k_1=-\frac{1}{10}=k_2$, and $l_1=-\frac{17}{20}$, $l_2=-\frac{13}{20}$. One sees that $k_1>-1$, $\alpha-m+l_1>0$, $2+k_1+l_1>0$, $k_2>-1$, $\beta-n+l_2>0$, $2+k_2+l_2>0$. Hence, (i)-(iv) defined in Sect. 1 hold.

By direct calculation using Matlab7, we find that

$$\mu_1=\frac{1}{2}\left(\frac{1}{2}\right)^{\frac{9}{10}}, \quad \upsilon_1=1-\frac{1}{2}\left(\frac{1}{2}\right)^{-\frac{1}{10}},$$
$$\omega_1=1-\frac{1}{2}\left(\frac{3}{4}\right)^{\frac{9}{10}}, \quad \lambda_1=1-\frac{1}{2}\left(\frac{3}{4}\right)^{-\frac{1}{10}},$$
$$\mu_2=\frac{1}{2}\left(\frac{1}{2}\right)^{\frac{19}{20}}, \quad \upsilon_2=1-\frac{1}{2}\left(\frac{1}{2}\right)^{-\frac{1}{20}},$$
$$\omega_2=1-\frac{1}{2}\left(\frac{3}{4}\right)^{\frac{19}{20}}, \quad \lambda_2=1-\frac{1}{2}\left(\frac{3}{4}\right)^{-\frac{1}{20}}$$

and

$$\Delta=\mu_1\lambda_1+\upsilon_1\omega_1>0, \; 0\le a<\frac{1}{\xi^{\alpha-2}(1-\xi)}, \quad 0\le b<\frac{1}{\eta^{\alpha-1}},$$
$$\nabla=\mu_2\lambda_2+\upsilon_2\omega_2>0, \; 0\le c<\frac{1}{\xi^{\beta-2}(1-\xi)}, \quad 0\le d<\frac{1}{\eta^{\beta-1}}.$$

(i) By

$$|f(t,t^{-\frac{1}{20}}u,t^{-\frac{41}{40}}v)| \le H, \; t\in(0,1), u,v\in \mathbb{R},$$
$$|g(t,t^{-\frac{1}{10}}u,t^{-\frac{21}{20}}v)| \le H, \; t\in(0,1), u,v\in \mathbb{R},$$
$$\begin{aligned}\phi_1(t,t^{-\frac{1}{20}}u,t^{-\frac{41}{40}}v) &= \psi_1(t,t^{-\frac{1}{20}}u,t^{-\frac{41}{40}}v)\\ &= \phi_2(t,t^{-\frac{1}{10}}u,t^{-\frac{21}{20}}v)\\ &= \psi_2(t,t^{-\frac{1}{10}}u,t^{-\frac{21}{20}}v)=0,\end{aligned}$$

It follows from Theorem 3.2 that BVP(13) has at least one positive solution.

(ii) One sees that (B1) holds with

$$\Phi(u,v)=c_1+b_1|u|^{\epsilon_1}+a_1v^{\delta_1},$$
$$\Psi(u,v)=c_2+b_2u^{\sigma_1}+a_2v^{\gamma_1},$$
$$\Phi_i(u,v)=\Psi_i(u,v)=0 \; (i=1,2)$$

. Furthermore, we have by direct computation (use Mathlab7.0) that

$$\begin{aligned} Q_1 = {} & \left[1+\frac{\upsilon_1+\mu_1}{\Delta}+\frac{b\upsilon_1+b\mu_1}{\Delta}\eta^{\alpha+k_1+l_1} + \frac{a\lambda_1+a\omega_1}{\Delta}\xi^{\alpha+k_1+l_1}\right]\frac{\mathbf{B}(\alpha+l_1,k_1+1)}{\Gamma(\alpha)}\\ & +\frac{\mathbf{B}(\alpha-m+l_1,k_1+1)}{\Gamma(\alpha-m)}\\ & +\frac{\upsilon_1\frac{\Gamma(\alpha)}{\Gamma(\alpha-m)}+\mu_1\frac{\Gamma(\alpha-1)}{|\Gamma(\alpha-m-1)|}}{\Delta} \times\frac{\mathbf{B}(\alpha+l_1,k_1+1)}{\Gamma(\alpha)}\\ & +\frac{b\upsilon_1\frac{\Gamma(\alpha)}{\Gamma(\alpha-m)}+b\mu_1\frac{\Gamma(\alpha-1)}{|\Gamma(\alpha-m-1)|}}{\Delta} \times\frac{\eta^{\alpha+k_1+l_1}\mathbf{B}(\alpha+l_1,k_1+1)}{\Gamma(\alpha)}\\ & +\frac{a\lambda_1\frac{\Gamma(\alpha)}{\Gamma(\alpha-m)}+a\omega_1\frac{\Gamma(\alpha-1)}{|\Gamma(\alpha-m-1)|}}{\Delta} \times\frac{\xi^{\alpha+k_1+l_1}\mathbf{B}(\alpha+l_1,k_1+1)}{\Gamma(\alpha)}\\ & \simeq 67.8769\le 68,\end{aligned}$$

and

$$
\begin{aligned}
Q_2 = & \left[1 + \frac{\upsilon_2 + \mu_2}{\nabla} + \frac{c\upsilon_2 + d\mu_2}{\nabla}\eta^{\beta+k_2+l_2}\right. \\
& \left. + \frac{c\lambda_2 + c\omega_2}{\nabla}\xi^{\beta+k_2+l_2}\right]\frac{\mathbf{B}(\beta+l_2,k_2+1)}{\Gamma(\beta)} \\
& + \frac{\mathbf{B}(\beta-n+l_2,k_2+1)}{\Gamma(\beta-n)} \\
& + \frac{\upsilon_2\frac{\Gamma(\beta)}{\Gamma(\beta-n)} + \mu_2\frac{\Gamma(\beta-1)}{|\Gamma(\beta-n-1)|}}{\nabla} \\
& \times \frac{\mathbf{B}(\beta+l_2,k_2+1)}{\Gamma(\beta)} \\
& + \frac{d\upsilon_2\frac{\Gamma(\beta)}{\Gamma(\beta-n)} + d\mu_2\frac{\Gamma(\beta-1)}{|\Gamma(\beta-n-1)|}}{\nabla} \\
& \times \frac{\eta^{\beta+k_2+l_2}\mathbf{B}(\beta+l_2,k_2+1)}{\Gamma(\beta)} \\
& + \frac{c\lambda_2\frac{\Gamma(\beta)}{\Gamma(\beta-n)} + c\omega_2\frac{\Gamma(\beta-1)}{|\Gamma(\beta-n-1)|}}{\nabla} \\
& \times \frac{\xi^{\beta+k_2+l_2}\mathbf{B}(\beta+l_2,k_2+1)}{\Gamma(\beta)} \\
& \simeq 56.4653 \leq 57.
\end{aligned}
$$

One sees that inequality system (13) has positive solutions if

$$
\begin{aligned}
& 68[c_1 + b_1 r_2^{\epsilon_1} + a_1 r_2^{\delta_1}] \leq r_1, \\
& 57[c_2 + b_2 r_1^{\sigma_1} + a_2 r_1^{\gamma_1}] \leq r_2
\end{aligned}
\tag{14}
$$

has positive solutions. One sees that if

$$
\begin{aligned}
& 68[c_1 + (b_1 + a_1) r_2^{\max\{\epsilon_1,\delta_1\}}] \leq r_1, \\
& 57[c_2 + (b_2 + a_2) r_1^{\max\{\sigma_1,\gamma_1\}}] \leq r_2
\end{aligned}
\tag{14$'$}
$$

has positive solution (r_1, r_2) with $r_1 > 1$, $r_2 > 1$, then (14) has positive solution $(\max\{1, r_1\}, \max\{1, r_2\}$.

(ii)-(a) $\max\{\epsilon_1, \delta_1\}\max\{\sigma_1, \gamma_1\} < 1$. It is easy to see that (14) has positive a positive solution (r_1, r_2) with $r_1 > 0, r_2 > 0$. It follows from Theorem 3.1 that BVP(13) has at least one solution if one of the followings holds:

(ii)-(b) $\max\{\epsilon_1, \delta_1\}\max\{\sigma_1, \gamma_1\} = 1$. One sees that $(14)'$ becomes

$$68[c_1 + (b_1 + a_1) r_2] \leq r_1, \quad 57[c_2 + (b_2 + a_2) r_1] \leq r_2.$$

It is easy to see that the latest inequality system holds for sufficiently large $r_1', r_2' > 0$ if $68 \times 57(a_1 + b_1)(a_2 + b_2) < 1$. Hence (15) has positive solution $(\max\{1, r_1'\}, \max\{1, r_2'\}$. Then BVP(1) has positive solution by Theorem 3.1.

(ii)-(c) $\max\{\epsilon_1, \delta_1\}\max\{\sigma_1, \gamma_1\} > 1$. By

$$
\begin{aligned}
& \lim_{(a_1,b_1,c_1)\to(0,0,0)} Q_1[c_1 + b_1|u|^{\epsilon_1} + a_1 v^{\delta_1}] \\
& = \lim_{(a_2,b_2,c_2)\to(0,0,0)} Q_2[c_2 + b_2 u^{\sigma_1} + a_2 v^{\gamma_1}] = 0,
\end{aligned}
$$

we know that (15) has positive solution (r_1, r_2) with $r_i > 0$. Then Theorem 3.1 implies that BVP(1) has at least one positive solution if $a_1, b_1, c_1, a_2, b_2, c_2$ are sufficiently small. The proof is completed.

□

Example 4.2 We consider the following boundary value problem

$$
\begin{cases}
D_{0^+}^{\frac{19}{10}} u(t) + t^{-\frac{1}{2}}(1-t)^{-\frac{1}{5}} f(t, v(t), D_{0^+}^{\frac{39}{40}} v(t)) = 0, & t \in (0,1), \\
D_{0^+}^{\frac{39}{20}} v(t) + t^{-\frac{1}{2}}(1-t)^{\frac{1}{10}} g(t, u(t), D_{0^+}^{\frac{19}{20}} u(t)) = 0, & t \in (0,1), \\
\lim\limits_{t\to 0} t^{\frac{1}{5}} u(t) - \frac{1}{2}u(1/2) = A, \quad u(1) - \frac{1}{2}u(3/4) = B, & \\
\lim\limits_{t\to 0} t^{\frac{1}{9}} v(t) - \frac{1}{2}v(1/2) = C, \quad v(1) - \frac{1}{2}v(3/4) = D, &
\end{cases}
\tag{15}
$$

where

$$
\begin{aligned}
f(t,u,v) &= t^2 + \frac{b_1 t^{\frac{1}{20}} u^{\epsilon_1} + a_1 t^{\frac{41}{40}} v^{\delta_1}}{\sqrt{2}\sqrt{b_1^2 t^{\frac{1}{10}} u^{2\epsilon_1} + a_1^2 t^{\frac{41}{20}} v^{2\delta_1} + 1}}, {}_1, \\
& \quad b_1 \geq 0,\ \epsilon_1, \delta_1 > 0, \\
g(t,u,v) &= 4t^5 + \frac{b_2 t^{\frac{1}{10}} u^{\sigma_1} + a_2 t^{\frac{21}{20}} v^{\gamma_1}}{\sqrt{2}\sqrt{b_2^2 t^{\frac{1}{5}} u^{2\sigma_1} + a_2^2 t^{\frac{21}{10}} v^{2\gamma_1} + 1}}, \ a_2, \\
& \quad b_2 \geq 0,\ \sigma_1, \gamma_1 > 0.
\end{aligned}
$$

Then BVP(15) has at least one positive solution for sufficiently small $a_i, b_i (i = 1, 2)$.

Proof Corresponding to BVP(1), we have $\alpha = \frac{19}{10}, \beta = \frac{39}{20}$, $m = \frac{19}{20}$ and $n = \frac{39}{40}$, $a = b = c = d = \frac{1}{2}$ and $\phi_1(t,u,v) = A, \psi_1(t,u,v) = B, \phi_2(t,u,v) = C, \psi_2(t,u,v) = D$ and $p(t) = t^{-\frac{1}{2}}(1-t)^{-\frac{1}{5}}$, $q(t) = t^{-\frac{1}{2}}(1-t)^{\frac{1}{10}}$.

It is easy to see that (i)–(iv) hold with $k_1 = -\frac{1}{10} = k_2$, and $l_1 = -\frac{1}{5}$, $l_2 = -\frac{1}{10}$. One sees that $k_1 > -1$, $\alpha - m + l_1 > 0$, $2 + k_1 + l_1 > 0$, $k_2 > -1$, $\beta - n + l_2 > 0$, $2 + k_2 + l_2 > 0$. One sees $m > \alpha - 1$, $n > \beta - 1$.

Then similar to Example 4.1, we know that BVP(15) has at least one positive solution by Theorem 3.2. □

Conclusions

In this paper, we establish sufficient conditions for the existence of positive solutions of four-point integral type boundary value problems for singular fractional differential systems. We allow the nonlinearities $p(t)f(t, x, y)$ and $q(t)g(t, x, y)$ in fractional differential equations to be singular at $t = 0$. Both f and g may be super-linear and sub-linear. The analysis relies on some well known fixed point theorems. This paper contributes within the domain of fractional differential equations. The methods can be applied to solve other kinds of four-point integral type boundary value problems for singular fractional differential systems.

In [12, 22], authors studied the existence of positive solutions of two-point boundary value problems for fractional order elastic beam equations. One can discuss the following boundary value problem for nonlinear singular coupled fractional order elastic beam equations of the form

$$
\begin{cases}
D_{0^+}^\alpha u(t) = f(t, v(t), v'(t), v''(t)), \ t \in (0,1), \\
D_{0^+}^\beta v(t) = g(t. u(t), u'(t), u''(t)), \ t \in (0,1), \\
\lim_{t \to 0} t^{4-\alpha} u(t) = \lim_{t \to 0} t^{4-\alpha} u'(t) = 0, \\
u(1) = u'(1) = 0, \\
\lim_{t \to 0} t^{4-\beta} v(t) = \lim_{t \to 0} t^{4-\beta} v'(t) = 0, \\
v(1) = v'(1) = 0,
\end{cases}
\tag{16}
$$

where $3 < \alpha, \beta \le 4$, $D_{0^+}^*$ (D^* for short) is the Riemann–Liouville fractional derivative of order $*$, and $f, g : (0, 1) \times [0, \infty) \times \mathbb{R}^2 \to [0, \infty)$ is continuous. f, g depend on the lower order fractional derivatives u', v' and u'', v'' and may be singular at $t = 0$ and $t = 1$, f, g are non-Carathéodory functions.

Acknowledgements The author would like to thank the referees and the editors for their careful reading and some useful comments on improving the presentation of this paper.

References

1. Ahmad, B., Nieto, J.J.: Existence results for a coupled system of nonlinear fractional differential equations with three-point boundary conditions. Comput. Math. Appl. **58**, 1838–1843 (2009)
2. Avery, R.I., Peterson, A.C.: Three positive fixed points of nonlinear operators on ordered banach spaces. Comput. Math. Appl. **42**, 313–322 (2001)
3. Basset, A.B.: On the descent of a sphere in a vicous liquid. Q. J. Pure Appl. Math. **41**, 369–381 (1910)
4. Bai, C.Z., Fang, J.X.: The existence of a positive solution for a singular coupled system of nonlinear fractional differential equations. Appl. Math. Comput. **150**(3), 611–621 (2004)
5. Caponetto, R., Dongona, G., Fortuna, L.: Fractional Differential Systems, Modeling and control applications, World Scientific Series on Nonlinear Science, Ser. A, vol. 72. World Scientific Publishing Co. Pte. Ltd. (2010)
6. Duan, J., Temuer, C.: Solution for system of linear fractional differential equations with constant coefficients. J. Math. **29**, 599–603 (2009)
7. Gaber, M., Brikaa, M.G.: Existence results for a coupled system of nonlinear fractional differential equation with four-point boundary conditions. ISRN Math. Anal. Article ID 468346, pp 14 (2011)
8. Goodrich, C.S.: Existence of a positive solution to systems of differential equations of fractional order. Comput. Math. Appl. **62**, 1251–1268 (2011)
9. Liu, Y.: Existence and non-existence of positive solutions of BVPs for fractional order elastic beam equations with a non-Caratheodory nonlinearity. Appl. Math. Model. **38**(2), 620–640 (2014)
10. Liu, Y.: Existence of positive solutions of fractional order elastic beam equation with a non-Carathodory nonlinearity. Math. Methods Appl. Sci. **39**(6), 1311–1324 (2015)
11. Liu, Y.: New existence results for positive solutions of boundary value problems for coupled systems of multi-term fractional differential equations. Hacet. Univ. Bull. Nat. Sci. Eng. **2**(45), 391–416 (2016)
12. Liang, S., Zhang, J.: Positive solutions for boundary value problems of nonlinear frac- tional differential equations. Nonlinear Anal. **71**, 5545–5550 (2009)
13. Liu, L., Zhang, X., Wu, Y.: On existence of positive solutions of a two-point boundary value problem for a nonlinear singular semipositone system. Appl. Math. Comput. **192**, 223–232 (2007)
14. Mainardi, F.: Fraction Calculus: Some basic problems in continuum and statistical machanics. In: Carpinteri, A., Mainardi, F. (eds.) Fratals and Fractional Calculus in Continuum Machanics, pp. 291–348. Springer, Vien (1997)
15. Miller, K.S., Ross, B.: An Introduction to the Fractional Calculus and Fractional Differential Equation. Wiley, New York (1993)
16. Mamchuev, M.O.: Boundary value problem for a system of fractional partial differential equations. Partial Differ. Equ. **44**, 1737–1749 (2008)
17. Rehman, M., Khan, R.: A note on boundaryvalueproblems for a coupled system of fractionaldifferential equations. Comput. Math. Appl. **61**, 2630–2637 (2011)
18. Su, X.: Boundary value problem for a coupled system of nonlinear fractional differential equations. Appl. Math. Lett. **22**(1), 64–69 (2009)
19. Torvik, P.J., Bagley, R.L.: On the appearance of the fractional derivative in the behavior of real materials. J. Appl. Mech. **51**, 294–298 (1984)
20. Trujillo, J.J., Rivero, M., Bonilla, B.: On a Riemann–Liouville generalized Taylor's formula. J. Math. Anal. Appl. **231**, 255–265 (1999)
21. Wang, J., Xiang, H., Liu, Z.: Positive solution to nonzero boundary values problem for a coupled system of nonlinear fractional differential equations. Int. J. Differ. Equ. Article ID 186928, pp 12. (2010). doi:10.1155/2010/186928
22. Xu, X., Jiang, D., Yuan, C.: Multiple positive solutions for the boundary value problem of a nonlinear fractional differential equation. Nonlinear Anal. **71**, 4676–4688 (2009)
23. Yuan, C.: Multiple positive solutions for (n-1, 1)-type semipositone conjugate boundary value problems for coupled systems

of nonlinear fractional differential equations, E. J. Qual. Theory of Differ. Equ. **13**, 1–12 (2011)
24. Yang, A., Ge, W.: Positive solutions for boundary value problems of N-dimension nonlinear fractional differential systems, Boundary Value Problems, article ID 437453. (2008). doi:10.1155/2008/437453
25. Yuan, C., Jiang, D., O'Regan, D., Agarwal, R.P.: Existence and uniqueness of positive solutions of boundary value problems for coupled systems of singular second-order three-point non-linear differential and difference equations. Appl. Anal. **87**, 921–932 (2008)
26. Yuan, C., Jiang, D., O'regan, D., Agarwal, R.P.: Multiple positive solutions to systems of nonlinear semipositone fractional differential equations with coupled boundary conditions. Eur. J. Qual. Theory Diff. Equ. **13**, 1–17 (2012)

16

Nonlocal fractional functional differential equations with measure of noncompactness in Banach space

Junfei Cao[1] · **Qian Tong** [1] · **Xianyong Huang**[1]

Abstract In this paper, we are concerned with the following fractional functional differential equations with nonlocal initial conditions in Banach space

$$\begin{aligned} D^{\alpha}x(t) &= Ax(t) + f(t, x(t), x_t), \quad t \in [0, T], \\ x(0) &= \phi + g(x). \end{aligned}$$

By virtue of the theory of measure of noncompactness associated with Darbo's fixed point theorem, upon making some suitable assumptions, some existence results of mild solutions are obtained. Moreover the results obtained are utilized to study the existence of solutions to fractional parabolic equations as an illustrative example to show the practical usefulness of the analytical results.

Keywords Fractional functional differential equation · Nonlocal initial condition · Hausdorff measure of noncompactness · Mild solution · Darbo's fixed point theorem

Introduction

In this paper, we are concerned with the nonlocal initial value problem

$$\begin{aligned} D^{\alpha}x(t) &= Ax(t) + f(t, x(t), x_t), \quad t \in [0, T], \\ x(0) &= \phi + g(x), \end{aligned} \tag{1.1}$$

where A is the infinitesimal generator of a strongly continuous semigroup of bounded linear operators $T(t)$ in a separable Banach space X,

$$f : [0, T] \times X \times C \to X, \quad g : L^p([0, T], X) \to X,$$

are given X-valued functions. The fractional derivative is understood in the Riemann–Liouville sense. The aim of this paper is to study the existence of mild solutions for the fractional functional differential Eq. (1.1) in a separable Banach space. The technique used here is the measure of noncompactness associated with Darbo's fixed point theorem.

The fractional derivative is understood in the Riemann–Liouville sense. The origin of fractional calculus goes back to Newton and Leibnitz in the seventieth century. One observes that fractional order can be very complex in viewpoint of mathematics and they have recently proved to be valuable in various fields of science and engineering. In fact, one can find numerous applications in electrochemistry, electromagnetism, viscoelasticity, biology and hydrogeology. For example space-fractional diffusion equations have been used in groundwater hydrology to model the transport of passive tracers carried by fluid flow in a porous medium [1, 2] or to model activator–inhibitor dynamics with anomalous diffusion [3]. For details, see [4–7] and the references therein.

Differential equations of fractional order have appeared in many branches of physics and technical sciences [8, 9]. It has seen considerable development in the last decade, see [3–29] and the references therein. Recently, the existence and uniqueness problem for various fractional differential equations were considered by Ahmad [10], Bhaskarc[11], Lakshmikantham and Leela[12] et al. The nonlocal Cauchy problem was considered by Anguraj, Karthikeyan and N'Guérékata [13], and the importance of nonlocal initial

✉ Junfei Cao
jfcaomath@163.com

1 Department of Mathematics, Guangdong University of Education, Guangzhou 510310, People's Republic of China

conditions in different fields has been discussed in [6, 7] and the references therein.

The nonlocal problem (1.1) was motivated by physical problems. Indeed, the nonlocal initial condition $x(0) = \phi + g(x)$ can be applied in physics with better effect than the classical initial condition $x(0) = \phi$. For this reason, the problem (1.1) has gotten considerable attention in recent years, see [30–32] and the references therein. See also [33–35] and the references therein for recent generalizations of problem (1.1) to various kinds of differential equations.

To the best of our knowledge, the existence of mild solutions for the fractional functional differential Eq. (1.1) with nonlocal initial conditions using the theory of measure of noncompactness is a subject that has not been treated in the literature. Our purpose in this paper is to establish some results concerning the existence of mild solutions for equations that can be modeled in the form (1.1) by virtue of the theory of measure of noncompactness associated with Darbo's fixed point theorem. Upon making some appropriate assumptions, some sufficient conditions for the existence of mild solutions for the fractional functional differential Eq. (1.1) are given. It is worthwhile mentioning that the cases of $T(t)$ or f compact and of f Lipschitz are special cases of our conditions. Also we hope that the concept of measure of noncompactness considered here may be a stimulant for further investigations concerning solutions of fractional differential equations of other types.

The rest of this paper is organized as follows. In "Notations, definitions and auxiliary facts" section, we give some notations, definitions and auxiliary facts. "Main results" section contains the main results of this paper with two existence theorems. An example is given to illustrate our results in "Applications" section.

Notations, definitions and auxiliary facts

Let $(X, \|\cdot\|)$ be a real separable Banach space. Denote by $C([0,T],X)$ the space of X-valued continuous functions on $[0,T]$ and by $L^p([0,T],X)$ the space of X-valued measurable functions on $[0,T]$ with

$$\int_0^T \|x(t)\|^p \mathrm{d}t < \infty,$$

provided with norm

$$\|x\|_p = \left(\int_0^T \|x(t)\|^p \mathrm{d}t\right)^{\frac{1}{p}}.$$

Let r be a given positive real number, if $x : [-r, T] \to X$, define $x_t \in C([-r,0],X)$ by

$$x_t(\theta) = x(t+\theta), \quad \text{for} \quad -r \le \theta \le 0,$$

and denote

$$\|x\|_C = \sup\left\{\|x(t)\| : t \in [-r,0]\right\}, \quad \text{for} \quad x_t \in C([-r,0],X).$$

We need some basic definitions and properties of the fractional calculus theory which are used further in this paper.

Definition 2.1 [36] The fractional integral of order $\alpha > 0$ with the lower limit t_0 for a function f is defined as

$$I^\alpha f(t) = \frac{1}{\Gamma(\alpha)} \int_{t_0}^t (t-s)^{\alpha-1} f(s)\mathrm{d}s, \quad t > t_0, \quad \alpha > 0,$$

provided the right-hand side is pointwise defined on $[t_0, \infty)$, where f is an abstract continuous function and $\Gamma(\alpha)$ is the Gamma function [36].

Definition 2.2 [36] Riemann–Liouville derivative of order $\alpha > 0$ with the lower limit t_0 for a function $f : [t_0, \infty) \to \mathbb{R}$ can be written as

$$D_t^\alpha f(t) = \frac{1}{\Gamma(n-\alpha)} \frac{\mathrm{d}^n}{\mathrm{d}t^n} \int_{t_0}^t (t-s)^{-\alpha} f(s)\mathrm{d}s,$$
$$t > t_0, \quad n-1 < \alpha < n.$$

The first and maybe the most important property of Riemann–Liouville fractional derivative is that for $t > t_0$ and $\alpha > 0$, we have

$$D_t^\alpha (I^\alpha f(t)) = f(t),$$

which means that Riemann–Liouville fractional differentiation operator is a left inverse to the Riemann–Liouville fractional integration operator of the same order α.

Let Y be a Banach space, for bounded set $B \subset Y$, the Hausdorff's measure of noncompactness χ_Y is defined by

$$\chi_Y(B) = \inf\Big\{r > 0, B \text{ can be covered by finite number of balls with radii } r\Big\}.$$

In this paper, we denote χ by the Hausdorff's measure of noncompactness of X and denote χ_p by the Hausdorff's measure of noncompactness of $L^p([0,T],X)$. To discuss the problem in this paper, we need the following lemmas.

Lemma 2.1 [37] Let $B, C \subset Y$ be bounded, the following properties are satisfied

(1) B is precompact if and only if $\chi_Y(B) = 0$;
(2) $\chi_Y(B) = \chi_Y(\overline{B}) = \chi_Y(\mathrm{conv}B)$, where $\overline{B}$ and $\mathrm{conv}B$ mean the closure and convex hull of B, respectively;
(3) $\chi_Y(B) \le \chi_Y(C)$ when $B \subseteq C$;
(4) $\chi_Y(B+C) \le \chi_Y(B) + \chi_Y(C)$, where $B + C = \{x + y : x \in B, y \in C\}$;

(5) $\chi_Y(B \cup C) \leq \max\{\chi_Y(B), \chi_Y(C)\}$;
(6) $\chi_Y(\lambda B) = |\lambda|\chi_Y(B)$ for any $\lambda \in \mathbb{R}$;
(7) If the map $Q : D(Q) \subseteq Y \to Z$ is Lipschitz continuous with constant k, then

$$\chi_Z(QB) \leq k\chi_Y(B),$$

for any bounded subset $B \subseteq D(Q)$, where Z is a Banach space;
(8) $\chi_Y(B) = \inf\{d_Y(B, C) : C \subseteq Y \text{ be precompact}\} = \inf\{d_Y(B, C) : C \subseteq Y \text{ be finite valued}\}$, where $d_Y(B, C)$ means Hausdorff distance between B and C in Y;
(9) If $\{W_n\}_{n=1}^{\infty}$ is a decreasing sequence of bounded closed nonempty subsets of Y and

$$\lim_{n\to\infty} W_n = 0,$$

then $\cap_{n=1}^{\infty} W_n$ is nonempty and compact in Y. The map $Q : W \subseteq Y \to Y$ is said to be a χ_Y-contraction if there exists a positive constant $k < 1$ such that

$$\chi_Y(Q(C)) \leq k\chi_Y(C),$$

for any bounded closed subset $C \subseteq W$.

In 1955, Darbo [38] proved the fixed point property for α-set contraction (i.e., $\alpha(S(A)) \leq k\alpha(A)$ with $k \in [0, 1]$) on a closed, bounded and convex subset of Banach spaces. Since then many interesting works have appeared. For example, in 1972, Sadovskii [39] proved the fixed point property for condensing functions (i.e., $\alpha(S(A)) < \alpha(A)$ with $\alpha(A) \neq 0$) on closed, bounded and convex subset of Banach spaces. It should be noted that any α-set contraction is a condensing function, but the converse is not true (see [40]). In 2007, Hajji and Hanebaly [41] proved the existence of a common fixed point for commuting mappings satisfying

$$\alpha(S(A)) \leq k \sup_{i\in I}(\alpha(T_i(A))), \quad \alpha(S(A)) < \sup_{i\in I}(\alpha(T_i(A)), \alpha(A)),$$

where α is the measure of noncompactness on a closed, bounded and convex subset Ω of a locally convex space X, T_i and S are continuous functions from Ω to Ω with T_i, and in addition, are affine or linear. Furthermore, for every $i \in I$, T_i are equal to the identity function, moreover the obtain in particular Darbo's (see [38]) as well as Sadovskii's (see [39]) fixed point theorems, which are used to study the existence of solutions of one equation. Recently, Hajji [42] present common fixed point theorems for commuting operators which generalize Darbo's and Sadovskii's fixed point theorems, furthermore, as examples and applications, they study the existence of common solutions of equations in Banach spaces using measure of noncompactness. Our purpose in this paper is to establish some results concerning the existence of mild solutions for equations that can be modeled in the form (1.1) by virtue of the theory of measure of noncompactness associated with Darbo's fixed point theorem.

Lemma 2.2 ([37], Darbo–Sadovskii) If $W \subseteq Y$ is bounded closed and convex, the continuous map $Q : W \to W$ is a χ_Y-contraction, then Q has at least one fixed point in W.

We call $B \subset L([0, T], X)$ uniformly integrable if there exists $\eta \in L([0, T], \mathbb{R}^+)$ such that

$$\|u(s)\| \leq \eta(s), \quad \text{for all } u \in B \text{ and a.e. } s \in [0, T].$$

Lemma 2.3 [43] If $\{u_n\}_{n=1}^{\infty} \in L([0, T], X)$ is uniformly integrable, then $t \to \chi\left(\{u_n(t)\}_{n=1}^{\infty}\right)$ is measurable and

$$\chi\left(\left\{\int_0^t u_n(s)\mathrm{d}s\right\}_{n=1}^{\infty}\right) \leq \int_0^t \chi\left(\{u_n(s)\}_{n=1}^{\infty}\right)\mathrm{d}s.$$

Lemma 2.4 [44] Let $B \subset C([0, T], X)$ be bounded and equicontinuous on $[0, T]$. Then

$$\chi_p = \left(\int_0^T \chi^p(B(t))\mathrm{d}t\right)^{\frac{1}{p}},$$

where

$$B(t) = \{u(t) : u \in B\} \subset X.$$

A C_0 semigroup $T(t)$ is said to be equicontinuous if

$$t \to \{T(t)x : x \in B\}$$

is equicontinuous for all bounded set B in X and $t > 0$. It is known that the analytic semigroup is equicontinuous.

The following lemma is obvious.

Lemma 2.5 If the semigroup $T(t)$ is equicontinuous, $\eta \in L([0, T], \mathbb{R}^+)$, then the set

$$\left\{\frac{1}{\Gamma(\alpha)}\int_0^t (t-s)^{\alpha-1}T(t-s)u(s)\mathrm{d}s : \|u(s)\| \leq \eta(s)\right\},$$

for a.e. $s \in [0, T]$, is equicontinuous for all $t \in [0, T]$.

Proof Note that,

$$\left\|\frac{1}{\Gamma(\alpha)}\int_0^{t+h} (t+h-s)^{\alpha-1}T(t+h-s)u(s)\mathrm{d}s - \frac{1}{\Gamma(q)}\int_0^t (t-s)^{\alpha-1}T(t-s)u(s)\mathrm{d}s\right\|$$
$$\leq \frac{1}{\Gamma(\alpha)}\left\|\int_0^t \left[(t+h-s)^{\alpha-1} - (t-s)^{\alpha-1}\right] T(t+h-s)u(s)\mathrm{d}s\right\|$$
$$+ \frac{1}{\Gamma(\alpha)}\left\|\int_t^{t+h} (t+h-s)^{\alpha-1}T(t+h-s)u(s)\right\|\mathrm{d}s$$

$$+\frac{1}{\Gamma(\alpha)}\left\|\int_0^t (t-s)^{\alpha-1}[T(t+h-s)-T(t-s)]u(s)\right\|\mathrm{d}s$$
$$=I+II+III, \tag{2.1}$$

where

$$I=\frac{1}{\Gamma(\alpha)}\left\|\int_0^t\left[(t+h-s)^{\alpha-1}-(t-s)^{\alpha-1}\right]\right.$$
$$T(t+h-s)u(s)\mathrm{d}s\|$$
$$II=\frac{1}{\Gamma(\alpha)}\left\|\int_t^{t+h}(t+h-s)^{\alpha-1}T(t+h-s)u(s)\right\|\mathrm{d}s$$
$$III=\frac{1}{\Gamma(\alpha)}\left\|\int_0^t (t-s)^{\alpha-1}[T(t+h-s)-T(t-s)]u(s)\right\|\mathrm{d}s.$$

Estimating the terms on the right-hand side of (2.1) yields

$$I\leq\frac{M}{\Gamma(\alpha)}\int_0^t|(t+h-s)^{\alpha-1}-(t-s)^{\alpha-1}|\eta(s)\mathrm{d}s$$
$$\leq\frac{M}{\Gamma(\alpha)}\int_0^{t-\varepsilon}[(t-s)^{\alpha-1}-(t+h-s)^{\alpha-1}]\eta(s)\mathrm{d}s$$
$$+\frac{M}{\Gamma(q)}\int_{t-\varepsilon}^t(t-s)^{\alpha-1}\eta(s)\mathrm{d}s$$
$$=I'+II',$$

where

$$I'=\frac{M}{\Gamma(\alpha)}\int_0^{t-\varepsilon}[(t-s)^{\alpha-1}-(t+h-s)^{\alpha-1}]\eta(s)\mathrm{d}s$$
$$II'=\frac{M}{\Gamma(q)}\int_{t-\varepsilon}^t(t-s)^{\alpha-1}\eta(s)\mathrm{d}s,$$

with

$$M=\sup\{\|T(t)\|:t\in[0,T]\}.$$

It follows from the assumption of $\eta(s)$ that $I'\to 0$ as $h\to 0$. Using Hölder inequality, one obtains $II'\to 0$ as $h\to 0$ and $\varepsilon\to 0$.

For II, one has

$$II\leq\frac{1}{\Gamma(\alpha)}\int_t^{t+h}(t+h-s)^{\alpha-1}\|T(t+h-s)u(s)\|\mathrm{d}s$$
$$\leq\frac{M}{\Gamma(\alpha)}\int_t^{t+h}(t+h-s)^{\alpha-1}\eta(s)\mathrm{d}s\to 0 \text{ as } h\to 0.$$

As to III, one gets

$$III\leq\frac{1}{\Gamma(\alpha)}\left\|\int_0^{t-\varepsilon}(t-s)^{\alpha-1}[T(t+h-s)-T(t-s)]u(s)\right\|\mathrm{d}s$$
$$+\frac{1}{\Gamma(\alpha)}\left\|\int_{t-\varepsilon}^t(t-s)^{\alpha-1}[T(t+h-s)-T(t-s)]u(s)\right\|\mathrm{d}s$$
$$\leq\frac{1}{\Gamma(\alpha)}\int_0^{t-\varepsilon}(t-s)^{\alpha-1}\left\|T\left(\frac{h}{2}+\frac{t+h-s}{2}\right)-T\left(\frac{t-s}{2}\right)\right\|$$
$$\left\|T\left(\frac{t-s}{2}\right)u(s)\right\|\mathrm{d}s$$
$$+\frac{2M}{\Gamma(\alpha)}\int_{t-\varepsilon}^t(t-s)^{\alpha-1}\eta(s)\mathrm{d}s$$
$$\leq\frac{M}{\Gamma(\alpha)}\int_0^{t-\varepsilon}(t-s)^{\alpha-1}\left\|T\left(\frac{h}{2}+\frac{t+h-s}{2}\right)\right.$$
$$\left.-T\left(\frac{t-s}{2}\right)\right\|\eta(s)\mathrm{d}s+\frac{2M}{\Gamma(\alpha)}\int_{t-\varepsilon}^t(t-s)^{\alpha-1}\eta(s)\mathrm{d}s$$
$$=I''+II'',$$

where

$$I''=\frac{M}{\Gamma(\alpha)}\int_0^{t-\varepsilon}(t-s)^{\alpha-1}\left\|T\left(\frac{h}{2}+\frac{t+h-s}{2}\right)\right.$$
$$\left.-T\left(\frac{t-s}{2}\right)\right\|\eta(s)\mathrm{d}s$$
$$II''=\frac{2M}{\Gamma(\alpha)}\int_{t-\varepsilon}^t(t-s)^{\alpha-1}\eta(s)\mathrm{d}s.$$

Using the assumption that $T(t)$ is equicontinuous in X, integrating with $s\to\eta(s)\in L([0,T],\mathbb{R}^+)$, one sees that $I''\to 0$ as $h\to 0$. From the assumption of $\eta(s)$ and Hölder inequality, it is easy to see that $II''\to 0$ as $h\to 0$ and $\varepsilon\to 0$. Therefore, the family of functions

$$\left\{\frac{1}{\Gamma(\alpha)}\int_0^{\cdot}(\cdot-s)^{\alpha-1}T(\cdot-s)u(s)\mathrm{d}s:\|u(s)\|\leq\eta(s)\right\},$$

is equicontinuous. □

Main results

In this section, we use the measure of noncompactness of $L^p([0,T],X)$ to consider the following functional differential equations of fractional order $0<\alpha>1$ when g is continuous in the norm of $L^p([0,T],X)$

$$D^\alpha x(t)=Ax(t)+f(t,x(t),x_t),\quad t\in[0,T],\ x(0)=\phi+g(x). \tag{3.1}$$

Eq. (3.1) will be considered under the following assumptions:

(H₁) The C_0-semigroup $\{T(t)\}_{t\geq 0}$ generated by A is equicontinuous;

(H₂)

(1) $f:[0,T]\times X\times C([-r,0],X)\to X$ satisfies the Carathéodory-type condition, i.e., $f(\cdot,x,\varphi):[0,T]\to X$

is measurable for all $(t,x,\varphi)\in[0,T]\times X\times C([-r,0],X)$ and $f(t,\cdot):X\times C([-r,0],X)\to X$ is continuous for a.e. $t\in[0,T]$;

(2) there exists $d_2,e_2\in L^p([0,T],\mathbb{R}^+)$ such that for all $(t,x,\varphi)\in[0,T]\times X\times C([-r,0],X)$

$$\|f(t,x,\varphi)\|\le d_2(t)(\|x\|+\|\varphi\|_C)+e_2(t);$$

(3) there exists $c_2\in L^q([0,T],\mathbb{R}^+)$ such that for a.e. $t,s\in[0,T]$ and any bounded subset $D_1\subseteq X$, $D_2\subseteq C([-r,0],X)$

$$\chi\Big(T(t)f(s,D_1,D_2)\Big)\le c_2(t)\Big(\chi(D_1)+\sup_{\theta\in[-r,0]}\chi(D_2(\theta))\Big),$$

where

$$D_2(\theta)=\{v(\theta):v\in D_2\}\ \text{ and }\ \frac{1}{p}+\frac{1}{q}=1;$$

(H$_3$)

(1) The function $g:L^p([0,T],X)\to X$ is continuous;

(2) there exist positive constants d_1,e_1 such that

$$\|g(x)\|\le d_1\|x\|_p+e_1,\ \text{ for any } x\in L^p([0,T],X);$$

(3) there exists a positive constant c_1 such that for any $B\subset C([0,T],X)$ which is bounded and equicontinuous on [0,T],

$$\chi\Big(T(t)g(B)\Big)\le c_1\chi_p(B),\ \text{ for any a.e. } t\in[0,T];$$

(H$_4$)

$$Md_1T^{\frac{1}{p}}+\frac{4MT^{\alpha}\|d_2\|_p}{\Gamma(\alpha)}\left(\frac{p-1}{p\alpha-1}\right)^{\frac{p-1}{p}}<1.$$

Definition 3.1 A continuous function $x:[-r,T]\to X$ satisfying the integral equation

$$x(t)=T(t)(x(0)-g(x))+\frac{1}{\Gamma(\alpha)}\int_0^t(t-\eta)^{\alpha-1}T(t-\eta)f(\eta,x(\eta),x_\eta)\mathrm{d}\eta,$$

is called a mild solution for Eq.(3.1).

Now, we are prepared to state and prove our main theorems of this section.

Theorem 3.1 Let (H_1)–(H_4))be satisfied. Then Eq.(3.1) has at least one mild solution whenever

$$T^{\frac{1}{p}}\left(c_1+\frac{2\|c_2\|_qT^{\alpha-\frac{1}{p}}}{\Gamma(\alpha)}\left(\frac{p-1}{p\alpha-1}\right)^{\frac{p-1}{p}}\right)<1.$$

Proof For each $k\in\mathbb{N}$, denote by

$$B_k=B_k(L^p([-r,T],X))=\Big\{x\in L^p([-r,T],X):\|x(s)\|\le k,s\in[-r,T]\Big\}.$$

Obviously $B_k\subset L^p([-r,T],X)$ is uniformly integrable, closed and convex. For each $x\in B_k$, the restriction of x on $[0,T]$ denoted by $x|_{[0,T]}$ is an element of $B_k(L^p([0,T],X))$. For simplicity, we also write $g(x|_{[0,T]})$ as $g(x)$.

Define $F:L^p([-r,T],X)\to L^p([-r,T],X)$ by $F=F_1+F_2$, where

$$(F_1x)(t)=\begin{cases}\phi(t), & t\in[-r,0],\\ T(t)(x(0)-g(x)), & t\in[0,T],\end{cases}$$

$$(F_2x)(t)=\begin{cases}0, & t\in[-r,0],\\ \dfrac{1}{\Gamma(\alpha)}\displaystyle\int_0^t(t-\eta)^{\alpha-1}T(t-\eta)f(\eta,x(\eta),x_\eta)\mathrm{d}\eta, & t\in[0,T].\end{cases}$$

First, we show that F is well defined.

If $t\in[-r,0]$, then

$$\|Fx(t)\|\le\|\phi\|_C,$$

and if $t\in[0,T]$, one has

$$\begin{aligned}
&\|Fx(t)\|\le\|F_1x(t)+F_2x(t)\|\\
&\le M(\|\phi(0)\|+\|g(x(0))\|)+\frac{1}{\Gamma(\alpha)}\int_0^t\|(t-\eta)^{\alpha-1}T(t-\eta)f(\eta,x(\eta),x_\eta)\|\mathrm{d}\eta\\
&\le M\Big[\|\phi(0)\|+d_1\|x\|_p+e_1+\frac{1}{\Gamma(\alpha)}\int_0^t(t-\eta)^{\alpha-1}\big[d_2(\eta)(\|x\|+\|x\|_C)+e_2(\eta)\big]\mathrm{d}\eta\Big]\\
&\le M(\|\phi(0)\|+d_1\|x\|_p+e_1)+\frac{M}{\Gamma(\alpha)}\left(\int_0^t(t-\eta)^{\frac{(\alpha-1)p}{p-1}}\mathrm{d}s\right)^{\frac{p-1}{p}}\left(\int_0^t(d_2(\eta)(\|x\|+\|x\|_C\|)+e_2(\eta))^p\mathrm{d}\eta\right)^{\frac{1}{p}}\\
&\le M(\|\phi(0)\|+d_1\|x\|_p+e_1)+\frac{M}{\Gamma(\alpha)}\left(\frac{p-1}{p\alpha-1}\right)^{\frac{p-1}{p}}T^{\alpha-\frac{1}{p}}\left(\int_0^t(d_2(\eta)(\|x\|+\|x\|_C\|)+e_2(\eta))^p\mathrm{d}\eta\right)^{\frac{1}{p}}\\
&\le M(\|\phi(0)\|+d_1T^{\frac{1}{p}}k+e_1)+\frac{2M}{\Gamma(\alpha)}\left(\frac{p-1}{p\alpha-1}\right)^{\frac{p-1}{p}}T^{\alpha-\frac{1}{p}}\Big[2\|d_2\|_pT^{\frac{1}{p}}k+\|e_2\|_p\Big].
\end{aligned}$$

Thus, one has

$$\|Fx(t)\| \leq \max\Big\{\|\phi\|_C,\ M(\|\phi(0)\| + d_1 T^{\frac{1}{p}}k + e_1) + \frac{2MT^{\alpha-\frac{1}{p}}}{\Gamma(\alpha)}\left(\frac{p-1}{p\alpha-1}\right)^{\frac{p-1}{p}}\left[2\|d_2\|_p T^{\frac{1}{p}}k + \|e_2\|_p\right]\Big\}.$$

Thus, we conclude that Fx exists.

Second, we show that there is a $k \in \mathbb{N}$ such that $F(B_k) \subseteq B_k$.

Suppose contrary that for each $k \in \mathbb{N}$ there is $x^k \in B_k$ and $t^k \in [-r, T]$ such that

$$\|Fx(t^k)\| > k.$$

If $t^k \in [-r, 0]$, then

$$k \leq \|Fx(t^k)\| \leq \|\phi(t^k) + g(x(t^k))\| \leq \|\phi\|_C + d_1 T^{\frac{1}{p}}k + e_1,$$

and if $t^k \in [0, T]$, one has

$$\begin{aligned} k &\leq \|Fx(t^k)\| \leq \|F_1x(t^k) + F_2x(t^k)\| \\ &\leq M(\|\phi(0)\| + \|g(x(0))\|) \\ &\quad + \frac{1}{\Gamma(\alpha)}\int_0^{t^k} \|(t^k-\eta)^{\alpha-1}Q(t^k-\eta)f(\eta, x(\eta), x_\eta)\|\mathrm{d}\eta \\ &\leq M\Big[\|\phi(0)\| + d_1\|x\|_p + e_1 \\ &\quad + \frac{1}{\Gamma(\alpha)}\int_0^t (t-\eta)^{\alpha-1}\left[d_2(\eta)\left(\|x\| + \|x\|_C\right) + e_2(\eta)\right]\mathrm{d}\eta\Big] \\ &\leq M(\|\phi(0)\| + d_1 T^{\frac{1}{p}}k + e_1) + \frac{2M}{\Gamma(\alpha)}\left(\frac{p-1}{p\alpha-1}\right)^{\frac{p-1}{p}} \\ &\quad T^{\alpha-\frac{1}{p}}\left[\|d_2\|_p T^{\frac{1}{p}}k + \|e_2\|_p\right]. \end{aligned} \tag{3.2}$$

Divided by k on both sides of (3.2), one has

$$1 \leq Md_1T^{\frac{1}{p}} + \frac{2MT^{\alpha}\|d_2\|_p}{\Gamma(\alpha)}\left(\frac{p-1}{p\alpha-1}\right)^{\frac{p-1}{p}},$$

which contradicts the hypotheses (H_4). Therefore, there is a $k \in \mathbb{N}$ such that $F(B_k) \subseteq B_k$.

From now on, we will restrict F on such B_k.

Third, we will verify that F is a χ_C-contraction.

To this end, from the hypothesises (H_2) (1) and (3), one can prove that F is continuous by the continuity of g and of the operator f. The hypothesis (H_1) and Lemma 2.5 imply that $FB_k \subset C([0,T], X)$ is bounded and equicontinuous on [0, T], so is conv(FB_k). As X is separable, from Lemma 2.1 and Lemma 2.3–2.5 for any $B \subset \text{conv}(FB_k)$, one has

$$\begin{aligned} \chi(FB(t)) &\leq \chi(T(t)(x_0 - g(B))) \\ &\quad + \chi\left(\frac{1}{\Gamma(\alpha)}\int_0^t (t-\eta)^{\alpha-1}T(t-\eta)f(\eta, B(\eta), B_\eta)\right) \\ &\leq \chi(T(t)(x_0 - g(B))) \\ &\quad + \frac{1}{\Gamma(\alpha)}\int_0^t \chi((t-\eta)^{\alpha-1}T(t-\eta)f(\eta, B(\eta), B_\eta)\mathrm{d}\eta \\ &\leq \chi(T(t)(x_0 - g(B))) + \frac{1}{\Gamma(\alpha)} \\ &\quad \int_0^t (t-\eta)^{\alpha-1}\chi(T(t-\eta)f(\eta, B(\eta), B_\eta)\mathrm{d}\eta \\ &\leq c_1\chi_p(B) + \frac{2}{\Gamma(\alpha)}\int_0^t (t-\eta)^{\alpha-1}c_2(\eta)\chi(B)\mathrm{d}\eta \\ &\leq \left(c_1 + \frac{2\|c_2\|_q T^{\alpha-\frac{1}{p}}}{\Gamma(\alpha)}\left(\frac{p-1}{p\alpha-1}\right)^{\frac{p-1}{p}}\right)\chi_p(B), \end{aligned}$$

for a.e. $t \in [0, T]$, where

$$B(t) = \{x(t) : x \in B\} \subseteq X,\ B_t = \{x_t : x \in B\} \subseteq C([-r, 0], X).$$

By Lemma 2.3, this implies that

$$\chi_p(FB) \leq T^{\frac{1}{p}}\left(c_1 + \frac{2\|c_2\|_q T^{\alpha-\frac{1}{p}}}{\Gamma(\alpha)}\left(\frac{p-1}{p\alpha-1}\right)^{\frac{p-1}{p}}\right)\chi_p(B). \tag{3.3}$$

Note that, by Lemma 2.4, the inequality (3.3) may not remain valid in the case of $B \subset B_k$ as B_k is not equicontinuous on [0,T]. So one must look for another closed convex and bounded subset of $L^p([0,T], X)$ such that F is a χ_p-contraction on it.

Let

$$U = L^p - \text{conv}(FB_k),$$

where $L^p - \text{conv}$ means closure of convex hull in $L^p([0,T], X)$. Then

$$FU \subset U \quad \text{as} \quad FB_k \subset B_k,$$

and B_k is closed and convex in $L^p([0,T], X)$. For any closed subset $V \subset U$, let

$$B = V \cap \text{conv}(FB_k).$$

Then

$$V = L^p - \text{cl}(B),$$

where $L^p - \text{cl}$ means closure in $L^p([0,T], X)$. Furthermore

$$FV \subset L^p - \text{cl}(FB),$$

as F is continuous on $L^p([0,T], X)$. By (3.3) this implies that

$$\begin{aligned} \chi_p(FV) &\leq \chi_p(L^p - \text{cl}(FB)) = \chi_p(FB) \\ &\leq T^{\frac{1}{p}}\left(c_1 + \frac{2\|c_2\|_q T^{\alpha-\frac{1}{p}}}{\Gamma(\alpha)}\left(\frac{p-1}{p\alpha-1}\right)^{\frac{p-1}{p}}\right)\chi_p(B) \end{aligned}$$

$$\leq T^{\frac{1}{p}}\left(c_1+\frac{2\|c_2\|_q T^{\alpha-\frac{1}{p}}}{\Gamma(\alpha)}\left(\frac{p-1}{p\alpha-1}\right)^{\frac{p-1}{p}}\right)\chi_p(V).$$

Since

$$T^{\frac{1}{p}}\left(c_1+\frac{2\|c_2\|_q T^{\alpha-\frac{1}{p}}}{\Gamma(\alpha)}\left(\frac{p-1}{p\alpha-1}\right)^{\frac{p-1}{p}}\right)<1,$$

so $F: U \to U$ is a continuous χ_p-contraction. By Darbo–Sadovskii's fixed point theorem, there is a fixed point x of F on B_k, which is a mild solution of the Eq. (3.1). This completes the proof due to Lemma 2.2.

Remark 3.1 Clearly the conclusion of Theorem 3.1 remains valid if the hypotheses (H_2) (3) and (H_4) (3) are replaced by the following (H_2) (3') and (H_4) (3'), respectively: (H_2) (3') There exists $c_2 \in L^q([0,T],\mathbb{R}^+)$ such that

$$\chi(f(t,D_1,D_2)) \leq \frac{c_2(t)}{M}\left(\chi(D_1)+\sup_{\theta\in[-r,0]}\chi(D_2(\theta))\right),$$

for a.e. [0,*T*] and any bounded subset $D_1 \subseteq X$, $D_2 \subseteq C([-r,0],X)$;

(H_4) (3') There exists a positive constant c_1 such that for any $B \subset C([0,T],X)$ which is bounded and equicontinuous on [0,*T*],

$$\chi(g(B)) \leq c_1\chi_p(B)/M, \text{ for any a. e. } t\in[0,T].$$

Remark 3.2 The hypothesis (H_2) (3') (H_4) (3') holds if $T(t)$ is compact or f (g) is the sum of compact and Lipschitz functions with constant $c_2(s) = M$ ($c_1 = M$).

If X is a Hilbert space, and ϕ is a proper, convex and lower semicontinuous function from X into $(-\infty,+\infty)$, then its subdifferential $\partial\Phi$ is m-accretive. Let $A = \partial\Phi$ then A generates an equicontinuous nonlinear contraction semigroup (cf. [45, 46]). From above we can get the following existence result.

Corollary 3.1 If X is a separable Hilbert space, the hypotheses (H_2)–(H_4) are true, and $A = \partial\Phi$ with ϕ is proper, convex and lower semicontinuous from X into $(-\infty,+\infty)$. Then the nonlocal Eq. (3.1) has at least one integral solution provided that

$$T^{\frac{1}{p}}\left(c_1+\frac{2\|c_2\|_q T^{\alpha-\frac{1}{p}}}{\Gamma(\alpha)}\left(\frac{p-1}{p\alpha-1}\right)^{\frac{p-1}{p}}\right)<1.$$

Let us now formulate an existence result when g is uniformly bounded.

Theorem 3.2 Assume that (H_1), (H_3) and (H_4) are true with $d_1 = 0$, and f satisfies (H_2) (2), (3). In addition, suppose that there exists $\theta \in L^p([0,T],\mathbb{R}^+)$, an increasing function $\Omega:\mathbb{R}^+\to\mathbb{R}^+$ such that

$$\|f(t,x,\varphi)\| \leq \theta(t)\Omega(\|x\|+\|\varphi\|_C), \tag{3.4}$$

for $(t,x,\varphi)\in[0,T]\times X\times C([-r,0],X)$. Then Eq.(3.1) has at least one mild solution whenever

$$T^{\frac{1}{p}}\left(c_1+\frac{3\|c_2\|_q T^{q-1}}{\Gamma(q)}\right)<1,$$

$$M(\|\phi\|+e_1)+\frac{MT^{\alpha-\frac{1}{p}}\|\theta(s)\|_p\Omega(2T^{\frac{1}{p}}k)}{\Gamma(\alpha)}\left(\frac{p-1}{p\alpha-1}\right)^{\frac{p-1}{p}}\leq k.$$

Proof Define

$$W_0=\left\{x\in L^p([0,T],X): \|x(t)\|\leq k \text{ for a.e. } t\in[0,T]\right\}.$$

For any $x \in W_0$,

$$\begin{aligned}
\|Fx(t)\| &\leq \|F_1x(t)+F_2x(t)\|\\
&\leq M(\|\phi(0)\|+\|g(x(0))\|)+\frac{1}{\Gamma(\alpha)}\\
&\quad \int_0^t \|(t-\eta)^{\alpha-1}T(t-\eta)f(\eta,x(\eta),x_\eta)\|d\eta\\
&\leq M\left[\|\phi(0)\|+e_1+\frac{1}{\Gamma(\alpha)}\right.\\
&\quad \left.\int_0^t (t-\eta)^{\alpha-1}\theta(\eta)\Omega(\|x\|+\|x\|_C)d\eta\right]\\
&\leq M(\|\phi(0)\|+e_1)+\frac{MT^{\alpha-\frac{1}{p}}}{\Gamma(\alpha)}\left(\frac{p-1}{p\alpha-1}\right)^{\frac{p-1}{p}}\\
&\quad \left(\int_0^t \theta^p(s)\Omega^p(\|x\|+\|x\|_C\|)ds\right)^{\frac{1}{p}}\\
&\leq M(\|\phi(0)\|+e_1)\\
&\quad +\frac{MT^{\alpha-\frac{1}{p}}\|\theta(s)\|_p\Omega(2T^{\frac{1}{p}}k)}{\Gamma(\alpha)}\left(\frac{p-1}{p\alpha-1}\right)^{\frac{p-1}{p}}.
\end{aligned} \tag{3.5}$$

for any $t\in[0,T]$. This means that $FW_0 \subset W_0$.

Let

$$W_{n+1}=FW_n \quad \text{for} \quad n=0,1,2,\ldots.$$

Then

$$FW_n \subset C([0,T],X),$$

is bounded and equicontinuous on [0, T]. Furthermore, $W_{n+1}\subset W_n$, because $W_1 \subset W_0$. Hence,

$$\begin{aligned}
\chi(W_{n+1}(t)) &\leq \chi(T(t)(x_0-g(W_n)))+\chi\left(\frac{1}{\Gamma(\alpha)}\int_0^t (t-\eta)^{\alpha-1}\right.\\
&\quad \left. T(t-\eta)f(\eta,W_n(\eta),(W_n)_\eta)d\eta\right)\\
&\leq \chi(T(t)(x_0-g(W_n)))+\frac{1}{\Gamma(\alpha)}\\
&\quad \int_0^t \chi((t-\eta)^{\alpha-1}T(t-\eta)f(\eta,W_n(\eta),(W_n)_\eta))d\eta
\end{aligned}$$

$$\leq c_1\chi_p(W_n)+\frac{2}{\Gamma(\alpha)}\int_0^t (t-\eta)^{\alpha-1}c_2(\eta)\chi(W_n)\mathrm{d}\eta$$
$$\leq\left(c_1+\frac{2\|c_2\|_q T^{\alpha-\frac{1}{p}}}{\Gamma(\alpha)}\left(\frac{p-1}{p\alpha-1}\right)^{\frac{p-1}{p}}\right)\chi_p(W_n),$$

for $n=0,1,2,\dots$ and a.e. $t\in[0,T]$. By Lemma 2.4, this implies that

$$\chi_p(W_{n+1})\leq T^{\frac{1}{p}}\left(c_1+\frac{2\|c_2\|_q T^{\alpha-\frac{1}{p}}}{\Gamma(\alpha)}\left(\frac{p-1}{p\alpha-1}\right)^{\frac{p-1}{p}}\right)\chi_p(W_n),$$
$$\text{for } n=0,1,2,\dots.$$

Define

$$\widehat{W}_n=L^p-\overline{\mathrm{conv}}(W_n),\quad \text{for}\quad n=0,1,2,\dots.$$

Then $\widehat{W}_{n+1}\subset\widehat{W}_n$ because $W_{n+1}\subset W_n$, and furthermore one has

$$\chi_p(\widehat{W}_{n+1})=\chi_p(W_{n+1})$$
$$\leq T^{\frac{1}{p}}\left(c_1+\frac{2\|c_2\|_q T^{\alpha-\frac{1}{p}}}{\Gamma(\alpha)}\left(\frac{p-1}{p\alpha-1}\right)^{\frac{p-1}{p}}\right)\chi_p(W_n)$$
$$=T^{\frac{1}{p}}\left(c_1+\frac{2\|c_2\|_q T^{\alpha-\frac{1}{p}}}{\Gamma(\alpha)}\left(\frac{p-1}{p\alpha-1}\right)^{\frac{p-1}{p}}\right)\chi_p(\widehat{W}_n),$$

for $n=0,1,2,\dots$. Lemma 2.1 shows that

$$\widehat{W}=\cap_{n=1}^{\infty}\widehat{W}_n,$$

is nonempty, convex and compact in $L^p([0,T],X)$ and $F\widehat{W}\subset\widehat{W}$. Let

$$U=\overline{\mathrm{conv}}(F\widehat{W}).$$

Then $U\subset C([0,T],X)$ and $FU\subset U$, since

$$U=\overline{\mathrm{conv}}(F\widehat{W})\subset\overline{\mathrm{conv}}(\widehat{W})\subset L^p-\overline{\mathrm{conv}}(\widehat{W})=\widehat{W}.$$

Now we prove that $U\subset C([0,T],X)$ is compact. First, by the hypothesis (H_1) and Lemma 2.5, $F\widehat{W}$ is equicontinuous on [0, T], as

$$g:L^p([0,T],X)\to X,$$

is continuous and $\widehat{W}\subset L^p([0,T],X)$ is compact. Furthermore,

$$\chi(\widehat{W}(t))\leq\chi(T(t)(x_0-g(\widehat{W})))+\chi\left(\frac{1}{\Gamma(\alpha)}\int_0^t(t-\eta)^{\alpha-1}\right.$$
$$\left.T(t-\eta)f(\eta,\widehat{W}(\eta),\widehat{W}_\eta)\mathrm{d}\eta\right)$$
$$\leq\chi(T(t)(x_0-g(\widehat{W})))+\frac{1}{\Gamma(\alpha)}$$
$$\int_0^t\chi((t-\eta)^{\alpha-1}T(t-\eta)f(\eta,\widehat{W}(\eta),\widehat{W}_\eta))\mathrm{d}\eta$$
$$\leq c_1\chi_p(\widehat{W})+\frac{2}{\Gamma(\alpha)}\int_0^t(t-\eta)^{\alpha-1}c_2(\eta)\chi(\widehat{W})\mathrm{d}\eta$$
$$\leq\left(c_1+\frac{2\|c_2\|_q T^{\alpha-\frac{1}{p}}}{\Gamma(\alpha)}\left(\frac{p-1}{p\alpha-1}\right)^{\frac{p-1}{p}}\right)\chi_p(\widehat{W})=0,$$

for any $t\in[0,T]$. Hence

$$F\widehat{W}\subset C([0,T],X),$$

is precompact, and hence so is

$$U\subset C([0,T],X).$$

The proof is complete by Schauder's fixed point theorem.

Remark 3.3 Without hypothesis (H_2) the map F, defined above, may not be continuous from $L^p([0,T],X)$ to itself, since the operator f may fail to be continuous under the growth condition (3.4) above. So we use the fixed point theorem on $C([0,T],X)$ rather than on $L^p([0,T],X)$, as F is obviously continuous from $C([0,T],X)$ to itself.

Applications

In this section, we give an example to illustrate the above results.

Consider the following nonlinear fractional parabolic systems of the form

$$\begin{aligned}
&\frac{\partial^\alpha}{\partial t^\alpha}u(t,x)=\mu_1\Delta u(t,x)+F_1(t,u(t,x),u_t(x),v(t,x),v_t(x)),\\
&\qquad t\in(0,T),\ x\in\Omega,\\
&\frac{\partial^\alpha}{\partial t^\alpha}v(t,x)=\mu_2\Delta v(t,x)+F_2(t,u(t,x),u_t(x),v(t,x),v_t(x)),\\
&\qquad t\in(0,T),\ x\in\Omega,\\
&u(0,x)=\varphi_1+g_1(u(t,x),v(t,x)),\quad x\in\Omega,\\
&v(0,x)=\varphi_2+g_2(u(t,x),v(t,x)),\quad x\in\Omega,
\end{aligned}\tag{4.1}$$

where Ω is a bounded domain of $\mathbb{R}^p$, $p\geq 1$, with smooth boundary Γ, μ_1, μ_2 are positive constants,

$$F_1,F_2:\mathbb{R}\times\mathbb{R}\times C([-q,0],\mathbb{R})\to\mathbb{R},$$

are given mappings. Here

$$F_2(t,u(t,x),u_t(x),v(t,x),v_t(x)) = \int_\Omega h_1(t,x,z,u(z),u_t(z))\mathrm{d}z,$$

$$g_2(u(t,x),v(t,x)) = \int_\Omega \int_0^T h_2(t,x,z,u(z))\mathrm{d}t\mathrm{d}z.$$

Let

$$X = L^2(\Omega) \times L^2(\Omega)$$

be endowed with the inner product $<\cdot,\cdot>$ defined by

$$<(u,v),(\bar{u},\bar{v})> \;=\; <u,\bar{u}>_{L^2(\Omega)} + <v,\bar{v}>_{L^2(\Omega)},$$

for each $(u,v),(\bar{u},\bar{v}) \in X$. Obviously X is a separable real Hilbert space. Define $A : D(A) \subset X \to X$ given by

$$A(u,v) = (\mu_1 \Delta u, \mu_2 \Delta v), \quad \text{for each} \quad (u,v) \in D(A)$$

with the domain

$$D(A) = \left\{ (u,v) \in X : \frac{\partial^\alpha}{\partial t^\alpha} u, \frac{\partial^\alpha}{\partial t^\alpha} v \in X, \Delta u, \Delta v \in X \right\}.$$

Now define

$$F : [0,T] \times X \to X \quad \text{and} \quad g : C([0,T],X) \to X$$

by

$$F(t,(u,v),(u_t,v_t)) = \Big(F_1(t,u(t,x),u_t(x),v(t,x),v_t(x)),$$
$$F_2(t,u(t,x),u_t(x),v(t,x),v_t(x))\Big),$$
$$g(t,(u,v)) = \Big(g_1(t,(u,v),(u_t,v_t)), g_2(t,(u,v),(u_t,v_t))\Big),$$

for

$$(u,v) \in X, (u_t,v_t) \in C([-q,0],X) \times C([-q,0],X),$$

where F_i and g_i are superposition mappings associated with F_i and g_i defined by

$$F_i(t,(u,v)) = \{h \in L^2(\Omega), h(x) = F_i(t,(u(x),v(x)),$$
$$(u_t(x),v_t(x))), \quad \text{a.e. for } x \in \Omega\},$$
$$g_i(t,(u,v)) = \{h \in L^2(\Omega), h(x) = g_i((u(x),v(x)),$$
$$(u_t(x),v_t(x))), \quad \text{a.e. for } x \in \Omega\}.$$

Observe that Eq.(4.1) may be rewritten as

$$\begin{aligned} &\frac{\mathrm{d}^\alpha U}{\mathrm{d}t^\alpha} = Au + F(t,U(t),U_t), \quad t \in (0,T), \quad \text{a. e.} \\ &U(0) = \varphi + g(U), \end{aligned} \tag{4.2}$$

where

$$U(t) = (u(t),v(t)), \quad \varphi = (\varphi_1,\varphi_2),$$

while A, F and g are as above.

Suppose that:

(1) There exists $k_2(t) \in L(0,T)$ such that $F_1 : [0,T] \times \mathbb{R} \to \mathbb{R}$ is a Carathédory-type function and, for $u,u',v,v' \in \mathbb{R}$,

$$|F_1(t,u,v) - F_1(t,u',v')| \le k_2(t)(|u-u'|_2 + |v-v'|^2).$$

(2) There exists a constant k_1 such that $g_1 : \mathbb{R} \to \mathbb{R}$ is a Carathédory-type function and, for $u,u',v,v' \in \mathbb{R}$,

$$|g_1(u,v) - g_1(u',v')| \le k_1(|u-u'|_2 + |v-v'|^2).$$

(3) For $i = 1,2$, $h_i : [0,T] \times \Omega \times \Omega \times \mathbb{R} \times \mathbb{R} \to \mathbb{R}$ satisfies Carathédory conditions. In addition:

(i)

$$|h_i(t,x,z,r,s) - h_i(t,x',z,r,s)| \le \omega_i^k(t,x,x',z),$$

for $(t,x,z),(t,x',z) \in [0,T] \times \Omega \times \Omega$ and $|r|,|s| \le k$, where $\omega_i^k \in L([0,T] \times \Omega^3)$ are such that

$$\lim_{x' \to x} \int_\Omega \int_0^T \omega_i^k(t,x,x',z)\mathrm{d}t\mathrm{d}z = 0,$$

uniformly for $x \in \Omega$, $i = 1,2$, and for $t \in [0,T]$

$$\lim_{x' \to x} \int_\Omega \omega_i^k(t,x,x',z)\mathrm{d}z = 0,$$

uniformly for $x \in \Omega$;

(ii)

$$|h_i(t,x,z,r,s| \le \rho_i(t)(|r|^2 + |s|^2)^{\frac{1}{2}} + \omega_i(t,x,z),$$

where $\rho_i \in L(0,T)$ and

$$\delta_i = \int_0^T \int_{\Omega \times \Omega} (\omega_i(t,x,z))^2 \mathrm{d}x\mathrm{d}z\mathrm{d}t < +\infty, \quad i = 1,2.$$

Adapting the arguments given in [47] it is not difficult to show that g satisfies the hypothesis (H_3) with

$$c_1 = k_1, \quad d_1^2 = k_1^2 + 2m(\Omega)\|\rho_1\|_1^2,$$
$$e_1^2 = 2m(\Omega)\delta_1,$$

and f satisfies the hypothesis (H_2) with

$$c_2(t) = k_2(t), \quad d^2(t) = k_2^2(t) + 2m(\Omega)\rho_2^2(t),$$
$$\|e_2\|_1^2 = 2m(\Omega)\delta_2,$$

where $m(\Omega)$ means the Lebesgue measure of Ω in $\mathbb{R}^p$. Using Corollary 3.1, we conclude that Eq.(4.1) has at least one generalized solution

$$(u_1, u_2) \in C([0,T], L^2(\Omega) \times L^2(\Omega)),$$

provided that

$$M\left(k_1^2+2m(\Omega)\|\rho_1\|_1^2\right)^{\frac{1}{2}} T^{\frac{1}{p}} + \frac{4MT^{\alpha}}{\Gamma(\alpha)}\left(\frac{p-1}{p\alpha-1}\right)^{\frac{p-1}{p}}$$

$$\left(\int_0^T \left(k_2^2(t)+2m(\Omega)\rho_2^2(t)\right)^{\frac{p}{2}} \mathrm{d}t\right)^{\frac{1}{p}} < 1,$$

and

$$T^{\frac{1}{p}}\left(k_1 + \frac{2\|k_2\|_q T^{\alpha-\frac{1}{p}}}{\Gamma(\alpha)}\left(\frac{p-1}{p\alpha-1}\right)^{\frac{p-1}{p}}\right) < 1.$$

Acknowledgments This work is supported by the National Natural Science Foundation of China (No.11301090), Appropriative Researching Fund for Professors and Doctors, Guangdong University of Education (No. 2013ARF02).

References

1. Ahn, V., McVinisch, R.: Fractional differential equations driven by Lévy noise. J. Appl. Math. Stoch. Anal. **16**, 97–119 (2003)
2. D. Benson, The Fractional Advection-Dispersion Equation, Ph.D. Thesis, University of Nevada, Reno, NV, 1998
3. Schumer, R., Benson, D.: Eulerian derivative of the fractional advection-dispersion equation. J. Contam. Hydrol. **48**, 69–88 (2001)
4. Sayed, A.: Nonlinear functional-differential equations of arbitrary orders. Nonlinear Anal. Theory Methods Appl. **33**, 181–186 (1998)
5. Ling, Y., Ding, S.: A class of analytic functions defined by fractional derivation. J. Math. Anal. Appl. **186**, 504–513 (1994)
6. N'Guérékata, G.: Cauchy problem for some fractional abstract differential equation with nonlocal conditions. Nonlinear Anal. Theory Methods Appl. **70**, 1873–1876 (2009)
7. Lahshmikantham, V., Devi, J.: Theory of fractional differential equations in Banach spaces. Eur. J. Pure Appl. Math. **1**, 38–45 (2008)
8. Glockle, W., Nonnemacher, T.: A fractional calculus approach of self-similar protein dynamics. Biophys. J. **68**, 46–53 (1995)
9. Metzler, F., Schick, W., Kilian, H., Nonnemacher, T.: Relaxation in filled polymers: A fractional calculus approach. J. Chem. Phys. **103**, 7180–7186 (1995)
10. Ahmad, B., Sivasundaram, S.: Some existence results for fractional integrodifferential equations with nonlinear conditions. Commmun. Math. Anal. **12**, 107–112 (2008)
11. Bhaskar, T., Lakshmikantham, V., Leela, S.: Fractional differential equations with Krasnoselskii-Krein-type condition. Nonlinear Anal. Hybrid Syst. **3**, 734–737 (2009)
12. Lakshmikantham, V., Leela, S.: Nagumo-type uniqueness result for fractional differential equations. Nonlinear Anal. Theory Methods Appl. **71**, 2886–2889 (2009)
13. Anguraj, A., Karthikeyan, P., N'Guérékata, G.: Nonlocal Cauchy problem for some fractional abstract integro-differential equations in Banach spaces. Commmun. Math. Anal. **6**, 31–35 (2009)
14. Cao, J., Yang, Q., Huang, Z.: Existence of anti-periodic mild solutions for a class of semilinear fractional differential equations. Commun. Nonlinear Sci. Numer. Simulat. **17**, 277–283 (2012)
15. Cao, J., Yang, Q., Huang, Z.: Optimal mild solutions and weighted pseudo-almost periodic classical solutions of fractional integro-differential equations. Nonlinear Anal. Theory Methods Appl. **74**, 224–234 (2011)
16. Rida, S., Sayed, A., Arafa, A.: On the solutions of time-fractional reaction-diffusion equations. Commun. Nonlinear Sci. Numer. Simulat. **15**, 3847–3854 (2010)
17. Wang, R., Chen, D., Xiao, T.: Abstract fractional Cauchy problems with almost sectorial operators. J. Differ. Equations **252**, 202–235 (2012)
18. Obukhovskii, V., Yao, J.: Some existence results for fractional functional differential equations. Fixed Point Theory **11**, 85–96 (2010)
19. Eidelman, S., Kochubei, A.: Cauchy problem for fractional diffusion equations. J. Differ. Equations **199**, 211–255 (2004)
20. Balachandran, K., Kiruthika, S., Trujillo, J.: Existence results for fractional impulsive integrodifferential equations in Banach spaces. Commun. Nonlinear Sci. Numer. Simulat. **16**, 1970–1977 (2011)
21. Wang, J., Zhou, Y., Wei, W.: A class of fractional delay nonlinear integrodifferential controlled systems in Banach spaces. Commun. Nonlinear Sci. Numer. Simulat. **16**, 4049–4059 (2011)
22. Agrawal, O.: Solution for a fractional diffusion-wave equation defined in a bounded domain. J. Nonlinear Dynam. **29**, 145–155 (2002)
23. Liang, J., Yang, H.: Controllability of fractional integro-differential evolution equations with nonlocal conditions. Appl. Math. Comput. **254**, 20–29 (2015)
24. Zhang, X.: Positive solutions for a class of singular fractional differential equation with infinite-point boundary value conditions. Appl. Math. lett. **39**, 22–27 (2015)
25. Agarwal, R., Lupulescu, V., O'Regan, D., Rahman, G.: Fractional calculus and fractional differential equations in nonreflexive Banach spaces. Commun. Nonlinear Sci. Numer. Simulat. **20**, 59–73 (2015)
26. Henderson, J., Luca, R.: Positive solutions for a system of fractional differential equations with coupled integral boundary conditions. Appl. Math. Comput. **249**, 182–197 (2014)
27. Wang, R., Xiang, Q., Zhu, P.: Existence and approximate controllability for systems governed by fractional delay evolution inclusions. Optimization **63**, 1191–1204 (2014)
28. Wang, R., Yang, Y.: On the Cauchy problems of fractional evolution equations with nonlocal initial conditions. Results. Math. **63**, 15–30 (2013)
29. Wang, R., Xiang, Q., Zhou, Y.: Fractional delay control problems: Topological structure of solution sets and its applications. Optimization **63**, 1249–1266 (2014)
30. Zhou, Y., Jiao, F.: Existence of mild solutions for fractional neutral evolution equations. Comput. Math. Appl. **59**, 1063–1077 (2010)
31. Bëleanu, D., Mustafa, O.: On the global existence of solutions to a class of fractional differential equations. Comput. Math. Appl. **59**, 1835–1841 (2010)
32. Mophou, G., N'Guérékata, G.: Existence of the mild solution for some fractional differential equations with nonlocal conditions. Semigroup Forum **79**, 315–322 (2009)
33. Li, C., Luo, X., Zhou, Y.: Existence of positive solutions of the boundary value problem for nonlinear fractional differential equations. Comput. Math. Appl. **59**, 1363–1375 (2010)
34. Zhang, S.: Positive solutions to singular boundary value problem for nonlinear fractional differential equation. Comput. Math. Appl. **59**, 1300–1309 (2010)
35. Agarwal, R., Zhou, Y., He, Y.: Existence of fractional neutral functional differential equations. Comput. Math. Appl. **59**, 1095–1100 (2010)
36. Podlubny, I.: Fractional Differential Equations. Academic Press, San Diego (1999)

37. J. Banaś, K. Goebel, Measures of Noncompactness in Banach Spaces, Lect. Notes Pure Appl. Math., vol. 60, Dekker, New York, 1980
38. Darbo, G.: Punti unitti in transformazioni a condominio non compatto. Rend. Semin. Mat. Univ. Padova **24**, 84–92 (1955)
39. Sadovski, B.: On a fixed point principle. Funct. Anal. Appl. **1**, 74–76 (1967)
40. Istratescu, V.: Fixed Point Theory. Reidel, Boston (1981)
41. Hajji, A., Hanebaly, E.: Commuting mappings and -compact type fixed point theorems in locally convex spaces. Int. J. Math. Anal. **1**, 661–680 (2007)
42. Hajji, A.: A generalization of Darbos fixed point and common solutions of equations in Banach spaces. Fixed Point Theory and Applications **2013**, 62 (2013)
43. Kisielewicz, M.: Multivalued differential equations in separable Banach spaces. J. Optim. Theory Appl. **37**, 231–249 (1982)
44. Xue, X.: L^p theory for semilinear nonlocal problems with measure of noncompactness in separable Banach spaces. J. Fixed Point Theory Appl. **5**, 129–144 (2009)
45. Barbu, V.: Nonlinear Semigroups and Differential Equations in Banach Spaces. Noordhoff, Leyden (1976)
46. Pavel, N.: Nonlinear evolution operators and semigroup. Lecture Notes in Math, vol. 1260. Springer-Verlag, New York (1987)
47. Martin, R.: Nonlinear Operators and Differential Equations in Banach Spaces. Wiley, New York (1976)

17

Two-dimensional Bernoulli wavelets with satisfier function in the Ritz–Galerkin method for the time fractional diffusion-wave equation with damping

Z. Barikbin[1]

Abstract In this paper, the two-dimensional Bernoulli wavelets (BWs) with Ritz–Galerkin method are applied for the numerical solution of the time fractional diffusion-wave equation. In this way, a satisfier function which satisfies all the initial and boundary conditions is derived. The two-dimensional BWs and Ritz–Galerkin method with satisfier function are used to transform the problem under consideration into a linear system of algebraic equations. The proposed scheme is applied for numerical solution of some examples. It has high accuracy in computation that leads to obtaining the exact solutions in some cases.

Keywords Two-dimensional Bernoulli wavelets basis · Fractional diffusion-wave equation · Ritz–Galerkin method · Satisfier function · Caputo derivative

Introduction

Many phenomena in various field of the science, can be modeled very successfully by time-fractional differential equations. In this paper we focus on the following fractional diffusion-wave equation (FDWE) with damping [1]:

$$\frac{\partial^q u(x,t)}{\partial t^q}+\frac{\partial u(x,t)}{\partial t}=\frac{\partial^2 u(x,t)}{\partial x^2}+h(x,t), \quad 0<x<L, \quad 0<t<T, \tag{1.1}$$

with the initial conditions:

$$u(x,0)=f_0(x), \qquad \frac{\partial u(x,0)}{\partial t}=f_1(x), \qquad 0\le x\le L, \tag{1.2}$$

and the boundary conditions:

$$u(0,t)=g_0(t), \qquad u(L,t)=g_1(t), \qquad 0\le t\le T, \tag{1.3}$$

where $L>0, T>0, 1<q\le 2$ is the order of the fractional derivative in the Caputo sense, f_0,f_1, g_0 and g_1 are known and sufficiently smooth functions, while the function u is to be determined. In the case $q=2$, this equation is named telegraph equation.

Recently, considerable amount of papers have been proposed methods for solving the FDWE [2–13]. Chen et al. [1] obtained the analytical solution by the method of separation of variables and proposed the numerical solution with finite difference method. In [2], Bhrawy et al. applied a spectral tau method based on the Jacobi operational matrix to solve the problem. Liu et al. [3] proposed the fractional predictor–corrector method to solve this problem. In [4], Mainardi derived the fundamental solutions for the FDWE. The combination of the compact difference method and alternating direction implicit method are used for solving two-dimensional fractional Cattaneo equation in [5]. A fully discrete difference scheme is recommended for a diffusion-wave system by Wess [6]. Heydari et al. [7] applied fractional operational matrix (FOM) of integration for the Legendre wavelets (LWs) to solve the problem. In [8] a compact finite-difference scheme is used for the fourth-order fractional diffusion-wave system. In [9] finite difference schemes of second-order are proposed for the time-fractional diffusion-wave equation. In [10], Hu and Zhang used finite-difference methods for fourth-order fractional diffusion-wave. Sumudu transform method for solving fractional

✉ Z. Barikbin
barikbin@ikiu.ac.ir

[1] Department of Mathematics, Imam Khomeini International University, Qazvin 34149-16818, Iran

differential equations and fractional diffusion-wave equation applied by Darzi et al. [11]. Hosseini et al. [12] employed the meshless local radial point interpolation method which is based on the Galerkin weak form and radial point interpolation approximation for solving FDWE.

The Ritz–Galerkin method is the method to transform a continuous problem to a discrete problem. Several partial differential equations are numerically solved by Ritz–Galerkin method, but using of the appropriate satisfier function in the Ritz–Galerkin method is taken into consideration recently, see for instance [14–20]. The satisfier function fulfills all the problem conditions. In conclusion, employing of it in Ritz–Galerkin method provides the facility to satisfy the problem conditions, also leads to a system of algebraic equations of smaller size and hence reduces the computation time.

In mathematical research, wavelet theory is a relatively new and growing area. It has been used in a wide range of engineering; for instance, wavelets are very successfully applied in signal analysis for waveform representations and segmentations [21]. Wavelets allow the accurate representation of many types of functions and operators [22, 23]. Furthermore, wavelets make a connection with fast numerical algorithms [24].

The shifted Legendre polynomials $p_n(x), n = 0, 1, 2, \ldots$, that $0 \leq x \leq 1$, are more efficient in approximation theory, among other orthogonal polynomials [25, 26]. The Bernoulli polynomials are not orthogonal functions. In [27], the superiority of Bernoulli polynomials $\beta_n(x)$, $n = 0, 1, 2, \ldots$, so that $0 \leq x \leq 1$, to shifted Legendre polynomials in approximation of functions is proposed.

In this paper, we define two-dimensional BWs for the first time. Moreover, this is the first time the Ritz–Galerkin method in the two-dimensional BWs basis and with utilizing the satisfier function is employed to give an approximate solution of FDWE. We also compare our results with those results obtained by [3] and [7]. Comparison for the numerical examples shows the more accuracy and less computations of our scheme in comparison to other published methods.

This paper is separated in to the following sections: In Sect. 2, we introduce basic formulation of wavelets and the Bernoulli wavelets. In Sect. 3, we construct two different satisfier functions and apply the Ritz–Galerkin method in the two-dimensional BWs basis for numerically discretize the problem. Section 4 presents and discusses the numerical results for two test examples, whilst Sect. 5 includes the conclusions of this paper.

Properties of Bernoulli wavelets

Wavelets and Bernoulli wavelets

Dilation and translation of a function (mother wavelet) construct a family of functions called wavelets and is defined as follows [28]

$$\varphi_{a,b}(t) = \mid a \mid^{\frac{-1}{2}} \varphi\left(\frac{t-b}{a}\right), \quad a \neq 0,$$

where $a, b \in \mathbb{R}$ are dilation and translation parameters that vary continuously. The family of discrete wavelets that form a basis for $L^2(\mathbb{R})$ is defined as follows

$$\varphi_{k,n}(t) = \mid a_0 \mid^{\frac{k}{2}} \varphi(a_0^k t - n b_0),$$

where n and k are positive integers and $a_0 > 1, b_0 > 1$.

Bernoulli wavelets are obtained, when we choose Bernoulli polynomial $\beta_m(t)$ as mother wavelet. BWs $\varphi_{n,m}(t) = \varphi(k, \hat{n}, m, t)$ have four arguments; $\hat{n} = n - 1, n = 1, 2, 4, \ldots, 2^{k-1}, m$ is the order for Bernoulli polynomials and k can assume any positive integer. They are defined on the interval [0, 1) for $m = 1, 2, \ldots, M - 1$, that $M > 0$ is a fixed integer by [29]

$$\varphi_{n,m}(t) = \begin{cases} 2^{\frac{k-1}{2}} \dfrac{1}{\sqrt{\dfrac{(-1)^{m-1}(m!)^2}{(2m)!}\alpha_{2m}}} \beta_m(2^{k-1}t - \hat{n}), & \dfrac{\hat{n}}{2^{k-1}} \leq t < \dfrac{\hat{n}+1}{2^{k-1}}, \\ 0, & \text{otherwise}, \end{cases}$$

and

$$\varphi_{n,0}(t) = \begin{cases} 2^{\frac{k-1}{2}}, & \dfrac{\hat{n}}{2^{k-1}} \leq t < \dfrac{\hat{n}+1}{2^{k-1}}, \\ 0, & \text{otherwise}, \end{cases}$$

where $n = 1, 2, \ldots, 2^{k-1}$. The coefficient $\frac{1}{\sqrt{\frac{(-1)^{m-1}(m!)^2}{(2m)!}\alpha_{2m}}}$ is applied to normalize the Bernoulli wavelets. Bernoulli polynomials form a complete basis over the interval [0, 1] [30] and are defined by [31]

$$\beta_m(t) = \sum_{i=0}^{m} \binom{m}{i} \alpha_{m-i} t^i, \tag{2.1}$$

that α_i, $i = 0, 1, \ldots, m$ are Bernoulli numbers.

For Bernoulli polynomials we have [32]

$$\int_a^\tau \beta_m(t) \mathrm{d}t = \frac{\beta_{m+1}(\tau) - \beta_{m+1}(a)}{m+1}, \tag{2.2}$$

$$\int_0^1 \beta_{m_1}(t)\beta_{m_2}(t) \mathrm{d}t = (-1)^{m_1 - 1} \frac{m_1! m_2!}{(m_1 + m_2)!} \alpha_{m_1 + m_2}, \quad m_1, m_2 \geq 1. \tag{2.3}$$

Now, let

$$\tilde{\beta}_m(t) = \begin{cases} 1, & m = 0, \\ \dfrac{1}{\sqrt{\dfrac{(-1)^{m-1}(m!)^2}{(2m)!}\alpha_{2m}}}\beta_m(t), & m > 0. \end{cases}$$

We define, for the first time, the two-dimensional Bernoulli wavelets $\varphi_{n_1 m_1 n_2 m_2}(x,y)$ as

$$\begin{cases} 2^{\frac{(k_1-1)+(k_2-1)}{2}}\tilde{\beta}_{m_1}(2^{k_1-1}x-\hat{n}_1)\tilde{\beta}_{m_2}(2^{k_2-1}y-\hat{n}_2), & \text{for } \dfrac{\hat{n}_1}{2^{k_1-1}} \le x < \dfrac{\hat{n}_1+1}{2^{k_1-1}}, \quad \dfrac{\hat{n}_2}{2^{k_2-1}} \le y < \dfrac{\hat{n}_2+1}{2^{k_2-1}}, \\ 0, & \text{otherwise} \end{cases}$$

where $m_1 = 0, 1, 2, ..., M_1 - 1, m_2 = 0, 1, 2, ..., M_2 - 1$. Here, $\hat{n}_1$ and $\hat{n}_2$ are defined similarly to $\hat{n}$, k_1 and k_2 can be any positive integer, $\tilde{\beta}_{m_1}$ and $\tilde{\beta}_{m_2}$ are defined similarly to $\tilde{\beta}_m$ of order m_1 and m_2, respectively.

Function approximation

A function $f(x,y) \in L^2([0,1) \times [0,1))$ may be expanded as in terms of two-dimensional Bernoulli wavelets as

$$f(x,y) = \sum_{n=1}^{\infty}\sum_{i=0}^{\infty}\sum_{l=1}^{\infty}\sum_{j=0}^{\infty} c_{nilj}\varphi_{nilj}(x,y). \tag{2.4}$$

If the infinite series for f is truncated, then Eq. (2.4) can be written as

$$f(x,y) \simeq \sum_{n=1}^{2^{k_1-1}}\sum_{i=0}^{M_1-1}\sum_{l=1}^{2^{k_2-1}}\sum_{j=0}^{M_2-1} c_{nilj}\varphi_{nilj}(x,y) = C^T\Phi(x,y) = \Phi_1^T(x)F\Phi_2(y), \tag{2.5}$$

where $\Phi_1(x)$ and $\Phi_2(y)$ are $2^{k_1-1}M_1 \times 1$ and $2^{k_2-1}M_2 \times 1$ matrices, respectively, given by

$$\begin{aligned} \Phi_1(x) =& [\varphi_{10}(x), \varphi_{11}(x), \ldots, \varphi_{1M_1-1}(x), \varphi_{20}(x), \ldots, \\ & \varphi_{2M_1-1}(x), \ldots, \varphi_{2^{k_1-1}0}(x), \ldots, \varphi_{2^{k_1-1}M_1-1}(x)]^T, \\ \Phi_2(y) =& [\varphi_{10}(y), \varphi_{11}(y), \ldots, \varphi_{1M_2-1}(y), \varphi_{20}(y), \ldots, \\ & \varphi_{2M_2-1}(y), \ldots, \varphi_{2^{k_2-1}0}(y), \ldots, \varphi_{2^{k_2-1}M_2-1}(y)]^T. \end{aligned}$$

In Eq. (2.5), F is a $2^{k_1-1}M_1 \times 2^{k_2-1}M_2$ matrix that can be calculated from [33]

$$F = D_1^{-1} < \Phi_1(x), < f(x,y), \Phi_2(y) > > D_2^{-1},$$

where $\langle . \rangle$ denotes the inner product, $D_1 = \langle \Phi_1(x), \Phi_1(x) \rangle = \int_0^1 \Phi_1(x)\Phi_1^T(x)dx$ and $D_2 = \langle \Phi_2(y), \Phi_2(y) \rangle = \int_0^1 \Phi_2(y)\Phi_2^T(y)dy$ are $2^{k_1-1}M_1 \times 2^{k_1-1}M_1$ and $2^{k_2-1}M_2 \times 2^{k_2-1}M_2$ matrices, respectively.

Satisfier function

In the Ritz–Galerkin method with the two-dimensional BWs basis, the approximation $\widetilde{u}(x,t)$ of the solution $u(x, t)$ in (1.1) is sought in the form of the truncated series

$$\widetilde{u}(x,t) = \sum_{n=1}^{2^{k_1-1}}\sum_{i=0}^{M_1-1}\sum_{l=1}^{2^{k_2-1}}\sum_{j=0}^{M_2-1} c_{nilj}\sigma_{nilj}(x,t) + w(x,t), \quad (x,t) \in [0,L] \times [0,T], \tag{3.1}$$

where $\sigma_{nilj}(x,t) = x(x-L)t^2\varphi_{ni}(x)\varphi_{lj}(t)$ and $w(x, t)$ is a satisfier function. The most important point in using the Ritz–Galerkin method is finding the satisfier function, which satisfies all the problem conditions [16]. Interpolation is one of the methods that is usually used to derive Satisfier functions. On the other hand, we have seen from experience, when in constructing of the satisfier function we use only the problem's data, we get a satisfier function that is closer to the exact solution. Therefore, we obtain the cost-effective computational results [14, 16]. Here, we construct two different satisfier functions for the initial and boundary conditions:

$$u(x,0) = f_0(x), \quad 0 \le x \le L, \tag{3.2}$$

$$\frac{\partial u(x,0)}{\partial t} = f_1(x), \quad 0 \le x \le L, \tag{3.3}$$

$$u(0,t) = g_0(t), \quad 0 \le t \le T, \tag{3.4}$$

$$u(L,t) = g_1(t), \quad 0 \le t \le T. \tag{3.5}$$

It is worth pointing out that $f_0(x), f_1(x), g_0(t)$ and $g_1(t)$ satisfy the following compatibility conditions:

$$f_0(0) = g_0(0), \quad f_0(L) = g_1(0), \tag{3.6}$$

$$f_1(0) = g_0'(0), \quad f_1(L) = g_1'(0). \tag{3.7}$$

The first technique for obtaining the satisfier function is as follows:

We set

$$w(x,t) = k_1(x)g_0(t) + k_2(x)g_1(t), \tag{3.8}$$

then we construct $k_1(x)$ and $k_2(x)$, such that (3.8) fulfils the conditions (3.2)–(3.5).

Clearly, if $k_1(x)$ and $k_2(x)$ satisfy the following conditions:

$$k_1(0) = k_2(L) = 1, \quad k_2(0) = k_1(L) = 0, \tag{3.9}$$

$$k_1(x)g_0(0) + k_2(x)g_1(0) = f_0(x), \quad k_1(x)g_0'(0) + k_2(x)g_1'(0) = f_1(x), \tag{3.10}$$

then (3.8) can be a satisfier function for (3.2)–(3.5).

Equations (3.10) form a system of linear equations, which can be solved for $k_1(x)$ and $k_2(x)$ when $g_0(0)g_1'(0) - g_1(0)g_0'(0) \neq 0$. By solving this system, we obtain

$$k_1(x) = \frac{f_0(x)g_1'(0) - g_1(0)f_1(x)}{g_0(0)g_1'(0) - g_1(0)g_0'(0)}, \tag{3.11}$$

$$k_2(x) = \frac{f_1(x)g_0(0) - g_0'(0)f_0(x)}{g_0(0)g_1'(0) - g_1(0)g_0'(0)}. \tag{3.12}$$

From compatibility conditions (3.6) and (3.7), it is easy to see that $k_1(x)$ and $k_2(x)$ have properties (3.9). Therefore, we introduce the satisfier function $w(x, t)$ which satisfies the initial conditions (3.2) and (3.3), and the boundary conditions (3.4) and (3.5) when

$$g_0(0)g_1'(0) - g_1(0)g_0'(0) \neq 0,$$

as:

$$w(x,t) = k_1(x)g_0(t) + k_2(x)g_1(t), \tag{3.13}$$

where $k_1(x)$ and $k_2(x)$ are obtained from (3.11) and (3.12).

The second satisfier function is constructed as follows:

We firstly transform the nonhomogeneous boundary condition into a homogeneous boundary condition. Let

$$u(x,t) = v(x,t) + \phi(x,t),$$

where

$$\phi(x,t) = \left(1 - \frac{x}{L}\right)g_0(t) + \frac{x}{L}g_1(t).$$

The function $v(x, t)$ then satisfies the problem with homogeneous boundary conditions:

$$v(x,0) = F_0(x), \quad 0 \leq x \leq L, \tag{3.14}$$

$$\frac{\partial v(x,0)}{\partial t} = F_1(x), \quad 0 \leq x \leq L, \tag{3.15}$$

$$v(0,t) = 0, \quad 0 \leq t \leq T, \tag{3.16}$$

$$v(L,t) = 0, \quad 0 \leq t \leq T, \tag{3.17}$$

that $F_0(x) = f_0(x) - \phi(x,0)$, and $F_1(x) = f_1(x) - \phi_t(x,0)$. From conditions (3.14)–(3.17), we drive compatibility conditions:

$$F_0(0) = F_0(L) = F_1(0) = F_1(L) = 0.$$

Therefore, $W(x,t) = F_0(x) + tF_1(x)$ satisfies conditions (3.14)–(3.17) and eventually, we introduce the satisfier function for (3.2)–(3.5) as:

$$w(x,t) = W(x,t) + \phi(x,t). \tag{3.18}$$

It is worth to mention that if $g_0(0)g_1'(0) - g_1(0)g_0'(0) \neq 0$, we prefer the first attained satisfier function in (3.13), since in constructing of (3.13) we used only the problem's data.

Returning now to the Ritz–Galerkin approximation (3.1), the expansion coefficients c_{nilj} are determined by the Galerkin equations:

$$\langle F(\tilde{u}), \varphi_{ni}(x)\varphi_{lj}(t)\rangle = 0, \tag{3.19}$$

where

$$F(u) = \frac{\partial^q u(x,t)}{\partial t^q} + \frac{\partial u(x,t)}{\partial t} - \frac{\partial^2 u(x,t)}{\partial x^2} - h(x,t),$$

and

$$\langle F(\tilde{u}), \varphi_{ni}(x)\varphi_{lj}(t)\rangle = \int_0^L \int_0^T F(\tilde{u})\varphi_{ni}(x)\varphi_{lj}(t)\mathrm{d}t\mathrm{d}x, \tag{3.20}$$

where $\varphi_{ni}(x), \varphi_{lj}(t)$ are BWs. Equations (3.19) form a linear system of equations which can be solved for the elements of $c_{nilj}, n = 1, ..., 2^{k_1-1}, i = 0, ..., M_1 - 1,\ l = 1, ..., 2^{k_2-1}, j = 0, ..., M_2 - 1$ using mathematical softwares.

Numerical results and comparisons

In this section, we apply the numerical scheme in the previous section for finding the approximation solutions of two examples of FDWE. We compare our results with obtained results in [3] and [7]. In all examples the package of Mathematica ver. 10.4 has been used. The approximate norm-2 of absolute error is given as

$$\|e(x,t)\|_{L^2}^2 = \int_0^L \int_0^T e^2(x,t)\mathrm{d}t\mathrm{d}x = \int_0^L \int_0^T (u(x,t) - \tilde{u}(x,t))^2 \mathrm{d}t\mathrm{d}x$$

Example 1

Notice the following FDWE [7]:

$$\frac{\partial^q u(x,t)}{\partial t^q} + \frac{\partial u(x,t)}{\partial t} = \frac{\partial^2 u(x,t)}{\partial x^2} + h(x,t), \quad (x,t) \in [0,1] \times [0,1], \quad 1 < q \leq 2, \tag{4.1}$$

with the homogenous initial and boundary conditions, and let

$$h(x,t) = \frac{2x(1-x)t^{2-q}}{\Gamma(3-q)} + 2tx(1-x) + 2t^2.$$

Now, we apply the numerical method presented in this paper for $k_1 = k_2 = 1$ and $M_1 = M_2 = 3$. From Eq. (3.18) we have $w(x,t) = 0$ and from Eq. (3.19), for $q = 1.1, 1.3, 1.5, 1.7, 1.9$ we obtain

$$c_{1010} = -1, \quad c_{1i1j} = 0, \quad i,j = 1,2.$$

Thus, from (3.1) we have

$$\tilde{u}(x,t) = t^2 x(1-x),$$

Table 1 Comparison of absolute errors of our scheme with scheme in [7], for Example 2

(x, t)	FOM of the LWs [7]				
	$q = 1.1$	$q = 1.3$	$q = 1.5$	$q = 1.7$	$q = 1.9$
(0.1, 0.1)	6.7028E−5	6.2270E−5	6.0407E−5	4.9516E−5	2.0243E−5
(0.2, 0.2)	1.8718E−4	1.6817E−5	1.5683E−4	1.3074E−4	5.9155E−5
(0.3, 0.3)	3.0913E−4	2.8256E−4	2.7985E−4	2.3269E−4	1.0947E−4
(0.4, 0.4)	4.0221E−4	3.6211E−4	3.7035E−4	3.4739E−4	1.6790E−4
(0.5, 0.5)	4.5801E−4	3.8782E−4	3.8089E−4	3.2553E−4	1.6277E−4
(0.6, 0.6)	4.5260E−4	3.7198E−4	3.6309E−4	3.1089E−4	1.9284E−4
(0.7, 0.7)	4.0597E−4	3.0859E−4	3.2603E−4	4.6656E−5	6.2825E−5
(0.8, 0.8)	3.1039E−4	7.9174E−5	6.5594E−4	1.2388E−4	1.0181E−5
(0.9, 0.9)	1.7283E−4	2.1787E−4	7.1269E−3	1.6600E−3	2.0918E−5
	Two-dimensional BWs with Ritz–Galerkin				
(0.1, 0.1)	1.15559E−8	2.74255E−8	4.98128E−8	7.94344E−8	1.15539E−7
(0.2, 0.2)	5.1508E−7	4.34347E−7	3.22397E−7	1.7431E−7	1.02961E−8
(0.3, 0.3)	2.63293E−6	2.4816E−6	2.27573E−6	2.00433E−6	1.66006E−6
(0.4, 0.4)	4.60055E−6	4.45255E−6	4.25572E−6	3.99849E−6	3.66878E−6
(0.5, 0.5)	5.4117E−7	5.28932E−7	5.14574E−7	4.98345E−7	4.80654E−7
(0.6, 0.6)	1.5061E−5	1.48076E−5	1.44822E−5	140612E−5	1.35084E−5
(0.7, 0.7)	3.84166E−5	3.78505E−5	3.71462E−5	3.62581E−5	3.5106E−5
(0.8, 0.8)	5.0541E−5	4.97745E−5	4.88533E−5	4.77303E−5	4.63092E−5
(0.9, 0.9)	2.89942E−5	2.83482E−5	2.7603E−5	2.6739E−5	2.57007E−5

which is the exact solution.

In [7], Heydari et al. used fractional operational matrix of integration for the LWs to solve this problem for $q = 1.1, 1.3, 1.5, 1.7, 1.9$. The best absolute error of the approximate solutions at some different points, in [7], with $k_1 = k_2 = 3$ and $M_1 = M_2 = 3$ is 1.6695×10^{-6}.

Example 2

Consider the following FDWE [3, 7]:

$$\frac{\partial^q u(x,t)}{\partial t^q} + \frac{\partial u(x,t)}{\partial t} = \frac{\partial^2 u(x,t)}{\partial x^2} + h(x,t), \quad (x,t) \in [0,1] \times [0,1], \quad 1 < q \leq 2, \tag{4.2}$$

where

$$h(x,t) = \frac{6t^{3-q}e^x}{\Gamma(4-q)} + 3t^2 e^x - t^3 e^x.$$

The initial and boundary conditions are determined correspondingly to the exact solution $u(x,t) = e^x t^3$.

From (3.18), we obtain $w(x,t) = t^3(1 - x + ex)$, then apply the numerical method presented in this paper for $k_1 = k_2 = 1$ and $M_1 = M_2 = 3$. Tables 1 and 2 present, respectively, the absolute error and the L^2 norm error for $u(x,t) - \tilde{u}(x,t)$ with different values of q. In Table 3, the absolute error of $u(x, t)$ and CPU times for $q = 1.5$ with $k_1 = k_2 = 1$ and different values of $M_1 = M_2$ are given. It is seen from Table 3 that, with increase in the number of the two-dimensional Bernoulli wavelets basis, the approximate values of $u(x, t)$ converge to the exact solutions. In Figs. 1 and 2, the exact and approximate solutions, and also the absolute difference between exact and approximate solutions of $u(x, t)$ with $q = 1.1$ and $q = 1.9$ are plotted, respectively. Also, the graphs of the absolute errors for $q = 1.5$ at $t = 0.5$ and $x = 0.5$ are shown in Fig. 3.

Table 2 The L^2 norm error, for Example 2

q	$\|u(x,t) - \tilde{u}(x,t)\|_{L^2}^2$
1.1	8.30025E−10
1.3	8.10785E−10
1.5	7.88518E−10
1.7	7.62722E−10
1.9	7.32127E−10

Liu et al. [3] employed the fractional predictor–corrector method and solved this problem with $q = 1.85$ and different values of time and space step sizes. They obtained 1.6341×10^{-3} for the best maximum absolute error. Moreover, in Table 1 we compare our results with obtained results in [7]. These comparisons show the more accuracy and less computations of our technique in comparison to other published methods.

Table 3 Absolute error for different values of M_1 and M_2 in $q = 1.5$, for Example 2

(x,t)	$M_1 = M_2 = 1$	$M_1 = M_2 = 2$	$M_1 = M_2 = 3$	$M_1 = M_2 = 4$	$M_1 = M_2 = 5$
(0, 0)	0	0	0	0	0
(0.1, 0.1)	6.18035E−4	2.90292E−6	4.98128E−8	5.21312E−10	$7.07305E-11$
(0.2, 0.2)	3.89089E−3	1.27728E−5	3.22397E−7	2.39923E−8	3.82387E−10
(0.3, 0.3)	9.90664E−3	3.18197E−6	2.27573E−6	3.31857E−8	2.504E−9
(0.4, 0.4)	1.67023E−2	7.22114E−5	4.25572E−6	1.44036E−7	7.99214E−9
(0.5, 0.5)	2.12456E−2	2.34008E−4	5.14574E−7	5.13346E−7	5.66433E−10
(0.6, 0.6)	2.06188E−2	4.30503E−4	1.44822E−5	5.85401E−7	2.76078E−8
(0.7, 0.7)	1.34408E−2	5.26881E−4	3.71462E−5	2.95966E−7	4.00415E−8
(0.8, 0.8)	1.57144E−3	3.72958E−4	4.88533E−5	1.78359E−6	1.23163E−8
(0.9, 0.9)	7.85399E−3	6.6203E−7	2.7603E−5	1.7876E−6	6.36963E−8
(1, 1)	0	0	0	0	0
CPU times (s)	0.312	0.342	0.374	0.390	0.405

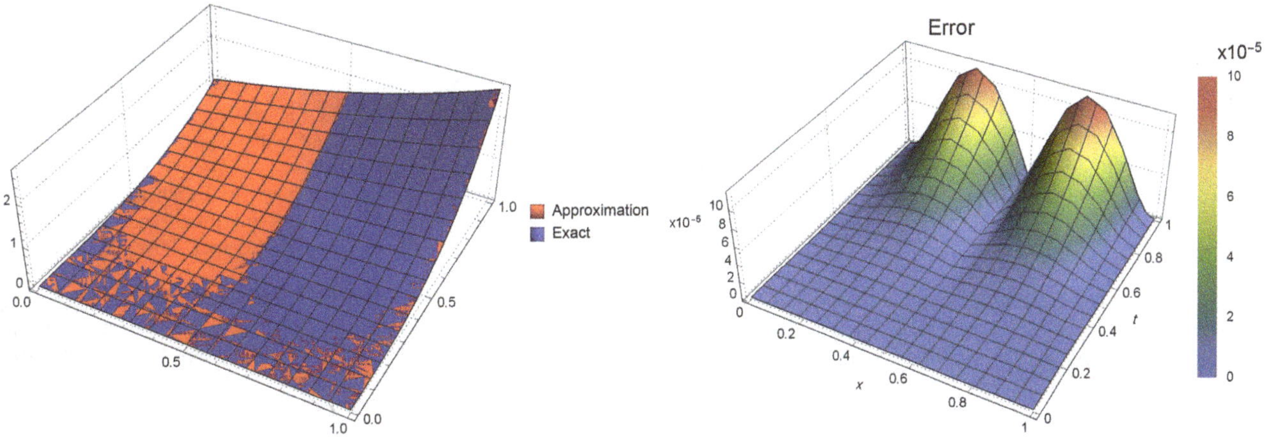

Fig. 1 The graphs of the Exact (*blue*) and approximate (*red*) solutions (*left side*) and the absolute error (*right side*) of $u(x, t)$ in $q = 1.1$, for Example 2

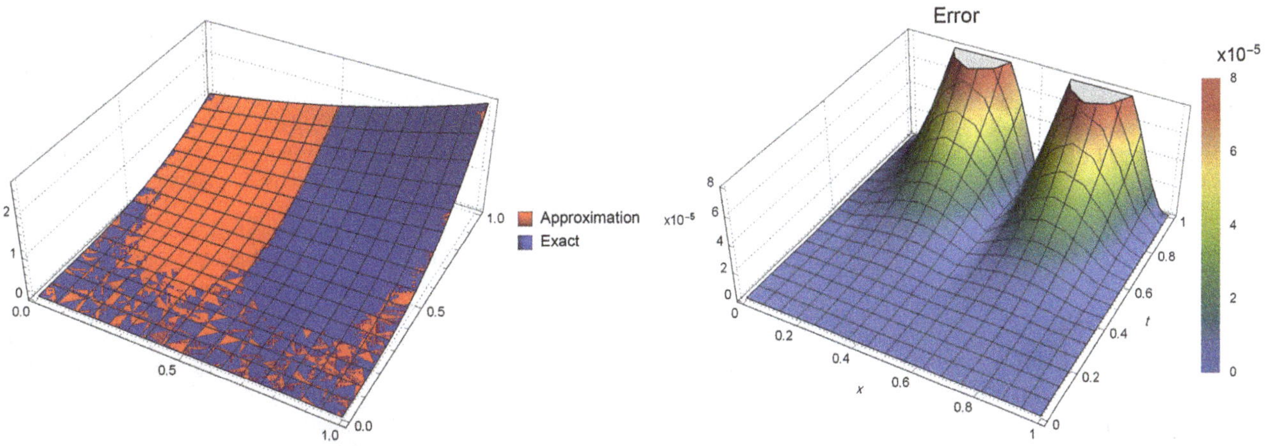

Fig. 2 The graphs of the Exact (*blue*) and approximate (*red*) solutions (*left side*) and the absolute error (*right side*) of $u(x, t)$ in $q = 1.9$, for Example 2

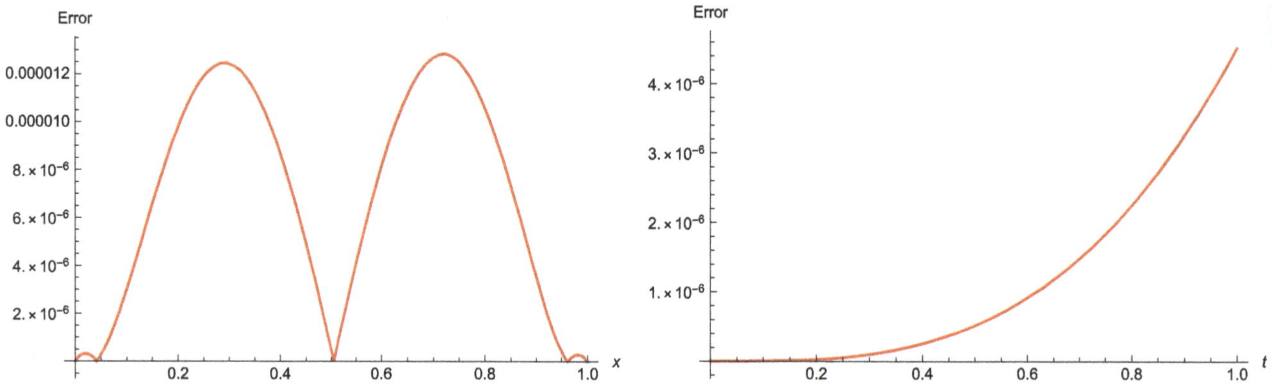

Fig. 3 The graphs of the absolute errors at $t = 0.5$ (*left*) and $x = 0.5$ (*right*) in $q = 1.5$, for Example 2

Conclusion

In this paper, the two-dimensional BWs was defined. Then the satisfier function in Ritz–Galerkin method with the two-dimensional BWs basis was successfully applied to solve the second-order time FDWE. Using of satisfier function in the Ritz–Galerkin method is an efficient tool to put on the initial and boundary conditions. Furthermore, a small number of basis elements were sufficient to derive accurate numerical solutions. Also, our results were compared with obtained results in [3] and [7]. Comparison for the numerical examples shows the more accuracy and less computations of the proposed method in comparison to other published methods.

References

1. Chen, J., Liu, F., Anh, V., Shen, S., Liu, Q., Liao, C.: The analytical solution and numerical solution of the fractional diffusion-wave equation with damping. Appl. Math. Comput. **219**, 1737–1748 (2012)
2. Bhrawy, A.H., Dohac, E.H., Baleanud, D., Ezz-Eldien, S.S.: A spectral tau algorithm based on Jacobi operational matrix for numerical solution of time fractional diffusion-wave equations. J. Comput. Phys. **293**, 142–156 (2015)
3. Liu, F., Meerschaert, M.M., McGough, R.J., Zhuang, P., Liu, Q.: Numerical methods for solving the multi-term time-fractional wave-diffusion equation. Fract. Calc. Appl. Anal. **16**, 9–25 (2013)
4. Mainardi, F.: The fundamental solutions for the fractional diffusion-wave equation. Appl. Math. Lett. **9**, 23–28 (1996)
5. Ren, J., Gao, G.: Efficient and stable numerical methods for the two-dimensional fractional Cattaneo equation. Numer. Algorithms **1**(24), 876–895 (2014)
6. Wess, W.: The fractional diffusion equation. J. Math. Phys. **27**, 2782–2785 (1996)
7. Heydari, M., Hooshmandasl, M.R., Ghaini, F.M., Cattani, C.: Wavelets method for the time fractional diffusion-wave equation. Phys. Lett. A **379**, 71–76 (2015)
8. Hu, X., Zhang, L.: A compact finite difference scheme for the fourth-order fractional diffusion-wave system. Comput. Phys. Commun. **182**, 1645–1650 (2011)
9. Zeng, F.: Second-order stable finite difference schemes for the time-fractional diffusion-wave equation. J. Sci. Comput. 1–20 (2014)
10. Hu, X., Zhang, L.: On finite difference methods for fourth-order fractional diffusion-wave and subdiffusion systems. Appl. Math. Comput. **218**, 5019–5034 (2012)
11. Darzi, R., Mohammadzade, B., Mousavi, S., Beheshti, R.: Sumudu transform method for solving fractional differential equations and fractional diffusion-wave equation. J. Math. Comput. Sci. **6**, 79–84 (2013)
12. Hosseini, V., Shivanian, E., Chen, W.: Local radial point interpolation (MLRPI) method for solving time fractional diffusion-wave equation with damping. J. Comput. Phys. **312**, 307–332 (2016)
13. Cui, M.: Convergence analysis of high-order compact alternating direction implicit schemes for the two-dimensional time fractional diffusion equation. Numer. Algorithms **62**, 383–409 (2013)
14. Yousefi, S.A., Barikbin, Z.: Ritz Legendre multiwavelet method for the damped generalized regularized long-wave equation. J. Comput. Nonlinear Dyn. **7**, 1–4 (2011)
15. Yousefi, S.A., Barikbin, Z., Dehghan, M.: Ritz–Galerkin method with Bernstein polynomial basis for finding the product solution form of heat equation with non-classic boundary conditions. Int. J. Numer. Methods Heat Fluid Flow **22**, 39–48 (2012)
16. Yousefi, S.A., Lesnic, D., Barikbin, Z.: Satisfier function in Ritz–Galerkin method for the identification of a time-dependent diffusivity. J. Inverse Ill Posed Probl. **20**, 701–722 (2012)
17. Barikbin, Z., Ellahi, R., Abbasbandy, S.: The Ritz–Galerkin method for MHD Couette flow of non-Newtonian fluid. Int. J. Ind. Math. **6**, 235–243 (2014)
18. Lesnic, D., Yousefi, S.A., Ivanchov, M.: Determination of a time-dependent diffusivity from nonlocal conditions. J. Appl. Math. Comput. **41**, 301–320 (2013)
19. Rashedi, K., Adibi, H., Dehghan, M.: Application of the Ritz–Galerkin method for recovering the spacewise-coefficients in the wave equation. Comput. Math. Appl. **65**, 1990–2008 (2013)
20. Rashedi, K., Adibi, H., Dehghan, M.: Determination of space-time dependent heat source in a parabolic inverse problem via the Ritz–Galerkin technique. Inverse Probl. Sci. Eng. **22**, 1077–1108 (2014)
21. Chui, C.K.: Wavelets: A Mathematical Tool for Signal Analysis. SIAM, Philadelphia (1997)
22. Shamsi, M., Razzaghi, M.: Solution of Hallen's integral equation using multiwavelets. Comput. Phys. Commun. **168**, 187–197 (2005)
23. Lakestani, M., Razzaghi, M., Dehghan, M.: Semi orthogonal spline wavelets approximation for Fredholm integro-differential equations. Math. Probl. Eng. 1–12 (2006)

24. Beylkin, G., Coifman, R., Rokhlin, V.: Fast wavelet transforms and numerical algorithms I. Commun. Pure Appl. Math. **44**, 141–183 (1991)
25. Marzban, H., Razzaghi, M.: Hybrid functions approach for linearly constrained quadratic optimal control problems. Appl. Math. Model. **27**, 471–485 (2003)
26. Razzaghi, M., Elnagar, G.: Linear quadratic optimal control problems via shifted Legendre state parametrization. Int. J. Syst. Sci. **25**, 393–399 (1994)
27. Mashayekhi, S., Ordokhani, Y., Razzaghi, M.: Hybrid functions approach for optimal control of systems described by integro-differential equations. Appl. Math. Model. **37**, 3355–3368 (2013)
28. Guf, J.S., Jiang, W.S.: The Haar wavelets operational matrix of integration. Int. J. Syst. Sci. **27**, 623–628 (1996)
29. Keshavarz, E., Ordokhani, Y., Razzaghi, M.: Bernoulli wavelet operational matrix of fractional order integration and its applications in solving the fractional order differential equations. Appl. Math. Model. **38**(24), 6038–6051 (2014)
30. Kreyszig, E.: Introductory Functional Analysis with Applications. Wiley, New York (1978)
31. Costabile, F., Dellaccio, F., Gualtieri, M.I.: A new approach to Bernoulli polynomials. Rend. Mat. **26**, 1–12 (2006)
32. Arfken, G.: Mathematical Methods for Physicists, 3rd edn. Academic press, San Diego (1985)
33. Maleknejad, K., Hashemizadeh, E., Basirat, B.: Computational method based on Bernstein operational matrices for nonlinear Volterra–Fredholm–Hammerstein integral equations. Commun. Nonlinear Sci. Numer. Simulat. **17**, 52–61 (2012)

18

On a fractional multi-agent cloud computing system based on the criteria of the existence of fractional differential equation

Rabha W. Ibrahim[1] · Yass K. Salih[2]

Abstract In the current work, we deal with the dynamic of multi-agent cloud computing system. We introduce a new criteria to minimize the cost with high accuracy of product description, contingent upon a class of fractional differential equations. By employing the fractional difference method on fractional Poisson's equation, we demonstrate that the solution is bounded for some domain. Simulation results are illustrated in the sequel. Outcomes obviously imply that the proposed method exhibits the highest performance compared with other methods.

Keywords Fractional calculus · Multi-agent system · Cloud computing · Fractional dynamic system

Introduction

A multi-agent system (MAS) is a self-possessed system of multiple networking, intelligent agents contained by an environment (cloud system CS). Agent environments can be allocated into: discrete environment; virtual environment; and continuous environment. These environments provide backward compatibility to guarantee easy mobility allowance. More data and information are required to cope with the increasing need for service delivery. This need is particularly obvious when considering high-bitrate multimedia applications that demand high quality of service (QoS) levels. These systems can be employed to resolve problems that are demanding or difficult for a separate agent or a uniform system to solve. Aptitude may involve some well-designed, functional method. Moreover, there is significant overlap, and MAS is not continuously the same as an agent-based system. Subjects where MAS investigation may carry a fitting method to solve scientific and social problems. MAS can be designed in many ways. Discrete methods (whether the total of potential actions in the environment is finite); dynamic methods (how many agents guidance the environment at the moment); periodicity (whether agent travels in confident time periods); and dimensional spaces (whether features are significant factors of the environment, such that each agent in MAS reflects space in its decision making).

During the last five decades, a huge amount of investigators has been worked in almost all fields of sciences and engineering in fractional calculus (fractional derivative and integral) of arbitrary calculus. This field is a major concept in the mathematical analysis to describe the non-linear. This indicates the importance of fractional calculus as an exciting mathematical method for solving various problems in science and engineering. Nowadays, the fractional calculus is considered as a key for opining the generalizations, modifications, and extensions in all sciences. The most well known of these operators that have been promoted in the world of fractional calculus are the Riemann–Liouville, Caputo (continuous and discrete fractional differential and integral operators), and Grunwald–Letnikov (discrete fractional differential operator) (see [1]). There are many types of fractional operators in real and complex planes, which are named by their finders.

✉ Rabha W. Ibrahim
rabhaibrahim@yahoo.com

Yass K. Salih
yass19752000@yahoo.com

1 Center of Mobile Cloud Computing Research, University Malaya, 50603 Kuala Lumpur, Malaysia

2 College of Engineering, Universiti Tenaga Nasional, 43000 Kajang, Selangor, Malaysia

Recently, the author suggested different classes of fractional multi-agent computing systems (see [2–7]). The first author showed the advantages of utilizing fractional calculus. Specifically, when it was detected that the explanation of some complex systems is more accurate when the fractional derivative is utilized. In addition, this subject deals with integro-differential equations, where the integrals are of convolution category and display kernels of power-law form. Hence, the fractional calculus discoveries usage in many categories of science and engineering, containing fluid flow, diffusive transport theory, electrical networks, electromagnetic theory, probability and statistics, theory of chaos and fractals, viscoelasticity, image processing, signal processing, food processing, chemical processing, and information theory.

In this study, we suggest a new fractional multi-agent cloud computing system. We introduce a new criteria to minimize the cost with high accuracy of product description, contingent upon a class of fractional differential equations. By employing the fractional difference method on fractional Poisson's equation, we demonstrate that the solution is bounded for some domain. Simulation results are illustrated in the sequel. Outcomes obviously imply that the proposed method exhibits the highest performance compared with other methods.

The paper is devoted as follows: "Formulation" deals with the preliminary and mathematical formulation; "Proposed algorithm" contains the proposed algorithm; "Applications" involves applications modeling systems; "Stability" studies the stability of the proposed algorithm; and "Conclusion" concludes our work.

Formulation

We aim to improve the dynamic of the computing model change between the choice of the traditional on-site computing model and the cloud computing model. We investigate a large economy in a continuous time situation; therefore, there are an appropriately big number of companies, businesses, and firms permitting to be close to importance. The planning pressures on the effect of the general trend of choices of technology transition times. In such a state, a continuum of agents having non-homogeneous references pay a cost to transfer from one point to another point in the state space. Using the usual financial derivatives is caused many disadvantageous. The most renowned disadvantage that financial derivatives could imply some financial difficulties. These difficulties can be recognized when the MAS may frustrate about some operations on derivative tools, thus losing morale for attempting another innovative financial tool. Therefore, here, we avoid using the usual derivative to keep our system in the stable case.

In general, for integer systems, the stability properties of any dynamic system are completely calculated from the location of the Laplacian eigenvalues of the network [8]. In this work, we suggest another method, based on the fractional Gâteaux derivative (FGD) [9]. This type of derivative connects two agents χ_i and χ_j in the formula as follows:

Definition 2.1 Let $(\Xi, \|.\|)$ be a real Banach space (the agreement space) with the dual space Ξ^*. A function $\psi : \Xi \to \mathbb{R}$ has a fractional Gâteaux derivative, of the order $0<\wp<1$ at $\chi_i \in \Xi$ if $\psi^{(\wp)}(\chi_i) \in \Xi^*$ exists, such that for a constant $\hbar > 0$, the forward operator $\mathrm{FW}_{\chi_j}(\hbar)$, $\chi_j \in \Xi$ is introduced by the equality

$$\mathrm{FW}_{\chi_j}(\hbar)\psi(\chi_i) := \psi(\chi_i + \hbar\,\chi_j),$$

with the fractional difference on the right

$$\Delta^{\wp}_{\chi_j}\psi(\chi_i) := \left(\mathrm{FW}_{\chi_j}(\hbar) - 1\right)^{\wp}\psi(\chi_i),$$

and its fractional derivative on the right

$$\lim_{\hbar\to 0^+} \frac{\Delta^{\wp}_{\chi_j}[\psi(\chi_i) - \psi(0)]}{\hbar^{\wp}} = \psi^{(\wp)}(\chi_i)(\chi_j), \quad \forall \chi_j \in \Xi.$$

The Sobolev spaces involve a natural norm:

$$\|\phi\|_{\kappa,\wp} = \left(\sum_{n=0}^{\kappa}\left\|\phi^{(n)}\right\|^{\wp}_{L^{\wp}}\right)^{\frac{1}{\wp}} = \left(\sum_{n=0}^{\kappa}\int\left|\phi^{(n)}(\chi)\right|^{\wp}\mathrm{d}\chi\right)^{\frac{1}{\wp}}.$$

Subjected to the norm $\|.\|_{\kappa,\wp}$, $W_{\kappa,\wp}$ yields a Banach space. A generalization of the space is imposed for an open set $\Omega \in \mathbb{R}^n$, $\kappa \in \mathbb{N}$, and $1 \le p < \infty$. The Sobolev space $W^{\kappa,p}(\Omega)$ is presented to be the set of all functions ϕ defined on Ω, such that for every multi-index ι with $|\iota| = \kappa$, the mixed partial derivative

$$\phi^{(\wp)} = \frac{\partial^{|\wp|}\phi}{\partial\chi_1^{\wp_1}\ldots\partial\chi_n^{\wp_n}}$$

is both locally integrable and in $L_p(\Omega)$, that is

$$\left\|\phi^{(\wp)}\right\|_{L^p} < \infty.$$

That is, the Sobolev space $W^{\kappa,p}(\Omega)$ is introduced as

$$W^{\kappa,p}(\Omega) = \{v \in L^p(\Omega) : v \text{ is absolutely continuos and } D^{\wp}v \in L^p(\Omega)\ \forall|\wp| \le \kappa\}.$$

The advantages of suggesting the FGD are included the distribution, stabilizing, and increasing the number of agents in the system and the purchase cost is fixed in contractual format.

The agent i has the cost function $F_i(\chi_i, u_i, \mu_i, u^{(\wp)}(\chi_i)(\chi_j), u^{(2\wp)}(\chi_i)(\chi_j))$ representing what an agent pays to have the characteristics χ_i (i.e., the level of cloud computing at any time t) under the controller input u_i and the density μ_i of the population for a given level of χ_i. Thus, we may present the fractional equation:

$$F_i(\chi_i, u_i, \mu_i, u_i^{(\wp)}(\chi_i)(\chi_j), u_i^{(2\wp)}(\chi_i)(\chi_j)) = \vartheta_i(\chi), \quad i = 1, \ldots, n$$
$$= \sum_{j \in \mathbb{N}} \alpha_{ij}(\chi_j - \chi_i) + u_i + \mu_i, \tag{1}$$
$$\left(D_i = \chi_i\right),$$

where D_i is the outcome at any time t, $\chi_i^{(\wp)}$ is the fractional $\wp$-order derivative, such that $\wp \in (0, 1)$, $\alpha_{ij} > 0$ is the (i, j) (agent i can receive information from agent j; otherwise, $\alpha_{ij} = 0$) element of the adjacency matrix. In matrix form, we can extend Eq. (1) as follows:

$$F(\chi, u, \mu, \nabla^{(\wp)}u(\chi), \nabla^{(2\wp)}u(\chi)) = \vartheta(\chi), \tag{2}$$
$$(u(\chi) = D(\chi)),$$

where $\chi = (\chi_1, \ldots, \chi_n)^\top$, $u = (u_1, \ldots, u_n)^\top$, $\mu = (\mu_1, \ldots, \mu_n)^\top$, $\vartheta(t) = (\vartheta_1, \ldots, \vartheta_n)^\top$ and $\nabla^{(\wp)}$ is the fractional gradient. It is clear that the $u(\chi)$ is a function of independent variable $\chi \in \mathbb{R}^n$. In operating form, we have the system:

$$\mathcal{L}u(\chi) = \vartheta(\chi), \quad \chi \in \Omega \subset \mathbb{R}^n, \tag{3}$$
$$(\Theta u(\chi) = \phi(\chi) \text{ on } \partial\Omega),$$

where $\mathcal{L}$ and Θ are differential operators in $\mathbb{R}^n$ and Ω is bounded in $\mathbb{R}^n$. Our aim is to minimize the problem:

$$\int_\Omega \left(\mathcal{L}u(\chi) - \vartheta(\chi)\right)^2 d\chi, \quad \chi \in \Omega \subset \mathbb{R}^n, \tag{4}$$

such that

$$\int_{\partial\Omega} \left(\Theta u(\chi) - \phi(\chi)\right)^2 d\chi = 0,$$

which the objectives and constraint are functional with $W^{\kappa,2}(\Omega)$. We shall solve the problem (4) approximately.

Proposed algorithm

To solve problem (4), we need the following facts:

Objective function Our chief objective is to investigate a company's optimal flat of usage of cloud computing. With the cost proposed in "Formulation", all rational companies make the decision on the flat of switch to the cloud computing pattern to minimize its estimated discount cost with respect to the effort cost. The objective function provides how much each variable contributes to the value to be optimized in the problem. Moreover, the negative of this function is called the utility function, in which to be maximized. Therefore, it is a function that maps all variables onto a real number intuitively representing some evolution with the minimization of the cost. For this purpose, we define a suitable objective function as follows:

$$\widehat{F}(\chi) = F(\chi) + \sum_{i=1}^{m} f(\alpha_i, \rho_i(\chi)) + \sum_{i=1}^{n} [f(\beta_i, \sigma_i(\chi)) + f(\beta_i, -\sigma_i(\chi))] \tag{5}$$

satisfying the conditions

$$\left(\rho_i(\chi) \le 0,\ i = 1, \ldots, m,\ \sigma_j(\chi) = 0,\ j = 1, \ldots, n\right),$$

where α_i and β_i are positive constants, f is the penalty function, which modifies the original objective function, and ρ_i and σ_i are the constraints. Our method is based on the finite-difference techniques. The penalty strategy makes the use of of finite-difference techniques on any boundary conditions on a boundary as fitness function in the cloud.

Fitness function (FF) The fitness function of the cloud, between agent χ_i and χ_j, is employed to control and summarize how close a given design solution to achieve the set aims of the process during the evolution of the system. Equation (4) can be reduced to minimize the problem:

$$\widehat{E}_\Omega(\widehat{u}(\chi)) = \sum_{\chi_i \in \Omega} \left(\mathcal{L}\widehat{u}(\chi_i) - \vartheta(\chi_i)\right)^2 \tag{6}$$

satisfying the conditions:

$$\widehat{E}_{\partial\Omega}(\widehat{u}(\chi)) = \sum_{\chi_i \in \partial\Omega} \left(\Theta\widehat{u}(\chi_i) - \phi(\chi_i)\right)^2 = 0, \tag{7}$$

where $\widehat{u}$ is the estimated value of u. Problems (6)–(7) can be solved by assuming the fitness function:

$$\Phi(\chi) = \widehat{E}_\Omega(\widehat{u}(\chi)) + \tau\widehat{E}_{\partial\Omega}(\widehat{u}(\chi)), \tag{8}$$

where τ is the penalty parameter.

Fractional Poisson's equation (FPE) It is a generalization of Laplace's equation, which is used widely in the cloud computing systems [8]. Poisson's equation arises to describe the potential connection caused by a given charge distribution of the agent i, with the agent j. Moreover, Poisson's equation is employed to reconstruct a smooth 3D system based on a large number of cloud agents χ_i, $i = 1, \ldots, n$, where each agent carries an estimate of the local cost. By applying the fractional derivative in "Formulation", we may generalize the Poisson's equation on two-dimensional rectangular domains:

$$u_{\chi_i}^{(2\wp)} + u_{\chi_j}^{(2\wp)} = \vartheta(\chi_i, \chi_j), \quad (a<\chi_i<b, c<\chi_j<d, 0<\wp\leq 1), \tag{9}$$

satisfying

$$u(a,\chi_j) = u_1(\chi_j),\ u(b,\chi_j) = u_2(\chi_j),\ u(\chi_i, c) = u_3(\chi_i),\ u(\chi_i, d) = u_4(\chi_i),$$

where u_{ij} is the approximated value of $u(\chi_i, \chi_j)$. By applying the finite-difference method for fractional order, then Poisson's equation implies

$$\frac{u_{i+1,j} - 2\wp u_{i,j} + \wp(2\wp - 1)u_{i-1,j}}{h^{2\wp}} + \frac{u_{i,j+1} - 2\wp u_{i,j} + \wp(2\wp-1)u_{i,j-1}}{k^{2\wp}} = \vartheta_{i,j}. \tag{10}$$

Note that, when $\wp = 1$, Eq. (10) becomes in normal form. This method is admitted consistency, stability, accuracy ($u \approx \widehat{u}$), and convergence for fractional differential equations. These properties of the method help us to understand how well a numerical approximation can be schemed for various classes of differential equations. We introduce a formal relation of the consistency that can be utilized for any partial differential equation defined on any domain. Stability deals with the behavior of solution $|u_{i,j} - u(ih, jk)|$ as numerical calculation progresses for fixed discrete steps.

Based on Eq. (10), the fitness function can be calculated as follows:

$$\begin{aligned}\Phi(\chi_i, \chi_j) =& \widehat{E}_\Omega(\widehat{u}(\chi)) + \tau \widehat{E}_{\partial\Omega}(\widehat{u}(\chi)) \\ =& \sum_{\Omega} \left[2\wp(1+\kappa)u_{ij} - (u_{i-1,j} + u_{i+1,j}) - (u_{i,j-1} + u_{i,j+1}) - h^{2\wp}\vartheta_{ij}\right]^2 \\ &+ \tau \sum_{\partial\Omega} \left[(u_{1,j} - u_1(\chi_j))^2 + (u_{m,j} - u_2(\chi_j)) + (u_{i,1} - u_3(\chi_i))^2 + (u_{i,n} - u_4(\chi_i))^2\right],\end{aligned} \tag{11}$$

where $i = 1, \ldots, m$, $j = 1, \ldots, n$, $\kappa = h^{2\wp}/k^{2\wp}$. The function is in fact an evaluation of good and bad connecting results in the cloud system.

Examples

We proceed to illustrate two examples to describe the proposed algorithm.

The above examples are shown the domain of the MAS. The first example is described a convex domain, while the second is suggested in a concave domain.

Table 1 shows two examples of the proposed method. Figure 1 shows the solution for the two systems for different $\wp \in (0, 1]$ values. The approximate solution of the first problem is convex type. The error is calculated utilizing the minimum of the fitness function of these outcomes given across domains of various measures ($h = k = 0.1, 0.2$). The comparison is imposed by well-known techniques, such as the Genetic Algorithm (GA), LMI, and PSOA. For example, the minimum error for the first problem for $\wp = 0.75$, is equal to $E_{\mathrm{m}} = 2.35e - 005$, while $E_{\mathrm{GA}} = 2.12$ and $E_{\mathrm{PSOA}} = 8.1e - 004$. The second problem has a concave solution. It is evident from the above fractional systems that the proposed technique is a very powerful to solve not only initial value problem, but boundary value problems too.

Table 1 Fractional multi-agent system

$\nabla^{(2\wp)}u =$	Initial conditions	Exact solution	Error ($\wp = 0.75$)	Error ($\wp = 0.5$)
Example 1				
4	$u(0,\chi_j) = \chi_j^2 + \chi_j + 1$	$u = \chi_i^2 + \chi_j^2 + \chi_i + \chi_j + 1$		
$\chi_{i,j} \in [0,1]$	$u(1,\chi_j) = \chi_j^2 + \chi_j + 3$		2.35e−005	2.35e−01
	$u(\chi_i, 0) = \chi_i^2 + \chi_i + 1$			
	$u(\chi_i, 1) = \chi_i^2 + \chi_i + 3$			
Example 2				
$(\chi_i - 2)e^{-\chi_i} + \chi_i e^{-\chi_j}$	$u(0,\chi_j) = 0$	$u = \chi_i(e^{-\chi_i} + e^{-\chi_j})$		
$\chi_{i,j} \in [0,1]$	$u(1,\chi_j) = e^{-\chi_j} + \frac{1}{e}$		6.3e−0066	6.3e−026
	$u(\chi_i, 0) = \chi_i(e^{-\chi_i} + 1)$			
	$u(\chi_i, 1) = \chi_i(e^{-\chi_i} + \frac{1}{e})$			

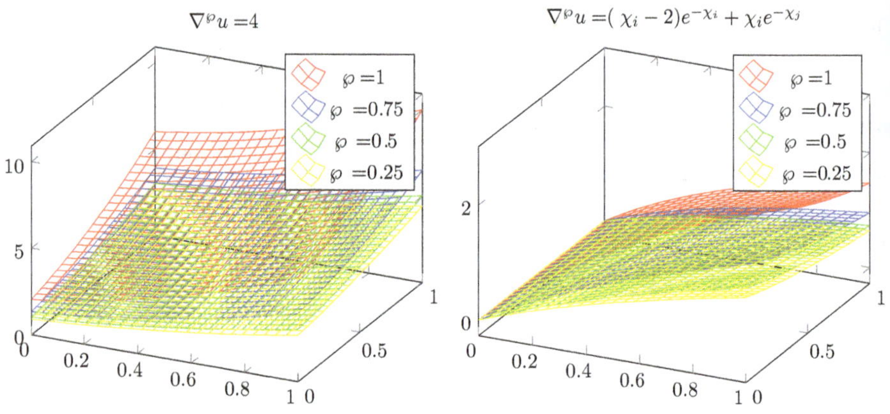

Fig. 1 Convergence of solutions of systems $\nabla^{\wp} = 4$ and $\nabla^{\wp}u = (\chi_i - 2)e^{-\chi_i} + \chi_i e^{-\chi_j}$

Table 2 Utility distribution

Request χ_i	Utility (Example 1)	Utility (Example 2)	Utility [10]
1	0.89	0.65	0.5
2	1.34	1.1	1
3	2.8	1.33	1.5
4	3.66	2.9	2
5	4.78	3.1	2.5

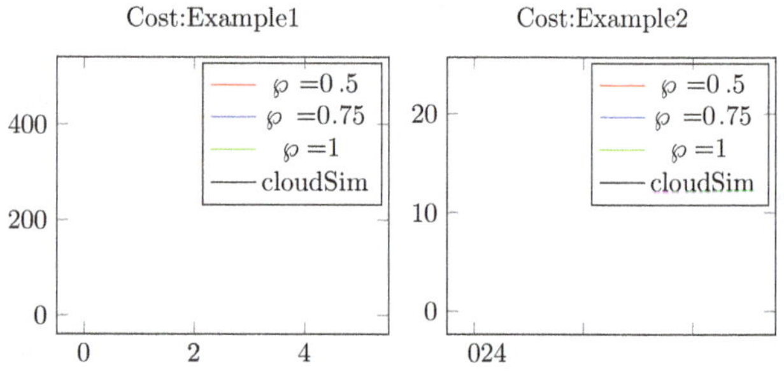

Fig. 2 Cost function with respect to utility function u_i

Table 2 shows five-agent system dynamic of its utility in a convex domain (example 1) and concave domain (example 2) comparing with the utility function which is given in [10] as follows:

$$U_{ij} = \sum_{j=1}^{n} \alpha_{ij}\chi_i.$$

Convexity has provided a good utility rather than the concavity domain. The distribution is stable and gives best-connected on satisfying a diversity of radio access technologies that collaborate to construct an integrated cloud computing system that reunites the requests of agents. Thus, users are provided the optimum service delivery through selecting the most appropriate network among different available wireless networks. The MAS utility is utilized to analytically process [11]. A hybrid system can be realized depending on the weight and quality levels for scoring [12]. Numerous investigations which used cloud computing systems to solve various problems are summarized in [13].

Applications

To faithfully model customer forms agreeing to their preferences, we utilize the above multi-agent systems, each with a characteristic form of the utility function (convex and concave forms). Our setting closely looks like a public game, where customers acquire costs for providing incomes to the public, and in this case, it is often presumed that self-interest is the most related parameters that illustrate the behavior of customers. Moreover, the utility function is capable to seizure various procedures of utility functions, such as the convex utility functions, which makes it very flexible. Therefore, all customers will choose their separate optimal value subject to their level of self-sacrifice and the convexity of their preferences (see Fig. 2). In this system model, agents request resources service by service. A service is composed of multiple tasks, such as data downloading and computing. Agent i wants to complete the jobs involved in the service as soon as possible. To process jobs, agents request to utilize their resources. The set of the sizes of jobs that agent i has to use is introduced by $\mu_i = \{\mu^{\ell}, \ell \text{ type of the job}\}$. The proposed method which allows two different jobs may require the same type of resources in which case the specific type of resource has to be connected between them. In addition, agent i requests extra resources and agent j shares its resources with agent i. A job of μ^{ℓ} can be divided into smaller parts (μ^{ℓ}), and when agent j shares resources with agent i, agent i outsources some parts of the job to agent j. Therefore, the mechanism of the process is as follows:

Demand $(\chi_{i,j}) \implies$ Controller of utility $(u) \implies$ outcomes (mini of the cost function).

Hence, we have

$$\text{Cost } = \alpha_{ij}|i-j| * |\chi_i - \chi_j|^{\wp} * u_{i,j} * \mu^{\ell}. \quad (12)$$

Thus, the total cost of the agent i, connecting with n agents, is calculated by

$$\text{Total Cost } = \sum_{j=1}^{n} \alpha_{ij}|i-j| * |\chi_i - \chi_j|^{\wp} * u_{i,j} * \mu^{\ell}. \quad (13)$$

Note that the distance between agents i and j takes the maximal value in the rectangle, which is determined by the diameter of the rectangle, i.e, $|i-j| = 5.83$ (for the first example) and $|i-j| = 1$ (for the second example).

Figure 2 shows the cost for the convex and concave systems. The convex system approaches to cloudSim at the value $\wp = 0.75$, while the concave system converges to the exact solution $\wp = 1$.

Stability

The mathematical term well-posed problem (Hadamard well-posed) is defined that mathematical platforms of physical phenomena should have the following properties:

(i) A solution exists.
(ii) The solution is unique.
(iii) The solution's behavior changes continuously with the initial conditions.

Otherwise, it is called ill-posed. It is well known that if the problem is well-posed, then it can be viewed as a good algorithm of outcome on a computer employing a stable algorithm. Since the system (1) is formulated for multi-agent users, it is enough to show that (1) has multi-solutions. In this case, we satisfy part (i) of the well-posed problem.

The discrete-time collective dynamics of the network under this algorithm can be read as follows:

$$\Delta^{\wp}_{\chi_j}\chi_i(k-1) + \aleph(k)\chi_i(k) = \Pi_i(k,\chi_i), \quad k \in [1,N],\ i = 1,\ldots,n, \quad (14)$$

$$\chi_i(0) = \chi_i(N+1) = 0,$$

where $\Pi_i : [1,N] \to \mathbb{R}$ is a continuous function, $\aleph := \sum \alpha_{ij} \geq 0$.

Introduce the Banach space as [14]

$$\mathcal{B} := \{\chi_i : [0,N+1] \to \mathbb{R} : \chi_i(0) = \chi_i(N+1) = 0\}$$

endowed with a discrete norm:

$$\|\chi_i\| := \left(\sum_{k=1}^{N+1} |\Delta^{\wp}_{\chi_j}\chi_i(k-1)|^p + \aleph(k)|\chi_i|^p\right)^{1/p},$$

such that

$$\max_{k\in[1,N]} |\chi_i| \leq \frac{(N+1)^{(p-1)/p}}{2}\|\chi_i\|, \quad \forall \chi_i \in \mathcal{B},\ i = 1,\ldots,n.$$

Let

$$\Phi(\chi_i) := \frac{\|\chi_i\|^p}{p}, \quad \Psi(\chi_i) := \sum_{k=1}^{N} \Pi_i(k),$$

$$\Theta(\chi_i) := \Phi(\chi_i) - \Psi(\chi_i), \quad \forall \chi_i \in \mathcal{B},$$

$$\overline{\Pi}_i(k,u) := \int_0^u \Pi_i(k,w)dw.$$

Note that $\Theta \in C^1(\mathcal{B},\mathbb{R})$, $\Theta(0) = 0$ and that all the critical points of Θ are the solutions of (12).

Definition 5.1 [15] A function $\theta : \Xi \to \mathbb{R}$ is standard as a Gâteaux differentiable and verifies the Palais–Smale (PS) condition if any bounded sequence $\{\chi_n\}$ with $\lim_{n\to\infty} \|\theta'(\chi_n)\|_{\Xi^*} = 0$ has a convergent subsequence.

We need the following result in the sequel [15].

Lemma 5.1 *Consider three positive constants C, C_1 and C_2, such that*

$$C_1 < \frac{(2+\overline{\sigma})^{1/p}(N+1)^{(p-1)/p}C}{2} < C_2, \quad (15)$$

where $\overline{\sigma} := \sum_{k=1}^{N}\sigma_k$. If

$$\beta_1 := \frac{\sum_{k=1}^{N}\max_t H(k,t) - \sum_{k=1}^{N}H(k,C)}{(2C_1)^p - (2+\overline{\sigma})(N+1)^{p-1}C^p} \leq \frac{1}{p(N+1)^{p-1}}$$

and

$$\beta_2 := \frac{\sum_{k=1}^{N}\max_t H(k,t) - \sum_{k=1}^{N}H(k,C)}{(2C_2)^p - (2+\overline{\sigma})(N+1)^{p-1}C^p} \leq \frac{1}{p(N+1)^{p-1}},$$

where $(2C_i)^p \neq (2+\overline{\sigma})(N+1)^{p-1}C^p$, $i = 1,2$, then (12) has at least one non-trivial solution μ^*, such that

$$\frac{2C_1}{(N+1)^{(p-1)/p}} < \|\mu^*\| < \frac{2C_2}{(N+1)^{(p-1)/p}}.$$

Lemma 5.2 [16] *Let the Φ satisfied (PS)-condition. Then there is a sequence u_n of pairwise distinct critical points (local minima) with $\lim_n\Phi(u_n) = \inf\Phi$ which weakly converges to a global minimum of Φ.*

In view of Lemmas 5.1 and 5.2, we conclude the following result.

Theorem 5.1 *Assume that there exist two real positive sequences ξ_n and ζ_n with $\lim_{n\to\infty}\zeta_n = 0$, such that*

$$\xi_n < \frac{2}{(\aleph+2)^{1/p}(N+1)^{(p-1)/p}}\zeta_n, \quad n \in \mathbb{N},$$

$$\ell := \frac{\sum_{k=1}^{N} \max_u \overline{\Pi}_i(k,u) - \sum_{k=1}^{N} \overline{\Pi}_i(k,K)}{(2\zeta_n)^p - (2+\aleph)(N+1)^{p-1}\xi_n^p} \leq \frac{\beta}{(\aleph+2)(N+1)^{p-1}}$$

and

$$\beta := \limsup_{u\to 0} \frac{\sum_{k=1}^{N} \overline{\Pi}_i(k,u)}{u^p}.$$

Then, (12) *admits a sequence of non-zero solutions which converges to zero.*

Conclusion

We discussed the question as how to optimize resources, scheduling based on an algorithm in cloud computing. The technique was additional capable and surpasses those of mathematical software design and reproducing the definite benefit of saving with the overall cost as well as task distribution. In addition, the objective function is taken in the sense of the fractional Poisson's equation. Utilizing this equation, we imposed a modification of the fitness function. This algorithm is proposed to evaluate and solve well-posed problems, such as initial and boundary value problems on finite domain. The method admitted two advantages: transforms the problem of constrained optimization into unconstrained, and with a suitable choice of the fractional order, the method is a good approximation. We suggested special type of fractional calculus named the fractional Gâteaux derivative. This class allows us to connect two agents in a multi-agent system. Moreover, one may suggests multi-connection, utilizing the above method. We studied the stability in view of Hadamard well-posed strategy. We proved that for some special class of fractional differential equations, the problem is stable, by approximating the solution to a convergent sequence. Finally, the method can be extended to higher dimension when the number of agents in the multi-agent system becomes large.

References

1. Tarasov, V.: Fractional Dynamics: Applications of Fractional Calculus to Dynamics of Particles, Fields and Media. Springer, New York (2010)
2. Ibrahim, R.W., Jalab, H.A., Gani, A.: Cloud entropy management system involving a fractional power. Entropy **18**(1), 14 (2015)
3. Ibrahim, R.W., Gani, A.: A new algorithm in cloud computing of multi-agent fractional differential economical system. Computing **98**(11), 1061–1074 (2016)
4. Ibrahim, R.W., Jalab, H.A., Gani, A.: Entropy solution of fractional dynamic cloud computing system associated with finite boundary condition. Bound. Value Probl. **2016**(1), 1–12 (2016)
5. Ibrahim, R.W., Jalab, H.A., Gani, A.: Perturbation of fractional multi-agent systems in cloud entropy computing. Entropy **18**(1), 31 (2016)
6. Ibrahim, R.W., Gani, A.: Hybrid cloud entropy systems based on Wiener process. Kybernetes **45**(7), 1072–1083 (2016)
7. Ibrahim, R.W.: Measurement of the communication possibility of service requests for multiservers in parallel connection in cloud computing systems. In: Thomas, C. (eds.) Complex Systems, Sustainability and Innovation. InTech. doi:10.5772/65285
8. Olfati-Saber, R., Fax, J.A., Murray, R.M.: Consensus and cooperation in networked multi-agent systems. Proc. IEEE **95**(1), 215–233 (2007)
9. Ibrahim, R.W., Jalab, H.A.: Discrete boundary value problem based on the fractional Gteaux derivative. Bound. Value Probl. **2015**(1), 23 (2015)
10. Salih, Y.K., Ong, H.S., Ibrahim, R.W.: An intelligent selection method based on game theory in heterogeneous wireless networks. Trans. Emerg. Telecommun. Technol. **27**(12), 1641–1652 (2016)
11. Fei, W., Tian, H., Lian, R.: Utility-based dynamic multi-service bandwidth allocation in heterogeneous wireless networks. Veh. Technol. Conf. **60971125**, 1–5 (2012)
12. Dyer, J.S., Fishburn, P.C., Steuer, R.E., Wallenius, J., Zionts, S., Science, M., May, N.: Multiple criteria decision making, multi-attribute utility theory. JSTOR Manag. Sci. **38**(5), 645–654 (1992)
13. Qi, H., Gani, A.: Research on mobile cloud computing: Review, trend and perspectives. In: Digital Information and Communication Technology and it's Applications (DICTAP), 2012 Second International Conference on. IEEE (2012)
14. Bonanno, G.: A critical points theorem and nonlinear differential problems. J. Glob. Optim. **28**(3–4), 249–258 (2004)
15. Motreanu, D., Râdulescu, V.: Variational and non-variational methods in nonlinear analysis and boundary value problems. In: Non-convex Optimization and Applications. Springer, US (2003)
16. Bonanno, G., Giovanni, M.: Infinitely many solutions for a boundary value problem with discontinuous non-linearities. Bound. Value Probl. **2009**(1), 1–20 (2009)

19

Wavelet methods for solving three-dimensional partial differential equations

Inderdeep Singh[1] · **Sheo Kumar**[1]

Abstract We present, a collocation method based on Haar wavelet and Kronecker tensor product for solving three-dimensional partial differential equations. The method is based on approximating a sixth-order mixed derivative by a series of Haar wavelet basis functions. The present method is suitable for numerical solution of all kinds of three-dimensional Poisson and Helmholtz equations. Numerical examples are solving to establish the efficiency and accuracy of the present method. Numerical results obtained are better as compared to numerical results obtained in past.

Keywords Haar wavelet method · Three-dimensional partial differential equations · Linear systems · Kronecker tensor product

Mathematics Subject Classification 65N99

Introduction

In many applications of engineering and science, there are various boundary value problems which involve three-dimensional partial differential equations. Only a few of these equations can be solved by analytical methods. In most cases, we depend on numerical solutions of such partial differential equations. There are several numerical methods available for solving these equations; the most common method used for solving such equations is finite difference method; but this method is slow. In the last few years, other numerical techniques were developed, which are more accurate, efficient and faster than previous numerical algorithms, such as: (a) Jacobi pseudospectral approximation for solving nonlinear complex generalized Zakharov system in [2], (b) A highly accurate collocation algorithm for solving $1 + 1$ and $2 + 1$ fractional percolation equations in [3], (c) Spectral-Galerkin algorithms using Jacobi polynomials for solving second- and fourth-order differential equations in [8] and [9] respectively, and (d) Legendre spectral-Galerkin method for solving multi-dimensional elliptic Robin boundary value problems in [11], (e) Jacobi spectral-Galerkin method for the integrated forms of fourth-order elliptic differential equations in [10]. In the last few decades, methods based on wavelet basis functions have been used abruptly. These methods are more efficient and give more accurate numerical results as compared to other well known methods. Wavelet methods are more interesting, accurate and reliable for solving integral and differential equations. Wavelets are a powerful and efficient mathematical tool that divides the data functions or operators into distinct frequency constituents and each constituent is analyzed or investigated with a resolution matching on its scale. Nowadays, wavelet methods are becoming a favorite choice of researchers for solving differential and integral equations. Haar [12] discovered a function, later known as Haar wavelet, in 1909. Such Haar functions are rectangular pair pulses and these are known as Daubechies wavelet of order 1. Also, it is a simplest orthonormal wavelet with compact support. The main disadvantage of Haar wavelets is their discontinuity and, therefore, derivatives do not exist at the points of discontinuities. Due to this, it is impossible to obtain the numerical solution of differential and integral equations.

✉ Sheo Kumar
sheoks53@gmail.com

Inderdeep Singh
inderdeeps.ma.12@nitj.ac.in

1 Department of Mathematics, Dr. B.R. Ambedkar National Institute of Technology, Jalandhar, Punjab 144011, India

There are two possibilities for overcoming these shortcomings. First, to regularize the piecewise constant Haar functions with interpolation splines; this technique has been applied by Cattani in [4, 5]. But, by this technique, it is difficult to find the solution easily and simplicity of Haar wavelets gets lost. Another possibility, which is proposed by Chen and Hsiao in [6, 7] is that they recommended to expand the highest derivative appearing in the differential equation into the Haar series, instead of the function itself. The other derivatives (and the functions) are obtained through integrations.

Numerical solutions of differential and integral equations using Haar wavelet have been presented by Lepik in [14, 15]. Numerical solutions of two-dimensional PDEs using Haar wavelet have been presented in Lepik [16]. The fundamental idea behind the Haar wavelet method is to convert the given problem into a system of equations which involves finite number of variables. In numerical analysis, because of the property of localization, wavelet based algorithms have become an important tools for solving ordinary and partial differential equations. A new approach of the Chebyshev wavelets method for partial differential equations with boundary conditions of the telegraph type equation has been presented in [13]. Haar wavelet collocation method has been presented in [19], for solving boundary layer fluid flow problems. A numerical assessment of parabolic partial differential equations using Haar and Legendre wavelets has been presented in [20]. Numerical solution of two-dimensional elliptic PDEs with nonlocal boundary conditions has been presented in [21]. In the present paper, we use Haar wavelet collocation method for solving three-dimensional Poisson and Helmholtz equations because Haar wavelet method has sparse representation, fast transformation, and possibility of implementation of fast algorithms. The general linear partial differential equation of the second order in three independent variables is of the form

$$A_{11}(x,y,z)\frac{\partial^2 u}{\partial x^2}+B_{11}(x,y,z)\frac{\partial^2 u}{\partial y^2}+C_{11}(x,y,z)\frac{\partial^2 u}{\partial z^2} + F_{11}\left[x,y,z,u,\frac{\partial u}{\partial x},\frac{\partial u}{\partial y},\frac{\partial u}{\partial z},\frac{\partial^2 u}{\partial x^2\partial y},\dots\right]=0, \tag{1}$$

on $K=\{(x,y,z): a<x<b, c<y<d, e<z<f\}$, with boundary conditions

$$u(0,y,z)=g_0(y,z), u(x,0,z)=g_1(x,z), u(x,y,0)=g_2(x,y), \tag{2}$$

and

$$u(1,y,z)=g_3(y,z), u(x,1,z)=g_4(x,z), u(x,y,1)=g_5(x,y), \tag{3}$$

where $A_{11}, B_{11}, C_{11}, g_0, g_1, g_2, g_3, g_4, g_5$ and f are known functions. The Poisson equation in three-dimensional Cartesian coordinates system plays an important role due to its wide range of application in areas like ideal fluid flow, heat conduction, elasticity, electrostatics, gravitation and other science fields especially in physics and engineering. Poisson and Helmholtz equations are arising in different branches of science and engineering such as fluid mechanics, electricity and magnetism and torsion problems.

Our main aim is to develop an accurate and efficient collocation method using Haar wavelet and Kronecker product for solving three-dimensional partial differential equations such as Poisson and Helmholtz equations, by approximating a sixth-order mixed derivative by a series of Haar wavelet basis functions. In Sect. 2, Haar wavelet method has been discussed. Error analysis has been described in Sect. 3. In Sect. 4, numerical examples have been solved using the present method and compared with the exact solutions.

Kronecker product of two matrices

For saving calculation time, we use the concept of Kronecker product of matrix A with matrix B of orders $p \times q$ respectively and is defined as:

$$A \otimes B = \begin{pmatrix} a_{11}B & a_{12}B & \cdots & a_{1q}B \\ a_{21}B & a_{22}B & \cdots & a_{2q}B \\ \vdots & \vdots & \vdots & \vdots \\ a_{p1}B & a_{p2}B & \cdots & a_{pq}B \end{pmatrix}. \tag{4}$$

The first documented work on Kronecker products was written by Johann Georg Zehfuss between 1858 and 1868. In MATLAB, the Kronecker product of two matrices A and B is directly calculated with the command *kron(A, B)*.

Kronecker product of three matrices

The Kronecker product of three A, B and C matrices each of orders $p \times q$ can be calculated as:

$$A \otimes B \otimes C = \begin{pmatrix} a_{11}E & a_{12}E & \cdots & a_{1q}E \\ a_{21}E & a_{22}E & \cdots & a_{2q}E \\ \vdots & \vdots & \vdots & \vdots \\ a_{p1}E & a_{p2}E & \cdots & a_{pq}E \end{pmatrix}, \tag{5}$$

where E is of the form:

$$E = B \otimes C = \begin{pmatrix} b_{11}C & b_{12}C & \cdots & b_{1q}C \\ b_{21}C & b_{22}C & \cdots & b_{2q}C \\ \vdots & \vdots & \vdots & \vdots \\ b_{p1}C & b_{p2}C & \cdots & b_{pq}C \end{pmatrix}. \tag{6}$$

Haar wavelet method

Consider $x \in [\sigma_1, \sigma_2]$, $y \in [\sigma_3, \sigma_4]$ and $z \in [\sigma_5, \sigma_6]$ where σ_1, σ_2, σ_3, σ_4, σ_5 and σ_6 are given constants. We shall define the quantities $M_1 = 2^{J_1}$, $M_2 = 2^{J_2}$ and $M_3 = 2^{J_3}$ where J_1, J_2 and J_3 are the maximal levels of resolution. Now, divide the interval $[\sigma_1, \sigma_2]$, $[\sigma_3, \sigma_4]$ and $[\sigma_5, \sigma_6]$ respectively into $2M_1$, $2M_2$ and $2M_3$ subintervals, each of length $\Delta x = (\sigma_2 - \sigma_1)/2M_1$, $\Delta y = (\sigma_4 - \sigma_3)/2M_2$ and $\Delta z = (\sigma_6 - \sigma_5)/2M_3$ respectively. Now, we introduce parameters : dilatation parameter $j_1 = 0, 1, 2, \ldots, J_1$; $j_2 = 0, 1, 2, \ldots, J_2$ and $j_3 = 0, 1, 2, \ldots, J_3$ and translation parameter $k_1 = 0, 1, 2, \ldots, m_1 - 1$; $k_2 = 0, 1, 2, \ldots, m_2 - 1$ and $k_3 = 0, 1, 2, \ldots, m_3 - 1$, where $m_1 = 2^{j_1}$, $m_2 = 2^{j_2}$ and $m_3 = 2^{j_3}$. The wavelet numbers i_1, i_2 and i_3 are calculated according the formula $i_1 = m_1 + k_1 + 1$, $i_2 = m_2 + k_2 + 1$ and $i_3 = m_3 + k_3 + 1$ respectively. Therefore, we have

$$h_{i_1}(x) = \begin{cases} 1, & \alpha_1 \le x < \alpha_2, \\ -1, & \alpha_2 \le x < \alpha_3, \\ 0, & \text{otherwise}, \end{cases} \tag{7}$$

$$h_{i_2}(y) = \begin{cases} 1, & \beta_1 \le y < \beta_2, \\ -1, & \beta_2 \le y < \beta_3, \\ 0, & \text{otherwise}, \end{cases} \tag{8}$$

and

$$h_{i_3}(z) = \begin{cases} 1, & \gamma_1 \le z < \gamma_2, \\ -1, & \gamma_2 \le z < \gamma_3, \\ 0, & \text{otherwise}, \end{cases} \tag{9}$$

where $\alpha_1 = \sigma_1 + 2k_1\omega_1\Delta x$, $\alpha_2 = \sigma_1 + (2k_1 + 1)\omega_1\Delta x$, $\alpha_3 = \sigma_1 + 2(k_1 + 1)\omega_1\Delta x$, $\omega_1 = M_1/m_1$,
$\beta_1 = \sigma_3 + 2k_2\omega_2\Delta y$, $\beta_2 = \sigma_3 + (2k_2 + 1)\omega_2\Delta y$, $\beta_3 = \sigma_3 + 2(k_2 + 1)\omega_2\Delta y$, $\omega_2 = M_2/m_2$,
$\gamma_1 = \sigma_5 + 2k_3\omega_3\Delta z$, $\gamma_2 = \sigma_5 + (2k_3 + 1)\omega_3\Delta z$, $\gamma_3 = \sigma_5 + 2(k_3 + 1)\omega_3\Delta z$, $\omega_3 = M_3/m_3$.

The collocation points are obtained as:

$$x_{l_1} = \alpha_1 + \frac{(2l_1 - 1)}{2M_1}, l_1 = 1, 2, 3, 4, \ldots, 2M_1; \tag{10}$$

$$y_{l_2} = \beta_1 + \frac{(2l_2 - 1)}{2M_2}, l_2 = 1, 2, 3, 4, \ldots, 2M_2; \tag{11}$$

and

$$z_{l_3} = \gamma_1 + \frac{(2l_3 - 1)}{2M_3}, l_3 = 1, 2, 3, 4, \ldots, 2M_3. \tag{12}$$

Consider the approximate wavelet solution of the form

$$u_{xxyyzz}(x, y, z) = \sum_{i_1=1}^{2M_1}\sum_{i_2=1}^{2M_2}\sum_{i_3=1}^{2M_3} W_{i_1i_2i_3} h_{i_1}(x) h_{i_2}(y) h_{i_3}(z). \tag{13}$$

Integrating (13), twice with respect to x, from 0 to x, we obtain

$$u_{yyzz}(x, y, z) = u_{yyzz}(0, y, z) + x u_{xyyzz}(0, y, z) + \sum_{i_1=1}^{2M_1}\sum_{i_2=1}^{2M_2}\sum_{i_3=1}^{2M_3} W_{i_1i_2i_3} P_{2,i_1}(x) h_{i_2}(y) h_{i_3}(z). \tag{14}$$

Putting $x = 1$ in (14), we obtain

$$u_{xyyzz}(0, y, z) = u_{yyzz}(1, y, z) - u_{yyzz}(0, y, z) - \sum_{i_1=1}^{2M_1}\sum_{i_2=1}^{2M_2}\sum_{i_3=1}^{2M_3} W_{i_1i_2i_3} P_{2,i_1}(1) h_{i_2}(y) h_{i_3}(z). \tag{15}$$

From (14) and (15), we obtain

$$u_{yyzz}(x, y, z) = u_{yyzz}(0, y, z) + x(u_{yyzz}(1, y, z) - u_{yyzz}(0, y, z)) + \sum_{i_1=1}^{2M_1}\sum_{i_2=1}^{2M_2}\sum_{i_3=1}^{2M_3} W_{i_1i_2i_3} (P_{2,i_1}(x) - xP_{2,i_1}(1)) h_{i_2}(y) h_{i_3}(z). \tag{16}$$

Again, integrating (16), twice with respect to y, from 0 to y, we obtain

$$u_{zz}(x, y, z) = \psi_{11}(x, y, z) + y\psi_{12}(x, z) + x\psi_{13}(y, z) + \sum_{i_1=1}^{2M_1}\sum_{i_2=1}^{2M_2}\sum_{i_3=1}^{2M_3} W_{i_1i_2i_3} (P_{2,i_1}(x) - xP_{2,i_1}(1)) P_{2,i_2}(y) h_{i_3}(z). \tag{17}$$

where

$$\psi_{11}(x, y, z) = u_{zz}(x, 0, z) + u_{zz}(0, y, z) - u_{zz}(0, 0, z), \tag{18}$$

$$\psi_{12}(x, z) = (u_{yzz}(x, 0, z) - u_{yzz}(0, 0, z) + x(u_{yzz}(0, 0, z) - u_{yzz}(1, 0, z))), \tag{19}$$

$$\psi_{13}(y, z) = (u_{zz}(1, y, z) - u_{zz}(1, 0, z) - u_{zz}(0, y, z) + u_{zz}(0, 0, z)). \tag{20}$$

Putting $y = 1$ in (17), we obtain

$$\psi_{12}(x, z) = u_{zz}(x, 1, z) - \psi_{11}(x, 1, z) - x\psi_{13}(1, z) - \sum_{i_1=1}^{2M_1}\sum_{i_2=1}^{2M_2}\sum_{i_3=1}^{2M_3} W_{i_1i_2i_3} (P_{2,i_1}(x) - xP_{2,i_1}(1)) P_{2,i_2}(1) h_{i_3}(z). \tag{21}$$

From (17), using (21), we obtain

$$u_{zz}(x,y,z) = \psi_{11}(x,y,z) + y(u_{zz}(x,1,z) - \psi_{11}(x,1,z) - x\psi_{13}(1,z)) + x\psi_{13}(y,z) + \sum_{i_1=1}^{2M_1}\sum_{i_2=1}^{2M_2}\sum_{i_3=1}^{2M_3} W_{i_1i_2i_3}(P_{2,i_1}(x) - xP_{2,i_1}(1))(P_{2,i_2}(y) - yP_{2,i_2}(1))h_{i_3}(z). \tag{22}$$

Now, integrating (13) twice with respect to z from 0 to z, we obtain

$$u_{xxyy}(x,y,z) = u_{xxyy}(x,y,0) + zu_{xxyyz}(x,y,0) + \sum_{i_1=1}^{2M_1}\sum_{i_2=1}^{2M_2}\sum_{i_3=1}^{2M_3} W_{i_1i_2i_3}h_{i_1}(x)h_{i_2}(y)P_{2,i_3}(z). \tag{23}$$

Putting $z = 1$ in (23), we obtain

$$u_{xxyyz}(x,y,0) = u_{xxyy}(x,y,1) - u_{xxyy}(x,y,0) - \sum_{i_1=1}^{2M_1}\sum_{i_2=1}^{2M_2}\sum_{i_3=1}^{2M_3} W_{i_1i_2i_3}h_{i_1}(x)h_{i_2}(y)P_{2,i_3}(1). \tag{24}$$

From (23), using (24), we obtain

$$u_{xxyy}(x,y,z) = u_{xxyy}(x,y,0) + z[u_{xxyy}(x,y,1) - u_{xxyy}(x,y,0)] + \sum_{i_1=1}^{2M_1}\sum_{i_2=1}^{2M_2}\sum_{i_3=1}^{2M_3} W_{i_1i_2i_3}h_{i_1}(x)h_{i_2}(y)(P_{2,i_3}(z) - zP_{2,i_3}(1)). \tag{25}$$

Again, integrating (25) twice with respect to x, from 0 to x, we obtain

$$u_{yy}(x,y,z) = \psi_{21}(x,y,z) + x\psi_{22}(y,z) + z\psi_{23}(x,y) + \sum_{i_1=1}^{2M_1}\sum_{i_2=1}^{2M_2}\sum_{i_3=1}^{2M_3} W_{i_1i_2i_3}P_{2,i_1}(x)h_{i_2}(y)(P_{2,i_3}(z) - zP_{2,i_3}(1)), \tag{26}$$

where

$$\psi_{21}(x,y,z) = u_{yy}(0,y,z) + u_{yy}(x,y,0) - u_{yy}(0,y,0), \tag{27}$$

$$\psi_{22}(y,z) = (u_{xyy}(0,y,z) - u_{xyy}(0,y,0) + z(u_{xyy}(0,y,0) - u_{xyy}(0,y,1))), \tag{28}$$

$$\psi_{23}(x,y) = (u_{yy}(x,y,1) - u_{yy}(0,y,1) - u_{yy}(x,y,0) + u_{yy}(0,y,0)) \tag{29}$$

Putting $x = 1$ in (26), we obtain:

$$\psi_{22}(y,z) = u_{yy}(1,y,z) - \psi_{21}(1,y,z) - z\psi_{23}(1,y) - \sum_{i_1=1}^{2M_1}\sum_{i_2=1}^{2M_2}\sum_{i_3=1}^{2M_3} W_{i_1i_2i_3}P_{2,i_1}(1)h_{i_2}(y)(P_{2,i_3}(z) - zP_{2,i_3}(1)). \tag{30}$$

Substituting (30) in (26), we obtain

$$u_{yy}(x,y,z) = \psi_{21}(x,y,z) + x(u_{yy}(1,y,z) - \psi_{21}(1,y,z) - z\psi_{23}(1,y)) + z\psi_{23}(x,y) + \sum_{i_1=1}^{2M_1}\sum_{i_2=1}^{2M_2}\sum_{i_3=1}^{2M_3} W_{i_1i_2i_3}(P_{2,i_1}(x) - xP_{2,i_1}(1))h_{i_2}(y)(P_{2,i_3}(z) - zP_{2,i_3}(1)). \tag{31}$$

Now, integrating (13) twice with respect to y, from 0 to y, we obtain:

$$u_{xxzz}(x,y,z) = u_{xxzz}(x,0,z) + yu_{xxyzz}(x,0,z) + \sum_{i_1=1}^{2M_1}\sum_{i_2=1}^{2M_2}\sum_{i_3=1}^{2M_3} W_{i_1i_2i_3}h_{i_1}(x)P_{2,i_2}(1)h_{i_3}(z). \tag{32}$$

Putting $y = 1$ in (32), we obtain

$$u_{xxyzz}(x,0,z) = u_{xxzz}(x,1,z) - u_{xxzz}(x,0,z) - \sum_{i_1=1}^{2M_1}\sum_{i_2=1}^{2M_2}\sum_{i_3=1}^{2M_3} W_{i_1i_2i_3}h_{i_1}(x)P_{2,i_2}(y)h_{i_3}(z). \tag{33}$$

From (32) and (33), we obtain

$$u_{xxzz}(x,y,z) = u_{xxzz}(x,0,z) + y[u_{xxzz}(x,1,z) - u_{xxzz}(x,0,z)] + \sum_{i_1=1}^{2M_1}\sum_{i_2=1}^{2M_2}\sum_{i_3=1}^{2M_3} W_{i_1i_2i_3}h_{i_1}(x)(P_{2,i_2}(y) - yP_{2,i_2}(1))h_{i_3}(z). \tag{34}$$

Again, integrating (34), twice with respect z, from 0 to z, we obtain

$$u_{xx}(x,y,z) = \psi_{31}(x,y,z) + z\psi_{32}(x,y) + y\psi_{33}(x,z) + \sum_{i_1=1}^{2M_1}\sum_{i_2=1}^{2M_2}\sum_{i_3=1}^{2M_3} W_{i_1i_2i_3}h_{i_1}(x)(P_{2,i_2}(y) - yP_{2,i_2}(1))P_{2,i_3}(z), \tag{35}$$

where

$$\psi_{31}(x,y,z) = u_{xx}(x,y,0) + u_{xx}(x,0,z) - u_{xx}(x,0,0), \tag{36}$$

$$\psi_{32}(x,y) = (u_{xxz}(x,y,0) - u_{xxz}(x,0,0) + y(u_{xxz}(x,0,0) - u_{xxz}(x,1,0))), \tag{37}$$

$$\psi_{33}(x,z) = (u_{xx}(x,1,z) - u_{xx}(x,1,0) - u_{xx}(x,0,z) + u_{xx}(x,0,0)). \quad (38)$$

Putting $z = 1$ in (35), we obtain

$$\psi_{32}(x,y) = u_{xx}(x,y,1) - \psi_{31}(x,y,1) - y\psi_{33}(x,1) - \sum_{i_1=1}^{2M_1}\sum_{i_2=1}^{2M_2}\sum_{i_3=1}^{2M_3} W_{i_1i_2i_3} h_{i_1}(x)(P_{2,i_2}(y) - yP_{2,i_2}(1))P_{2,i_3}(1). \quad (39)$$

From (35) and (39), we obtain

$$u_{xx}(x,y,z) = \psi_{31}(x,y,z) + z(u_{xx}(x,y,1) - \psi_{31}(x,y,1) - y\psi_{33}(x,1)) + y\psi_{33}(x,z) + \sum_{i_1=1}^{2M_1}\sum_{i_2=1}^{2M_2}\sum_{i_3=1}^{2M_3} W_{i_1i_2i_3} h_{i_1}(x)(P_{2,i_2}(y) - yP_{2,i_2}(1)) (P_{2,i_3}(z) - zP_{2,i_3}(1)). \quad (40)$$

Again, integrating (40), twice with respect to x, from 0 to x, we obtain

$$u(x,y,z) = \varphi_0(x,y,z) + x\varphi_1(y,z) + z\varphi_2(x,y) - yz\varphi_3(x) + y\varphi_4(x,z) + \sum_{i_1=1}^{2M_1}\sum_{i_2=1}^{2M_2}\sum_{i_3=1}^{2M_3} W_{i_1i_2i_3} P_{2,i_1}(x)(P_{2,i_2}(y) - yP_{2,i_2}(1)) (P_{2,i_3}(z) - zP_{2,i_3}(1)), \quad (41)$$

where

$$\varphi_0(x,y,z) = u(0,y,z) + u(x,y,0) - u(0,y,0) + u(x,0,z) - u(0,0,z) - u(x,0,0) + u(0,0,0), \quad (42)$$

$$\begin{aligned}\varphi_1(y,z) = {} & (u_x(0,y,z) - u_x(0,y,0) - u_x(0,0,z) + u_x(0,0,0)) \\ & + z(u_x(0,y,0) - u_x(0,y,1) + u_x(0,0,1) - u_x(0,0,0)) \\ & - yz(u_x(0,1,0) - u_x(0,1,1) + u_x(0,0,1) - u_x(0,0,0)) \\ & + y(u_x(0,1,0) - u_x(0,1,z) + u_x(0,0,z) - u_x(0,0,0)),\end{aligned} \quad (43)$$

$$\varphi_2(x,y) = u(x,y,1) - u(0,y,1) - u(x,y,0) + u(0,y,0) - u(x,0,1) + u(0,0,1) + u(x,0,0) - u(0,0,0), \quad (44)$$

$$\varphi_3(x) = u(x,1,1) - u(0,1,1) - u(x,1,0) + u(0,1,0) - u(x,0,1) + u(0,0,1) + u(x,0,0) - u(0,0,0), \quad (45)$$

$$\varphi_4(x,z) = u(x,1,z) - u(0,1,z) - u(x,1,0) + u(0,1,0) - u(x,0,z) + u(0,0,z) + u(x,0,0) - u(0,0,0). \quad (46)$$

Putting $x = 1$ in (41), we obtain

$$\varphi_1(y,z) = u(1,y,z) - \varphi_0(1,y,z) - z\varphi_2(1,y) + yz\varphi_3(1) - y\varphi_4(1,z) - \sum_{i_1=1}^{2M_1}\sum_{i_2=1}^{2M_2}\sum_{i_3=1}^{2M_3} W_{i_1i_2i_3} P_{2,i_1}(1)(P_{2,i_2}(y) - yP_{2,i_2}(1)) (P_{2,i_3}(z) - zP_{2,i_3}(1)). \quad (47)$$

From (41) and (47), we obtain

$$u(x,y,z) = \varphi_0(x,y,z) + x\{u(1,y,z) - \varphi_0(1,y,z) - z\varphi_2(1,y) + yz\varphi_3(1) - y\varphi_4(1,z)\} + z\varphi_2(x,y) - yz\varphi_3(x) + y\varphi_4(x,z) + \sum_{i_1=1}^{2M_1}\sum_{i_2=1}^{2M_2}\sum_{i_3=1}^{2M_3} W_{i_1i_2i_3}(P_{2,i_1}(x) - xP_{2,i_1}(1)) (P_{2,i_2}(y) - yP_{2,i_2}(1))(P_{2,i_3}(z) - zP_{2,i_3}(1)). \quad (48)$$

Substituting the values from (22), (31) and (40) in (1), we obtain

$$\sum_{i_1=1}^{2M_1}\sum_{i_2=1}^{2M_2}\sum_{i_3=1}^{2M_3} W_{i_1i_2i_3}[R_{i_1i_2i_3}(x,y,z) + S_{i_1i_2i_3}(x,y,z) + T_{i_1i_2i_3}(x,y,z)] = F(x,y,z), \quad (49)$$

where

$$R_{i_1i_2i_3}(x,y,z) = \{h_{i_1}(x)\}\{(P_{2,i_2}(y) - yP_{2,i_2}(1))\}\{(P_{2,i_3}(z) - zP_{2,i_3}(1))\} \approx \{H_1(i_1,x)\}\{P_{11}(i_2,y)\}\{P_{12}(i_3,z)\}, \quad (50)$$

$$S_{i_1i_2i_3}(x,y,z) = \{(P_{2,i_1}(x) - xP_{2,i_1}(1))\}\{h_{i_2}(y)\}\{(P_{2,i_3}(z) - zP_{2,i_3}(1))\} \approx \{P_{21}(i_1,x)\}\{H_2(i_2,y)\}\{P_{22}(i_3,z)\}, \quad (51)$$

$$T_{i_1i_2i_3}(x,y,z) = \{(P_{2,i_1}(x) - xP_{2,i_1}(1))\}\{(P_{2,i_2}(y) - yP_{2,i_2}(1))\}\{h_{i_3}(z)\} \approx \{P_{31}(i_1,x)\}\{P_{32}(i_2,y)\}\{H_3(i_3,z)\}, \quad (52)$$

and

$$\begin{aligned}F(x,y,z) = {} & f(x,y,z) - (\psi_{31}(x,y,z) + z(u_{xx}(x,y,1) - \psi_{31}(x,y,1) - y\psi_{33}(x,1)) + y\psi_{33}(x,z)) \\ & - (\psi_{21}(x,y,z) + x(u_{yy}(1,y,z) - \psi_{21}(1,y,z) - z\psi_{23}(1,y)) + z\psi_{23}(x,y)) \\ & - (\psi_{11}(x,y,z) + y(u_{zz}(x,1,z) - \psi_{11}(x,1,z) - x\psi_{13}(1,z)) + x\psi_{13}(y,z)).\end{aligned} \quad (53)$$

Expressions for $\{h_{i_1}(x)\}$, $\{h_{i_2}(y)\}$ and $\{h_{i_3}(z)\}$ are given below:

$$\{h_{i_1}(x)\} = \{H_1(i_1,x)\}, \{(P_{2,i_2}(y) - yP_{2,i_2}(1))\} = \{P_{11}(i_2,y)\},$$
$$\{(P_{2,i_3}(z) - zP_{2,i_3}(1))\} = \{P_{12}(i_3,z)\}, \tag{54}$$

$$\{h_{i_2}(y)\} = \{H_2(i_2,y)\}, \{(P_{2,i_1}(x) - xP_{2,i_1}(1))\} = \{P_{21}(i_1,x)\},$$
$$\{(P_{2,i_3}(z) - zP_{2,i_3}(1))\} = \{P_{22}(i_3,z)\}, \tag{55}$$

$$\{h_{i_3}(z)\} = \{H_3(i_3,z)\}, \{(P_{2,i_1}(x) - xP_{2,i_1}(1))\} = \{P_{31}(i_1,x)\},$$
$$\{(P_{2,i_2}(y) - yP_{2,i_2}(1))\} = \{P_{32}(i_2,y)\}. \tag{56}$$

Discretising (49)–(53) using (10)–(12), we obtain the following system of equations in matrix form

$$[W][R+S+T] = [F], \tag{57}$$

where W represents the wavelet coefficient matrix. The value of matrix R can be calculated as:

$$R = H_1 \otimes P_{11} \otimes P_{12}, \tag{58}$$

where

$$H_1 = \begin{pmatrix} H_1(1,x_1) & H_1(1,x_2) & \cdots & H_1(1,x_{2M_1}) \\ H_1(2,x_1) & H_1(2,x_2) & \cdots & H_1(2,x_{2M_1}) \\ \vdots & \vdots & \vdots & \vdots \\ H_1(2M_1,x_1) & H_1(2M_1,x_2) & \cdots & H_1(2M_1,x_{2M_1}) \end{pmatrix}, \tag{59}$$

$$P_{11} = \begin{pmatrix} P_{11}(1,y_1) & P_{11}(1,y_2) & \cdots & P_{11}(1,y_{2M_2}) \\ P_{11}(2,y_1) & P_{11}(2,y_2) & \cdots & P_{11}(2,y_{2M_2}) \\ \vdots & \vdots & \vdots & \vdots \\ P_{11}(2M_2,y_1) & P_{11}(2M_2,y_2) & \cdots & P_{11}(2M_2,y_{2M_2}) \end{pmatrix}, \tag{60}$$

and

$$P_{12} = \begin{pmatrix} P_{12}(1,z_1) & P_{12}(1,z_2) & \cdots & P_{12}(1,z_{2M_3}) \\ P_{12}(2,z_1) & P_{12}(2,z_2) & \cdots & P_{12}(2,z_{2M_3}) \\ \vdots & \vdots & \vdots & \vdots \\ P_{12}(2M_3,z_1) & P_{12}(2M_3,z_2) & \cdots & P_{12}(2M_3,z_{2M_3}) \end{pmatrix}. \tag{61}$$

The value of matrix S can be calculated as:

$$S = P_{21} \otimes H_2 \otimes P_{22}, \tag{62}$$

where

$$P_{21} = \begin{pmatrix} P_{21}(1,x_1) & P_{21}(1,x_2) & \cdots & P_{21}(1,x_{2M_1}) \\ P_{21}(2,x_1) & P_{21}(2,x_2) & \cdots & P_{21}(2,x_{2M_1}) \\ \vdots & \vdots & \vdots & \vdots \\ P_{21}(2M_1,x_1) & P_{21}(2M_1,x_2) & \cdots & P_{21}(2M_1,x_{2M_1}) \end{pmatrix}, \tag{63}$$

$$H_2 = \begin{pmatrix} H_2(1,y_1) & H_2(1,y_2) & \cdots & H_2(1,y_{2M_2}) \\ H_2(2,y_1) & H_2(2,y_2) & \cdots & H_2(2,y_{2M_2}) \\ \vdots & \vdots & \vdots & \vdots \\ H_2(2M_2,y_1) & H_2(2M_2,y_2) & \cdots & H_2(2M_2,y_{2M_2}) \end{pmatrix}, \tag{64}$$

and

$$P_{22} = \begin{pmatrix} P_{22}(1,z_1) & P_{22}(1,z_2) & \cdots & P_{22}(1,z_{2M_3}) \\ P_{22}(2,z_1) & P_{22}(2,z_2) & \cdots & P_{22}(2,z_{2M_3}) \\ \vdots & \vdots & \vdots & \vdots \\ P_{22}(2M_3,z_1) & P_{22}(2M_3,z_2) & \cdots & P_{22}(2M_3,z_{2M_3}) \end{pmatrix}. \tag{65}$$

The value of matrix T can be calculated as:

$$T = P_{31} \otimes P_{32} \otimes H_3, \tag{66}$$

where

$$P_{31} = \begin{pmatrix} P_{31}(1,x_1) & P_{31}(1,x_2) & \cdots & P_{31}(1,x_{2M_1}) \\ P_{31}(2,x_1) & P_{31}(2,x_2) & \cdots & P_{31}(2,x_{2M_1}) \\ \vdots & \vdots & \vdots & \vdots \\ P_{31}(2M_1,x_1) & P_{31}(2M_1,x_2) & \cdots & P_{31}(2M_1,x_{2M_1}) \end{pmatrix}, \tag{67}$$

$$P_{32} = \begin{pmatrix} P_{32}(1,y_1) & P_{32}(1,y_2) & \cdots & P_{32}(1,y_{2M_2}) \\ P_{32}(2,y_1) & P_{32}(2,y_2) & \cdots & P_{32}(2,y_{2M_2}) \\ \vdots & \vdots & \vdots & \vdots \\ P_{32}(2M_2,y_1) & P_{32}(2M_2,y_2) & \cdots & P_{32}(2M_2,y_{2M_2}) \end{pmatrix}, \tag{68}$$

and

$$H_3 = \begin{pmatrix} H_3(1,z_1) & H_3(1,z_2) & \cdots & H_3(1,z_{2M_3}) \\ H_3(2,z_1) & H_3(2,z_2) & \cdots & H_3(2,z_{2M_3}) \\ \vdots & \vdots & \vdots & \vdots \\ H_3(2M_3,z_1) & H_3(2M_3,z_2) & \cdots & H_3(2M_3,z_{2M_3}) \end{pmatrix}. \tag{69}$$

Each component of F can be evaluated as:

$$F(x,y,z) = \{F_1(x)\}\{F_2(y)\}\{F_3(z)\} = F_1 \otimes (F_2 \otimes F_3), \tag{70}$$

where

$$F_1 = [F_1(x_1), F_1(x_2), \ldots, F_1(x_{2M_1})], \tag{71}$$

$$F_2 = [F_2(y_1), F_2(y_2), \ldots, F_2(y_{2M_2})], \tag{72}$$

$$F_3 = [F_3(z_1), F_3(z_2), \ldots, F_3(z_{2M_3})]. \tag{73}$$

The numerical solution of given problem is obtained by substituting the values of wavelet coefficients into (48).

Error analysis for three-dimensional PDEs

In this section we present the error analysis for our proposed scheme. In order to analyze the convergence of our method, we state and prove the following convergence theorem:

Theorem Suppose that $u(x, y, z)$ satisfies a Lipschitz condition on $D = [0,1) \times [0,1) \times [0,1)$, that is there exist a positive constant L_1, L_2, L_3 and L_4, such that for all (x_1, y, z), (x_2, y, z), (x_3, y, z), (x_4, y, z), (x_5, y, z), (x_6, y, z), (x_7, y, z) and (x_8, y, z) in D, we have

$$\begin{cases} | u(x_2, y, z) - u(x_1, y, z) |= L_1| x_2 - x_1 |, \\ | u(x_4, y, z) - u(x_3, y, z) |= L_2| x_4 - x_3 |, \\ | u(x_6, y, z) - u(x_5, y, z) |= L_3| x_6 - x_5 |, \\ | u(x_8, y, z) - u(x_7, y, z) |= L_4| x_8 - x_7 |. \end{cases} \tag{74}$$

Then, the error bound $\| E_m \|_2$ obtained from above is

$$\| E_m \|_2 \approx O\left(\frac{1}{m}\right)^4. \tag{75}$$

Here, the order of convergence is of the order 4.

Proof Consider $M_1 = M_2 = M_3 = M$. Let $u_{exact}(x, y, z)$ and $u_{approximate}(x, y, z)$ be the exact and approximate solutions of the partial differential equation. The error at the Jth level of resolution is defined as:

$$\begin{aligned} E_m &= u_{exact}(x, y, z) - u_{approximate}(x, y, z) \\ &= \sum_{i_1=2M+1}^{\infty} \sum_{i_2=2M+1}^{\infty} \sum_{i_3=2M+1}^{\infty} C_{i_1 i_2 i_3} h_{i_1}(x) h_{i_2}(y) h_{i_3}(z) \\ &= \sum_{i_1, i_2, i_3=2M+1}^{\infty} C_{i_1 i_2 i_3} h_{i_1}(x) h_{i_2}(y) h_{i_3}(z), \end{aligned} \tag{76}$$

where

$$u_{approximate}(x, y, z) = \sum_{i_1=1}^{2M} \sum_{i_2=1}^{2M} \sum_{i_3=1}^{2M} C_{i_1 i_2 i_3} h_{i_1}(x) h_{i_2}(y) h_{i_3}(z), \tag{77}$$

and the wavelet coefficients are calculated as:

$$\begin{aligned} C_{i_1 i_2 i_3} &= \int_0^1 \int_0^1 \int_0^1 u(x, y, z) h_{i_1}(x) h_{i_2}(y) h_{i_3}(z) \mathrm{d}x\mathrm{d}y\mathrm{d}z \\ &= <h_{i_1}(x), <h_{i_2}(y), <u(x, y, z), h_{i_3}(z)> > >. \end{aligned} \tag{78}$$

Here $<.>$ shows the inner product. Define $\| \, . \, \|_2$ as:

$$\| E_m \|_2^2 = \int_0^1 \int_0^1 \int_0^1 (u_{exact}(x, y, z) - u_{approximate}(x, y, z))^2 dxdydz. \tag{79}$$

From (76) and (79), we obtain

$$\| E_m \|_2^2 = \int_0^1 \int_0^1 \int_0^1 \left(\sum_{i_1, i_2, i_3=2M+1}^{\infty} C_{i_1 i_2 i_3} h_{i_1}(x) h_{i_2}(y) h_{i_3}(z) \right)^2 \mathrm{d}x\mathrm{d}y\mathrm{d}z. \tag{80}$$

Using definition of inner product, from (80), we obtain

$$\begin{aligned} \| E_m \|_2^2 = & \sum_{i_1, i_2, i_3=2M+1}^{\infty} \sum_{p,q,r=2M+1}^{\infty} C_{i_1 i_2 i_3} C_{pqr} \\ & \left(\int_0^1 h_{i_1}(x) h_p(x) \mathrm{d}x \right) \left(\int_0^1 h_{i_2}(y) h_q(y) \mathrm{d}y \right) \\ & \left(\int_0^1 h_{i_3}(z) h_r(z) \mathrm{d}z \right). \end{aligned} \tag{81}$$

Using orthogonality conditions, from (81), we obtain

$$\| E_m \|_2^2 = \frac{1}{m^3} \sum_{i_1, i_2, i_3=2M+1}^{\infty} (C_{i_1 i_2 i_3})^2. \tag{82}$$

According to (9), we can write

$$\begin{aligned} <u(x, y, z), h_{i_3}(z)> &= \int_0^1 u(x, y, z) h_{i_3}(z) \mathrm{d}z \\ &= \int_{k/m}^{k+0.5/m} u(x, y, z) dz - \int_{k+0.5/m}^{k+1/m} u(x, y, z) \mathrm{d}z. \end{aligned} \tag{83}$$

Applying mean value theorem, that is there exist $z_1 \in [\frac{k}{m}, \frac{k+0.5}{m}]$ and $z_2 \in [\frac{k+0.5}{m}, \frac{k+1}{m}]$, such that

$$\begin{aligned} & <u(x, y, z), h_{i_3}(z)> \\ & = \left[\left(\frac{k+0.5}{m} - \frac{k}{m} \right) u(x, y, z_1) - \left(\frac{k+1}{m} - \frac{k+0.5}{m} \right) u(x, y, z_2) \right] \\ & = \frac{1}{2m} [u(x, y, z_1) - u(x, y, z_2)]. \end{aligned} \tag{84}$$

Again,

$$\begin{aligned} & <h_{i_2}(y), <u(x, y, z), h_{i_3}(z)> > \\ & = <h_{i_2}(y), \frac{1}{2m}(u(x, y, z_1) - u(x, y, z_2))>, \end{aligned} \tag{85}$$

From (85), using the definition of inner product, we obtain

$$<h_{i_2}(y), <u(x, y, z), h_{i_3}(z)> > = \int_0^1 \frac{1}{2m} [u(x, y, z_1) - u(x, y, z_2)] h_{i_2}(y) dy, \tag{86}$$

Using (8), from (86), we obtain

$$\begin{aligned} & <h_{i_2}(y), <u(x, y, z), h_{i_3}(z)> > \\ & = \frac{1}{2m} \left[\int_{\frac{k}{m}}^{\frac{k+0.5}{m}} u(x, y, z_1) dy - \int_{\frac{k+0.5}{m}}^{\frac{k+1}{m}} u(x, y, z_1) dy \right] \\ & - \frac{1}{2m} \left[\int_{\frac{k}{m}}^{\frac{k+0.5}{m}} u(x, y, z_2) dy + \int_{\frac{k+0.5}{m}}^{\frac{k+1}{m}} u(x, y, z_2) dy \right]. \end{aligned} \tag{87}$$

Again, applying mean value theorem, we obtain:

$$<h_{i_2}(y),<u(x,y,z),h_{i_3}(z)>>=\frac{1}{2m}\left[\left(\frac{k+0.5}{m}-\frac{k}{m}\right)u(x,y_1,z_1)-\left(\frac{k+1}{m}-\frac{k+0.5}{m}\right)u(x,y_2,z_1)\right]+\frac{1}{2m}\left[\left(\frac{k+1}{m}-\frac{k+0.5}{m}\right)u(x,y_4,z_2)-\left(\frac{k+0.5}{m}-\frac{k}{m}\right)u(x,y_3,z_2)\right] \tag{88}$$

After simplifications, from (88), we obtain

$$<h_{i_2}(y),<u(x,y,z),h_{i_3}(z)>>=\frac{1}{2^{j+2}m}[u(x,y_1,z_1)-u(x,y_2,z_1)-u(x,y_3,z_2)+u(x,y_4,z_2)]. \tag{89}$$

Hence,

$$\mathcal{C}_{i_1i_2i_3}=<h_{i_1}(x),\frac{1}{2^{j+2}m}[u(x,y_1,z_1)-u(x,y_2,z_1)-u(x,y_3,z_2)+u(x,y_4,z_2)]>=\frac{1}{2^{j+2}m}\int_0^1[u(x,y_1,z_1)-u(x,y_2,z_1)-u(x,y_3,z_2)..+u(x,y_4,z_2)]h_i(x)dx. \tag{90}$$

From (90), using (7), we obtain

$$\mathcal{C}_{i_1i_2i_3}=\frac{1}{2^{j+2}m}\left[\int_{\frac{k}{m}}^{\frac{k+0.5}{m}}u(x,y_1,z_1)dx-\int_{\frac{k+0.5}{m}}^{\frac{k+1}{m}}u(x,y_1,z_1)dx\right]-\frac{1}{2^{j+2}m}\left[\int_{\frac{k}{m}}^{\frac{k+0.5}{m}}u(x,y_2,z_1)dx+\int_{\frac{k+0.5}{m}}^{\frac{k+1}{m}}u(x,y_2,z_1)dx\right]+\frac{1}{2^{j+2}m}\left[\int_{\frac{k}{m}}^{\frac{k+0.5}{m}}u(x,y_4,z_2)dx-\int_{\frac{k+0.5}{m}}^{\frac{k+1}{m}}u(x,y_4,z_2)dx.-\int_{\frac{k}{m}}^{\frac{k+0.5}{m}}u(x,y_3,z_2)dx+\int_{\frac{k+0.5}{m}}^{\frac{k+1}{m}}u(x,y_3,z_2)dx\right] \tag{91}$$

Applying mean value theorem, from (91), we obtain

$$\mathcal{C}_{i_1i_2i_3}=\frac{1}{2^{j+2}m}\left[\left(\frac{k+0.5}{m}-\frac{k}{m}\right)u(x_1,y_1,z_1)-\left(\frac{k+1}{m}-\frac{k+0.5}{m}\right)u(x_2,y_1,z_1)\right]+\frac{1}{2^{j+2}m}\left[-\left(\frac{k+0.5}{m}-\frac{k}{m}\right)u(x_3,y_2,z_1)+\left(\frac{k+0.5}{m}-\frac{k}{m}\right)u(x_4,y_2,z_1)\right]+\frac{1}{2^{j+2}m}\left[\left(\frac{k+0.5}{m}-\frac{k}{m}\right)u(x_5,y_4,z_2)-\left(\frac{k+1}{m}-\frac{k+0.5}{m}\right)u(x_6,y_4,z_2)\right]+\frac{1}{2^{j+2}m}\left[-\left(\frac{k+0.5}{m}-\frac{k}{m}\right)u(x_7,y_3,z_2)+\left(\frac{k+1}{m}-\frac{k+0.5}{m}\right)u(x_8,y_3,z_2)\right]. \tag{92}$$

After simplifications, from (92), we obtain

$$|\mathcal{C}_{i_1i_2i_3}|\leq\frac{1}{2^{2j+3}m}[|u(x_1,y_1,z_1)-u(x_2,y_1,z_1).+u(x_4,y_2,z_1)-u(x_3,y_2,z_1)|].+\frac{1}{2^{2j+3}m}[|u(x_5,y_4,z_2)-u(x_6,y_4,z_2)+u(x_8,y_3,z_2)-u(x_7,y_3,z_2)|]. \tag{93}$$

Using (74), from (93), we obtain

$$|\mathcal{C}_{i_1i_2i_3}|\leq\frac{1}{2^{2j+3}m}\frac{4L}{2m}, \tag{94}$$

where $L=max\{L_1,L_2,L_3,L_4\}$. After simplifications, from (94), we obtain

$$|\mathcal{C}_{i_1i_2i_3}|\leq\frac{4L}{2^{2j+4}m^2}\leq\frac{L}{2^{2j+2}}\frac{1}{m^2}. \tag{95}$$

Squaring both sides, from (95), we obtain

$$(\mathcal{C}_{i_1i_2i_3})^2\leq\frac{L^2}{2^{4j+4}}\frac{1}{m^4}. \tag{96}$$

By substituting (96) in (82), we obtain

$$\| E_m \|_2^2\leq\sum_{i_1,i_2,i_3=2M+1}^{\infty}\frac{L^2}{2^{4j+4}m^4}\frac{1}{m^3}. \tag{97}$$

After simplifications, from (97), we obtain

$$\| E_m \|_2^2\leq\frac{L^2}{m^7}\frac{1}{2^4}\sum_{i_1,i_2,i_3=2M+1}^{\infty}\frac{1}{2^{4j}}. \tag{98}$$

Expanding (98), we obtain

$$\| E_m \|_2^2\leq\frac{L^2}{m^7}\frac{1}{2^4}\sum_{j=J+1}^{\infty}\left(\sum_{i_1=0}^{2^j-1}\sum_{i_2=0}^{2^j-1}\sum_{i_3=0}^{2^j-1}\frac{1}{2^{4j}}\right). \tag{99}$$

From (99), after simplification, we obtain

$$\| E_m \|_2^2\leq\frac{L^2}{m^7}\frac{1}{2^4}\sum_{j=J+1}^{\infty}\left(\frac{1}{2^j}\right). \tag{100}$$

From (100), after series summation, we obtain

$$\| E_m \|_2^2\leq\frac{L^2}{m^8}. \tag{101}$$

After taking square root, we obtain

$$\| E_m \|_2\approx O\left(\frac{1}{m^4}\right). \tag{102}$$

This shows that the convergence is of the order 4. □

Table 1 Maximum absolute errors of Example 1

J1 = J2 = J3	Maximum absolute error
0	8.9227E−004
1	6.4125E−004
2	2.0157E−004
3	5.3267E−005

Numerical examples and discussion

We have applied our method on some numerical examples, to observe the accuracy and efficiency of the present method for solving three-dimensional Poisson equations.

Example 1 Consider the following linear three-dimensional Poisson equation

$$\nabla^2 u = \frac{\partial^2 u}{\partial x^2} + \frac{\partial^2 u}{\partial y^2} + \frac{\partial^2 u}{\partial z^2} = f(x,y,z), \quad (103)$$

on $K = \{(x,y,z) : 0<x<1, 0<y<1, 0<z<1\}$, with boundary conditions:

$$u(0,y,z) = u(x,0,z) = u(x,y,0) = u(1,y,z) = u(x,1,z) = u(x,y,1) = 0, \quad (104)$$

where

$$f(x,y,z) = \sin \pi x.\sin \pi y.\sin \pi z. \quad (105)$$

The exact solution is $u(x,y,z) = -\frac{1}{3\pi^2}.\sin(\pi x).\sin(\pi y).\sin(\pi z)$. Table 1 shows the maximum absolute errors of Example 1.

The proposed method is more simplest and different from the method presented in [23]. For $J = 3$, the maximum absolute error obtained by [23] is $1.3730E - 004$, where as in our research paper maximum absolute error is $5.3267E - 005$.

Example 2 Consider the following linear three-dimensional Poisson equation

$$\nabla^2 u = \frac{\partial^2 u}{\partial x^2} + \frac{\partial^2 u}{\partial y^2} + \frac{\partial^2 u}{\partial z^2} = f(x,y,z), \quad (106)$$

on $K = \{(x,y,z) : 0<x<1, 0<y<1, 0<z<1\}$, with boundary conditions:

$$u(0,y,z) = u(x,0,z) = u(x,y,0) = u(1,y,z) = u(x,1,z) = u(x,y,1) = 0, \quad (107)$$

where

$$f(x,y,z) = 6xyz[y^2z^2(1-2x)(1-y)(1-z) + x^2z^2(1-x)(1-2y)(1-z) + x^2y^2(1-x)(1-y)(1-2z)]. \quad (108)$$

The exact solution is $u(x,y,z) = x^3y^3z^3(1-x)(1-y)(1-z)$. Table 2 shows the maximum absolute errors of Example 2.

Example 3 Consider the following linear three-dimensional Helmholtz equation

$$\nabla^2 u = \frac{\partial^2 u}{\partial x^2} + \frac{\partial^2 u}{\partial y^2} + \frac{\partial^2 u}{\partial z^2} + 2500u = f(x,y,z), \quad (109)$$

on $K = \{(x,y,z) : 0<x<1, 0<y<1, 0<z<1\}$, with boundary conditions:

$$u(0,y,z) = u(x,0,z) = u(x,y,0) = u(1,y,z) = u(x,1,z) = u(x,y,1) = 0, \quad (110)$$

where

$$f(x,y,z) = (2500 - 3\pi^2).\sin \pi x.\sin \pi y.\sin \pi z. \quad (111)$$

The exact solution is $u(x,y,z) = \sin(\pi x).\sin(\pi y).\sin(\pi z)$. Table 3 shows the maximum absolute errors of Example 3.

Example 4 Consider the following linear three-dimensional Helmholtz equation

$$\nabla^2 u = \frac{\partial^2 u}{\partial x^2} + \frac{\partial^2 u}{\partial y^2} + \frac{\partial^2 u}{\partial z^2} + 1500u = f(x,y,z), \quad (112)$$

on $K = \{(x,y,z) : 0<x<1, 0<y<1, 0<z<1\}$, with boundary conditions:

$$u(0,y,z) = u(x,0,z) = u(x,y,0) = u(1,y,z) = u(x,1,z) = u(x,y,1) = 0, \quad (113)$$

where

Table 2 Maximum absolute errors of Example 2

J1 = J2 = J3	Maximum absolute error
0	8.9892E−005
1	4.5356E−005
2	1.1495E−005
3	3.0247E−006

Table 3 Maximum absolute errors of Example 3

J1 = J2 = J3	Maximum absolute error
0	3.4255E−004
1	2.3325E−004
2	7.1992E−005
3	1.8931E−005

Table 4 Maximum absolute errors of Example 4

J1 = J2 = J3	Maximum absolute error
0	3.6587E−005
1	3.0419E−005
2	1.1531E−005
3	2.9835E−006

$$f(x,y,z) = (6x(1-2x) - (1500 - 2\pi^2)x^3(1-x)).\sin \pi y.\sin \pi z. \quad (114)$$

The exact solution is $u(x,y,z) = x^3(1-x).\sin(\pi y).\sin(\pi z)$. Table 4 shows the maximum absolute errors of Example 4.

Conclusion

It is concluded from here that the Haar wavelet method is a powerful mathematical tool for solving three-dimensional partial differential equations. As we increase the values of $2M_1$, $2M_2$ and $2M_3$, absolute errors decrease rapidly and the numerical solutions are much closer to the exact solutions. Also, proposed method gives better results as compared to numerical results presented in [23]. Also, proposed method is applicable to many types of three-dimensional Poisson equations (see, for example, in Sect. 4, Example 2) whereas method presented in [23] is applicable to only one type of three-dimensional Poisson equation. In view of numerical results, it is concluded that proposed method based on Haar wavelet is more efficient and accurate for solving three-dimensional partial differential equations.

Acknowledgements We are grateful to the anonymous reviewers for their valuable comments which let to the improvement of the manuscript.

References

1. Aziz, I., Sarler, B.: Wavelets collocation methods for the numerical solution of elliptic BV problems. Appl. Math. Model. **37**, 676–697 (2013)
2. Bhrawy, A.H.: An efficient Jacobi pseudospectral approximation for nonlinear complex generalized Zakharov system. Appl. Math. Comput. **247**, 30–46 (2014)
3. Bhrawy, A.H.: A highly accurate collocation algorithm for 1+1 and 2+1 fractional percolation equations. J. Vibrat. Control **22**, 2288–2310 (2016)
4. Cattani, C.: Haar wavelet splines, J. Interdiscipl. Math., 4, 35–47 (2001)
5. Cattani, C.: Haar wavelets based technique in evolution problems, Proc. Estonian Acad. Sci. Phys. Math., 53, 45–65 (2004)
6. Chen, C.F., Hsiao, C.H.: Haar wavelet method for solving lumped and distributed- parameter systems, IEE Proc. Control Theory Appl. **144**, 87–94 (1997)
7. Chen, C. F. Hsiao, C. H.: Wavelet approach to optimising dynamic systems, IEE Proc. Control Theory Appl., 146, 213–219 (1997)
8. Doha, E.H., Bhrawy, A.H.: Efficient spectral- Galerkin algorithms for direct solution of second-order differential equations using Jacobi polynomials. Num. Algorith. **42**, 137–164 (2006)
9. Doha, E.H., Bhrawy, A.H.: Efficient spectral- Galerkin algorithms for direct solution of fourth-order differential equations using Jacobi polynomials. Appl. Num. Math. **58**, 1224–1244 (2008)
10. Doha, E.H., Bhrawy, A.H.: A Jacobi spectral-Galerkin method for the integrated forms of fourth-order elliptic differential equations. Num. Methods Par. Diff. Equ. **25**, 712–739 (2009)
11. Doha, E.H., Bhrawy, A.H.: An efficient direct solver for multidimensional elliptic Robin boundary value problems using a Legendre spectral-Galerkin method. Comp. Math. Appl. **64**, 558–571 (2012)
12. Haar, A.: Zur theorie der orthogonalen Funktionsysteme. Math. Annal. **69**, 331–371 (1910)
13. Heydari, M.H., Hooshmandasl, M.R., Maalek, F.M., Ghaini, : A new approach of the Chebyshev wavelets method for partial differential equations with boundary conditions of the telegraph type. Appl. Math. Model. **38**, 1597–1606 (2014)
14. Lepik, Ü.: Application of Haar wavelet transform to solving integral and differential equations. Proc. Estonian Acad. Sci. Phys. Math. **56**(1), 28–46 (2007)
15. Lepik, Ü.: Solving differential and integral equations by Haar wavelet method, revisted. Int. J. Math. Comput. **1**, 43–52 (2008)
16. Lepik, Ü.: Solving PDEs with the aid of two-dimensional Haar wavelets. Comp. Math. Appl. **61**(7), 1873–1879 (2011)
17. Mittal, R.C., Bhatia, R.: A numerical study of two dimensional hyperbolic telegraph equation by modified B-spline differential quadrature method. Appl. Math. Comput. **244**, 976–997 (2014)
18. Mittal, R.C., Gahlaut, S.: High-order finite-differences schemes to solve Poissons equation in polar coordinates, IMA J. Numer. Anal., 11, 261–270 (1987)
19. Siraj ul Islam, Arler, B., Aziz, I., Haq, F.: Haar wavelet collocation method for the numerical solution of boundary layer fluid flow problems, Int. J. Therm. Sci., 50, 686–697 (2011)
20. Siraj-ul-Islam, Aziz, I., Al-Fhaid, A.S., Shah, A.: A numerical assessment of parabolic partial differential equations using Haar and Legendre wavelets, Applied Mathematical Modelling, 37, 9455–9481 (2013)
21. Siraj-ul-Islam, I., Aziz, M., Ahmad, : Numerical solution of two-dimensional elliptic PDEs with nonlocal boundary conditions. Comp. Math. Appl. **69**, 180–205 (2015)
22. Sutmann, G., Steffen, B.: High–order compact solvers for the three–dimensional Poisson equation, J. Comput. Appl. Math., 187, 142–170 (2006)
23. Shi, Zhi, Cao, Yong-yan, Chen, Qing-jiang: Solving 2D and 3D Poisson equations and biharmonic equations by the Haar wavelet method. Appl. Math. Modell. **36**, 5143–5161 (2012)

20

An efficient approach based on radial basis functions for solving stochastic fractional differential equations

N. Ahmadi[1] · A. R. Vahidi[2] · T. Allahviranloo[1]

Abstract In this paper, we present a collocation method based on Gaussian Radial Basis Functions (RBFs) for approximating the solution of stochastic fractional differential equations (SFDEs). In this equation the fractional derivative is considered in the Caputo sense. Also we prove the existence and uniqueness of the presented method. Numerical examples confirm the proficiency of the method.

Keywords Stochastic fractional differential equations · Radial basis functions

Introduction

Fractional calculus introduced because it can fill the existing gap for describing a large amount of work in engineering [1, 2], and different phenomena in nature such as biology, physics [3, 4]. Mathematicians and physicists have been created numerous articles about fractional differential equations(FDEs) for finding analytical and numerical methods, including Adomian Decomposition Method [5], Variational Iteration Method [6, 7], Homotopy perturbation Method [8] and homotopy analysis method [9]. H. Rezazadeh et al. have generalized the Floquet system to the fractional Floquet system in 2016 [10]. Stochastic Differential Equations (SDEs) models play a great role in various sciences such as physics, economics, biology, chemistry and finance [11–15]. The reader has at least knowledge about independence, expected values and variances and also basic definitions of stochastic, that is necessary to read articles in this field [16]. M.Khodabin et al. approximate solution of stochastic Volterra integral equations in 2014 [17], also R.Ezzati et al. work on a stochastic operational matrix based on block pulse functions in 2014 [18]. We introduce SFDEs [19]:

$$D^{\alpha}u(t) = f(t, u(t)) + \sigma \int_{t_0}^{t} g(t,s)dw(s), \qquad u(t_0) = u_{t_0}. \tag{1}$$

for $0 \leq \alpha \leq 1$ and $t \in [0, T]$, where D^{α} is the Caputo fractional derivative of order α which will be defined later. σ is Max amplitude of noise also $\int_{t_0}^{t} g(t,s)dw(s)$ is the stochastic term, that produce some noise in our result, throughout the paper we putting $\sigma = 1$. SFDEs play a remarkable role for physical applications in nature [20–22].

Using RBFs for solving partial differential equations (PDEs) are very popular among many researchers, during the last two decades [23, 24]. Also RBFs is applied in mechanics [25], Kdv equation [26], Klein-Gordon equation [27], then in 2012 Vanani et al. used RBF for solving fractional partial differential equations [28]. Gonzalez-Gaxiola and Gonzalez-Perez used Multi-Quadratic RBF for approximating the solution of the Black-Scholes equation in 2014 [29].

The motivation of this paper is to extend the application of the RBF to solve SFDEs.

✉ A. R. Vahidi
alrevahidi@yahoo.com

N. Ahmadi
nafiseh_ahmadi@ymail.com

T. Allahviranloo
tofigh@allahviranloo.com

1 Department of Mathematics, Science and Research Branch, Islamic Azad University, Tehran, Iran

2 Department of Mathematics, Yadegar-e-Imam Khomeini (RAH) Shahre Rey Branch, Islamic Azad University, Tehran, Iran

The layout of the paper is the following. In Sect. 2 some essential definitions of fractional calculus is proposed. In Sect. 3 we explain using RBFs method for SFDEs and prove the existence and uniqueness of the presented method. In Sect. 4 various examples are solved to illustrate the effectiveness of the proposed method. Also a conclusion is given in the last section.

Preliminaries and notations

In this section, we give some basic definitions and properties of fractional calculus which are defined as follow [4].

Definition 2.1 The Caputo fractional derivative of order v is defined as

$$D^{\vartheta}f(x) = J^{p-\vartheta}D^{p}f(x) = \frac{1}{\Gamma(p-\vartheta)}\int_0^x (x-t)^{p-\vartheta-1}\frac{d^p}{dt^p}f(t)dt, \quad p-1<\vartheta\leq p, \quad x>0$$

where D^p is the classical differential operator of order p.

Remark 2.2 For the Caputo derivative we have

$$D^{\vartheta}x^{\beta} = \begin{cases} 0, & \beta<\vartheta, \\ \dfrac{\Gamma(\beta+1)}{\Gamma(\beta+1-\vartheta)}, & \beta\geq\vartheta, \end{cases}$$

Remark 2.3 $D^{-\vartheta}$ is defined as $D^{-\vartheta}f(t) = \frac{1}{\Gamma(\vartheta)}\int_0^t f(t)(t-\zeta)^{\vartheta-1}d\zeta, \quad t>0, \quad 0<\vartheta\leq 1.$

Definition 2.4 Let (Ω, F, ρ) be a probability space with a normal filteration $(F_t)_{t\geq 0}$ and $w = \{w(t) : t\geq 0\}$ be a Brownian motion defined over this filtered probability space. Consider the following SFDE

$$D^{\alpha}u(t,v) = f(t,u(t,v)) + \int_0^t g(t,s,v)dw(s), \qquad u(0,v) = u_0$$

for $t\in[0,T]$, and $v\in\Omega$. For simplicity of notation we drop the variable v so we have the following equation

$$D^{\alpha}u(t) = f(t,u(t)) + \int_0^t g(t,s)dw(s), \qquad u(0) = u_0$$

from Remark (2.3) we can see

$$u(t) = u_0 + D^{-\alpha}f(t,u(t)) + D^{-\alpha}\int_0^t g(t,s)dw(s),$$

therefore, we have

$$u(t) = u_0 + \frac{1}{\Gamma(\alpha)}\int_0^t f(v,u(v))(t-v)^{\alpha-1}dv + \frac{1}{\Gamma(\alpha)}\int_0^t\int_0^v g(v,s)(t-v)^{\alpha-1}dw(s)dv. \tag{2}$$

Also we admit the following assumptions.

Assumption 2.5 Suppose f and g are L^2 measurable functions satisfying

$$|f(m,x)-f(m,y)|\leq K_1|x-y|, \quad |g(t,m)-g(t,n)|\leq K_2(|m-n|), \tag{3}$$

$$(|f(m,x)|)\leq K_3(1+|x|), \qquad (|g(t,m)|)\leq K_4, \tag{4}$$

for some constants K_1, K_2, K_3, K_4 and for every $x,y\in\mathbb{R}$ and $0\leq m,n\leq t\leq T=1$.

Stochastic integral 2.6 Now we should explain the approximation of the stochastic term. White noise is known as the derivative of the brownian motion $W(s)$ [30], so we approximating the term $\frac{dw_s}{dt}$. Let $t_0 = 0 \prec t_1 = \Delta t \prec \cdots \prec t_N = T = 1$, with $t_i = i\Delta t$, for $i = 0,\ldots,N$ be a partition of $[0,1]$. This method introduced in [31], we approximate $\frac{dw_s}{dt}$ by $\frac{d\hat{w}_s}{dt}$

$$\frac{d\hat{w}_s}{dt} = \frac{1}{\sqrt{\Delta t}}\sum_{i=1}^{N}\gamma_i\zeta_i(s), \tag{5}$$

where $\gamma_i\sim N(0,1)$ is introducing by

$$\gamma_i = \frac{1}{\sqrt{\Delta t}}\int_{t_i}^{t_{i+1}} dW(t), \quad i=1,\ldots,N$$

where

$$\zeta_i(s) = \begin{cases} 1, & t_i\leq s<t_{i+1}, \\ 0, & \text{otherwise}, \end{cases}$$

Collocation method based on RBFs for solving SFDEs

The Radial basis functions method has been known as a powerful tool for solving ordinary, partial and fractional differential equations and also integral equations and etc. So in this section we use this method for solving (2). Before that we consider some preliminaries.

Interpolation by RBFs

Let $\{t_1,\ldots,t_N\}$ be a given set of distinct points in $[0,T]\subseteq\mathbb{R}$. Then the approximation of a function $u(t)$ using RBFs $\varphi(t) = \varphi(\|t\|)$, can be written in the following form [32, 33]

$$u(t)\approx\pi_{N,m-1}u(t) = \sum_{k=1}^{N}c_k\varphi(\|t-t_k\|) + \sum_{l=0}^{m-1}d_l p_l(t), \quad x\in D$$

where $p_0,\ldots,p_m$ form a basis for m-dimensional linear space $P_m([0,T])$ of polynomials of total degree less than or equal to m on the [0, T]. Suppose $C_N^m([0,T]) = span\{\varphi_0,\ldots,\varphi_N,p_0,\ldots,p_m\}$ then $\pi_{N,m}: C([0,T]) \to C_N^m([0,T])$ is the collocation projector on the collocation points $X = \{t_0,\ldots,t_N\} \subset [0,T]$. Since enforcing the interpolation conditions $\pi_{N,m}(t_i) = u(t_i),\quad i = 1,\ldots,N$, leads to a system of N linear equations with $N+m$ unknown, usually we add m additional conditions:

$$\sum_{k=0}^{m-1} c_k p_l(t_k) = 0, \quad l = 0,\ldots,m-1.$$

Using RBFs for solving SFDEs (2)

Let $D = [0,T] \subseteq \mathbb{R}$ and $u : C([0,T]) \to \mathbb{R}$ and also suppose that u_N is the approximation of u based on these functions so we can write

$$u_N(t) = \sum_{i=1}^{N} \lambda_i \varphi_i(t) + \sum_{l=0}^{m-1} d_l p_l(t), \qquad t \in D \subset \mathbb{R}.$$

N is the number of nodal points within the domain D and c_i denotes the shape parameter. Also we know there are different kinds of RBFs, but in this research we need only one of them with titled Gaussian that we represented as follow:

$$\varphi_i(t) = e^{\frac{-\|t-t_i\|^2}{c_i^2}}.$$

The collocation method based on RBF basis for solving (2) can be written in the following form:

Lemma 3.1 (Existence and Uniqueness) Assume that there exists a constant $K_1 > 0$ such that

$$|f(u_1,t) - f(u_2,t)| \le K_1|u_1 - u_2|$$

and

$$\frac{K_1}{\alpha\Gamma(\alpha)} < 1$$

for each $t \in [0,T]$ and all $x, y \in \mathbb{R}^n$, then Eq. (2) has an unique solution on [0, T].

Proof First, we transform Eq. (2) in to a fixed point problem. For this purpose consider the operation

$$P : C([0,T],\mathbb{R}^n) \to C([0,T],\mathbb{R}^n)$$

defined by

$$P(u)(t) = u_0 + \frac{1}{\Gamma(\alpha)}\int_0^t f(v,u(v))(t-v)^{\alpha-1}dv + \frac{1}{\Gamma(\alpha)}\int_0^t\int_0^v g(v,s)d\hat{w}(s)dv$$

As we can see according to Eq. (2) we have

$$P(u)(t) = u(t)$$

If we show P is a contraction operator, using the Banach contraction principle we conclude P has a fixed point and we conclude Eq.(2) has a unique solution. Consequently applying Eq.(3) it is easy to see that

$$\begin{cases} u_N(t_1) = u(t_0) + \dfrac{1}{\Gamma(\alpha)}\displaystyle\int_0^{t_1} f(v,u_N(v))(t_1-v)^{\alpha-1}dv + \dfrac{1}{\Gamma(\alpha)}\int_0^{t_1}\int_0^v g(v,s)d\hat{w}(s)dv \\ u_N(t_2) = u(t_0) + \dfrac{1}{\Gamma(\alpha)}\displaystyle\int_0^{t_2} f(v,u_N(v))(t_2-v)^{\alpha-1}dv + \dfrac{1}{\Gamma(\alpha)}\int_0^{t_2}\int_0^v g(v,s)d\hat{w}(s)dv \\ \vdots \\ u_N(t_N) = u(t_0) + \dfrac{1}{\Gamma(\alpha)}\displaystyle\int_0^{t_N} f(v,u_N(v))(t_N-v)^{\alpha-1}dv + \dfrac{1}{\Gamma(\alpha)}\int_0^{t_N}\int_0^v g(v,s)d\hat{w}(s)dv \\ \displaystyle\sum_{k=0}^{m-1} c_k p_l(t_k) = 0, \quad l = 0,\ldots,m-1 \end{cases}$$

where

$$u_N(t) = \sum_{i=1}^{N} c_i\varphi_i(\|t-t_i\|) + \sum_{l=0}^{m-1} d_l p_l(t), \quad x \in D$$

$$|P(u_1)(t) - P(u_2)(t)| \le \frac{1}{\Gamma(\alpha)}\int_0^t |(f(u_1(v),v) - f(u_2(v),v))|(t-v)^{\alpha-1}dv \le$$

$$\frac{K_1}{\Gamma(\alpha)}|u_1-u_2|\int_0^t (t-v)^{\alpha-1}dv \leq \frac{K_1}{\alpha\Gamma(\alpha)}|u_1-u_2|.$$

On the other side using the assumption of the lemma we know that $\frac{K_1}{\alpha\Gamma(\alpha)}<1$, therefore the proof is completed. □

Illustrative example

In this section, we solve SFDEs using RBFs and Galerkin method [19], these equations don't have exact solution so we use numerical approximation for sufficiently small partition on t. While we have this relation

$$|e_{exact}-e_{RBFs}| = |e_{exact}-e_{Galerkin}+e_{Galerkin}-e_{RBFs}|$$
$$\leq |e_{exact}-e_{Galerkin}| + |e_{Galerkin}-e_{RBFs}|$$

then we approximate the solution in the form of RMSError [34] as follow

$$RMSError = \sqrt{\frac{\sum_{i=0}^{n}(U_{RBFs}(x_i,0)-U_{Galerkin}(x_i,0))^2}{n}}$$

In this paper, we introduce RMSError after 50 and 60 times run the program for different points and σ.

Example 4.1 Consider the following SFDE:

$$D^{\frac{3}{2}}u(t)+u(t)=t+1+\sigma\int_0^t dW(s), \qquad u(0)=1.$$

We have Tables 1, 2 and 3 after 60 times run the program with $n=17$.

Example 4.2 Consider the following SFDE:

$$D^{\alpha}u(t)+u(t)=\frac{2t^{2-\alpha}}{\Gamma(3-\alpha)}-\frac{t^{1-\alpha}}{\Gamma(2-\alpha)}+t^2-t+\sigma\int_0^t sdw(s), \qquad u(0)=0, \qquad 0<\alpha\leq 1.$$

Table 1 U_{RBFs} for Example (4.1) with $n=17$ and $\sigma=1$.

t	U_{RBFs}	t	U_{RBFs}	t	U_{RBFs}
0	0.6666	0.3750	1.1109	0.75	1.5894
0.0625	0.7391	0.4375	1.188	0.8125	1.6696
0.1250	0.8113	0.5	1.2666	0.875	1.7478
0.1875	0.8852	0.5625	1.3445	0.9375	1.8201
0.25	0.9604	0.6250	1.4233	1	1.9076
0.3125	1.0356	0.6875	1.5059		

For $\alpha=\frac{1}{2}$ and various σ we have Table 4 after 50 times run the program with nodal points $n=12$ and for $\alpha=\frac{3}{2}$ and different value for σ we have Table 5 after 50 times run the program with $n=11$.

Table 2 $U_{Galerkin}$ for example (4.1) with $n=17$ and $\sigma=1$

t t	U_{RBFs} $U_{Galerkin}$	t t	U_{RBFs} $U_{Galerkin}$	t t	U_{RBFs} $U_{Galerkin}$
0	0.866	0.3750	2.1326	0.75	0.5876
0.0625	1.2892	0.4375	0.5363	0.8125	0.3984
0.1250	0.3070	0.5	1.7974	0.875	0.0539
0.1875	1.6771	0.5625	1.9590	0.9375	2.6604
0.25	1.1019	0.6250	2.6902	1	1.9019
0.3125	0.7147	0.6875	2.7538		

Table 3 RMS Error for example (4.1), $n=17$

σ	RMS
1	0.058
0.8	0.0464
0.6	0.0229
0.3	0.0251
0.2	0.0107
0.09	0.0038
0.009	4.287×10^{-4}
0.001	7.1078×10^{-5}

Table 4 RMS Error for example (4.2) with $\alpha=\frac{1}{2}$, $n=12$

σ	RMS
0.7	0.0308
0.3	0.0230
0.1	0.0248
0.05	0.0107
0.03	0.0103
0.01	0.0140
0.006	0.0104

Table 5 RMS Error for Example (4.2) with $\alpha=\frac{3}{2}$, $n=11$

σ	RMS
0.6	0.0470
0.5	0.0450
0.4	0.0614
0.2	0.0422
0.1	0.0512
0.05	0.0503
0.03	0.0498
0.008	0.04995

Table 6 RMS Error for Example 4.3 for different values of n at points t

$t\backslash n$	8	11	15	20
0	0.005	0.0006	0.0007	6 (−7)
0.2	0.242	0.023	3.2341 (−3)	8.234 (−4)
0.4	0.453	0.067		
0.6	0.483	0.034	5.327 (−3)	3.792 (−4)
0.8	0.129	0.022	5.901 (−3)	5.325 (−4)
1	0.321	0.049	6.383 (−3)	7.235 (−4)

Table 7 RMS error for Example 4.4 for different values of n at points t

$t\backslash n$	10	13	15	18
0	6.321 (−5)	3.289 (−6)	8.389 (−6)	5.329 (−7)
0.3	9.321 (−2)	5.736 (−2)	1.719 (−2)	6.132 (−3)
0.6	7.068 (−2)	6.368 (−2)	3.102 (−2)	9.618 (−3)
0.9	6.932 (−2)	5.971 (−2)	1.001 (−2)	8.320 (−3)
1	7.180 (−2)	8.143 (−2)	6.128 (−2)	9.731 (−3)

Example 4.3 Consider the following SFDE:

$$D^{0.5}u(t) + u(t) = (t+1)^5 + \int_0^t \cos(s)dw(s), \qquad u(0) = 1.$$

Example 4.4 Consider the following SFDE:

$$D^{0.75}u(t) + u^3(t) = t^3 + 1 + \int_0^t (s^2+1)^3 dw(s), \qquad u(0) = 0.$$

In this work, the accuracy of approximate solution, when taking larger n and smaller σ, is expected that more accurate the approximate results.

Conclusion

The main goal of this work was to purpose an efficient algorithm for the stochastic fractional differential equations. In this paper, while we don't have exact solution for SFDEs we used RBFs to approximate the solution of these kind of equations. In addition, we discussed about existence and uniqeness of the presented method. The present RMS Error in the tables shows that the results are highly accurate in comparison with another method using by Galerkin algorithm (Tables 6, 7).

References

1. Maimardi, F.: Fractional calculus and waves in linear viscoelasticity. Imperial College Press, London (2010)
2. Alvelid, M., Enelund, M.: Modeling of constrained thin rubber layer with emphasis on damping. J. Sound Vib. **300**, 662–675 (2007)
3. Atanackovich, T.M., Stankovic, B.: On a system of differential equations with fractional derivatives arising in rod theory. J. Phys. A **37**(4), 1241–1250 (2004)
4. A. A. Kilbas, H.M. Srivastava, J. J. Trujillo, Theory and applications of fractional differential equations, vol.204 of North-Holland Mathematics studies. Elsevier, Amsterdam(2006)
5. El-Tawil, M.A., Bahnasawi, A.A., Abdel-Naby, A.: solving Riccati Differential Equation using Adomian Decomposition Method, Appl. Math. Comput. 157(2), 503–514 (2004)
6. He, J. H.: Variation iteration method-a kind of nonlinear analytical technique: some examples. Int. J Non-linear Mech. 34, 699–708 (1999)
7. Inc, M.: The approximate and exact solutions of the space -and time-fractional burgers equations with initial conditions by VIM. J. Math. Anal. Appl. **345**, 476–484 (2008)
8. Abbasbandy, S.: homotopy perturbation method for quadratic Riccati Differential Equation and comparison with Adomian Decomposition Method, Appl. Math. Comput. 172, 485-490 (2006)
9. Abbasbandy, S.: The application of homotopy analysis method to nonlinear equations arising in heat transfer, Phys. Lett. A 360, 109-113 (2006)
10. Rezazadeh, H., Aminikhah, H., Refahi Sheikhani, A. H.: Analytical studies for linear periodic systems of fractional order, Math Sci 10, 1321 (2016)
11. Glasserman, P.: Monte Carlo Method in Financial Engineering. Applications of Mathematics, Vol. 53 (Springer, New York, 2004)
12. Kloeden, P., Platen, E.: Numerical solution of stochastic differential equations(springer, Berlin/NewYork, 1992)
13. Öhinger,: Stochastic Processes in Polymeric Fluids (Springer, Berlin, 1996). Tools and examples for developing simulation algorithms
14. P. Rué. villá-Freixa, K. Burrage, Simulation methods with extended stability for stiff biochemical kinetics. BMC Syst. Biol. 4(110), 1–13 (2010)
15. Dung, N.T.: Fractional stochastic differential equations with applications of finance. J. Math. Anal. Appl. **397**, 334–348 (2013)
16. Higham, D.J.: An algorithmic introduction to numerical simulation of Stochastic Differential Equations. Soc. Indust. Appl. Math. 43(3), 525–546
17. Khodabin, M., Maleknejad, K., Damercheli, T.: Approximate solution of the stochastic Volterra integral equations via expansion method. IJIM **6**(1), 41–48 (2014)
18. Ezzati, R., Khodabin, M., Sadati, Z.: Numerical Implementation of Stochastic Operational Matrix Driven by a Fractional Brownian Motion for Solving a Stochastic Differential Equation, Abstract and Applied Analysis Volume 2014, Article ID 523163, 11 pages
19. Kamrani, M.: Numerical solution of stochastic differential equations. Numer Algor. 68, 81–93 (2015)
20. Enelund, M. Josefson, B.L.: Time-domain finite element analysis of viscoelastic structures with fractional derivatives constitutive relations. AIAAJ. 35(10), 1630–1637 (1997)
21. Friedrich, C.: linear viscoelastic behavior of branched polybutadiens: a fractional calculus approach. Acta Polym. 385–390 (1995)
22. Govindan, T.E., Josh, M.C.: Stability and optimal control of stochastic functional differential equations with memory. Numer. Funct. Anal. Optim. **13**(3–4), 249–265 (1992)
23. Frank, C., Schaback, R.: solving partial differential equations by collocation using radial basis functions. Appl. Comut. **93**, 73–82 (1998)

24. Kansa, E.J.: multiquadric a scattered data approximation scheme with applications to computational fluid dynamics. Comput. Math. Appl. 19, 147–161 (1990)
25. Tran-Cong, T., Mai-Dug, N., Phan-Thien, N.: BEM-RBF approch for viscoelastic flow analysis. Eng Anal Boundary Element **26**, 757–762 (2002)
26. Shen, Q.: A meshless method of lines for the numerical solution of Kdv equation using radial basis functions. Eng. Anal. Boundary Element. **33**, 1171–1180 (2009)
27. Dehghan, M., Shokri, A.: Numerical solution of the nonlinear Klein-Gordon equation using radial basis functions, J.of. Comput. Appl. Math. **230**, 400–410 (2009)
28. Vanani, S.K., Aminataei, A.: On the numerical solution of fractional partial differential equations. Math. Compute. Appl. **17**(2), 140–151 (2012)
29. Gonzalez-Gaxiola, O.: Pedro pablo Gonzalez-Perez, Nonlinear black-scholes equation thorough Radial Basis Functions, J.of Appl. Math. Bioinformatics **4**, 75–86 (2014)
30. Oksendal, B.: Stochastic Differential Equations, An Introduction with Applications, Springer; 6th edn
31. Allen, E.J., Novosel, S.J., Zhang, Z.: Finite element and difference approximation of some linear stochastic partial differential equations. Stochast. Stochast. Rep. **64**(12), 117142 (1998)
32. Bohmann, M.D.: Radial Basis Functions, Theory and implementations. Cambridge University Press, Cambridge (2003)
33. Wendland, Scattered Data Approximation, Cambrige University Press (2005)
34. Chai, T., Draxler, R. R.: Root mean square error(RMSE) or mean absolute error (MAE)?- Argument against avoiding RMSE in the literature . Geosci. Model Dev. 1247–1250, 2014

21

Analytical studies for linear periodic systems of fractional order

Hadi Rezazadeh[1,3] · Hossein Aminikhah[1] · Amir Hossein Refahi Sheikhani[2]

Abstract In this paper, new periodic fractional trigonometric functions with the period $2\pi_\alpha$ are presented. We have generalized the Floquet system to the fractional Floquet system. The fractional derivatives are described with the use of modified Riemann–Liouville derivative. Moreover, the stability analysis of fractional Floquet system is introduced.

Keywords Floquet system · Stability · Fractional derivative · Modified Riemann–Liouville

Introduction

The study of systems governed by ordinary differential equation with period coefficients is of basic importance in many branches such as mathematics, physics, chemistry, biology, mechanics and finance, such systems are known as Floquet systems [1, 2]. The Floquet systems are defined with the $n \times n$ matrix function A as $x' = A(t)x$, where the components in matrix A are continuous and periodic function with smallest positive period w, that is, $A(t + w) = A(t)$ for all ts. Although the coefficient matrix in $x' = A(t)x$ is periodic, in general solutions they are not considered as periodic. The idea of Floquet systems has been stated by Gaston Floquet in the early 1880s, and later he established his celebrated theorem on the structure of solutions of periodic differential equations [3]. In this paper, we first focus our attention on fractional Floquet system, and then we consider the stability analysis for this class system.

The fractional order calculus establishes the branch of mathematics dealing with differentiation and integration under an arbitrary order of the operation, that is the order can be any real or even complex number, not only the integer one. Although the history of fractional calculus is more than three centuries old, it only has received much attention and interest in the past 20 years; the reader may refer to [4–6] for the theory and applications of fractional calculus. The generalization of dynamical equations using fractional derivatives proved to be useful and more accurate in mathematical modeling related to many interdisciplinary areas. Applications of fractional order differential equations include: electrochemistry [7], porous media [8] and so on [9–11]. It is worth noting that recently much attention has been paid to the distributed-order differential equations and their applications in engineering fields that both integer-order systems and fractional order systems are special cases of distributed-order systems. The reader may refer to [12–14]. The analytic results on the existence and uniqueness of solutions to the fractional differential equations have been investigated by many authors [5, 6].

✉ Amir Hossein Refahi Sheikhani
ah_refahi@liau.ac.ir

Hadi Rezazadeh
rezazadehadi1363@gmail.com

Hossein Aminikhah
aminikhah@guilan.ac.ir

1 Department of Applied Mathematics, Faculty of Mathematical Sciences, University of Guilan, P.O. Box 1914, Rasht, Iran

2 Department of Applied Mathematics, Faculty of Mathematical Sciences, Islamic Azad University, Lahijan Branch, P.O. Box 1616, Lahijan, Iran

3 Faculty of Engineering Technology, Amol University of Special Modern Technologies, P.O. Box 46168-49767, Amol, Iran

Preliminaries and notations

Basic definitions

We give some basic definitions and properties of the fractional calculus theory used in this work.

Definition 1 Let $f:\mathbb{R}\to\mathbb{R}$, $t\to f(t)$ denote a continuous (but not necessarily differentiable) function and let partition $h>0$ in the interval [0, 1]. The Jumarie'Derivative is defined through the fractional difference [15]:

$$\Delta^{\alpha}f(t)=(\mathrm{FW}-1)^{\alpha}f(t)=\sum_{k=0}^{\infty}(-1)^{k}\binom{\alpha}{k}f[t+(\alpha-k)h], \quad (1)$$

where $\mathrm{FW}\,f(t)=f(t+h)$. Then the fractional derivative is defined as the following limit

$$f^{(\alpha)}(t)=D_t^{\alpha}f(t)=\frac{\mathrm{d}^{\alpha}f(t)}{\mathrm{d}t^{\alpha}}=\lim_{h\to 0}\frac{\Delta^{\alpha}[f(t)-f(0)]}{h^{\alpha}}. \quad (2)$$

This definition is close to the standard definition of derivatives, and as a direct result, the αth derivative of a constant $0<\alpha\leq 1$ is zero.

Definition 2 The Riemann–Liouville fractional integral operator of order $\alpha>0$ is defined as [16]:

$$I_t^{\alpha}f(t)=\frac{1}{\Gamma(\alpha)}\int_0^t(t-\varepsilon)^{\alpha-1}f(\varepsilon)\mathrm{d}\varepsilon, \quad \alpha>0. \quad (3)$$

Definition 3 The modified Riemann–Liouville derivative is defined as [16]:

$$D_t^{\alpha}f(t)=\left(f^{(\alpha-1)}(t)\right)'=\frac{1}{\Gamma(1-\alpha)}\frac{\mathrm{d}}{\mathrm{d}t}\int_0^t(t-\varepsilon)^{-\alpha}(f(\varepsilon)-f(0))\mathrm{d}\varepsilon,$$
$$0<\alpha\leq 1, \quad (4)$$

and

$$D_t^{\alpha}f(t)=\left(f^{(\alpha-n)}(t)\right)^{(n)}, \quad n\leq\alpha<n+1, \ n\geq 1.$$

The proposed modified Riemann–Liouville derivative as shown in Eq. (4) is strictly equivalent to Eq. (2).

Definition 4 Fractional derivative of compounded functions is defined as [16]:

$$d^{\alpha}f\approx\Gamma(1+\alpha)\mathrm{d}f, \quad 0<\alpha<1. \quad (5)$$

Definition 5 The integral with respect to $(\mathrm{d}t)^{\alpha}$ is defined as the solution of fractional differential equation [17]:

$$\mathrm{d}y=f(t)(\mathrm{d}t)^{\alpha}, \quad t\geq 0, \ y(0)=0, \quad 0<\alpha\leq 1. \quad (6)$$

Lemma 1 Let $f(t)$ denotes a continuous function then the solution of the Eq. (6) is defined as [17]:

$$y=\int_0^t f(\varepsilon)(\mathrm{d}\varepsilon)^{\alpha}=\alpha\int_0^t(t-\varepsilon)^{\alpha-1}f(\varepsilon)\mathrm{d}\varepsilon, \quad 0<\alpha\leq 1. \quad (7)$$

Definition 6 Function $f(t)$ is αth differentiable then the following equalities holds:

$$f^{(\alpha)}(t)=\lim_{h\to 0}\frac{\Delta^{\alpha}f(t)}{h^{\alpha}}=\Gamma(1+\alpha)\lim_{h\to 0}\frac{\Delta f(t)}{h^{\alpha}}, \quad 0<\alpha\leq 1. \quad (8)$$

Mittag-Leffler function

The Mittag-Leffler function which plays a very important role in the fractional differential equations was in fact introduced by Mittag-Leffler in 1903 [18]. The Mittag-Leffler function $E_{\alpha}(t)$ is defined by the power series:

$$E_{\alpha}(t)=\sum_{n=0}^{\infty}\frac{t^n}{\Gamma(n\alpha+1)}, \quad \alpha>0, \quad (9)$$

which

$$D_t^{\alpha}E_{\alpha}(\lambda t^{\alpha})=\lambda E_{\alpha}(\lambda t^{\alpha}). \quad (10)$$

As further result of the above formula

$$E_{\alpha}(\lambda t^{\alpha})E_{\alpha}(\lambda(\pm s)^{\alpha})\approx E_{\alpha}(\lambda(t\pm s)^{\alpha}), \quad \lambda\in\mathbb{C}. \quad (11)$$

The matrix extension of the mentioned Mittag-Liffler function for $A\in M_m$ is defined as in the following representation:

$$E_{\alpha}(At^{\alpha})=\sum_{n=0}^{\infty}\frac{A^n t^{\alpha n}}{\Gamma(n\alpha+1)}, \quad \alpha>0. \quad (12)$$

If $A,B\in\mathbb{R}^{n\times n}$ and $\alpha>0$, then it is easy to prove the following nice properties of Mittag-Leffler matrix $E_{\alpha}(At^{\alpha})$:

(i) $E_{\alpha}^{-1}(At^{\alpha})\approx E_{\alpha}(-At^{\alpha})$,
(ii) If P is a non-singular matrix, then $E_{\alpha}(P^{-1}AP)=P^{-1}E_{\alpha}(A)P$,
(iii) $E_{\alpha}((A+B)t^{\alpha})\approx E_{\alpha}(At^{\alpha})E_{\alpha}(Bt^{\alpha})$ if and only if $AB=BA$,
(iv) $E_{\alpha}^{-1}(At^{\alpha})\approx E_{\alpha}(A(-t)^{\alpha})$.

Corollary 1 [19] *If the matrix A is diagonalizable, that is, there exists an invertible matrix T such that*

$$\Lambda=T^{-1}AT=diag(\lambda_1,\lambda_2,\ldots,\lambda_n),$$

then, we have

$$E_{\alpha}(At^{\alpha})=T\,E_{\alpha}(\Lambda t^{\alpha})T^{-1}=T\,diag(E_{\alpha}(\lambda_1 t^{\alpha}), E_{\alpha}(\lambda_2 t^{\alpha}),\ldots,E_{\alpha}(\lambda_n t^{\alpha}))T^{-1}.$$

Next, suppose the matrix A is similar to a Jordan canonical form, that is there exists an invertible matrix T such that

$$J = T^{-1}AT = diag(J_1, J_2, \ldots, J_n),$$

where j_i, $1 \le i \le r$ has the following form

$$\begin{bmatrix} \lambda_i & 1 & 0 & \cdots & 0 \\ 0 & \lambda_i & 1 & \ddots & \vdots \\ \vdots & 0 & \ddots & \ddots & 0 \\ 0 & \ddots & \ddots & \lambda i & 1 \\ 0 & 0 & \cdots & 0 & \lambda_i \end{bmatrix}_{n_i \times n_i},$$

and $\sum_{i=1}^{r} n_i = n$. Obviously,

$$E_\alpha(At^\alpha) = Tdiag(E_\alpha(J_1 t^\alpha), E_\alpha(J_2 t^\alpha), \ldots, E_\alpha(J_r t^\alpha))T^{-1},$$

and

$$E_\alpha(J_i t^\alpha) = \sum_{k=0}^{\infty} \frac{(J_i t^\alpha)^k}{\Gamma(\alpha k + 1)} = \sum_{k=0}^{\infty} \frac{(t^\alpha)^k}{\Gamma(\alpha k + 1)} \begin{pmatrix} \lambda_i^k & C_k^1 \lambda_i^{k-1} & \cdots & C_k^{n_i-1} \lambda_i^{k-n_i+1} \\ 0 & \lambda_i^k & \ddots & \vdots \\ \vdots & \ddots & \ddots & C_k^1 \lambda_i^{k-1} \\ 0 & \cdots & 0 & \lambda_i^k \end{pmatrix}$$

$$= \begin{pmatrix} \sum_{k=0}^{\infty} \frac{(t^\alpha)^k}{\Gamma(\alpha k + 1)} \lambda_i^k & \sum_{k=0}^{\infty} \frac{(t^\alpha)^k}{\Gamma(\alpha k + 1)} C_k^1 \lambda_i^{k-1} & \cdots & \sum_{k=0}^{\infty} \frac{(t^\alpha)^k}{\Gamma(\alpha k + 1)} C_k^{n_i-1} \lambda_i^{k-n_i+1} \\ 0 & \sum_{k=0}^{\infty} \frac{(t^\alpha)^k}{\Gamma(\alpha k + 1)} \lambda_i^k & \ddots & \vdots \\ \vdots & \ddots & \ddots & \sum_{k=0}^{\infty} \frac{(t^\alpha)^k}{\Gamma(\alpha k + 1)} C_k^1 \lambda_i^{k-1} \\ 0 & \cdots & 0 & \sum_{k=0}^{\infty} \frac{(t^\alpha)^k}{\Gamma(\alpha k + 1)} \lambda_i^k \end{pmatrix}$$

$$= \begin{pmatrix} E_\alpha(\lambda_i t^\alpha) & \frac{1}{1!} \frac{\partial}{\partial \lambda_i} E_\alpha(\lambda_i t^\alpha) & \cdots & \frac{1}{(n_i - 1)!} \left(\frac{\partial}{\partial \lambda_i} \right)^{n_i - 1} E_\alpha(\lambda_i t^\alpha) \\ 0 & E_\alpha(\lambda_i t^\alpha) & \ddots & \vdots \\ \vdots & \ddots & \ddots & \frac{1}{1!} \frac{\partial}{\partial \lambda_i} E_\alpha(\lambda_i t^\alpha) \\ 0 & \cdots & 0 & E_\alpha(\lambda_i t^\alpha) \end{pmatrix},$$

where C_k^j, $1 \le j \le n_i - 1, 1 \le i \le r$ are the binomial coefficients.

Fractional trigonometric functions and Mittag-Leffler logarithm function

The idea of the fractional trigonometric functions has been stated by Jumarie [20] asserting that these functions are not periodic. Now, we introduce new fractional trigonometric functions which are periodic with the period $2\pi_\alpha \approx 2\pi$. Analogous with the trigonometric function, we can write

$$E_\alpha((it)^\alpha) = \cos_\alpha(t^\alpha) + i \sin_\alpha(t^\alpha), \tag{13}$$

and

$$E_\alpha((-it)^\alpha) = \cos_\alpha(t^\alpha) - i \sin_\alpha(t^\alpha), \tag{14}$$

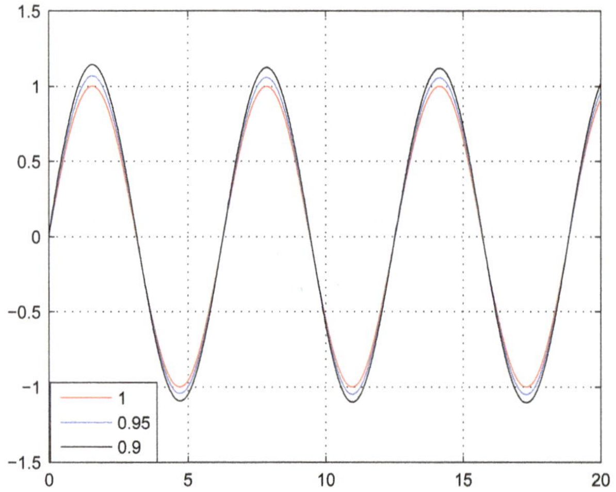

Fig. 1 Plot of $\sin_\alpha(t^\alpha)$ with respect to t for $\alpha = 1,\ 0.95,\ 0.9$

with

$$\cos_\alpha(t^\alpha) = \frac{E_\alpha((it)^\alpha) + E_\alpha((-it)^\alpha)}{2}, \quad \text{and}$$

$$\sin_\alpha(t^\alpha) = \frac{E_\alpha((it)^\alpha) - E_\alpha((-it)^\alpha)}{2i}.$$

These fractional functions have the period $2\pi_\alpha \approx 2\pi$. Figure 1 shows $\sin_\alpha(t^\alpha)$ for $\alpha = 1,\ 0.95,\ 0.9$ which is periodic with the period $2\pi_\alpha \approx 2\pi$.

Some properties of the fractional trigonometric functions are presented as follows:

$$\sin_\alpha^2 \theta^\alpha + \cos_\alpha^2 \theta^\alpha \approx 1,$$

$$\sin_\alpha(-t)^\alpha = -\sin_\alpha(t^\alpha),$$
$$\cos_\alpha(-t)^\alpha = \cos_\alpha(t^\alpha),$$

$$D_t^\alpha(\sin_\alpha(\omega^\alpha t^\alpha)) = \omega^\alpha (i)^{\alpha-1} \cos_\alpha(\omega^\alpha t^\alpha),$$
$$D_t^\alpha(\cos_\alpha(\omega^\alpha t^\alpha)) = \omega^\alpha (i)^{\alpha+1} \sin_\alpha(\omega^\alpha t^\alpha).$$

The fractional functions $\sin_\alpha(\omega^\alpha t^\alpha)$ and $\cos_\alpha(\omega^\alpha t^\alpha)$ both are periodic functions with the period $(2\pi_\alpha/\omega)$.

In addition Eq. (11) provides the equalities

$$\cos_\alpha(t+s)^\alpha \approx \cos_\alpha(t^\alpha)\cos_\alpha(s^\alpha) - \sin_\alpha(t^\alpha)\sin_\alpha(s^\alpha), \quad (15)$$

$$\sin_\alpha(t+s)^\alpha \approx \cos_\alpha(t^\alpha)\sin_\alpha(s^\alpha) + \cos_\alpha(s^\alpha)\sin_\alpha(t^\alpha). \quad (16)$$

There are similar formulas like for $\cos_\alpha(t-s)^\alpha$ and $\sin_\alpha(t-s)^\alpha$.

Substituting θ for both t and s in the addition formulas gives

$$\cos_\alpha 2\theta^\alpha \approx \cos_\alpha^2 \theta^\alpha - \sin_\alpha^2 \theta^\alpha, \qquad sin_\alpha 2\theta^\alpha \approx 2\sin_\alpha \theta^\alpha \cos_\alpha \theta^\alpha.$$

Additional formulas come from combining the equations

$$\sin_\alpha^2 \theta^\alpha + \cos_\alpha^2 \theta^\alpha \approx 1, \quad \cos_\alpha 2\theta^\alpha \approx \cos_\alpha^2 \theta^\alpha - \sin_\alpha^2 \theta^\alpha,$$

we add the two equations to get $\cos_\alpha 2\theta^\alpha \approx 2\cos_\alpha^2 \theta^\alpha - 1$ and subtract the second from the first to get $\cos_\alpha 2\theta^\alpha \approx 1 - 2\sin_\alpha^2 \theta^\alpha$.

Definition 7 $Ln_\alpha t$ denotes the inverse function of the $E_\alpha(t)$, referred to as Mittag-Leffler logarithm, clearly $E_\alpha(Ln_\alpha t) = t$ and the Mittag-Leffler logarithm function is defined as [20]:

$$\int_0^t \frac{d^\alpha \xi}{\xi} = \frac{1}{(1-\alpha)!}\int_0^t \frac{(d\xi)^\alpha}{\xi^\alpha} = Ln_\alpha(t). \quad (17)$$

Fractional linear system and its stability analysis

Here, we will consider the following linear fractional differential system with modified Riemann–Liouville fractional derivative

$$D_t^\alpha x = Ax, \quad (18)$$

with initial value $x(0) = x_0 = (x_{10}, x_{20}, \ldots, x_{no})^T$, where $x = (x_1, x_2, \ldots, x_n)^T$, $\alpha \in (0, 1]$ and $A \in \mathbb{R}^{n\times n}$. By implementation of the Laplace transform on the above system and using the initial condition, the general solution can be written as

$$x = x_0 E_\alpha(At^\alpha). \quad (19)$$

The stability of the equilibrium of system (18) was first defined and established by Matignon as follows [21].

Definition 8 The linear fractional differential system (18) is said to be

(i) stable if for any initial value x_0, there exists a $\varepsilon > 0$ such that for all $t \geq 0$,

(ii) asymptotically stable if at first it is stable and $\lim_{t\to\infty}\|x(t^\alpha)\| = 0$.

Theorem 1 *The linear fractional differential system* (18) *is asymptotically stable if all the eigenvalues of A satisfy*

$$|\arg(\lambda(A))| > \frac{\alpha\pi}{2}. \quad (20)$$

We can very easily prove Theorem 1 analogously using Proposition 3.1 in [21].

Now, we state the following important existence–uniqueness theorem for solutions of initial value problems (21).

Theorem 2 ([22]) *Let* $0 < \alpha \leq 1$, $(0, b) \subset \mathbb{R}$, *U be an open connected set in* $\mathbb{R}^{n+1}$, $\Delta = (0, b) \times U$ *and* $(t_0, x_0) \in \Delta$. *If*

$$A(t^\alpha) = \begin{bmatrix} a_{11}(t^\alpha) & a_{12}(t^\alpha) & \dots & a_{1n}(t^\alpha) \\ a_{21}(t^\alpha) & a_{22}(t^\alpha) & \dots & a_{2n}(t^\alpha) \\ \vdots & \vdots & \ddots & \vdots \\ a_{n1}(t^\alpha) & a_{n2}(t^\alpha) & \dots & a_{nn}(t^\alpha) \end{bmatrix} \quad \text{and}$$

$$B(t^\alpha) = \begin{bmatrix} b_1(t^\alpha) \\ \vdots \\ b_n(t^\alpha) \end{bmatrix},$$

are continuous matrices in $[0, b]$, then equation

$$D_t^\alpha x = A(t^\alpha)x + B(t^\alpha), \tag{21}$$

has a unique solution $x(t^\alpha)$, continuous in $(0, b]$, such that $x(t_0^\alpha) = x_0$.

Fractional Floquet system

In this Section, we will consider fractional order linear periodic differential equations involving modified Riemann–Liouville derivative that can be written in the form

$$D_t^\alpha x(t^\alpha) = f_\alpha(t^\alpha)x(t^\alpha), \tag{22}$$

where we assume that $f_\alpha : (a, b) \to \mathbb{R}$ is continuous periodic function with smallest positive periodic w_α, that is, $f_\alpha((t + w_\alpha)^\alpha) \approx f_\alpha(t^\alpha)$.

Also, the solution of Eq. (22) obtained with respect to the Mittag-Leffler function is as

$$x(t^\alpha) \approx C\, E_\alpha\left(\int_0^t f_\alpha(\tau^\alpha)(d\tau)^\alpha\right), \tag{23}$$

where C is a constant.

Example 1 Consider the fractional order linear periodic differential equation

$$D_t^\alpha x(t^\alpha) = \sin_\alpha(t^\alpha)x(t^\alpha). \tag{24}$$

Thus, by (23), $x(t^\alpha) \approx C\, E_\alpha(\frac{1}{j^\alpha+1}\cos_\alpha(t^\alpha))$, where for $t \in \mathbb{R}$ is a general solution on $\mathbb{R}$ for (24).

Definition 9 We say x is a solution of (22) on an interval $L \subset (0, b)$ if x is a continuously αth differentiable function on L and for $t \in L$, x satisfies (22).

In the rest of this section, we will generalize linear periodic systems to fractional periodic systems involving modified Riemann–Liouville derivative form

$$\begin{aligned} D_t^\alpha x_1 &= a_{\alpha_{11}}(t^\alpha)x_1 + a_{\alpha_{12}}(t^\alpha)x_2 + \dots + a_{\alpha_{1n}}(t^\alpha)x_n, \\ D_t^\alpha x_2 &= a_{\alpha_{21}}(t^\alpha)x_1 + a_{\alpha_{22}}(t^\alpha)x_2 + \dots + a_{\alpha_{2n}}(t^\alpha)x_n, \\ &\vdots \qquad\qquad \vdots \qquad\qquad\quad \vdots \qquad\qquad\qquad \vdots \\ D_t^\alpha x_n &= a_{\alpha_{n1}}(t^\alpha)x_1 + a_{\alpha_{n2}}(t^\alpha)x_2 + \dots + a_{\alpha_{nn}}(t^\alpha)x_n, \end{aligned} \tag{25}$$

where $a_{\alpha_{ij}}(t^\alpha)$, $(i, j = 1, 2, \dots, n)$ are given continuous periodic functions with smallest positive periodic w_α on an interval L.

This system can be transformed to a vector–matrix form as

$$D_t^\alpha x = A_\alpha(t^\alpha)x, \tag{26}$$

where

$$x = \begin{bmatrix} x_1 \\ \vdots \\ x_n \end{bmatrix}, \quad D_t^\alpha x = \begin{bmatrix} D_t^\alpha x_1 \\ \vdots \\ D_t^\alpha x_n \end{bmatrix},$$

and

$$A_\alpha(t^\alpha) = \begin{bmatrix} a_{\alpha_{11}}(t^\alpha) & a_{\alpha_{21}}(t^\alpha) & \dots & a_{\alpha_{1n}}(t^\alpha) \\ a_{\alpha_{21}}(t^\alpha) & a_{\alpha_{22}}(t^\alpha) & \dots & a_{\alpha_{2n}}(t^\alpha) \\ \vdots & \vdots & \ddots & \vdots \\ a_{\alpha_{n1}}(t^\alpha) & a_{\alpha_{n2}}(t^\alpha) & \dots & a_{\alpha_{nn}}(t^\alpha) \end{bmatrix},$$

where the components in matrix A_α are continuous and periodic functions with smallest positive period w_α (saying $A_\alpha((t + w_\alpha)^\alpha) \approx A_\alpha(t^\alpha)$).

Consider the matrix fractional differential equation

$$D_t^\alpha X = A_\alpha(t^\alpha)X, \tag{27}$$

where

$$X = \begin{bmatrix} x_{11} & x_{12} & \dots & x_{1n} \\ x_{21} & x_{22} & \dots & x_{2n} \\ \vdots & \vdots & \ddots & \vdots \\ x_{n1} & x_{n2} & \dots & x_{nn} \end{bmatrix}, \quad \text{and}$$

$$D_t^\alpha X = \begin{bmatrix} D_t^\alpha x_{11} & D_t^\alpha x_{12} & \dots & D_t^\alpha x_{1n} \\ D_t^\alpha x_{21} & D_t^\alpha x_{22} & \dots & D_t^\alpha x_{2n} \\ \vdots & \vdots & \ddots & \vdots \\ D_t^\alpha x_{n1} & D_t^\alpha x_{n2} & \dots & D_t^\alpha x_{nn} \end{bmatrix},$$

are $n \times n$ matrix variables and A_α is an $n \times n$ continuous matrix function on L.

Theorem 3 (Existence–Uniqueness Theorem) *If the entries of the square matrix* A_α *are continuous on an interval L containing* t_0, then the initial value problem

$$D_t^\alpha X = A_\alpha(t^\alpha)X, \qquad X(t_0) = X_0 \in \mathbb{R}^{n\times n},$$

has one and only one solution X on the whole interval L.

Proof The proof is similar to that of Theorem 2.21 in [2]. □

Definition 10 An $n \times n$ matrix fractional function Φ_α, defined on an interval L, is called a fractional fundamental matrix of the linear system (3.5) if Φ_α is a solution of the fractional matrix equation (27) on L and $\det \Phi_\alpha(t^\alpha) \neq 0$ on L.

Theorem 4 *If Φ_α is a fractional fundamental matrix for $D_t^\alpha x = A_\alpha(t^\alpha)x$, then, for an arbitrary $n \times n$ non-singular constant matrix C, $\Psi_\alpha = \Phi_\alpha C$ is a general fractional fundamental matrix of $D_t^\alpha x = A_\alpha(t^\alpha)x$.*

Proof Since Φ_α is a fractional fundamental matrix solution to $D_t^\alpha x = A_\alpha(t^\alpha)x$ and setting $\Psi_\alpha = \Phi_\alpha C$, we have

$$D_t^\alpha \Psi_\alpha(t^\alpha) = D_t^\alpha \Phi_\alpha(t^\alpha)C = A_\alpha(t^\alpha)\Phi_\alpha(t^\alpha)C = A_\alpha(t^\alpha)\Psi_\alpha(t^\alpha),$$

and also Ψ_α is continuously αth differentiable function on L. Thus, $\Psi_\alpha = \Phi_\alpha C$ is a solution of the matrix fractional equation (3.6). Since Φ_α is a fractional fundamental matrix solution to (26), Definition 10 implies that $\det[\Phi_\alpha(t^\alpha)] \neq 0$. As well, since, $\det[C] \neq 0$. Hence,

$$\det[\Psi_\alpha(t^\alpha)] = \det[\Phi_\alpha(t^\alpha)C] = \det[\Phi_\alpha(t^\alpha)]\det[C] \neq 0,$$

for $t \in L$, and by Definition 10, $\Psi_\alpha = \Phi_\alpha C$ is a fractional fundamental matrix of (27). □

Theorem 5 *If C is an $n \times n$ non-singular matrix, then there is a matrix B such that $E_\alpha(B) = C$.*

Proof To avoid some tedious calculations, we prove this theorem for a 2×2 matrices. For the eigenvalues μ_1, μ_2 / = 0 of nonsingular matrix C. We consider two special cases:

Case I Let

$$C = \begin{bmatrix} \mu_1 & 0 \\ 0 & \mu_2 \end{bmatrix},$$

then, in this case we are looking for a diagonal matrix

$$B = \begin{bmatrix} b_1 & 0 \\ 0 & b_2 \end{bmatrix},$$

so that $E_\alpha(B) = C$. For this purpose, according to the definition of the Mittag-Leffler function, we pick b_1 and b_2 so that

$$E_\alpha(B) = \begin{bmatrix} E_\alpha(b_1) & 0 \\ 0 & E_\alpha(b_2) \end{bmatrix} = \begin{bmatrix} \mu_1 & 0 \\ 0 & \mu_2 \end{bmatrix}.$$

Hence, the matrix B can be taken as

$$B = \begin{bmatrix} \ln_\alpha(\mu_1) & 0 \\ 0 & \ln_\alpha(\mu_2) \end{bmatrix}.$$

Case II Let

$$C = \begin{bmatrix} \mu_1 & 1 \\ 0 & \mu_1 \end{bmatrix},$$

then, we seek a matrix B of the form

$$B = \begin{bmatrix} a_1 & a_2 \\ 0 & a_1 \end{bmatrix},$$

so that $E_\alpha(B) = C$. We choose the parameters a_1 and a_2 so that

$$E_\alpha(B) = \begin{bmatrix} E_\alpha(a_1) & a_2 E_\alpha(a_1) \\ 0 & E_\alpha(a_1) \end{bmatrix} = \begin{bmatrix} \mu_1 & 1 \\ 0 & \mu_1 \end{bmatrix}.$$

Hence, in the view of the inverse function derivative, the matrix B can be taken as

$$B = \begin{bmatrix} \ln_\alpha(\mu_1) & \dfrac{1}{\mu_1} \\ 0 & \ln_\alpha(\mu_1) \end{bmatrix}.$$

Case III When $C \in \mathbb{R}^{2\times 2}$ is an arbitrary matrix such that $\det[C] \neq 0$. By the Corollary 1, there is a non-singular matrix P such that $C = PJP^{-1}$, where

$$C = \begin{bmatrix} \mu_1 & 0 \\ 0 & \mu_2 \end{bmatrix}, \quad or \quad C = \begin{bmatrix} \mu_1 & 1 \\ 0 & \mu_1 \end{bmatrix}.$$

Now, by the previous two cases there is a matrix B_1 so that $E_\alpha(B_1) = J$.

If we set the matrix B as

$$B = PB_1P^{-1},$$

then, we see that

$$E_\alpha(B) = E_\alpha(PB_1P^{-1}) = PE_\alpha(B_1)P^{-1} = C.$$

Similarly, for the higher order of n, the matrix B can be easily found. □

Example 2 For example, consider

$$D_t^\alpha x = \begin{bmatrix} 1 & 1 \\ 0 & \dfrac{\Gamma(\alpha+1)i^{\alpha-1}\cos_\alpha(t^\alpha) - \Gamma(\alpha+1)\sin_\alpha(t^\alpha)}{(2 + \Gamma(\alpha+1)\sin_\alpha(t^\alpha) - \dfrac{\Gamma(\alpha+1)}{i^{\alpha+1}}\cos_\alpha(t^\alpha))} \end{bmatrix} x.$$

Here, we know that the solution is in general

$$x_1(t^\alpha) \approx \mu E_\alpha(t^\alpha) + \beta\left(\frac{\Gamma(\alpha+1)}{i^{\alpha+1}}\cos_\alpha(t^\alpha) - 2\right),$$

$$x_2(t^\alpha) \approx \beta\left(2 + \Gamma(\alpha+1)\sin_\alpha(t^\alpha) - \frac{\Gamma(\alpha+1)}{i^{\alpha+1}}\cos_\alpha(t^\alpha)\right),$$

for $t \in \mathbb{R}$, where $\beta, \mu \in \mathbb{R}$ denote two constants. Using all the above definitions, the fractional fundamental matrix is

$$\Phi_\alpha(t^\alpha) \approx \begin{bmatrix} \frac{\Gamma(\alpha+1)}{i^{\alpha+1}}\cos_\alpha(t^\alpha) - 2 & E_\alpha(t^\alpha) \\ 2+\Gamma(\alpha+1)\sin_\alpha(t^\alpha) - \frac{\Gamma(\alpha+1)}{i^{\alpha+1}}\cos_\alpha(t^\alpha) & 0 \end{bmatrix}$$
$$= \begin{bmatrix} \frac{\Gamma(\alpha+1)}{i^{\alpha+1}}\cos_\alpha(t^\alpha) - 2 & 1 \\ 2+\Gamma(\alpha+1)\sin_\alpha(t^\alpha) - \frac{\Gamma(\alpha+1)}{i^{\alpha+1}}\cos_\alpha(t^\alpha) & 0 \end{bmatrix} \begin{bmatrix} 1 & 0 \\ 0 & E_\alpha(t^\alpha) \end{bmatrix}.$$

Theorem 6 (Fractional Floquet's Theorem) *Every fractional fundamental matrix solution $\Phi_\alpha(t^\alpha)$ of (26) has the form*

$$\Phi_\alpha(t^\alpha) \approx P_\alpha(t^\alpha)E_\alpha(Bt^\alpha), \tag{28}$$

where $P_\alpha(t^\alpha)$, B are $n \times n$ matrices, $P_\alpha((t+w_\alpha)^\alpha) \approx P_\alpha(t^\alpha)$ for all t and B is a constant.

Proof Assume that $\Phi_\alpha(t^\alpha)$ is a fractional fundamental matrix solution of (26). Then $\Phi_\alpha((t+w_\alpha)^\alpha)$ is also a fractional fundamental matrix solution, since $A_\alpha(t^\alpha)$ is periodic of period w_α. Therefore, there is a nonsingular matrix C such that

$$\Phi_\alpha((t+w_\alpha)^\alpha) = \Phi_\alpha(t^\alpha)C.$$

From Theorem 5, there is a matrix B so that $C = E_\alpha(w_\alpha B)$. For this matrix B, let $P_\alpha(t^\alpha) \approx \Phi_\alpha(t^\alpha)E_\alpha(B(-t)^\alpha)$. Then

$$P_\alpha((t+w_\alpha)^\alpha) \approx \Phi_\alpha((t+w_\alpha)^\alpha)E_\alpha(B(-t-w_\alpha)^\alpha)$$
$$\approx \Phi_\alpha(t^\alpha)E_\alpha(B(w_\alpha)^\alpha)E_\alpha(B(-t-w_\alpha)^\alpha) \approx P_\alpha(t^\alpha),$$

and the theorem is proved. □

Definition 11 The eigenvalues $\mu_1, \mu_2, \ldots, \mu_n$ of $C = \Phi_\alpha^{-1}(0)\Phi_\alpha(w_\alpha)$ are called the multipliers of the fractional Floquet system $D_t^\alpha x = A_\alpha(t^\alpha)x$, where $\Phi_\alpha(t^\alpha)$ is a fractional fundamental matrix of system $D_t^\alpha x = A_\alpha(t^\alpha)x$.

Example 3 Solving the following equation,

$$D_t^\alpha x = \sin_\alpha^2(t^\alpha)x, \tag{29}$$

we get that

$$\Phi_\alpha(t^\alpha) \approx E_\alpha\left(\frac{1}{2}t^\alpha - \frac{\Gamma(\alpha+1)}{4i^{\alpha-1}}\sin_\alpha(2t^\alpha)\right)$$
$$\approx E_\alpha(\frac{1}{2}t^\alpha)E_\alpha^{-1}\left(\frac{\Gamma(\alpha+1)}{4i^{\alpha-1}}\sin_\alpha(2t^\alpha)\right)$$

so that

$$C = \Phi_\alpha^{-1}(0)\Phi_\alpha(\pi_\alpha) = E_\alpha\left(\frac{(\pi_\alpha)^\alpha}{2}\right).$$

As a result $E_\alpha\left(\frac{(\pi_\alpha)^\alpha}{2}\right)$ is the multiplier for this fractional differential equation.

Theorem 7 *Let $\Phi_\alpha(t^\alpha) \approx P_\alpha(t^\alpha)E_\alpha(Bt^\alpha)$ be the fractional fundamental matrix in Theorem 6. Then, x is a solution of the fractional Floquet system $D_t^\alpha x = A_\alpha(t^\alpha)x$ if and only if the vector function y defined by $y(t^\alpha) = P_\alpha^{-1}(t^\alpha)x(t^\alpha)$ be a solution of*

$$D_t^\alpha y = By. \tag{30}$$

Proof Assume that x is a solution of the fractional Floquet system $D_t^\alpha x = A_\alpha(t^\alpha)x$. Then, for some vector $x_0 \in \mathbb{R}^{n\times 1}$ we have $x(t^\alpha) = \Phi_\alpha(t^\alpha)x_0$.

Now, by setting $y(t^\alpha) = P_\alpha^{-1}(t^\alpha)x(t^\alpha)$, we get

$$y(t^\alpha) = P_\alpha^{-1}(t^\alpha)\Phi_\alpha(t^\alpha)x_0 \approx P_\alpha^{-1}(t^\alpha)P_\alpha(t^\alpha)E_\alpha(Bt^\alpha)x_0$$
$$= E_\alpha(Bt^\alpha)x_0,$$

which is a solution of (30).

Conversely, assume that y is a solution of system (30) and set $x(t^\alpha) = P_\alpha(t^\alpha)y(t^\alpha)$. Since y is a solution of $D_t^\alpha y = By$, there is a vector $y_0 \in \mathbb{R}^{n\times 1}$ such that $y(t^\alpha) = E_\alpha(Bt^\alpha)y_0$.

It follows that

$$x(t^\alpha) = P_\alpha(t^\alpha)y(t^\alpha) = P_\alpha(t^\alpha)E_\alpha(Bt^\alpha)y_0$$
$$\approx \Phi_\alpha(t^\alpha)y_0,$$

which is a solution of the fractional Floquet system $D_t^\alpha x = A_\alpha(t^\alpha)x$. □

Theorem 8 *Two matrices A and B are called similar if there exists a nonsingular matrix A such that $A = TBT^{-1}$* [23].

Theorem 9 *A fractional Floquet system $D_t^\alpha x = A_\alpha(t^\alpha)x$ with the multipliers $\mu_1, \mu_2, \ldots, \mu_n$ is*

(i) asymptotically stable on $[0, \infty)$ if all multipliers satisfy $|\mu_i| < 1$, $1 \le i \le n$,
(ii) unstable on $[0, \infty)$, when there is an i_0, $1 \le i_0 \le n$, such that $|\mu_{i_0}| > 1$.

Proof Without loss of generality, we prove this theorem for 2×2 matrices A_α. Let $\Phi_\alpha(t^\alpha) \approx P_\alpha(t^\alpha)E_\alpha(Bt^\alpha)$ and C be the same as in the Theorem 6. Therefore, the matrix B can be chosen such that $E_\alpha(Bw^\alpha) = C$.

Suppose the matrix B is similar to a Jordan canonical form, i.e., there exists an invertible matrix M such that $B = MJM^{-1}$. Now by letting λ_1, λ_2 as the eigenvalues of B, we see that for the matrix C we have

$$C = E_\alpha(Bw^\alpha) = E_\alpha(MJM^{-1}w^\alpha) = ME_\alpha(Jw^\alpha)M^{-1} = MHM^{-1},$$

where either

$$H = \begin{bmatrix} E_\alpha(\lambda_1 w^\alpha) & 0 \\ 0 & E_\alpha(\lambda_2 w^\alpha) \end{bmatrix}, \quad or$$
$$H = \begin{bmatrix} E_\alpha(\lambda_1 w^\alpha) & w^\alpha E_\alpha(\lambda_1 w^\alpha) \\ 0 & E_\alpha(\lambda_1 w^\alpha) \end{bmatrix}.$$

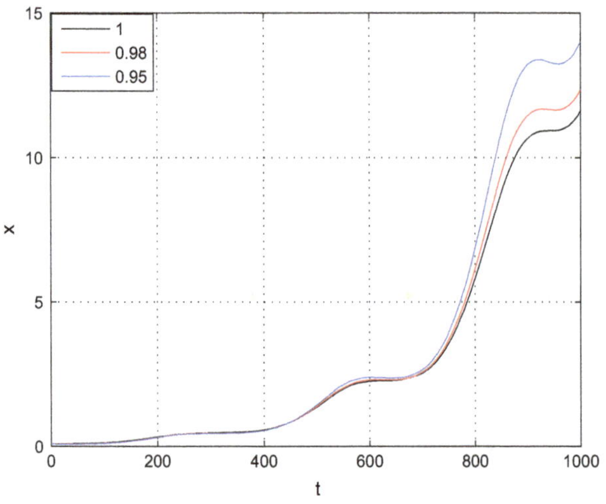

Fig. 2 The numerical approximations of equation (3.8) when $\alpha = 1,\ 0.98,\ 0.95$

Since the eigenvalues of H are the same as the eigenvalues of C, we take the multipliers μ_i as $\mu_i = E_\alpha(\lambda_i w^\alpha)$, $i = 1, 2$. Since $|\mu_i| = E_\alpha(Re(\lambda_i)w^\alpha)$, we have that

$$|\mu_i| < 1 \quad iff \quad Re(\lambda_i) < 0,$$

$$|\mu_i| > 1 \quad iff \quad Re(\lambda_i) > 0.$$

Since according to the Theorem 7, there is a one-to-one correspondence between solutions of the fractional Floquet system $D_t^\alpha x = A_\alpha(t^\alpha)x$ and system (30).

For constant $Q_1 > 0$, we have

$$\|x(t^\alpha)\| = \|P_\alpha(t^\alpha)y(t^\alpha)\| \le \|P_\alpha(t^\alpha)\|\|y(t^\alpha)\| \le Q_1\|y(t^\alpha)\|,$$

and for constant $Q_2 > 0$, we get

$$\|y(t^\alpha)\| = \left\|P_\alpha^{-1}(t^\alpha)x(t^\alpha)\right\| \le \left\|P_\alpha^{-1}(t^\alpha)\right\|\|x(t^\alpha)\| \le Q_2\|x(t^\alpha)\|.$$

Finally by Theorem 1 the results can be derived. □

In Example 3, we saw that the multiplier of the fractional differential equation (29) is $\mu = E_\alpha(\frac{(\pi_\alpha)^\alpha}{2})$. When $0 < \alpha \le 1$, then the solution fractional differential equation (29) is unstable, since we always have $|\mu| > 1$. Figure 2 indicates that equation (29) with parameters $\alpha = 1,\ 0.98,\ 0.95$ is unstable.

Example 4 We can show $\Phi_\alpha(t^\alpha)$ in the form

$$\Phi_\alpha(t^\alpha) \approx \begin{bmatrix} E_\alpha(-t^\alpha) & 0 \\ \frac{1}{i^{\alpha-1}} E_\alpha(-t^\alpha)\sin_\alpha(t^\alpha) & E_\alpha(-t^\alpha) \end{bmatrix},$$

is a fractional fundamental matrix for the fractional Floquet system

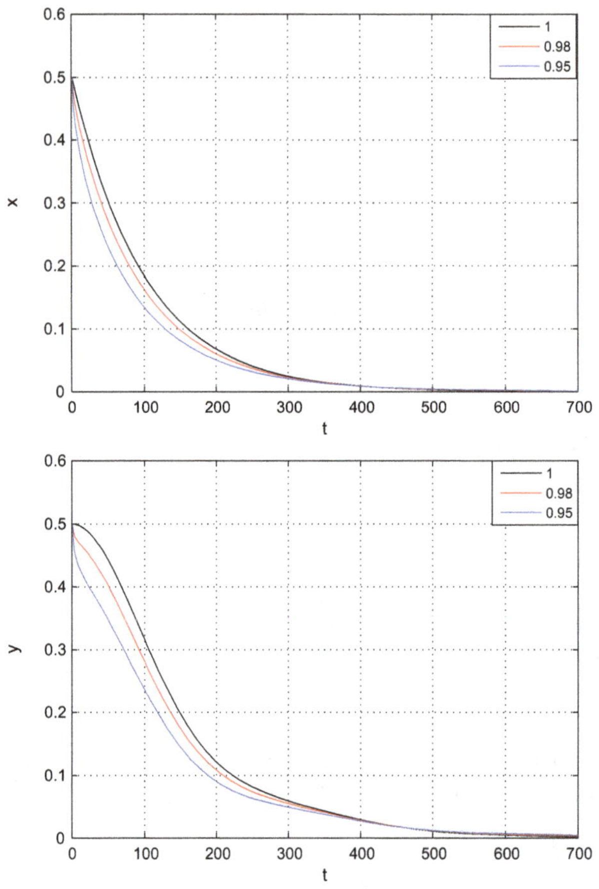

Fig. 3 The numerical approximations of fractional Floquet system (31) when $\alpha = 1,\ 0.98,\ 0.95$

$$\begin{bmatrix} D_t^\alpha x \\ D_t^\alpha y \end{bmatrix} = \begin{bmatrix} -1 & 0 \\ \cos_\alpha(t^\alpha) & -1 \end{bmatrix} \begin{bmatrix} x \\ y \end{bmatrix}. \tag{31}$$

Since

$$C = \Phi_\alpha^{-1}(0)\Phi_\alpha(2\pi_\alpha) = \begin{bmatrix} E_\alpha(-(2\pi_\alpha)^\alpha) & 0 \\ 0 & E_\alpha(-(2\pi_\alpha)^\alpha) \end{bmatrix},$$

the multipliers are $\mu_1 = \mu_2 = E_\alpha(-(2\pi_\alpha)^\alpha)$. When $0 < \alpha \le 1$, then the solution system (31) is asymptotically stable, as we always have $|\mu_i| < 1$ for $i = 1, 2$. Figure 3 indicates that the solution fractional Floquet system (3.10) with parameters $\alpha = 1,\ 0.98,\ 0.95$ is asymptotically stable.

Conclusion

In the present article, we have recalled some properties of the Mittag-Leffler function and Mittag-Leffler logarithm function as described in [20]. Then we have presented fractional trigonometric function and the fractional Floquet system based on the modified Riemann–Liouville derivation. Since, the study of stability for the fractional Floquet system is very important, the asymptotical stability for such systems has been investigated. We have shown the fractional Floquet system is asymptotically stable if all multipliers have real parts between -1 and 1. Finding the stability of nonlinear periodic fractional systems and delay linear periodic fractional systems can be an interesting topic for future research work.

Acknowledgments We are very grateful to anonymous referees for their careful reading and valuable comments which led to the improvement of this paper.

References

1. Hale, J.K.: Ordinary Differential Equations, Pure and Applied Mathematics, vol. XXI. Wiley-Interscience, New York (1969)
2. Kelley, W.G., Peterson, A.C.: The theory of differential equations: classical and qualitative. Springer Science, Business Media (2010)
3. Floquet, G.: Sur les equations differentielles lineaires a coeffcients peroidiques. Ann. Ecole Norm. Sup. **12**, 47–49 (1883)
4. Oldham, K.B., Spanier, J.: The Fractional Calculus. Acadmic Press, New York (1974)
5. Podlubny, I.: Fractional differential equations. Academic Press, San Diego (1999)
6. Kilbas, A.A., Srivastava, H.M., Trujillo, J.J.: Theory and Applications of Fractional Differential Equations. Elsevier, San Diego (2006)
7. Reyes-Melo, E., Martinez-Vega, J., Guerrero-Salazar, C., Ortiz-Mendez, U.: Application of fractional calculus to the modeling of dielectric relaxation phenomena in polymeric materials. J. Appl. Poly. Sci. **98**, 923–935 (2005)
8. Schumer, R., Benson, D.: Eulerian derivative of the fractional advection-dispersion equation. J. Contam. **48**, 69–88 (2001)
9. Ansari, A., Refahi Sheikhani, A., Saberi Najafi, H.: Math. Methods Appl. Sci. Solution to system of partial fractional differential equations using the fractional exponential operators **35**, 119–123 (2012)
10. Ansari, A., Refahi, A.: Sheikhani and S. Kordrostami, On the generating function $e^{xt+y\varphi(t)}$ and its fractional calculus. Cent. Euro. J. Phy. (2013). doi:10.2478/s11534013-0195-3
11. Ansari, A., Sheikhani, A.R.: New identities for the wright and the mittag-leffler functions using the laplace transform. Asian-Euro. J. Math. **7**, 1450038 (2014)
12. H. Saberi Najafi, A. Refahi Sheikhani, and A. Ansari, Stability Analysis of Distributed Order Fractional Differential Equations, Abstract and Applied Analysis, Vol. 2011, Article ID 175323, 12 pages, 2011. doi:10.1155/2011/175323
13. H. Aminikhah, A. Refahi Sheikhani, H. Rezazadeh, Stability Analysis of Distributed Order Fractional Chen System. Sci. World J. (2013). doi:10.1155/2013 /645080 **(Article ID 645080, 13 pages, 2013)**
14. H. Aminikhah, A. Refahi Sheikhani, and H.Rezazadeh, Stability analysis of linear distributed order fractional system with multiple time delays, U.P.B. Sci. Bull. **77** Series A, 207–218 (2015)
15. Jumarie, G.: Stochastic differential equations with fractional Brownian motion input. Int. J. Syst. Sci. **24**, 1113–1132 (1993)
16. Jumarie, G.: Modified Riemann–Liouville derivative and fractional Taylor series of non-differentiable functions further results. Comp. Math. Appl. **51**, 1367–1376 (2006)
17. Jumarie, G.: Oscillation of Non-Linear Systems Close to Equilibrium Position in the Presence of Coarse-Graining in Time and Space. Non. Anal. Modell. Control **14**, 177–197 (2009)
18. Mittag-Leffler, M.G.: Sur la nouvelle fonction $E_\alpha(x)$, Comptes Rendus. Acad. Sci. Paris **137**, 554–558 (1903)
19. Qian, D., Li, C., Agarwal, R.P., Wongd, P.J.Y.: Stability analysis of fractional differential system with Riemann–Liouville derivative. Math. Comp. Model. **52**, 862–874 (2010)
20. Jumarie, G.: Laplace's transform of fractional order via the Mittag-Leffler function and modified Riemann–Liouville derivative. Appl. Math. Lett. **22**, 1659–1664 (2009)
21. Matignon, D.: Stability results for fractional differential equations with applications to control processing. Comp. Eng. Syst. Appl. **2**, 963–968 (1996)
22. Bonilla, B., Rivero, M., Trujillo, J.J.: On systems of linear fractional differential equations with constant coefficients. Appl. Math. Comp. **187**, 68–78 (2007)
23. B. N. Datta, Numerical Linear Algebra and Applications, SIAM, 2nd Edition (2010)

Permissions

All chapters in this book were first published in MS, by Springer International Publishing AG.; hereby published with permission under the Creative Commons Attribution License or equivalent. Every chapter published in this book has been scrutinized by our experts. Their significance has been extensively debated. The topics covered herein carry significant findings which will fuel the growth of the discipline. They may even be implemented as practical applications or may be referred to as a beginning point for another development.

The contributors of this book come from diverse backgrounds, making this book a truly international effort. This book will bring forth new frontiers with its revolutionizing research information and detailed analysis of the nascent developments around the world.

We would like to thank all the contributing authors for lending their expertise to make the book truly unique. They have played a crucial role in the development of this book. Without their invaluable contributions this book wouldn't have been possible. They have made vital efforts to compile up to date information on the varied aspects of this subject to make this book a valuable addition to the collection of many professionals and students.

This book was conceptualized with the vision of imparting up-to-date information and advanced data in this field. To ensure the same, a matchless editorial board was set up. Every individual on the board went through rigorous rounds of assessment to prove their worth. After which they invested a large part of their time researching and compiling the most relevant data for our readers.

The editorial board has been involved in producing this book since its inception. They have spent rigorous hours researching and exploring the diverse topics which have resulted in the successful publishing of this book. They have passed on their knowledge of decades through this book. To expedite this challenging task, the publisher supported the team at every step. A small team of assistant editors was also appointed to further simplify the editing procedure and attain best results for the readers.

Apart from the editorial board, the designing team has also invested a significant amount of their time in understanding the subject and creating the most relevant covers. They scrutinized every image to scout for the most suitable representation of the subject and create an appropriate cover for the book.

The publishing team has been an ardent support to the editorial, designing and production team. Their endless efforts to recruit the best for this project, has resulted in the accomplishment of this book. They are a veteran in the field of academics and their pool of knowledge is as vast as their experience in printing. Their expertise and guidance has proved useful at every step. Their uncompromising quality standards have made this book an exceptional effort. Their encouragement from time to time has been an inspiration for everyone.

The publisher and the editorial board hope that this book will prove to be a valuable piece of knowledge for researchers, students, practitioners and scholars across the globe.

List of Contributors

Bhawna Kohli
Department of Mathematics, University of Delhi, Delhi 110007, India

Mohamed S. Al-luhaibi
Department of Mathematics, Faculty of Science, Kirkuk University, Kirkuk, Iraq

Déthié Dione, Salimata Gueye Diagne and Bakary Koné
Department of Mathematic and Computer Science, University of Cheikh Anta Diop, Dakar, Senegal

Youssou Gningue
University of Laurentia, Sudbury, Canada

Yuji Liu
Department of Mathematics, Guangdong University of Finance and Economics, Guangzhou 510320, People's Republic of China

G. J. Zalmai
Department of Mathematics and Computer Science, Northern Michigan University, Marquette, MI 49855, USA

Ram U. Verma
Department of Mathematics, University of North Texas, Denton, TX 76201, USA

Marwan Alquran
Department of Mathematics and Statistics, Jordan University of Science and Technology, P.O.Box(3030), Irbid 22110, Jordan
Department of Mathematics and Statistics, Sultan Qaboos University, P.O.Box(36), PC 123, Al-Khod, Muscat, Oman

NI˙hal Yilmaz Özgür and Nİhal Tas
Department of Mathematics, Balıkesir University, 10145 Balıkesir, Turkey

Adem Kılıçman
Department of Mathematics, Universiti Putra Malaysia (UPM), 43400 Serdang, Selangor, Malaysia

Rabha W. Ibrahim
Faculty of Computer Science and Information Technology, University of Malaya, 50603 Kuala Lumpur, Malaysia

Zainab E. Abdulnaby
Department of Mathematics, Universiti Putra Malaysia (UPM), 43400 Serdang, Selangor, Malaysia
Department of Mathematics College of Science, Al-Mustansiriyah University, Baghdad, Iraq

Saima Arshed
Department of Mathematics, University of the Punjab, Lahore 54590, Pakistan

M. De la Sen
Faculty of Science and Technology, Institute of Research and Development of Processes IIDP, University of the Basque Country, PO Box 644 de Bilbao. Barrio Sarriena, 48940 Leioa, Bizkaia, Spain

Antonio F. Roldán
Department of Mathematics, University of Jae´n, Paraje de las Lagunillas s/n, 23071 Jae´n, Spain

Kamel Al-Khaled and Marwan Alquran
Department of Mathematics and Statistics, Jordan University of Science and Technology, P.O. Box: 3030, Irbid 22110, Jordan
Department of Mathematics and Statistics, Sultan Qaboos University, Al-Khod, P.O. Box: 36, PC 123 Muscat, Oman

M. Mohamadi and E. Babolian
Department of Mathematics, Science and Research Branch, Islamic Azad University, Tehran, Iran

S. A. Yousefi
Department of Mathematics, Shahid Beheshti University, G. C. Tehran, Iran

Kazem Nouri, Behzad Abbasi, Farahnaz Omidi and Leila Torkzadeh
Department of Mathematics, Faculty of Mathematics, Statistics and Computer Sciences, Semnan University, P.O. Box 35195-363, Semnan, Iran

Khaled A. Gepreel
Mathematics Department, Faculty of Science, Taif University, Taif, Kingdom of Saudi Arabia
Mathematics Department, Faculty of Science, Zagazig University, Zagazig, Egypt

Taher A. Nofal
Mathematics Department, Faculty of Science, Taif University, Taif, Kingdom of Saudi Arabia
Mathematics Department, Faculty of Science, El-Minia University, El-Minia, Egypt

Xiaohui Yang
Department of Computer, Guangdong Police College, Guangzhou 510230, People's Republic of China

Yuji Liu
Department of Mathematics, Guangdong University of Finance and Economics, Guangzhou 510320, People's Republic of China

Junfei Cao, Qian Tong and Xianyong Huang
Department of Mathematics, Guangdong University of Education, Guangzhou 510310, People's Republic of China

Z. Barikbin
Department of Mathematics, Imam Khomeini International University, Qazvin 34149-16818, Iran

Rabha W. Ibrahim
Center of Mobile Cloud Computing Research, University Malaya, 50603 Kuala Lumpur, Malaysia

Yass K. Salih
College of Engineering, Universiti Tenaga Nasional, 43000 Kajang, Selangor, Malaysia

Inderdeep Singh and Sheo Kumar
Department of Mathematics, Dr. B.R. Ambedkar National Institute of Technology, Jalandhar, Punjab 144011, India

N. Ahmadi and T. Allahviranloo
Department of Mathematics, Science and Research Branch, Islamic Azad University, Tehran, Iran

A. R. Vahidi
Department of Mathematics, Yadegar-e-Imam Khomeini (RAH) Shahre Rey Branch, Islamic Azad University, Tehran, Iran

Hossein Aminikhah
Department of Applied Mathematics, Faculty of Mathematical Sciences, University of Guilan, P.O. Box 1914, Rasht, Iran

Amir Hossein Refahi Sheikhani
Department of Applied Mathematics, Faculty of Mathematical Sciences, Islamic Azad University, Lahijan Branch, P.O. Box 1616, Lahijan, Iran

Hadi Rezazadeh
Department of Applied Mathematics, Faculty of Mathematical Sciences, University of Guilan, P.O. Box 1914, Rasht, Iran
Faculty of Engineering Technology, Amol University of Special Modern Technologies, P.O. Box 46168-49767, Amol, Iran

Index